Handbook of Irrigation Hydrology and Management

Ever-increasing population growth has caused a proportional increased demand for water, and existing water sources are depleting day by day. Moreover, with the impact of climate change, the rates of rainfall in many regions have experienced a higher degree of variability. In many cities, government utilities have been struggling to maintain sufficient water for the residents and other users. The *Handbook of Irrigation Hydrology and Management* examines and analyzes irrigated ecosystems in which water storage, applications, or drainage volumes are artificially controlled in the landscape and the spatial domain of processes varies from micrometers to tens of kilometers, while the temporal domain spans from seconds to centuries. The continuum science of irrigation hydrology includes the surface, subsurface (unsaturated and groundwater systems), atmospheric, and plant subsystems. Further, the book includes coverage of environmental and economic impacts, water quality issues, water harvesting, satellite measurements for irrigation, and more.

Features:

- Offers water-saving strategies to increase the judicious use of scarce water resources
- Presents strategies to maximize agricultural yield per unit of water used for different regions
- Compares irrigation methods to offset changing weather patterns and impacts of climate change

Handbook of Irrigation Hydrology and Management
Irrigation Fundamentals

Edited by
Saeid Eslamian and Faezeh Eslamian

CRC Press
Taylor & Francis Group
Boca Raton London New York

CRC Press is an imprint of the
Taylor & Francis Group, an **informa** business

Designed cover image: Shutterstock

First edition published 2023
by CRC Press
6000 Broken Sound Parkway NW, Suite 300, Boca Raton, FL 33487-2742

and by CRC Press
4 Park Square, Milton Park, Abingdon, Oxon, OX14 4RN

CRC Press is an imprint of Taylor & Francis Group, LLC

© 2023 selection and editorial matter, Saeid Eslamian and Faezeh Eslamian; individual chapters, the contributors

Library of Congress Cataloging-in-Publication Data
Names: Eslamian, Saeid, editor. | Eslamian, Faezeh A., editor.
Title: Handbook of irrigation hydrology and management: irrigation
management and optimization / Edited by Saeid Eslamian and Faezeh Eslamian.
Description: Boca Raton, FL: CRC Press, 2023. | Includes bibliographical
references and index. | Also available online. | Description based on
print version record and CIP data provided by publisher; resource not viewed.
Identifiers: LCCN 2022050740 (print) | LCCN 2022050741 (ebook) | ISBN
9780429290152 (ebook) | ISBN 9780367258306 (hardback) | ISBN 9781032457468
(paperback) | ISBN 9780367258191(v. 1 ;hardback) | ISBN
9781032457451(v. 1 ;paperback) | ISBN 9781032457468(v. 2;paperback)
| ISBN 9780429290152(v. 2 ;ebook) | ISBN 9781032406077(v. 3 ;hardback)
| ISBN 9781032429106(v. 3 ;paperback) | ISBN 9781003353928(v. 3 ;ebook)
Subjects: LCSH: Irrigation. | Irrigation engineering.
Classification: LCC TC805 (ebook) | LCC TC805 .H36 2023 (print)
| DDC 627/.52 23/eng/20221–dc16
LC record available at https://lccn.loc.gov/2022050740

ISBN: 978-0-367-25819-1 (hbk)
ISBN: 978-1-032-45745-1 (pbk)
ISBN: 978-0-429-29011-4 (ebk)

DOI: 10.1201/9780429290114

Typeset in Minion
by codeMantra

To Prof. Mark Grismer (University of California, Davis), the best Agricultural Hydrologist that I have met in my 28 years academic life

Irrigation hydrology is at the nexus of water-energy-food (WEF) security and is a key element of Integrated Water Resources Management (IWRM) from the local to regional and in some cases transnational scales across the world.

(Grismer, 2022)

Contents

Preface

Water is known as the most important input required for plant growth in agricultural production. Irrigation can be defined as the completion of soil water storage in the root zone of the plant by methods other than natural rainfall. Irrigation seems to have its roots in human history from the beginning. This makes it possible to reduce uncertainties in agricultural practices, particularly climatic ones.

Archaeological research has found evidence of irrigation where natural rainfall was not sufficient to support farming. The ancient Egyptians, for example, practiced basin irrigation by using the Nile flood to flood land surrounded by dykes. Another example is the aqueduct, which was built in ancient Persia around 800 BC and is one of the oldest known methods of irrigation still in use.

Irrigation hydrology is constrained to analysis of irrigated ecosystems in which water storage, applications, or drainage volumes are artificially controlled in the landscape, and the spatial domain of processes varies from micrometers to tens of kilometers, while the temporal domain spans from seconds to centuries. The continuum science of irrigation hydrology includes the surface, subsurface (unsaturated and groundwater systems), atmospheric, and plant subsystems.

Irrigation management involves many different decisions: selection of economically viable cropping patterns, land allocation by crop, water resource allocation by crop, irrigation scheduling, deficit management irrigation, etc. Plants need adequate amounts of water, and its distribution throughout the growth cycle has a huge influence on the final yield of the crop. This means that the management of soil water content is crucial to obtain an optimal allocation of water resources, provided that the other production factors are adequate.

This book has several merits as follows:

- A comprehensive book on majority methods of Irrigation
- A new focus on Irrigation Hydrology, Landscape, Scales, and Social Context
- A robust tool for computational analysis of Irrigation Methods
- A updated book including global warming, adaptation, resilience, and sustainability associated with Irrigation
- Deficit and Over-irrigation Merits and Disadvantages
- Inclusion of Smart and Precision Irrigation
- Offering Solutions for Water Scarcity and Soil Salinity in Irrigation
- Satellite Measurements for Irrigation Management
- Selected case studies from different climates across world.

Volume I of this handbook, entitled *Irrigation Fundamentals*, includes the following topics:

- Earth and Satellite Measurements for Irrigation
- Environmental and Economical Impacts
- Evapotranspiration and Water Requirements
- Irrigation Hydrology

- Irrigation Water Quality Issues
- Water Harvesting for Irrigation.

The information contained in this *Handbook of Irrigation Hydrology and Management* can be beneficial to students at the following levels of the study: Undergraduate, Postgraduate, Research Students, and Short Course Programs, and could also be a useful resource for courses such as Surface Hydrology, Water Resources, Climatology, Agrometeorology, Irrigation Principals, Surface Irrigation, Irrigation System Design, Drip Irrigation, Sprinkler Irrigation, Water, Soils and Plants Relationships, Evapotranspiration and Water Requirement, Drainage Engineering, Irrigation and Drainage Networks, Irrigation Hydraulic Structures, and Special Problems in Irrigation.

All the scholars and students of Applied Geography, Geosciences, Environmental Engineering, Environmental Health, Natural Resources, Agricultural Engineering, Irrigation Engineering, and the related courses, as well as professionals, will find this handbook of great value.

Three-volume *Handbook of Irrigation Hydrology and Management* could be recommended not only for universities and colleges, but also for research centers, governmental departments, policy makers, engineering consultants, federal emergency management agencies, and the related bodies.

Saeid Eslamian
Isfahan University of Technology

Faezeh Eslamian
McGill University

Editors

Saeid Eslamian is a full professor of environmental hydrology and water resources engineering in the Department of Water Engineering at Isfahan University of Technology, where he has been serving since 1995. His research focuses mainly on statistical and environmental hydrology in a changing climate. In recent years, he has worked on modeling natural hazards, including floods, severe storms, wind, drought, pollution, water reuses, sustainable development and resiliency, etc. Formerly, he was a visiting professor at Princeton University, New Jersey, and the University of ETH Zurich, Switzerland. On the research side, he started a research partnership in 2014 with McGill University, Canada. He has contributed to more than 600 publications in journals, books, and technical reports. He is the founder and chief editor of both the *International Journal of Hydrology Science and Technology* (IJHST) and the *Journal of Flood Engineering* (JFE). Eslamian is now associate editor of three important publications: *Journal of Hydrology* (Elsevier), *Eco-Hydrology and Hydrobiology* (Elsevier), *Journal of Water Reuse and Desalination* (IWA), and *Journal of the Saudi Society of Agricultural Sciences* (Elsevier). Professor Eslamian is the author of approximately 65 books and Journal Special Issues, and also 350 chapter books.

Dr. Eslamian's professional experience includes membership on editorial boards, and he is a reviewer of approximately 100 Web of Science (ISI) journals, including the *ASCE Journal of Hydrologic Engineering*, *ASCE Journal of Water Resources Planning and Management*, *ASCE Journal of Irrigation and Drainage Engineering*, *Advances in Water Resources*, *Groundwater*, *Hydrological Processes*, *Hydrological Sciences Journal*, *Global Planetary Changes*, *Water Resources Management*, *Water Science and Technology*, *Eco-Hydrology*, *Journal of American Water Resources Association*, and *American Water Works Association Journal*. UNESCO has also nominated him for a special issue of the *Eco-Hydrology and Hydrobiology Journal* in 2015.

Professor Eslamian was selected as an outstanding reviewer for the *Journal of Hydrologic Engineering* in 2009 and received the EWRI/ASCE Visiting International Fellowship in Rhode Island (2010). He was also awarded outstanding prizes from the Iranian Hydraulics Association in 2005 and Iranian Petroleum and Oil Industry in 2011. Professor Eslamian has been chosen as a distinguished researcher of Isfahan University of Technology (IUT) and Isfahan Province in 2012 and 2014, respectively. In 2016, he was a candidate for national distinguished researcher in Iran.

He has also been the referee of many international organizations and universities. Some examples include the U.S. Civilian Research and Development Foundation (USCRDF), the Swiss Network for International Studies, the Majesty Research Trust Fund of Sultan Qaboos University of Oman, the Royal Jordanian Geography Center College, and the Research Department of Swinburne University of

Technology of Australia. He is also a member of the following associations: American Society of Civil Engineers (ASCE), International Association of Hydrologic Science (IAHS), World Conservation Union (IUCN), GC Network for Drylands Research and Development (NDRD), International Association for Urban Climate (IAUC), International Society for Agricultural Meteorology (ISAM), Association of Water and Environment Modeling (AWEM), International Hydrological Association (STAHS), and UK Drought National Center (UKDNC).

Professor Eslamian finished Hakimsanaei High School in Isfahan in 1979. After the Islamic Revolution, he was admitted to IUT for a BS in water engineering and graduated in 1986. After graduation, he was offered a scholarship for a master's degree program at Tarbiat Modares University, Tehran. He finished his studies in hydrology and water resources engineering in 1989. In 1991, he was awarded a scholarship for a PhD in civil engineering at the University of New South Wales, Australia. His supervisor was Professor David H. Pilgrim, who encouraged him to work on "Regional Flood Frequency Analysis Using a New Region of Influence Approach." He earned a PhD in 1995 and returned to his home country and IUT. In 2001, he was promoted to associate professor and in 2014 to full professor. For the past 26 years, he has been nominated for different positions at IUT, including university president consultant, faculty deputy of education, and head of department. Eslamian is now the director for center of excellence in Risk Management and Natural Hazards (RiMaNaH).

Professor Eslamian has made three scientific visits to the United States, Switzerland, and Canada in 2006, 2008, and 2015, respectively. In the first, he was offered the position of visiting professor by Princeton University and worked jointly with Professor Eric F. Wood at the School of Engineering and Applied Sciences for one year. The outcome was a contribution in hydrological and agricultural drought interaction knowledge by developing multivariate L-moments between soil moisture and low flows for northeastern U.S. streams.

Recently, Professor Eslamian has published fourteen handbooks by Taylor & Francis (CRC Press): the three-volume *Handbook of Engineering Hydrology* in 2014, *Urban Water Reuse Handbook* in 2016, *Underground Aqueducts Handbook* (2017), the three-volume *Handbook of Drought and Water Scarcity* (2017), *Constructed Wetlands: Hydraulic Design* (2019), *Handbook of Irrigation System Selection for Semi-Arid Regions* (2020), *Urban and Industrial Water Conservation Methods* (2020) and the three-volume *Flood Handbook* (2022)

An Evaluation of Groundwater Storage Potentials in a Semiarid Climate and *Advances in Hydrogeochemistry Research* by Nova Science Publishers are also his book publications in 2019 and 2020, respectively. The two-volume *Handbook of Water Harvesting and Conservation* (Wiley) and *Handbook of Disaster Risk Reduction and Resilience* (*New Frameworks for Building Resilience to Disasters*) are early 2021 book publications of Professor Eslamian. *Handbook of Disaster Risk Reduction and Resilience* (*Disaster Risk Management Strategies*) and the two-volume *Earth Systems Protection and Sustainability* are early 2022 handbooks of Professor Eslamian. The Handbook of Hydroinformatics (Elsevier) has been the latest book publication of him in early 2023.

Professor Eslamian has been appointed as World Top 2-Percent Researcher by Stanford University, USA, in 2019 and 2020. He has also been a Grant Assessor, Report Referee, Award Jury, and Invited Researcher for international organizations, namely, United States Civilian Research and Development Foundation (2006), Intergovernmental Panel on Climate Change (2012), World Bank Policy and Human Resources Development Fund (2021), and Stockholm International Peace Research Institute (2022), respectively.

Faezeh Eslamian is a PhD holder in bioresource engineering from McGill University. Her research focuses on the development of a novel lime-based product to mitigate phosphorus loss from agricultural fields. Faezeh completed her bachelor's and master's degrees in civil and environmental engineering from Isfahan University of Technology, Iran, where she evaluated natural and low-cost absorbents for the removal of pollutants such as textile dyes and heavy metals. Furthermore, she has conducted research on the worldwide water quality standards and wastewater reuse guidelines. Faezeh is an experienced multidisciplinary researcher with research interests in soil and water quality, environmental remediation, water reuse, and drought management.

Contributors

Hamid Zare Abyaneh
Department of Irrigation and Drainage,
  Faculty of Agriculture
Bu-Ali Sina University
Hamedan, Iran

Bashir Adelodun
Department of Agricultural and Biosystems
  Engineering
University of Ilorin
Ilorin, Kwara State, Nigeria
and
Department of Agricultural Civil Engineering
Kyungpook National University
Daegu, Korea

Adedayo Oreoluwa Adewole
Department of Geography
Obafemi Awolowo University
Ile-Ife, Osun State, Nigeria

Oluwaseyi A. Ajala
Department of Zoology and Environmental
  Sciences
Punjabi University
Patiala, Punjabi, India
and
Department of Chemistry, Faculty of Science
University of Ibadan
Ibadan, Oyo State, Nigeria

Fidelis O. Ajibade
Department of Civil and Environmental
  Engineering
Federal University of Technology
Akure, Ondo State, Nigeria
and
University of Chinese Academy of Sciences
Beijing, PR China
and
Research Centre for Eco-Environmental Sciences
Chinese Academy of Sciences
Beijing, PR China

Temitope F. Ajibade
University of Chinese Academy of Sciences
Beijing, PR China
and
Department of Civil and Environmental
  Engineering
Federal University of Technology
Akure, Ondo State, Nigeria
and
Institute of Urban Environment
Chinese Academy of Sciences
Xiamen, PR China

Nadhir Al-Ansari
Department of Civil, Environmental and
  Natural Resources Engineering
Lulea University of Technology
Luleå, Sweden

Laheab Abbas Al-Maliki
Department of Regional Planning,
    Faculty of Physical Planning
University of Kufa
Najaf, Iraq

Sohaib Kareem Al-Mamoori
Department of Environmental Planning,
    Faculty of Physical Planning
University of Kufa
Najaf, Iraq

Catariny Cabral Aleman
Department of Agricultural Engineering
Federal University of Viçosa
Viçosa, Minas Gerais, Brazil

Toju E. Babalola
Department of Water Resources Management &
    Agro-Meteorology
Federal University Oye-Ekiti
Oye, Ekiti State, Nigeria

Nelson Chanza
Department of Town and Regional Planning
University of Johannesburg
South Africa

Fadi G. Comair
Intergovernmental Hydrological
    Programme (IHP)
UNESCO IHP Council
Beirut, Lebanon

Alessia Corami
Geosciences Department
University of Louisiana at Lafayette
USA

T. C. Crusberg
Department of Biology and Biotechnology
Worcester Polytechnic Institute
Worcester, Massachusetts
USA

Nicolas R. Dalezios
Department of Civil Engineering
University of Thessaly
Volos, Greece

Nicholas Dercas
Department of Natural Resources Management
    and Agricultural Engineering
Agricultural University of Athens
Athens, Greece

Mohammad Mahdi Dorafshan
Department of Civil Engineering
Isfahan University of Technology
Isfahan, Iran

Khaled El-Tawil
Faculty of Engineering
Lebanese University
Beirut, Lebanon

Nabil I. Elsheery
Agriculture Botany Department,
    Faculty of Agriculture
Tanta University
Tanta, Egypt

Adebayo Oluwole Eludoyin
Department of Geography
Obafemi Awolowo University
Ile-Ife, Osun State, Nigeria

Saeid Eslamian
Department of Water Science and Engineering
College of Agriculture
Isfahan University of Technology
Isfahan, Iran
Excellence Center of Risk Management and
    Natural Hazards
Isfahan University of Technology
Isfahan, Iran

Ioannis N. Faraslis
Department of Environmental Sciences
University of Thessaly
Volos, Greece

Nashwa A. Fetian
Soils, Water and Environment Research Institute
Agricultural Research Center
Giza, Egypt

Nedjoud Grara
Faculté des Sciences de la Nature et de la Vie et
   Sciences de la Terre et de l'Univers, Université
Guelma, Guelma, Algeria

Mark E. Grismer
Department of Biological and Agricultural
   Engineering
University of California, Davis
Davis, California
USA

Mahbub Hasan
Department of Mechanical, Civil Engineering,
   and Construction Management
Alabama A&M University
Normal, Alabama
USA

Mir Bintul Huda
Consultant, Water Resources Management Centre
National Institute of Technology Srinagar
Srinagar, Jammu and Kashmir, India

Monzur A. Imteaz
Department of Civil and Construction
   Engineering
Swinburne University of Technology
Hawthorn, VC, Australia

Mehdi Jovzi
Soil and Water Research Department,
   Kermanshah Agricultural and Natural
   Resources Research and Education Center
Agricultural Research Education and Extension
   Organization
Kermanshah, Iran

Parth J. Kapupara
Agricultural engineering department
Parul University
Vadodara, Gujarat, India

Fadila Khaldi
Department of Biology, Laboratory of Sciences
   and Technology of Water and Environment,
   Faculty of Life and Natural Science
Mohamed Cherif Messaadia University of
   Souk-Ahras
Souk Ahras, Algeria

Mobushir Khan
School of Environmental Sciences
Charles Sturt University
Albury, NSW, Australia

Jahangir Abedi Koupai
Department of Water Science and Engineering
Isfahan University of Technology
Isfahan, Iran

Kayode H. Lasisi
University of Chinese Academy of Sciences
Beijing, PR China
and
Department of Civil and Environmental
   Engineering
Federal University of Technology
Akure, Ondo State, Nigeria
and
Institute of Urban Environment
Chinese Academy of Sciences
Xiamen, PR China

Mousa Maleki
Department of Civil Engineering
Illinois Institute of Technology
Chicago, Illinois

Nedjma Mamine
Department of Biology, Laboratory of Aquatic
   and Terrestrial Ecosystems, Faculty of Life
   and Natural Science,
Mohamed Cherif Messaadia University of
   Souk-Ahras
Souk Ahras, Algeria

Patricia Angélica A. Marques
Department of Biosystems Engineering (LEB)
University of São Paulo
Piracicaba, São Paulo, Brazil

Sushant Mehan
Department of Food, Agricultural, and Biological
   Engineering
The Ohio State University
Columbus, Ohio

Never Mujere
Geography Geospatial Sciences and Earth
    Observation Department
University of Zimbabwe
Harare, Zimbabwe

Nathaniel A. Nwogwu
Department of Agricultural and Bioresources
    Engineering
Federal University of Technology
Owerri, Imo State, Nigeria
and
Department of Biosystems Engineering
Auburn University
Auburn, AL

Ifeoluwa F. Omotade
Department of Agricultural and Environmental
    Engineering
Federal University of Technology
Akure, Ondo State, Nigeria

Mayowa Emmanuel Oyinloye
Department of Geography
Obafemi Awolowo University
Ile-Ife, Osun State, Nigeria

Nasir Ahmad Rather
Civil Engineering Department
Baba Ghulam Shah Badshah University
Rajouri, Jammu and Kashmir, India

Hosein Roknizadeh
Department of Water Science and Engineering
Isfahan University of Technology
Isfahan, Iran

Stavros Sakellariou
Department of Planning and Regional
    Development
University of Thessaly
Volos, Greece

Tamer A. Salem
Soils, Water and Environment Research Institute
Agricultural Research Center
Giza, Egypt

Pantelis Sidiropoulos
Department of Civil Engineering
University of Thessaly
Volos, Greece

Marios Spiliotopoulos
Department of Civil Engineering
University of Thessaly
Volos, Greece

Ahmad Aboueloyoun Taha
Soils, Water and Environment Research Institute
Agriculture Research Center
Giza, Egypt

Maryam Bayat Varkeshi
Department of Science and Soil Engineering,
    Faculty of Agriculture
Malayer University
Malayer, Iran

Irrigation
Hydrology

I

1

Irrigation Hydrology: Landscape, Scales and Social Context

Mark E. Grismer
University of California

Saeid Eslamian
*Isfahan University
of Technology*

1.1 Introduction

Development of irrigated agriculture across the globe and ages has generally required socially acceptable separation of the water from its landscape of origin and re-assignment as a resource, or economically tradable commodity. This is followed by alteration of existing local and regional hydrologic processes associated with taking water available at one location and/or time to another where it is less available and used for agricultural production or urban demands; thus, the transfer includes both spatial and temporal scales of consideration. Surface and groundwater reservoirs accumulate the water in particular locations at the regional scale and generally in periods prior (in the case of groundwater perhaps centuries prior) to when the water is allocated for irrigation or other uses elsewhere at the local scale. The continuum science of irrigation hydrology includes the surface, subsurface (unsaturated and groundwater systems), atmospheric and plant subsystems in which at least one part of the water storage, application or drainage systems is artificially controlled in the landscape (Wallender and Grismer, 2002).

In many semi-arid to arid regions, irrigated agriculture is the largest water user across the landscape and as such has a profound impact on water allocations for other uses critical to civilization (e.g. urban and industrial supplies, and hydropower) especially as affected by human-induced water scarcity and climate change (Lankford, 2013). In classical hydrology, physical processes interacting across space and time scales are very important (e.g. Blöschl, 2001; Wallender and Grismer, 2002; Merz and Blöschl, 2008; Montanari et al., 2010). However, considering the combined socio-hydrologic system, nonlinear feedbacks with human activities and unforeseen thresholds result in less predictability as appropriate incorporation of social processes is challenging and only now being studied directly. New variations in apparent processes brought about by feedbacks between the processes occurring at a range of scales, referred to as 'emergent behavior', may lead to exceedance of 'tipping points' through which the systems may evolve into new, perhaps previously unobserved, or unanticipated states. Presumably, including an understanding of resilience thresholds and unforeseen social behaviors should assist in formulating management decisions by accounting for these wider process dynamics, some of which are not as yet defined (Kumar, 2011; Melsen et al., 2018).

DOI: 10.1201/9780429290114-2

Historically, the redefinition of water (including treated wastewater) as a tradable commodity enabling its transfer was associated with irrigation of relatively dry lands in arid and semi-arid regions, though in the past several decades, irrigation has also developed in traditionally rain-fed agricultural areas to improve the crop yields and/or quality or enable transition to higher-value crops. While most irrigation practices affecting water use occur at the farm-local scale and some water conservation measures occur at the district scale associated with distribution system management, the transfer of waters from one locale to another is something of a regional-scale process affecting landscape-scale water resources (reservoirs or groundwater) as managed by county, state or national agencies. Moreover, there are parallel water quality impacts at the farm to regional scales associated with local application of fertilizers, herbicides and pesticides, and salt importation from regional water supplies (reservoirs or groundwater). Of course, no irrigated agricultural area is without urban areas that also require water resources and share concerns about water quality. Similarly, there are evolving concerns about meeting environmental water needs associated with preserving or maintaining fisheries and wildlife habitat. These needs are a reflection of current shifting social values and recognition of the broader impacts of irrigated agriculture perhaps considered in isolation previously of the larger landscape processes central to the quality of life. The failure to adequately recognize and address the hydrologic interconnections in irrigated agriculture at the landscape scale has led to the well-known demise of various civilizations of antiquity, or at least the loss of productive regions due to progressive soil salinization, wars or changing climate conditions limiting available water resources at the regional scale. Thus, contemporary discussion is associated with allocation of water resources within a social or environmental context, not unlike that of 'eco-hydrology'; some have referred to this view as 'Socio-Hydrology' (Sivapalan et al., 2012).

While integrated, or irrigation water resources management (IRWM) relies on economic or scenario-based approaches to investigate the human–water interactions, socio-hydrology is concerned with longer-term dynamics; that is, predicting possible trajectories of the system dynamics to be considered by policy developers and planners faced with making strategic, long-term decisions at the local to the national scales (Rath et al., 2021). In other words, socio-hydrology attempts to account for the co-evolutionary dynamics of the coupled human–water system, including spontaneous or unexpected behaviors, whereas the focus of IWRM is on controlling or managing the water system to reach desired outcomes for irrigated agriculture, society and possibly the environment that supports local communities and wildlife habitat. In essence, "the focus of socio-hydrology is on observing, understanding and predicting future trajectories of co-evolution of coupled human-water systems" (Sivapalan et al., 2012). In this sense, one could say that socio-hydrology is the fundamental science underpinning the practice of IWRM.

In the traditional water resources framework, society buffers against 'scarce' and variable surface water flows through regional basin water storage and transfers using surface water reservoirs, groundwater pumps and reuse schemes that capture excess diversions (urban wastewater and irrigation return flows). In this, what some would consider more narrow view of the landscape focused on water resources, riparian ecosystems experience unintended consequences of direct, often irreversible, impacts of water appropriation. Around the world, infrastructure expansion continues within and between basins, often resulting in surface- and groundwater depletions downstream affecting in-stream flows leading to intermittency and loss of ecosystem services followed by regime shifting as natural riparian systems become hydraulic conveyance systems. This is further complicated in the present where reservoir storage has declined to that of several decades past as dams impound increasing amounts of sediment not truly considered economically or in terms of the functional life of the structure, nor in the face of changing climate conditions leading to decreased wet season flows in many regions. Thus, as interconnected social-ecological systems, river basins respond to both internal and external climate and human drivers that vary with time. They are 'resilient' to the degree that they can absorb the disturbances and continue to provide the services and functionality to human and ecological communities alike. However, some question the moral-philosophical implied framework of this 'mechanical' view of the landscape and water.

Schmidt (2017) penned a provocative essay considering the sociology and the developing philosophy underpinning water resources management in the Western world and designation of water as a resource will ultimately lead to scarcity and other attributes typical of economic quantities. In their reviews, Williams (2017) and McCool (2018) noted that hydro-social scientists have been getting things wrong for years. Schmidt contends that the conceptual starting point of the transformation of naturally occurring and materially messy 'water', to the industrial product 'H$_2$O', delineated separate from nature is a false premise. He further argues that the contemporary view of water as the connecting element between things, people and politics is "at the confluence of claims about human impacts on Earth systems, economics, and governance" (p. 183). In other words, as conceptualized by society today, water has an ontological priority and is materially different from other resources or environmental elements. Schmidt considers that this false premise evolved from the old story about the modern separation of society and nature and the entrenchment of binary Enlightenment thinking. Further, that because this thinking does not really apply to water in the landscape has resulted in the ongoing abundance/scarcity aspects of IWRM worldwide.

Schmidt (2017) suggested that the contemporary approach to IWRM was premised on three assumptions; "that water was once abundant, that it has now become scarce, and, as an outcome of mismanaging scarcity, that water is now an issue of security" (p. 41). Such a view leads to the current concepts of the water-food-energy-security nexus and 'resiliency' largely framed in an economic sense. He argued that the other concepts of water in the landscape were available from the indigenous peoples or from the perspective of naturalists such as Aldo Leopold. Leopold viewed human relationships to the environment differently and that the 'land' included 'waters, plants and animals' and the ecological community as "everything assembled within the category of 'land'" (p. 216). As part of 'conservation', it was permissible to manage the 'land'; however, it was necessary to ensure that every part of the ecological system is considered as a whole. Essentially Leopold's holistic ecological thinking of the system is in contrast to judgments in which, for example, weeds, sinuous brooks or the number/type of fish or other wildlife became the primary reasons for the management/conservation efforts; instead, he believed that species and fauna were "entitled to share the land with us" (p. 218).

While several comment on the problems of how water is managed or ways to improve our water management systems, Schmidt (2017) traced the trajectory of water management philosophy as the possible source of current management actions and the associated problems encountered in IWRM. He advocated that "social scientists should refuse the notion that water resources are a neutral category", as this supports judgments that "foster unequal practices favoring one cultural understanding of water over others" (p. 229). Recently, the United Nations (UN) declared this decade (2020–2030) as that of 'Ecosystem Restoration' in an effort to highlight the need to massively scale up restoration efforts of degrade ecosystems so as to improve 'resiliency' in the face of changing climate and improve food security and biodiversity. As actual restoration of pre-alteration landscape conditions is not possible, the meaning applied here requires active participation and management of the landscape by society for the benefit of all. The UN noted that key ecosystems delivering numerous services essential to food and agriculture, including supply of freshwater, protection against hazards and provision of habitat for species such as fish and pollinators, are rapidly declining worldwide. Further, the UN defines ecosystem restoration within a social context as the process of reversing the ecosystem degradation, such that landscapes, lakes and oceans regain their ecological functionality; in other words, that they improve the productivity and capacity of ecosystems to meet the needs of society. However, Schmidt (2017) argued that new conceptual tools are required to interrogate philosophies of water management if we are to arrive at more holistic solutions to problems of water scarcity, security and food production within a thriving resilient environment. It seems that Schmidt's provocation is to have society revisit how we view water resources management at a fundamental level; perhaps such an effort will reveal greater opportunities and ideas needed for socio-hydrologic assessment and landscape restoration to a level of functionality beneficial to the quality of life.

Here, first taking the traditional water-as-a-resource perspective, the basic processes involved in irrigated hydrologic systems are briefly reviewed. Next, considering the spatial scales of farm to region and the associated political entities associated with its management, we outline some of the central knowledge deficiencies and obstacles in planning, or lack thereof at all of the scales. After describing some basic processes and their measurement, the discussion centers on the different evolving perspectives of irrigated hydrology in the broader socio-hydrologic landscape.

1.2 The Irrigated Hydrology System – Processes and Scales

Based on the water-as-a-resource perspective, Figure 1.1 illustrates a conceptual model of the irrigated agriculture hydrologic landscape focused on the transfer of water across the landscape-atmosphere continuum noting the range of spatial and implied temporal scales. While not an exhaustive sketch, this illustration is a starting point and enables definition of terms used in the following discussions of each process shown.

As in most hydrologic cycle conceptualizations, in the natural landscape water available for agricultural use begins with precipitation either historically preserved as groundwater storage, or more contemporarily as direct rainfall on crops, additions to surface storage and leaching, deep percolation, and seepage to shallow groundwater. Precipitation is generally considered as a regional process that is measured at the local (e.g. rain gages) to district/region scales (radar) and averaged to the regional scale by a variety of established methods. With respect to precipitation, the IWRM approach is generally to capture and store that water for future use by agriculture, cities and the environment to the extent practically, or economically possible in an effort to manage its wide-ranging temporal variability. Evaporation and evapotranspiration (ET) processes are similar to that of precipitation. Evaporation and ET rates are measured directly at the hourly and local spatial scale using collected micrometeorological data (e.g. Bowen ratio, eddy covariance flux and surface renewal methods) and typically summed to daily depths and averaged across regions of like ground cover. Regional assessment of ET is either through averaging values from micrometeorological stations with the region (e.g. CIMIS, AZMET and AgriMET), or indirectly possible through remote sensing methods, but usually corroborated or calibrated to local ground measurements (e.g. ALEXI model; see Neale, 2018). Daily evaporation and ET depths exhibit the known seasonal patterns that have much less year-to-year temporal variability than precipitation, though with changing climates these amounts are slowly increasing in many parts of the world. Evaporation 'losses' not associated with plant growth (indicated with red arrows in Figure 1.1) are not presently 'recoverable' for other uses, and they occur directly from precipitation, reservoirs, canals and soils at local to regional scales.

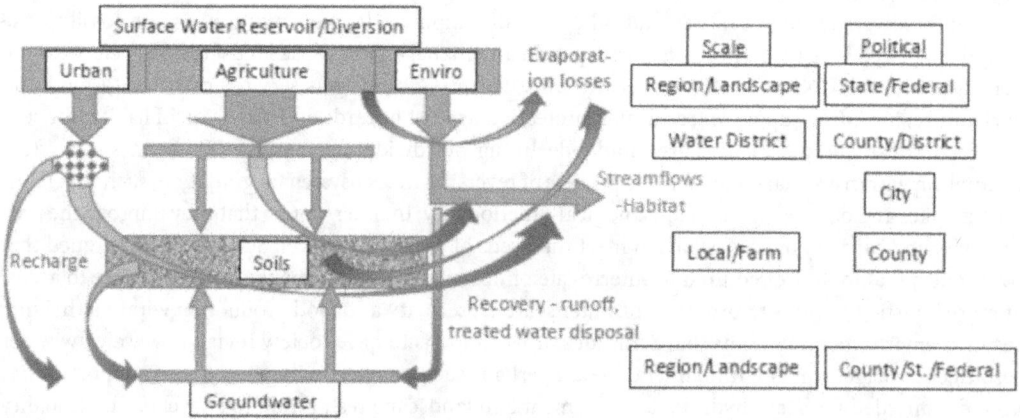

FIGURE 1.1 Conceptual model of hydrologic processes and associated scales operating in irrigated agriculture.

Soils are the mediating, temporary storage media for applied (irrigation or rainfall) water used by vegetation, crops and various other species for metabolism and growth. Their properties vary spatially across all of the scales (e.g. log-normally distributed hydraulic conductivity across the farm scale) and to some degree in the longer temporal scales associated with soil degradation (e.g. erosion and salinization) or restoration (improved soil health or tilth). They also vary in the microbial, animal and plant communities that they support, though some display common characteristics across the regional landscape. From the perspective of IWRM and proposed 'precision' or 'smart farming', the soil-dependent aspect of agriculture water applications is managed at the local scale using the locally derived precipitation and ET measurements. Soils also control the applied-water seepage rates and quality to deeper groundwater and possibly adjacent down-gradient streams and other surface water bodies in the landscape; however important as seepage fluxes measurements are, their direct measurement at any scale remains problematic and such fluxes are usually inferred as water-balance-driven closure terms in regional, or basin integrated surface-subsurface water modeling efforts. Seepage flows to surface or subsurface waters are considered 'recoverable' as indicated by the purple arrows in Figure 1.1. Groundwater storage as reflected by point measurements of groundwater levels across a region is generally considered to reflect district-wide or regional water supply as created by accumulated precipitation or applied-water seepage from years to centuries past and used in the present. In many regions, groundwater has only more recently been subject to regulation or management at the local scale or beyond.

Overlaying the landscape hydrology are the 'engineered' hydraulic systems that enable diversions of water from one locale to another in time and space at multiple scales. These socio-hydrologic systems are either directed at managing floodwaters to protect downstream urban and agricultural areas, or in terms of the focus here, intended to store and supply water for agricultural production at another location and later time. Both intentions alter the natural landscape and associated processes and are predicated on designation of 'water rights' that enable re-allocation of water resources from their source to another locale or region. How these water rights are created, viewed and enforced is a key element associated with IWRM across all scales and plays a critical role in the allocations of water to farms, cities and the 'environment'. Local and regional ET rates and areas of irrigated land drive water demands in the irrigated agricultural landscape. When associated with crop or animal production, these demands are 'beneficial' uses of water and efforts are made to limit their 'losses' at the farm and district distribution scales. On-farm seepage and tail-water recovery and district canal seepage and spillage 'losses' are potentially recoverable for future or downstream uses and if quantifiable may be included as beneficial water uses depending on the water right perspective adopted in the region. Consistent definition of the environmental benefit of these 'recoverable' losses, such as that for the value of in-stream flows, remains elusive in the agricultural landscape, but is under increasing scrutiny. Water evaporation from reservoirs and distribution canals remains 'losses' reducing water available for both the irrigated landscape as well as downstream flows and habitat that would have otherwise been available without the diversions. As alluded to in the introduction, within the social context the nature of the discussion about 'beneficial' uses and water 'losses' has been evolving and is captured in part by discussion of 'efficiencies', or paradoxes thereof, in the irrigated agricultural landscape.

1.3 Sustainable IWRM – Revised Perspectives?

Following establishment of water economic value or 'rights' that enable subsequent water diversions for irrigated agriculture and urban use, the trajectory toward eventual problems of water scarcity and security is well known, if not openly acknowledged. Contemporary examples sketched out below continue to evolve, that include loss of inland seas and related communities and fisheries to salinization of valleys and plains following irrigated agricultural development. Certainly, one aspect of the scarcity/security problem is growing population, but as this trajectory from defining water economic value to scarcity has occurred historically, it seems that it is rooted in the definition of water resources as an economic good rather than as a landscape property. Adequately defining this economic 'good' of water has been complicated by the range of water 'rights' definitions, especially in California where there is a mix of appropriative, riparian and historic water rights of differing value and related significance to IWRM.

For example, creation of the Salton Sea in southern California by uncontrolled Colorado River diversions in 1905 replaced a desert landscape with what became a thriving wildlife and recreational area, its survival and that of the associated communities depended on saline subsurface drainage flows from the irrigated Imperial Valley and urban wastewater flows from Mexico. Simultaneously, the Colorado River diversions depleted the thriving delta ecosystems in the Gulf of California, displacing several human and wildlife communities. As required by federal law, California reduced its take from the Colorado River and subsequently the Imperial Valley negotiated sale of 'conserved' water (seepage recovery) to the urban San Diego and Metropolitan (Los Angeles region) water districts. To do so, some lands were fallowed and application or irrigation efficiency in the Imperial Valley increased resulting in decreased subsurface flows to the Salton Sea. As a result, the Sea has been decreasing in area for the past couple decades, becoming increasingly saline and its wildlife and recreational values rapidly declining and the associated communities long gone. Failure to maintain some of the Sea will exacerbate poor air quality in the remaining communities of the Imperial and possibly adjacent Coachella Valleys.

In the western and southern San Joaquin Valley (SJV) of California, water diversions in northern California and from the southern Sierras enabled replacement of annual wetland/marsh ecosystems with irrigated agriculture and temporarily arrested decline of groundwater levels in the Valley associated with agricultural production during the middle half of the last century. With no effective Valley drainage outlet for subsurface drainage from irrigated lands, increasing soil salinization limited production, or resulted in fallowing lands in parts of the Valley, while market values for nut and other permanent crops increased water demands despite increasing application efficiencies. With more permanent, high-value crops in place, the recent (2010–2015) and current drought periods highlighted both perceived water scarcity concerns while depleting groundwater resources such that many groundwater-dependent rural communities went dry leading to real water security problems. Moreover, despite the wealth of food production in the SJV, the counties comprising the Valley are among those in the state and country with the greatest food insecurity (~21% of local population) that steadily increased during the past two decades (Pouslen, 2012). Recent formation of Groundwater Sustainability Agencies in several counties comprising the San Joaquin Valley, like many across the state, are tasked with developing pro-active groundwater sustainability plans (GSPs) by 2022. This was essentially a State government intervention toward encouraging local communities and stakeholders to come together to manage their groundwater for the benefit of all and plans to be implemented in the next two decades will likely involve some land fallowing and efforts to recharge groundwater locally (Hanak et al., 2019).

Irrigated hydrologic systems co-develop with the societies that create them and the resulting interdependency between the two creates reactionary feedback loops that are not always apparent. Zikalala (2018) considered the co-evolution of irrigation, water development and related policies or infrastructure in the Salinas Valley of the California central coast in terms of its impacts on the Salinas River and groundwater salinity from 1929 to 2017. The Salinas Valley is an interesting case study of this co-evolution because of the relatively long documented history of actions, and seemingly pro-active action of forming district-wide water agencies several decades ago. As Grismer and Zikalala (2019) noted, the economic drivers associated with national and international vegetable, lettuce and strawberry markets increased irrigated areas and year-round cropping within the Valley, while reactive water policies and infrastructure were developed to curb salinization of the surface- and ground-waters. Multiple reactive actions taken by the local water agency over time included adoption of on-farm water conservation measures, increased use of drip irrigation, replacement of groundwater for irrigation with treated wastewater, and construction of additional surface water storage and diversion structures. Interestingly, each major action or combination thereof resulted in a 1–3-year increase in groundwater levels and decreased salinized groundwater area followed by further declines in groundwater levels and quality 2–5 years after implementation. Currently, the water agency is pursuing additional surface water rights in the upper Valley as a partial replacement for groundwater pumping in the lower Valley. However, absent efforts to severely limit groundwater extractions in the lower Valley as part of their local GSP efforts, we expect from the current historical trajectory that further development of surface water supplies to address the current scarcity problems will likely only briefly relieve groundwater stress followed again by declining groundwater conditions. Clearly, the interplay of national and global interests with the temporal and spatial variations of the water resources at the local and regional scales, which are often the determining factors for water scarcity, leads to the complex systems dynamics (Savenije and Zaag, 2000). Increasingly, water policies and market forces tend to be the main drivers affecting local IWRM.

Considering unintended consequences of IWRM in the irrigated hydrology landscape as briefly described in the previous examples, Scott et al. (2014) and Grafton et al. (2018) identified efficiency, scale or sectoral 'paradoxes'. For example, paradoxically, requiring greater on-farm application efficiency or adoption of other water conservation measures appears to result in increased land area irrigated or district/regional water use (e.g. Salinas Valley and SJV). Thus, in the absence of effective policy to constrain irrigated area expansion using 'saved water', increased application efficiency can aggravate scarcity, deteriorate water/soil quality, and impair river basin resilience through loss of flexibility and redundancy. Recoverable irrigation seepage or return flows that support key basin ecological functions are not actually available for increased human water appropriation without adverse consequences on both the ecosystem and the communities it supports. The scale paradox may be described in part by the Imperial Valley–Salton Sea connection whereby water scarcity and salinity effects in the lower basin (Sea) may partly be offset in the short term through increased surface water flows from the Valley, an unlikely scenario presently. However, such diversions would reduce ecological flows and potentially increasing salinity (soils and surface waters) while likely degrading riparian and estuarine ecosystems (e.g. Gulf of California) and possibly non-irrigation human water uses for urban supply and energy generation, examples of the sectoral paradox. Overall, eliminating slack in the hydrologic system through stringent water conservation and re-allocation of recoverable water can result in demand 'hardening'. Such demand then implies future crop losses or limits imposed on irrigated areas under conditions of water shortage resulting from changing climate conditions or enforcement of senior water rights in over-appropriated basins. Furthermore, some have suggested that there is a perverse incentive associated with the fact that environmental costs resulting with water diversions to the profit of agricultural production are not directly passed on to the consumer, but rather appear as 'undesirable' taxes associated with environmental restoration efforts.

Within this current view of IWRM, as noted by Schmidt (2017), water management is a zero-sum game pitting human consumptive use (agricultural and urban use) against 'environmental' use as though these are competing interests. Should the thinking be about inevitable tradeoffs between irrigation-efficiency interventions, ecosystem services and agricultural production? If so, what are the policy frameworks that encourage or discourage efficiency, and in turn, mediate impacts on river basin resilience? Or, is there a Leopoldian holistic view possible that obviates the question of 'tradeoffs' to considering overall improvement of basin/community resilience in the face of changing climates and populations? Clearly, in most developed irrigated agricultural landscapes, returning to pre-development ecosystem conditions is no longer an option, though reclaiming or redefining water value or rights (e.g. Murray-Darling basin, Australia) may enable creative solutions to manage the water needs across the human-ecosystem landscape. For example, understanding the pre-alteration landscape hydrology may be helpful, that is, "natural streamflow patterns and the role they play in supporting ecosystem health, to develop the management strategies that balance human and ecosystem needs" (Grantham et al., 2019).

1.4 Irrigated Hydrology and the Landscape – Needs and Examples

Of the many interacting processes in the earth system, human processes are the dominant drivers of change in water, nutrient and energy cycles and in landscape evolution (Crutzen and Stoemer, 2000; Röckstrom et al., 2009; Zalasiewicz et al., 2010). Water extraction for human use unleashes several unintended consequences such as land subsidence, seawater intrusion, altered streamflow and loss of habitats and fisheries, and water quality degradation. Human management of the water cycle results in complex coupled human–hydrological systems spanning physical infrastructure, regional economics to policy and legal frameworks governing water availability and use. New theoretical and quantitative frameworks to understand and mediate the 'competition' for water between humans and the environment through generating new understanding of how they coexist and interact are thus critically needed (Gleick and Palaniappan, 2010; Grafton et al., 2013; Ostrom, 2009). Hydrological predictions are needed that can explicitly account for both changes in landscape structure as well as the possibility for new dynamics that might emerge from such human–environment interactions. A conceptual model that takes into account the changing structure of the coupled human–water systems and how it affects the resulting dynamics is likely required (Kallis, 2007, 2010).

Sivapalan et al. (2012, 2014) suggested that we thoroughly study a few, well-documented human–water systems to gain more detailed insights into causal relationships that then complement the typical temporal and spatial analyses. Since socio-hydrology is about co-evolution and feedbacks operating at multiple scales, the notions of optimality and goal functions are likely to be important and useful, just as they have been in eco-hydrology. Zikalala (2018) takes one possible approach to analysis of the coupled human–water system for the Salinas Valley. The main objective was to analyze the linkages between long-term temporal distributions of land uses, changes in water policy or technology and ecosystem services as represented by surface and groundwater quality and groundwater elevations. Advances in technology and associated irrigation systems have increased agricultural productivity in the Salinas Valley with attended consequences on Valley groundwater, surface water and soil resources. Technology adoption sets in motion a recursive process of technological adaptation, social institution formation and environmental change; thus, a long-term perspective on socio-environmental dynamics is essential. First considering the history of agricultural development in the Valley, Zikalala explores the dynamics between changes in the agricultural landscape, water management, use of commercial fertilizer and soil amendments, trade opportunities, environmental policies and/or initiatives and the development of the salinity problem. Then she deploys the mediation analysis to address the question of how and why the water management systems – water supply, water demand, environmental and policy – are related over time. Using a dynamic factor analysis, she also develops an ensemble of weighted multi-crop indices to predict the total irrigated area and to simulate historical total irrigated crop area based on historical cropping patterns and gross crop values. The proposed analysis framework provides a quantitative analysis of the social and hydrologic drivers of water resources developments. With the quantitative linkages in changes in cropping patterns, water supply infrastructure, irrigation and conservation technologies, and economics of agricultural production and policy, Zikalala determined the key drivers of agricultural development in this area and how they might be manipulated to enable the Valley agriculture to continue.

Sabzi et al. (2019) further suggested that developing models that explicitly account for socio-hydrologic process feedbacks at multiple scales facilitates the critical stakeholder participation required to advance broader applications of IWRM across the landscape. In terms of long-range model predictions, the goal is to provide multiple scenarios that include projection of alternative, plausible and co-evolving trajectories of the socio-hydrologic system. Such scenarios may include revised definitions of water rights. This approach may yield additional "insights into cause-effect relationships and help stakeholders identify safe or desirable operating space", or system thresholds to be avoided. Typically, adjustments or modifications of water rights and efforts to resolve the competing water demands in over-allocated basins have required some government or legal intervention but always in concert with stakeholder groups. In the heavily managed Oxnard Plain groundwater basin of Ventura County California, growers led an initiative authorized by the Fox Canyon Groundwater Management Agency to create a centralized market for the individual landowners to buy and sell groundwater after obtaining a US Department of Agriculture $1.9 million grant to expand the original pilot project in 2017. An independent university economic group spent more than a year guiding about 50 farmers, city representatives and environmental stewards in the development of recommendations for the market-based remedy for groundwater depletion.

Considering a larger stakeholder group in Texas, regional competition for Edwards Aquifer groundwater historically pitted municipal and industrial users against agricultural interests and surface water rights holders downstream of the springs. This resulted in disputes involving the 'rule of capture' in Texas, private property rights, and endangered species. Through a 1993 federal court ruling, the Texas Legislature created the Edwards Aquifer Authority and an associated water market. Uncertainties over legal rights to groundwater and over how to structure transfer agreements were the major barriers to willingness to engage in transfers, but these were not insurmountable hurdles preventing water transfers. By 2013, stakeholders developed a regional demand management plan that effectively ended the disputes and retained state rather than federal control of this common resource. As Kaiser and Phillips (1998) noted, "market transfers are not an elixir for all of the region's water problems. However, they provide a means to respond to changing economic, environmental and social water needs". The Edwards Aquifer Authority has several planning, regulatory and managerial tools that could be used in combination with market transfers to allocate water in the Edwards in ways that minimize economic, political and social instability.

An example at the local scale in northern California illustrates the importance of definitions of water rights and of the role that stakeholders play in sustaining the irrigated landscape. The ~1,750 ha Shasta Big Springs Ranch (SBSR) is nestled in the valleys below Mount Shasta where hay and cattle farming work together with senior water rights and several km of key cold-water salmon habitat. In October 2018, "Shasta Big Springs Ranch is nearly dry, except for a ribbon of silvery thistles along the riverbank. Today, it's a source and symbol of the polarizing divide between farmers and conservationists facing an increasingly water-scarce future. Water is what makes this land so special. Hundreds of thousands of threatened red coho and hook-jawed chinook salmon once swam here, nearly 200 miles from where the Klamath River meets the Pacific Ocean, to clear tributaries in Siskiyou County fed by icy springs and thawing snow. Fish in this mineral-rich oasis grow nearly twice as fast and gain three times as much weight as those in nearby streams. But by the 2000s, their numbers had dwindled to just a few dozen adults each year. Since size largely determines whether juvenile fish survive, conservation organizations have been interested in this particular property" (Petersen-Rockney, 2019). Together with the California Department of Fish and Wildlife (CDFW), the Nature Conservancy purchased the ranch in 2009 with the goal of demonstrating that salmon habitat and other wildlife conservation could coexist with hay production and cattle ranching. Efforts by the Conservancy to engage the other local ranchers were unsuccessful and the 2007–2008 recession left the organization facing a $14 million mortgage that forced transfer of the Ranch senior water rights to the CDFW in 2012 for use in restoration of downstream habitat and loss of irrigation at the Ranch. Without irrigation, what short-lived partnerships between ranchers and conservationists existed ended and the local communities remain suspicious of CDFW such that the local director penned an editorial letter explaining their actions and asking for community support in acquiring the ranch in May 2018.

In the next valley over from SBSR lies the Scotts River Ranch (SRR) in the Scotts Valley that has been successfully farming hay, cattle and salmon for the past 30 years on a similar size property. SRR contends that farming water is "both a way of thinking about the land and a set of practices with the goal of producing food, in this case, beef, with as little water as possible" (Petersen-Rockney, 2019). They consider stewardship of not only ranch pasture, but also the wider landscape employing many of the land management practices promoted by the Nature Conservancy. These practices included improving soil infiltration and soil-water storage capacity, while maintaining channel meanders and beavers to help promote seepage and groundwater recharge. Where some trees are cleared at SRR to better practically manage pastures, other riparian areas are planted, while stream bars are maintained, sediment removed and gravel bottoms restored for salmon habitat after floods. The SRR has been able to develop and implement such practices and continue hay production while also growing the SRR cattle herd as some neighboring herds and pastures languished during the recent drought period. The US Department of Fish and Wildlife (USFWS) recognized his efforts. In turn, they issued the SRR an exemption from the federal Endangered Species Act and used it to relocate some 30,000 juvenile coho salmon during the

2010–2015 drought. In 2010, SRR was the first in the Scotts Valley to obtain a Conservation Easement from the Siskiyou Land Trust that protects the land from development in perpetuity. "For all of the stakeholders involved in Siskiyou County, there's a consensus that in order to be successful, conservation efforts must be interwoven with local values. For most family farmers, saving money is more compelling than saving fish. On the other hand, 'farming water' also touches on deep-seated rural ideals of self-efficacy, independence, and private property rights" (Petersen-Rockney, 2019) and may be a way of advancing both the conservation and ranching community's interests. A key element in both is keeping applied and snowmelt waters in the local landscape and groundwater, thus improving or maintaining seepage throughout the year is central to the SRR IWRM. However, actual seepage rates from stream channels or pastures that are critical toward developing groundwater management plans remain largely unknown.

Central to annual IWRM locally and at the basin scales is an assessment of net rates of seepage to groundwater from canals, streams, lakes and rain or irrigation-driven deep percolation (and associated lag times) to groundwater supplies that enable quantification of the basin water balance (Grismer, 2012a; Grismer and Asato, 2012). In most basins considering groundwater management plans, such quantification has relied on modeling efforts in which annual basin seepage quantities are determined as closure terms in the water balance, as regional-scale direct measurement of seepage rates is problematic. Historically, seepage rates from ponds, conveyance structures and canals in irrigated regions were measured, and in some cases, deep percolation rates below irrigated lands were estimated from subsurface drainage flowrates (Grismer et al., 1988; Grismer and Tod, 1991; Tod and Grismer, 1991). McCullough-Sanden and Grismer (1988) and later Grismer and McCullough-Sanden (1989) used a combination of direct infiltrometer seepage measurements, lab-measured hydraulics conductivity of soil cores and water balance calculations to determine seepage rates from subsurface drainwater collection ponds of the SJV. They found that seepage rates from multiple infiltrometers left in the pond for several weeks were quite small, on the order of 1–10 mm/day, and log-normally distributed in the clay loam soils comprising the base of most ponds. The log-normal mean of the hydraulic conductivity of the matching soil cores from the infiltrometers was six to seven times smaller than the average seepage rates from the infiltrometers and water balance calculations. Similarly, using infiltrometers to measure the seepage rates from low-flow stream channels silty-sandy/gravel bottoms in Putah Creek of central California, Grismer (2021) again found the values to be log-normally distributed with a log-normal mean of ~660 mm/day. Landon et al. (2001) and Solder et al. (2016) obtained similar seepage results for several sandy bottom creeks in Nebraska with mean values on the order of 100 mm/day. In Iraq, Mohsen and Mohammed (2016) measured the canal seepage losses of roughly 400–4,000 mm/day in lined and un-lined distribution channels, respectively, – values far greater than recommended values of 2–20 mm/day by the United Nation Food & Agriculture Organization (FAO) or United States Bureau of Reclamation (USBR) for distribution canals. Historically, in the United States, Sonnichsen (1993) summarized the results of canal seepage studies and created Figure 1.2 illustrating the log-normal and grain-size dependence of measured seepage rates from the canals. Consistent with the stream infiltrometer results above, the values ranged from about 30 mm/day for concrete-lined canals to about 600 mm/day for canals having sandy-gravel bottoms. In all of these studies, seepage rates estimated from grain-size distributions had greater variabilities and were of limited value in predicting seepage rates. Such results indicate something of the complexity of estimating seepage rates and only apply to actual seepage rates in the top roughly half-meter of bed materials; presumably this seepage eventually reaches shallow groundwater within the next 1–100 m depth, but much may be distributed laterally augmenting stream flows or appearing as downslope seeps.

While estimates of seepage rates from canal, rivers, lakes and irrigated fields are critical to determination of groundwater recharge rates and water quality across the landscape, most groundwater planning studies rely on soil–water balance, or soil-texture-based estimates of deep percolation rates. Further, they then presume vertical transport to groundwater within the next year to depths of ~100 m below ground surface (Grismer, 2012a; Grismer and Asato, 2012). Only recently have some studies begun

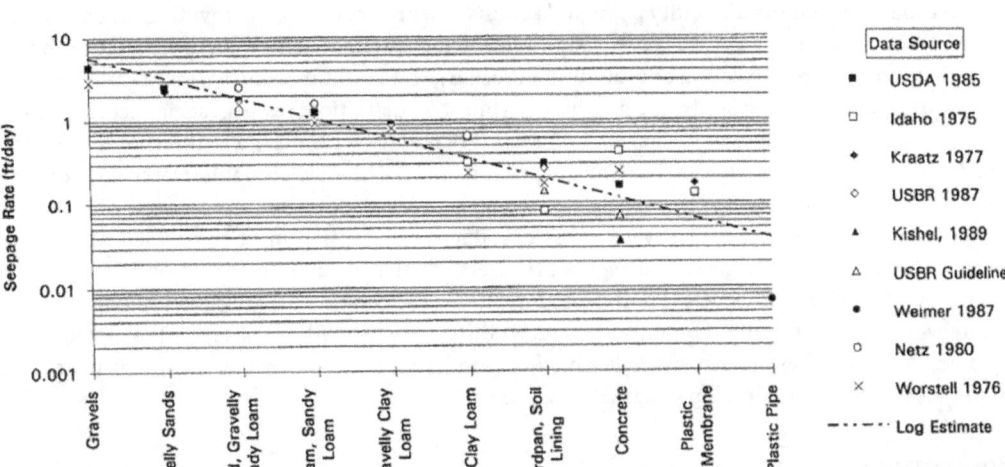

FIGURE 1.2   Historically measured canal seepage rates in the United States (taken from Sonnichsen, 1993).

considering groundwater augmentation through flooding of cropped fields, or orchards during winter rainy periods, though the approaches continue to rely on soil–water balance methodologies (e.g. Dahlke et al., 2018). Such practices may become an important management tool in irrigated basins subject to periodic drought, adding some resiliency to the landscape hydrology. Aside from those studies noted above, few contemporary direct measurements of seepage rates (Grismer, 2021) or volumes exist in the literature and this remains an important gap in understanding the irrigated hydrology landscape as well as socio-hydrologic planning.

1.5 Conclusions

Irrigation hydrology is at the nexus of water-energy-food security from the local to regional and in some cases the transnational scales in many agricultural regions of the world. Transnational considerations of irrigated hydrology include both the diversion of streamflows between upstream and downstream countries as well as the export of water as foodstuffs or forage from the poorer countries to richer, or from low-priced to high-priced water areas (e.g. alfalfa hay shipped from southern California to Saudi Arabia; see Markham, 2019). Consideration of irrigated agriculture as distinct from the regional landscape ecosystem of which it is a part has likely led to a number of counter-intuitive unintended consequences associated with water scarcity and food security. In the United States, it is surprising to find that many of the top agricultural production counties of the country also have the highest food insecurity, poor drinking water quality and related poverty rates, despite this abundance of production associated with irrigated agriculture. Designation of water as a tradable commodity separate from its landscape of origin when combined with market-driven agricultural systems has led to and will likely continue to provoke local issues of water and food security. Consideration of the landscape ecosystem as a whole within which water is an integral part and essential element critical toward meeting the basic needs of habitats and communities together with agricultural production likely will require reconsideration of the basic underpinnings of what water is to society. As irrigated agriculture in many basins has already altered the basin and perhaps adjoining basin hydrologic systems, landscape restoration takes on different meaning that is associated with developing sustainability and resilience within the habitat, human community and agricultural sectors. Generally, the redefinition of water requires government intervention within the basins affected, preferably working with the primary stakeholders to develop the water management sustainability plans and actions that are pro-active in providing robust broadly applied IWRM within the context of the socio-hydrologic system. Such IWRM will require careful determination of deep percolation or groundwater recharge rates/volumes and when they reach underlying aquifers critical to the landscape hydrology. Almost by definition, resilient social-ecological systems are expected to continually change and adapt yet remain within critical thresholds; however, climatic, hydrologic and societal drivers often appear as shocks (floods, droughts, wars and economic crises) that can push these systems beyond resilience thresholds leading to some type of 'collapse'. Possible programs that incentivize water users to reduce consumption must be coupled with strong policies, pre-determined triggers and regulatory oversight that include increased understanding of water users' behaviors and beliefs (Pfeiffer and Lin, 2014). Stakeholders associated with irrigated agriculture working with the local communities and wildlife managers will play a critical role in developing possible system trajectories that help anticipate undesirable exceedance of key socio-hydrologic thresholds as the regions develop sustainability plans.

References

Blöschl, G. (2001). Scaling in hydrology. *Hydrological Processes*, 15, 709–711.

Crutzen, P.J., E.F. Stoemer. (2000). The Anthropocene. *Global Change*, 41, 17–18.

Dahlke, H.E., A.G. Brown, S. Orloff, S. Putnam, A. O'Geen. (2018). Managed winter flooding of alfalfa recharges groundwater with minimal crop damage. *California Agriculture*, 72(1), 65–75. https://doi.org/10.3733/ca.2018a0001.

Gleick, P.H., M. Palaniappan. (2010). Peak water limits to freshwater withdrawal and use. *Proceedings National Academy of Science*, 107, 11155–11162.

Grafton, R.Q., Pittock, J., Davis, R., Williams, J.S., Fu, G., Warburton, M., Udall, B., McKenzie, R., Yu, X., Che, N., Connell, D., Jiang, Q., Kompas, T., Lynch, A., Norris, R., Possingham, H., Quiggin, J. (2013). Global insights into water resources, climate change and governance. *National Climate Change*, 3, 1–7.

Grafton, R.Q., Williams, J., Perry, C.J., Molle, F., Ringler, C., Steduto, P., et al. (2018). The paradox of irrigation efficiency. *Science*, 361(6404), 748–750. https://doi.org/10.1126/science.aat9314.

Grantham, T.E., Zimmerman, J.K.H, Carah, J.K., Howard, J.K. (2019). Stream flow modeling tools help inform environmental water policy in California. *California Agriculture*, 73(1), 33–39.

Grismer, M.E. (2012). Estimating agricultural deep percolation lag times to groundwater in the Antelope Valley, CA. *Hydrological Processes*, 27(3), 378–393, https://doi.org/10.1002/hyp.9249.

Grismer, M.E. (2021). Point- and reach-scale measurements are important to determine accurate seepage rates in controlled flow channels. *California Agriculture*, 75(2), 74–82. https://doi.org/10.3733/ca.2021a0013.

Grismer, M.E., B.L. McCullough-Sanden. (1988). Evaporation pond seepage. *California Agriculture*, 42(1), 4–5.

Grismer, M.E., B.L. McCullough-Sanden. (1989). Correlation of laboratory analyses of soil properties and infiltrometer seepage from drainwater evaporation ponds. *Transactions of the American Society of Agricultural Engineers*, 32(1), 173–176.

Grismer, M.E., C. Asato. (2012). Oak savanna versus vineyard soil-water use in Sonoma county, CA. *California Agriculture*, 66(4), 144–152.

Grismer, M.E., I.C. Tod. (1991). Drainage of clay overlaying artesian aquifer: I. Hydrologic Assessment. *Journal of Irrigation and Drainage Engineering*, 117(2), 255–270.

Grismer, M.E., I.C. Tod, F.E. Robinson. (1988). Drainage system performance after 20 years. *California Agriculture*, 42(3), 24–25.

Grismer, M.E., P.G. Zikalala. (2019). Recycled water use, water demand and infrastructure in the Salinas Valley, CA. *ASA California Plant and Soil Conference Proceedings*, pp. 18–28.

Hanak, E., A. Escriva-Bou, B. Gray, S. Green, T. Harter, J. Jezdimirovic, J. Lund, J. Medellín-Azuara, P. Moyle, N. Seavy. (2019). *Water and the Future of the San Joaquin Valley*. Report by the Public Policy Institute of California. https://www.ppic.org/wp-content/uploads/water-and-the-future-of-the-san-joaquin-valley-overview.pdf.

Kaiser, R.A., L.M. Phillips. (1998). Dividing the waters: Water marketing as a conflict resolution strategy in the Edwards Aquifer Region. *Natural Resources Journal*, 38, 411–444.

Kallis, G. (2007). When is it coevolution? *Ecological Economics,* 62, 1–6.

Kallis, G. (2010). Coevolution in water resource development: The vicious cycle of water supply and demand in Athens, Greece. *Ecological Economics,* 69, 796–809.

Kumar, P. (2011). Typology of hydrologic predictability. *Water Resources Research*, 47, W00H05, https://doi.org/10.1029/2010WR009769.

Landon, M.K., D.L. Rus, F.E. Harvey. (2001). Comparison of instream methods for measuring hydraulic conductivity in sandy streambeds. *Groundwater*, 39(6), 870–885.

Lankford, B.A. (2013). *Resource Efficiency Complexity and the Commons: The Paracommons and Paradoxes of Natural Resource Losses, Wastes and Wastages*, Routledge, London, UK.

Markham, L. (2019). Who keeps buying California's scarce water? Saudi Arabia. *The Guardian*, 25 March. https://www.theguardian.com/us-news/2019/mar/25/california-water-drought-scarce-saudi-arabia.

McCool, D. (2018). Book Review: Water: Abundance, scarcity, and security in the age of humanity. *The American Historical Review*, 123(1), 269–270, https://doi.org/10.1093/ahr/123.1.269.

McCullough-Sanden, B.L., M.E. Grismer. (1988). Field analysis of seepage from drainwater evaporation ponds. *Transactions of ASAE*, 31(6), 1710–1714.

Melsen, L.A., J. Vos, R. Boelens. (2018). What is the role of the model in socio-hydrology? Discussion of "Prediction in a socio-hydrological world". *Hydrological Sciences Journal*, 63(9), 1435–1443, https://doi.org/10.1080/02626667.2018.1499025.

Merz, R., G. Blöschl. (2008). Flood frequency hydrology: 1. Temporal, spatial, and causal expansion of information. *Water Resources Research*, 44(8), W08432.

Mohsen, M.H., O.M. Mohammed. (2016). Comparison between the measured seepage losses and estimation and evaluated the conveyance efficiency for part of the Hilla Main Canal and Three Distributary Canals (HC 4R, HC 5R and HC 6R) of Hilla-Kifil Irrigation Project. *Civil and Environmental Research*, 8(2), 1–10.

Montanari, A., G. Blöschl, M. Sivapalan, H.H.G. Savenije. (2010). Getting on target. *Public Service Review: Science and Technology*, 7, 167–169.

Neale, C.M.U. (2018). Satellite-based evapotranspiration estimates for irrigation water management. *Colorado Water*, 35(6), 38–39.

Ostrom, E. (2009). A general framework for analyzing sustainability of social-ecological systems. *Science*, 325, 419–422.

Petersen-Rockney, M. (2019). Farming water. *Grist*, February 26th, Cover Story.

Pfeiffer, L., C.-Y.C. Lin. (2014). Does efficient irrigation technology lead to reduced groundwater extraction? Empirical evidence. *Journal of Environmental Economics and Management*, 67(2), 189–208.

Pouslen, P. (2012). *Hunger and Food Insecurity. Update for the CCASSSC Region*. CSU-Fresno.

Rath, A., S. Samantaray, R.D. Raj, P.C. Swain, S. Eslamian. (2021). Impact assignment of integrated watershed management in the micro watersheds of Sambalpur District, Odisha, India, *Chapter 23 in Handbook of Water Harvesting and Conservation*, Vol. 2: Case Studies and Application Examples, Eds: Eslamian, S. & Eslamian, F., John Wiley & Sons, Inc., New Jersey, pp. 341–358.

Röckstrom, J., W. Steffen, K., Noone, K., Persson, Å., Chapin III, F.S., Lambin, E., Lenton, T.M., Scheffer, M., Folke, C., Schellnhuber, H., Nykvist, B., De Wit, C.A., Hughes, T., van der Leeuw, S., Rodhe, H., Sörlin, S., Snyder, P.K., Costanza, R., Svedin, U., Falkenmark, M., Karlberg, L., Corell, R.W., Fabry, V.J., Hansen, J., Walker, B., Liverman, D., Richardson, K., Crutzen, P., Foley, J. (2009). Planetary boundaries: Exploring the safe operating space for humanity. *Ecology and Society*, 14, 32. https://doi.org/10.5751/ES-03180-140232.

Sabzi, H.Z., S. Rezapour, R. Fovargue, H. Moreno, T.M. Neeson. (2019). Strategic allocation of water conservation incentives to balance environmental flows and societal outcomes. *Ecological Engineering*, 127, 160–169.

Savenije, H.G. and P. van der Zaag. (2000). Conceptual framework for the management of shared river basins with special reference to the SADC and EU. *Water Policy*, 2(1–2), 9–45.

Schmidt, J.J. (2017). *Water: Abundance, Scarcity, and Security in the Age of Humanity*. New York University Press, New York, 308 p.

Scott, C.A., S. Vicuna, I. Blanco Gutierrez, F. Meza, C. Varela Ortega. (2014). Irrigation efficiency and water-policy implications for river-basin resilience. *Hydrology and Earth System Sciences*, 18(4), 1339–1348.

Sivapalan, M., H.H.G. Savenije, G. Blöschl. (2012). Socio-hydrology: A new science of people and water. *Hydrological Processes*, 26, 1270–1276.

Sivapalan, M., M. Konar, V. Srinivasan, A. Chhatre, A. Wutich, C.A. Scott, J.L. Wescoat, I. Rodriguez-Iturbe. (2014). Socio-hydrology: User-inspired water sustainability science for the Anthropocene. *Earth's Future*, 2, 225–230.

Solder, J.E., T.E. Gilmore, D.P. Genereux, D.K. Solomon. (2016). A tube seepage meter for in situ measurement of seepage rate and groundwater sampling. *Groundwater*, 54(4), 588–595.

Sonnichsen, R.P. (1993). *Seepage Rates from Irrigation Canals*. Washington State Department of Ecology Water Resources Program Open-File Report 93-3, USA.

Srinivasan, V., M. Sanderson, M. Garcia, M. Konar, G. Blöschl, M. Sivapalan (2017). Prediction in a socio-hydrological world. *Hydrological Sciences Journal*, 62(3), 338–345.

Tod, I.C., M.E. Grismer. (1991). Drainage of clay overlaying artesian aquifer: II. Technical analysis. *Journal of Irrigation and Drainage Engineering*, 117(2), 271–284.

Wallender, W.W., M.E. Grismer. (2002). Irrigation hydrology: Crossing scales. *Journal of Irrigation and Drainage Engineering*, 128(4), 203–211.

Williams, J. (2017). Book Review: Water: Abundance, scarcity, and security in the age of humanity. *Water Alternatives*, 10(3), 934–936.

Zalasiewicz, J., M. Williams, W. Steffen, P. Crutzen. (2010). The new world of the Anthropocene. *Environmental Science and Technology*, 44, 2228–2231.

Zikalala, P.G. (2018). Social and hydrologic drivers of salinity in the Salinas Valley Agricultural Watershed, CA. *PhD Dissertation in Hydrologic Science*, UC Davis, Davis, CA, 189 p.

2

Infiltration and Irrigation Management

Mohammad Mahdi
Dorafshan and
Saeid Eslamian
*Isfahan University
of Technology*

2.1 Introduction

In recent years, water scarcity crisis has now become one of the most important problems in many countries. In this regard, competition over using freshwater in the development of urban engineering, industry, and agriculture has significantly reduced freshwater resources in the world (Amer, 2010; Kang et al., 2010). The issue of ensuring food security for the growing population of the world requires allocating a large part of freshwater resources (about 70%) to agriculture industry (FAO, 2002), as well as increasing the area under cultivation or adopting strategic approaches to enhance the level of performance per area unit (Deng et al., 2006). However, the global limitation of water resources limits increased production by increasing the area under cultivation (Douh et al., 2013). Hence, utilizing management strategies to properly consume these limited resources plays a crucial role in increasing the efficiency of water consumption in such circumstances (Rahil and Antonopoulos, 2007; WRI, 2005; Morison et al., 2008). Generally, the use of modern irrigation systems for the proper use of water seems to increase the area under cultivation and yield, in addition to reducing the water consumption (Wang et al., 2007). The use of water management and minimum leaching coefficient during the irrigation season for saving water consumption causes soil salinity and salinization, which are among the important challenges in arid and semi-arid regions (Abedi-Koupai et al., 2006; Jalali et al., 2008).

The issue of infiltration is one of the physical characteristics of soil that is affected by water quality and water management. In this way, infiltration and its variation, and ultimately, its effect on the efficiency of water are of great importance in designing and managing irrigation systems. The infiltration is regarded as the most important factor affecting surface irrigation. This parameter not only controls the amount of water infiltrated the soil but also affects the progress of surface water. The equation of water infiltration to soil plays an important role in evaluating, designing, or simulating irrigation systems.

DOI: 10.1201/9780429290114-3

In the process of water infiltration into the soil, soil moisture varies with time and location, and water redistribution in the soil depends heavily on the irrigation method, soil type, plant root distribution, and applied water content (Machiwal et al., 2006). Consequently, the dynamic properties of the soil taking all of these factors into account should be anticipated for the design and management of an efficient irrigation system (Fernández-Gálvez et al., 2006). The main drawbacks and challenges of using the surface irrigation methods are included, but not restricted to: (1) In case of inaccurate implementation and poor management, the efficiency of surface irrigation will be dramatically decreased and hereby lead to adverse effects on agriculture in the region of interest. (2) Developing the surface irrigation methods has to be done along with the full soil drainage; otherwise, the undesired issues like wetland and salinity will deprive the cultivability. Despite the high efficiency of pressure irrigation methods, these methods encounter various limitations that prevent their widespread usage and development, including the need for integration of agricultural land, investment, pipes clogging, and salinity. Therefore, the objectives of this chapter are: (1) introducing the mechanism of infiltration and the effective factors, (2) choosing the right irrigation method, and (3) sensitivity analysis of surface irrigation hydraulic models.

2.2 Mechanism of Infiltration

2.2.1 Infiltration

Infiltration is the most important physical property of soil from the viewpoint of agriculture. Infiltration refers to the downward entry of water into the soil surface. In this way, the cumulative infiltration is defined by the amount of the infiltrated water over a specific time (for example, a full day). On the other hand, the infiltration rate is determined by the average amount of water entered into the soil over a period while the instantaneous infiltration rate is defined as the velocity of water infiltration into the soil at any given time. Infiltration occurs in the one-dimensional way (vertical movement of water) in flooding and tape irrigations and a two-dimensional way in furrow irrigation (vertical and lateral movement of water). The fundamental relationship describing the movement of water into the soil was first obtained from Darcy's experiments in 1856. Darcy found that the rate of flow in porous mediums was directly related to the hydraulic gradient (Woolhiser and Brakensiek, 1982). Darcy's law can be shown as follows:

$$q = -K \frac{\partial H}{\partial S}, \tag{2.1}$$

where:

q is the flow in S direction ($L^3 L^{-2} T^{-1}$),

$\dfrac{\partial H}{\partial S}$ is the hydraulic gradient in S direction (m/m), and

K is the soil hydraulic conductivity in S direction (LT^{-1}).

Darcy conducted his experiments on the saturated soil columns and derived the law under above conditions. In another research, Richards extended Darcy's law to the unsaturated condition, except that the hydraulic conductivity was a function of the matrix potential or the amount of soil moisture in the unsaturated conditions (Richards, 1931). The speed of water entering the soil, or in other words, the velocity of water infiltration to rain, is one of the most important parameters in designing and managing the irrigation systems. Evaluating the amount of cumulative infiltration of water into the soil reveals the high speed of the infiltration at the beginning of the process. Figure 2.1 shows the variation curve of the cumulative infiltration relative to time. As observed, the cumulative infiltration curve gradually becomes smooth, indicating that a smaller amount of water is entered into the soil in a certain amount of time. Therefore, it can be concluded that the infiltration rate is reduced over time.

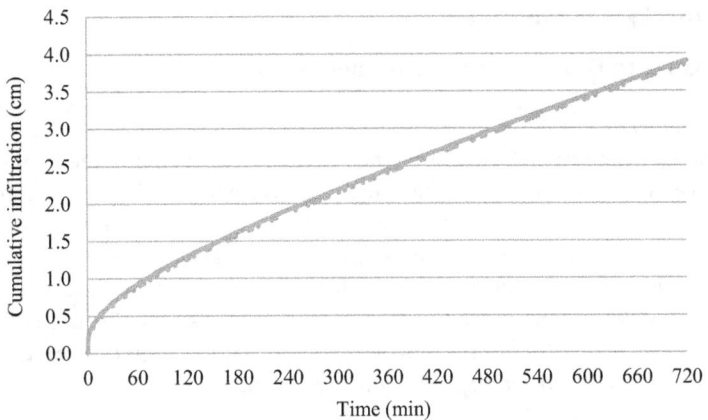

FIGURE 2.1 Cumulative infiltration curve for soil.

The infiltration rate curve of water to soil changes with the percentage of primary moisture in the soil. In general, an increase in the primary moisture decreases the infiltration rate of water into the soil (although the final infiltration rate will be the same eventually), representing that the level of surface runoff is higher during raining on the moist soil compared to when the soil is initially dry (Hillel, 2003; Sande and Chu, 2012). The entry of water into the soil occurs because of the simultaneous effect of all of the gravity force, capillary action, and surface absorption. The gravity force acts only in the vertical direction while the capillary action and surface absorption initially act both in the horizontal and vertical directions when the soil is dry and the capillary pores are free of water. As the pores gradually become saturated with water, only the gravity force intervenes and the infiltration flow becomes mainly vertical. For this reason, the infiltration rate is high at the beginning of the entry of water into the soil and then, gradually decreases and reaches a fixed level, which is the only result of the gravity force. Thus, the amount of water infiltrated the soil increases cumulatively over time, regardless of the soil surface condition. The water infiltration to the soil depends on factors such as the soil texture and structure, vegetation, land slope, and most importantly the dispersion capability of the soil surface particles (Vahabi and Mahdian, 2008). If the soil surface layer contains a large number of exchangeable sodium ions, the ions disperse the soil and prevent the water infiltration, implying that the infiltration of water into the soil is significantly low in clay and silty layers.

The spatial and temporal variability of infiltration rate has greatly complicated the management of the irrigation system. For example, researchers Oyonarte et al. (2002) and Childs et al. (1993) have shown that changing soil characteristics changes the characteristics of surface irrigation infiltration. Given that soil variability affects the distribution of infiltration depth, the use of only one infiltration equation for the entire farm significantly affects the efficiency of irrigation. Since infiltration properties are a function of time and place, a relatively large number of field measurements are required to represent the average farm conditions. In recent years, various methods have been used to reduce the need for measuring farm data required for representing the dynamic characteristics of water in the soil. Some of these examples include the infiltration process (Sharma et al., 1980), hydraulic conductivity (Tuli et al., 2001), and soil variability (Warrick, 1980). Scaling, as one of these methods, was first developed by (Miller and Miller, 1956) based on the theory of similar environments in water and soil knowledge. According to the theory of similar environments, two soils can be similar if there is a scale factor (such as γ) that can turn soil into another soil. Considering the concept of similar environments, Miller and Miller (1956) proposed the concept of similar flow to scale Darcy's and continuity equations, and stated that the flow of water in two similar soils is the same under boundary and scaled primary conditions. Researchers (Sadeghi et al., 2008; Rasoulzadeh and Sepaskhah, 2003) used scaling to express the characteristics of water dynamics in the soil.

2.2.2 Irrigation Water Quality

2.2.2.1 Water Quality Determination Parameters

a. *Electrical conductivity*

Electrical conductivity (EC) is a simplified index for determining the total concentration of water-soluble salts. This index states that pure water is a poor conductor of electricity while adding salt to the water can conduct electricity well because electrolytes decompose in the vicinity of water and become charged ions. These ions can also direct the flow of electricity, so that the higher the concentration of salts in the water, the higher the conductivity of the electric current will be. Regarding the inverse relationship between the EC and resistance, its unit is the reverse of the resistance unit (ohm), i.e. (mho). The unit of this parameter was mmho/cm or μmho/cm, although the unit decisiemens per meter (dS/m) is more used in current experiments (Chemura et al., 2014).

b. *Sodium absorption ratio*

The sodium absorption ratio (SAR) refers to the sodium concentration ratio in the soil to the total amount of calcium and magnesium. The SAR, as an effective index in evaluating the potential risk of sodium in solution, is in balance state with the soil phase, the relationship of which is as follows (Lesch and Suarez, 2009; Robbins, 1984):

$$SAR = \frac{Na^+}{\sqrt{\dfrac{Ca^{2+} + Mg^{2+}}{2}}}, \tag{2.2}$$

where:

Na^+ is the concentration of sodium (meq/L),
Ca^{+2} is the concentration of calcium (meq/L), and
Mg^{+2} is the concentration of magnesium (meq/L).

c. *Effect of irrigation water quality on infiltration*

Water salinity (total amount of salts) and the amount of sodium in water compared to the amount of calcium and magnesium are the most common quality factors that affect the infiltration rate of water into the soil. Water with low salinity or water with a high sodium to calcium ratio reduces permeability. Prolonging the time of each irrigation to optimize the infiltration of water into the soil can cause secondary problems, including crust creation in the seedbed, the spread of weeds, nutritional disorders, plant flooding, seeds rotting, and plants weakening in the low and wet places. Accordingly, a weakness in the stability of surface soil structure, as well as the small amount of calcium relative to the amount of sodium, leads to the problem of quality-caused permeability, which often occurs within a few centimeters of the soil surface. The irrigation of soil with high sodium content causes the sodiumization of the surface soil, which will be weak in terms of soil structure. In this case, the aggregates disperse the surface soil and turn it into much smaller particles, which block the soil's pores. Pal et al. (1980) studied the effects of SAR and salt concentration on the distribution of sandy and loamy soils and observed that an increase in the concentration decreased the distribution while a decrease in the SAR increased the distribution. Abu-Sharar et al. (1987) revealed the existence of a correlation relationship between the reduction of hydraulic conductivity and the distribution of aggregates in saline and sodium soils. Hanson et al. (1999) concluded that the distribution of salt in the soil followed the pattern of water flow in the soil. For example, the infiltration of water into the soil transfers the salt from the upper surface of the soil to the lower depths. In this way, the type of soil, type of salt or chemical compounds present in the soil, the amount of water, and the irrigation method all affect the pattern of salt distribution and movement in the soil. Ragab et al. (2008) observed that increasing the salinity of irrigation water decreased the accessible water in the soil and increased the amounts of soluble cations and anions in the soil, especially sodium. Wang et al. (2015) addressed "salinity distribution" in soil under "irrigation" with different levels of salinity. The salinity levels of irrigation water were 3.3, 5, and 6.8 dS/m, respectively. The results of the statistical analysis indicated the significant effect of treatments included the different salinity levels in irrigation water at the level of 5% on soil salinity at depths of 0–40 cm. However, there was no the significant difference (5%) for depths of 40–100 cm. Moreover, different levels of irrigation and the simultaneous effect of various levels of irrigation and salinity on EC_{soil} were not significant (5%).

2.2.2.2 Leaching

Leaching plays an important role in the regeneration of saline and sodium soils (Oster et al., 1999). Concerning each irrigation event, an amount of salt enters the soil so that the yield will decrease if the salts accumulate at harmful concentrations in the root depth. Although the most applied water is used for plant evaporation and transpiration, a major part of the salts remains in the soil at the same time, resulting in increasing their concentration in the soil water. In each irrigation, more salt is entered into the soil, which necessitates leaching a part of the salt and removing it from the root zone before the salt concentration in the soil reaches the limit affecting the yield of the crop. Leaching is performed by adding enough water to the soil and passing some of the accumulated salt in the area below the roots. Further, the washing of salts in unsaturated soil is more than saturated soil (Gardner and Fireman, 1958). By leaching, part of the salt accumulated in the root zone is transferred outside. A fraction of water that enters the soil passes through the root zone, and moves below this area is called a leaching fraction (LF).

$$LF = \frac{\text{The depth of leaching water to the root zone}}{\text{The depth of water applied on the soil surface}} \tag{2.3}$$

After continuous irrigation, salt accumulation in the soil tends to a balanced concentration depending on the salinity of applied water and LF. LF greater than 0.5 accumulates the less salt in the soil compared to the LF less than 0.1. The absence of leaching leads to a varied distribution of salts at different soil depths (van Schilfgaarde et al., 1974). If the salinity of irrigation water and LF are known or can be estimated, it is possible to estimate the average salinity of the root zone and the salinity of the drainage water infiltrated under the root zone. Drainage salinity can be estimated by the following equation (Rhoades et al., 1974):

$$EC_{dw} = \frac{EC_w}{LF}, \tag{2.4}$$

where:

EC_{dw} is the salinity of the drainage water that penetrates below the root zone, and
EC_w is the applied water salinity.

The salinity of the applied irrigation water (EC_w) and soil salinity threshold (EC_e) should be determined to estimate the need for leaching. The water salinity can be obtained in the laboratory by using its decomposition while EC_e should be estimated from the data of plant tolerance. The leaching requirement can be obtained from the following equation (Ayers and Westcot, 1985):

$$LR = \frac{EC_w}{5EC_e - EC_w}, \tag{2.5}$$

where:

LR is the minimum leaching required to control soil salts in the plant tolerance range (EC_e) via conventional surface irrigation methods,
EC_w is the salinity of applied irrigation water, and
EC_e is the mean plant tolerable salinity (derived from saturated soil extract).

Tagar et al. (2010) compared the continuous and intermittent leaching methods for leaching salts from a clay loam soil. After 2 months, 46.14% of the salts were removed from the layer of 0–60 cm of plot soil subjected to the intermittent leaching. The removal of salts was 61.59% in plots under continuous leaching. The researchers suggested that if time is a limiting factor, the continuous leaching is superior to intermittent leaching, although it is better to use the intermittent leaching to achieve the better results in the long term.

2.2.3 Infiltration Equations

The infiltration of water into the soil can be determined in two direct and indirect ways. The methods of small plots and double-ring are among the direct methods, the selection of each of which depends on the type of irrigation system. In this regard, the high cost and time-consuming nature of direct methods in determining the permeability of the soil has always led researchers to use indirect methods (Wösten et al., 2001). In this regard, the several infiltration models were described in different investigations (Green and Ampt, 1911; Philip, 1957; Horton, 1940; Kostiakov, 1932; US SCS, 1974). The infiltration equations under study are the experimental equations (Kostiakov, Kostiakov–Lewis, Horton and Soil Conservation Service [SCS]) and the physical equation (Philip and Green–Ampt).

2.2.4 Experimental Equations

Experimental equations relate the infiltration or the depth of infiltration, as the integration of infiltration rate, to the time through a nonlinear relationship. The effectiveness of spatial and temporal conditions and the inefficiency of the equations' parameters in explaining the physical concepts are the main disadvantages of this type of equations. The experimental functions of infiltration express the amount of infiltration only as a function of the time of the infiltration and assume constant the effect of factors such as initial moisture, flow rate (to furrow, plot, or tape), and various physical characteristics of the soil, including texture, structure, and saturation degree on the infiltration. These effects are reflected in the coefficients of the infiltration equation. Based on the results, increasing the flow rate through increasing the contact surface of the soil with water and reducing the initial soil moisture through decreasing the soil matric potential will increase the infiltration rate (Loáiciga and Huang, 2007). The modification of these equations based on the above factors allows applying the robust management during the cultivation season and in proportion to the situation.

2.2.4.1 Kostiakov and Kostiakov–Lewis Equations

Kostiakov presented the equation of cumulative infiltration for vertical infiltration in a given soil as follows:

$$Z = at^b,\tag{2.6}$$

where:

Z is the cumulative infiltration (cm),
t is the time of intrusion (min), and
a and b are empirical coefficients.

The units and dimensions a and b are completely meaningless in the Kostiakov equation such that the parameter a is dimensionless and b necessarily has a dimension, indicating that the equation is dimensionally correct. Although the units and dimensions of the coefficients in the Kostiakov equation are completely meaningless, these parameters are interpretable and important from a physical viewpoint (Taylor and Ashcroft, 1972; Dixon, 1976). Kostiakov equation provides the satisfactory results for the short-term periods (i.e. about 2–6 hours, which are common in irrigation systems). Therefore, this equation is widely used in water and soil planning. By deriving the equation to calculate the infiltration rate, since the power of t is a decimal and negative value, the final "infiltration" ratio will be zero if time limits to infinity. However, this issue does not occur in practice and the final infiltration rate, which is the same as the saturated soil hydraulic conductivity, is not zero. Thus, the Kostiakov equation is only applicable if the infiltration rate is higher than the saturated soil hydraulic conductivity (Dixon, 1976). To solve this problem, researchers have proposed corrections and developed other equations for infiltration, among which the Kostiakov–Lewis equation has the parameter of the final infiltration rate:

$$Z = At^B + f_0 t, \tag{2.7}$$

where:

f_0 is the final infiltration rate $(\text{cm}/\text{minute})$.

In long term, the infiltration rate is at least equal to the constant value of f_0, i.e. the value of final infiltration. The irrigation time may be such that irrigation is cut-off before the infiltration rate reaches a constant value, in which case the Kostiakov equation is sufficient for irrigation (Israelsen and Hansen, 1962).

2.2.4.2 Horton Equation

After numerous observations of water infiltration into the soil, Horton (1940) concluded that water infiltration into soil followed a declining exponential function. This exponential function includes a process in which the speed of the work is proportional to the remaining work. Accordingly, the infiltration rate of water into the soil, as a work that can be done at time t, is proportional to the infiltration rate at this time (f) minus the final infiltration rate (f_f) and thus, proportional to the $(f - f_f)$. If the work speed is written as $\dfrac{df}{dt}$, we have:

$$-\frac{df}{dt} = k(f - f_f), \tag{2.8}$$

where k represents a coefficient for the equality of the two proportions and the negative sign indicates the decreasing trend of the infiltration rate, which will be obtained after integrating and replacing the constant coefficient:

$$f(t) = f_f + (f_i - f_f)e^{-kt}. \tag{2.9}$$

If t is infinite, the infiltration rate tends to be positive and greater than zero (f_f), and the velocity is equal to (f_i) at zero time, which are the benefits of the equation. A disadvantage of the equation is that figures obtained for the infiltration velocity are initially less than the actual values. The amount of water infiltrated at time t is achieved by integrating the above equation (Horton, 1941):

$$Z(t) = f_f t + \frac{f_i - f_f}{k}(1 - e^{-kt}), \tag{2.10}$$

where:

Z is the cumulative infiltration (cm),
t is the time (minutes),
f_f is the final infiltration rate (cm/minute),
f_i is the initial infiltration rate (cm/minute), and
k is the constant coefficient.

2.2.4.3 Soil Conservation Service Equation

The experts of the US SCS conducted several experiments on farms based on the Kostiakov equation, which ultimately led to the development of a method for calculating the infiltration, which is known as the US SCS equation as follows:

$$Z = ct^d + 0.6985. \tag{2.11}$$

As observed, this equation is almost identical to the Kostiakov equation, except that the constant number is 0.6985. The coefficients c and d are related to the type of soil. To obtain these two coefficients, the US Soil Conservation Service has provided several numbered curves, which are the logarithmic relationship of cumulative infiltration and time for different soils (US SCS, 1974).

2.2.5 Physics-Based Equations

This category of equations involves the properties of the initial soil moisture and soil-water matric suction in determining the permeability, which will generally lead to complexities. Philip and Green–Ampt equations are among these equations.

2.2.5.1 Philip Equation

Philip (1957) assumed the infiltration process as a state of the general phenomenon of the water movement in the soil and solved Richards' equation numerically. Then, he considered the cumulative infiltration equal to the moisture stored in the moisture profile plus the water passed from the end of the moisture profile over time. Philip equation is as follows:

$$Z = St^{0.5} + k_s t, \tag{2.12}$$

where:

Z is the cumulative infiltration (cm),
t is the time (minute),
S is the soil absorption coefficient (cm/minute), and
k_s is the hydraulic conductivity (cm/minute).

The absorption coefficient of the soil and the hydraulic conductivity factor are dominant at the beginning and end of the infiltration, respectively. By deriving the above equation, the infiltration rate will be equal to:

$$Z = St^{-0.5} + k_s. \tag{2.13}$$

If time limits to infinity, the infiltration rate will tend to k_s . The two phrases of the Philip equation represent the suction load and the gravitational load. For a horizontal soil column, suction is the only force that draws water into the soil and thus, the equation in the horizontal state will be as follows (Kirkham and Powers, 1972):

$$Z = St^{0.5}. \tag{2.14}$$

2.2.5.2 Green–Ampt Equation

Green and Ampt (1911) examined the water movement in three soil types of loam, clay, and sandy inside the column and developed an approximation model according to Darcy's law. Their method can be used for specific cases of water infiltration in dry, coarse-textured soils with a certain moisture front, resulting in obtaining an estimation from the instantaneous infiltration rate. Green–Ampt equation is as follows:

$$f(t) = K\left(\frac{\Psi\Delta\theta}{F(t)} + 1\right), \tag{2.15}$$

$$F(t) = Kt + \Psi\Delta\theta\ln\left(\frac{F(t)}{\Delta\theta\Psi} + 1\right), \tag{2.16}$$

where:

K is the hydraulic conductivity (cm/hour),
Ψ is the wetting front soil suction head (cm),
$\Delta\theta$ is the difference between soil initial moisture and saturation (%),
t is the time (hour),
$f(t)$ is the infiltration rate (cm/hour), and
$F(t)$ is the cumulative infiltration (cm).

The original form of the Green–Ampt infiltration equation expresses the cumulative infiltration as a function of the moisture front. Later, other researchers evaluated the Green–Ampt model under various conditions and presented the modified forms of this equation. Some of the modified forms of the Green–Ampt model indicate the cumulative infiltration and infiltration rate as a function of time (Gowdish and Muñoz-Carpena, 2009).

By measuring the infiltration through the double-ring method in the Ohio region in the United States, Shukla et al. (2003) examined the performance of ten equations, including the experimental and physical equations, of water infiltration into the soil in different uses. The results indicated that the Horton equation has the best performance in the quantitative representation of the infiltration process, compared to other equations. By comparing the efficiency of the equations of Kostiakov, Kostiakov-Lewis, Philip, SCS, and Horton in estimating the cumulative infiltration and infiltration rate, Duan et al. (2011) revealed the higher accuracy of the equations of Kostiakov–Lewis and Horton. Zhang et al. (2012) compared four different equations of infiltration (Philip, Kostiakov–Lewis, Kostiakov, and Horton) in describing the phenomenon of infiltration in irrigation furrows. The results showed the better performance of the Kostiakov–Lewis infiltration equation in explaining the relationship between cumulative infiltration and time, compared to the other infiltration equations. Zolfaghari et al. (2012) examined the performance of seven infiltration equations in estimating the cumulative infiltration measured with double-acting cylinders in four classes of soil texture located in five provinces of Iran. Based on the results, the Kostiakov–Lewis and SCS equations were the most appropriate and inappropriate equations in all of the studied soils, respectively.

Turner (2006) studied the sensitivity of the coefficients of different infiltration equations to the changes in infiltration intensity and showed that the final infiltration rate of the Horton equation and the soil absorption coefficient of the Philip equation were more sensitive to the changes of infiltration intensity than the other coefficients of these equations. Hoyos and Cavalcante (2015) analyzed the sensitivity of the coefficients of the Kostiakov, Philip, Green–Ampt, and Horton equations in the sandy loam texture. Based on the results, the coefficients k and b of the Horton and Kostiakov equations were the least sensitive while the coefficient *a* of the Kostiakov equation and the saturated soil hydraulic conductivity of Green–Ampt equation were the most sensitive coefficients, respectively. Furthermore, Horton and Philip's equations were the most and least sensitive equations, respectively. Sepaskhah and Afshar-Chamanabad (2002) investigated the parameters of the Kostiakov–Lewis infiltration equation for conventional furrow irrigation (CFI) with the different discharge rates based on the progression phase, and recession and reserve phases. Their results showed that the infiltration equation should be modified for different discharge rates.

2.3 Choosing the Right Irrigation Method

2.3.1 Types of Irrigation

In many cases, irrigation method is not appropriate for the farm, due to the various reasons such as the light texture of the soil, poor topography, low depth of agricultural soil, low flow rate, high losses of evaporation and Aeolian processes, and unfamiliarity of the farmer in managing irrigation systems. Hence, the irrigation method should be selected according to the climatic, agricultural, water, and soil conditions of the farm, as well as the awareness level of farmers in using the irrigation system. However, the main reason for changing the irrigation system by farmers is not to reduce the water consumption, but to increase the income and economic efficiency, decrease labor costs, and reduce the water-pumping costs. On the other hand, limited water resources and the reduction of agricultural water rights have urged the farmers to use the available water resources effectively.

2.3.1.1 Surface Irrigation

Surface irrigation is one of the most common irrigation methods, in which water flows in a gravity way on the earth's surface and the earth's surface acts as a water absorber and transporter. Despite the high efficiency in pressurized irrigation systems, the increasing trend of energy costs has led many researchers to conduct the significant studies on increasing the efficiency of surface irrigation. Accordingly, many researchers have suggested the surface irrigation as a suitable alternative to pressurized irrigation methods. The surface irrigation method is more important than pressurized irrigation method, due to low initial investment costs, simplicity of repair and maintenance, lack of need for a skilled worker, and acceptability to the farmer (Walker and Skogerboe, 1987). The low efficiency of surface irrigation is the main problem in this method, which is caused by poor management of irrigation. Considering the high cost of pressurized irrigation systems, it is of great importance to improve and modify the traditional surface irrigation methods, which are early yielding and low cost and can be implemented by the farmer. The performance of surface irrigation is a function of farm design, soil permeability characteristics, and irrigation management method. However, the complexity of the interaction between these parameters causes problems for farmers in identifying the optimal design or management methods with current methods (Liu et al., 2018). Improvement of surface irrigation systems includes land leveling and proper selection of irrigation methods. Among the methods to increase the efficiency of surface irrigation are: surge, cutback, runoff recovery, and cablegation. These strategies are mainly possible in the furrow irrigation, which is regarded as one of the most compatible types of surface irrigation with the development of mechanized agriculture. In this method, water enters the soil through the wet environment and infiltrates laterally and vertically (Ramon et al., 1987).

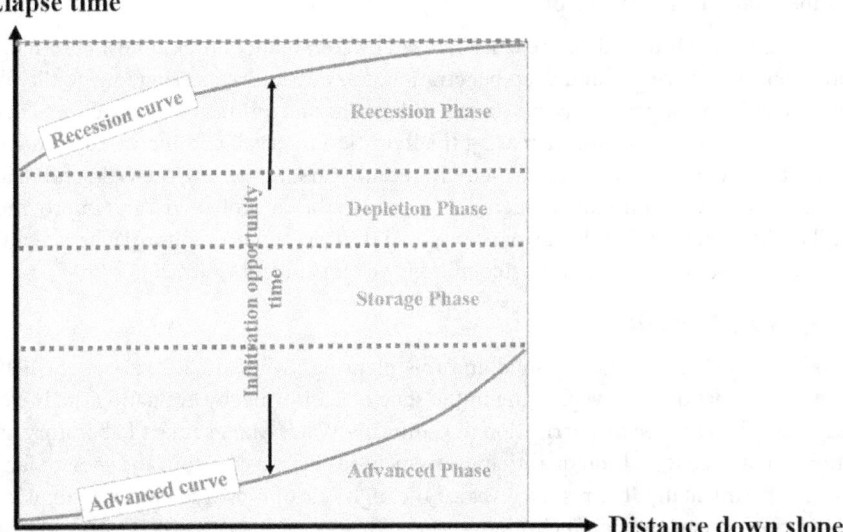

FIGURE 2.2 Surface irrigation process.

Reduced flow (cutback) management in surface irrigation reduces the runoff losses and increases the irrigation water productivity. By disconnecting and connecting the flow, wave irrigation increases uniformity and reduces water losses, especially in light soils (Bishop, 1980; Kemper et al., 1988). In cable irrigation, inlet discharge reduction can be managed over time, which results in a significant reduction in water losses. Both wave and cable irrigation methods are more commonly applied in research and less used in practice (Kemper et al., 1987). By conducting farm experiments, Ebrahimian and Liaghat (2011) showed that the efficiency of water for the corn crop led to a slight decrease in the yield of crop by one to two times. However, water consumption was reduced by up to 50% compared to the conventional irrigation methods.

In all surface irrigation systems, the water distribution on the soil follows a general rule and includes four phases (Figure 2.2):

i. *Advance phase:* The advance phase continues from when the water enters the field until the water covers the entire surface of the field. It is worth noting that the advance rate decreases over time.

ii. *Recession phase:* This phase begins when the discharge phase is over. A drying or recession front starts from the beginning until it reaches the end. The recession phase reaches the end when either the recession front reaches the end of the field or the flow depth becomes zero or a negligible amount across the field.

iii. *Storage phase:* The storage phase begins when the advance phase reaches the end of the field and continues until the flow is shut off.

iv. *Depletion phase:* The depletion phase starts when the flow is shut off and continues until the water depth at the beginning of the field reaches zero.

2.3.1.2 Pressure Irrigation

New methods of pressurized irrigation are among the technologies that have led to a significant increase in water efficiency. The two relatively new technologies of pressurized irrigation systems, namely subsurface drip irrigation (SDI) and Tape drip irrigation (Tape), are briefly introduced and described.

2.3.1.3 Subsurface Drip Irrigation

Significant benefits of SDI include more efficient use of water, reduction or almost elimination of surface evaporation, surface runoff and deep penetration, the possibility of using low-quality waters due to reduced transmission of pathogens, no contact of humans and animals with water in SDI, decreased moisture and osmotic stresses with increasing the irrigation interval, and increased evenness (uniformity) (Ayars et al., 2015; Woodrow et al., 2008). In recent years, increasing the yield and quality of the crop, decreasing water consumption, reducing crop costs to the control weeds, and improving fertilization and tillage have encouraged the farmers to spread this method on California farms, so that SDI is an available technology to improve the water efficiency (Johnson and Cody, 2015).

2.3.1.4 Tape Drip Irrigation

This method is suitable for irrigating solid and row planting. In research, Ayars et al. (1999) studied the effect of tape irrigation on row planting in the state of California, by evaluating the 15-year studies conducted on the field of tape drip irrigation in California Water Management Laboratory and studies on irrigation management and soil quality improvement in the corn, cotton, and tomato farms, which were planted experimentally. The results revealed the high rate of crop and efficiency of using water in the drip irrigation system compared to other systems. Brosz and Wiersma (1974) studied two drip and subsurface irrigation methods in corn planting in Oklahoma (United States) and reported that drip and subsurface irrigation systems increased crop yield by 5% and 15%, respectively, in addition to a 20% reduction in water consumption.

2.3.2 Irrigation Management Methods

Despite using new and appropriate irrigation systems on the farms, it is observed that the performance of the irrigation system has not been satisfactory in many cases, due to poor management of the irrigation on the farm. In the following, the important strategies are proposed to improve the water management in the farm, aiming to reduce water losses, increase crop yields, and enhance water productivity.

2.3.3 Irrigation Planning

Irrigation planning is one of the most important and basic measures in the correct use of irrigation water such that the irrigation system should follow the principles of this planning. It is possible to provide the accurate irrigation planning through methods such as using meteorological data to determine the time and depth of irrigation and applying the moisture and plant sensors to specify irrigation time and present precise irrigation planning (Irmak et al., 2012; Nolz and Loiskandl, 2017; Stieber and Shock, 1995; Müller et al., 2016). For example, the California Irrigation Management Information System was implemented in 1982 as a research project by the University of California and the California Department of Water Resources Conservation. This system included an integrated network of more than 249 automatic meteorological stations, located on many agricultural and urban sites across California. This information system helps the producers, farmers, and environmental managers to manage the irrigation time and water required by the plants in the best possible way, resulting in improving water and energy management through implementing the efficient irrigation methods (Eching, 2000).

2.3.4 Use of Deficit Irrigation

If irrigation is carried out when the soil moisture has reached a low level of readily available water (RAW), then the plant will use only the readily available moisture, which is called full irrigation in such cases. However, if the soil moisture is lower than the water level of RAW and the plant uses more energy to receive moisture from the soil, then it is said that the deficit irrigation has been applied. Accordingly, selecting an appropriate irrigation method requires performing the economic analysis and comparing the amount of crop to the water consumption. Given the limited resources of water, it is recommended to consider the deficit irrigation as an option to increase the water productivity in irrigation designs and projects. Deficit irrigation refers to the deliberate and wise consumption of less water, or in other words, more and better use of water volume per unit. Deficit irrigation is different from less irrigation. The main purpose of irrigation is to increase the water productivity, either by reducing the amount of irrigation water at each turn or by eliminating the irrigation methods with the lowest yields. Alternate furrow irrigation is less irrigation that can reduce the irrigation losses on the farm. There are two types of the irrigation management for this method, which are fixed alternate furrow irrigation (FAFI) and variable alternate furrow irrigation (VFFI; Ebrahimian, 2014; Kang et al., 2000a, b). Kang et al. (2000a, b) compared three irrigation systems, including the CFI, FAFI, and VFFI. All the three irrigation methods were evaluated for three different irrigation depths of 30, 45, and 22.5 mm in each irrigation of corn plant. According to the observations, a decrease in the level of irrigation water significantly decreased the level of crop yield in the CFI and FAFI treatments, although no significant reduction was observed in the crop yield in the VAFI treatment.

The partial root-zone drying irrigation was first established in Australia by studying the grape trees (Dry and Loveys, 1998). This method is regarded as one of the modern methods of crop management in recent years, which has greatly increased the efficiency of water consumption in many crops (de Lima et al., 2015; Romero et al., 2015; Yactayo et al., 2013). The specific theory governing the partial root-zone drying distinguishes this type of irrigation from the conventional less-irrigation methods, which mainly reduce the crop yield. In this technique, only half of the roots are irrigated each time and the other half remains dry. The remaining dry part of the root, as a physiological response to the water stress, produces a certain chemical hormone, called Abscisic acid, in the root, the transfer of which to the plant sap alkalinizes the sap and reduces the openness of the pores in the plant, resulting in reducing water loss (Stoll et al., 2000; Sepaskhah and Ahmadi, 2012). Deficit irrigation is an optimal approach to grow crops under the water deficit conditions, which is associated with reduced crop yield per unit area and increased crop yield with surface expansion. Extensive research has been conducted about the effect of deficit irrigation on crop yields (English, 1990; Doorenbos and Kassam, 1979).

In many plants, the effect of irrigation amount on yield highly relies on the location and climate of plant growth, such that the type of soil affects the water absorption by plants while the climate determines the amount of evapotranspiration of plants (Zheng et al., 2021). The results of previous studies indicate an increase in the water productivity and the lack of a significant decrease in yield, due to the utilization of partial root-zone drying (Zegbe and Serna-Pérez, 2011; Posadas, 2008; Shahnazari et al., 2007). Considering the results of research on the corn plant, Yazar et al. (2009) indicated the lack of significant difference between the weight values of one thousand seeds in the treatments of full irrigation and partial root-zone drying irrigation. Nevertheless, Kang and Zhang (2004) believed that the different environmental conditions of the farm, such as the weather conditions, soil, and plant type can affect the results of applying partial root-zone drying irrigation treatment, despite the advantages of the partial root-zone drying irrigation method compared to the full irrigation and the conventional deficit irrigation methods. The results of applying different percentages of deficit irrigation method in cultivating wheat and corn in the United States and Syria reveal that this method is the best treatment and reduces the water consumption by 33%, compared to the full irrigation conditions. Further, with a 0.4% and 9% reduction in the wheat yield and a 16% decrease in corn yield, this method leads to a 16% and 22% increase in the wheat productivity in the United States and Syria, respectively, and 7% increase in corn productivity (Kijne et al., 2003).

2.3.5 Reducing Water Losses

The use of mulches is one of the ways to increase the efficiency of irrigation and water productivity. Mulches can be generally found in two forms of organic (such as plant residues) and inorganic (chemical and mineral). The use of mulches prevents the growth of weeds and reduces the water consumption, in addition to retaining the soil moisture by reducing the evaporation from the soil surface (Brainard and Bellinder, 2004(. Furthermore, the utilization of mulch can increase the efficiency of water consumption by 60%, enhance the efficiency of nitrogen consumption, and greatly contribute to solving the environmental issues (Qin et al., 2015).

2.3.6 Agricultural Measures

In this regard, crop improvement measures such as transplanting, conservation agriculture (appropriate tillage), and changing planting date lead to the reduction of water consumption and maximum use of rainfall. Further, the management of plant nutrition and proper management of mechanization in planting, growing, and harvesting operations can increase the crop yield and productivity of agricultural water.

2.4 Sensitivity Analysis of Surface Irrigation Hydraulic Models

The flow of water on the soil surface is non-uniform and unstable. The flow rate changes over time at a particular point because the infiltration property is a function of time. At the endpoint, water is advancing and especially the depth changes with time and place. Therefore, the Saint-Venant equation, which includes the equations of continuity and motion (Akan, 1992), is used in the surface irrigation models to explain the process of water movement:

$$\frac{\partial Q}{\partial x} + \frac{\partial A}{\partial t} + I_x = 0, \tag{2.17}$$

$$\frac{1}{g}\frac{\partial V}{\partial t} + \frac{V}{g}\frac{\partial V}{\partial x} + \frac{\partial y}{\partial x} = S_0 - S_f + \frac{I_x V}{gA}, \tag{2.18}$$

where:

y is the depth of flow (m),

Q is the flow (m^3/s),

I_x is the infiltration intensity ($m^3/s/minute$),

S_0 is the bed slope (m/m),

S_f is the friction slope (m/m),

V is the flow velocity (m/s), and

g is the acceleration of gravity (m/s^2).

2.4.1 Hydrodynamic Model

Walker and Skogerboe (1987) and Bautista and Wallender (1993) proposed a hydrodynamic furrow irrigation model with the fixed spatial steps and the effect of wet environment on infiltration. However, because of using the Saint-Venant equation, this model is very complex and cannot be easily used in the normal situations. When the test data are uncertain and missing, this model allows calibrating data easily, which is considered as one of the applications of the hydrodynamic models. Later, Haie (1985), Strelkoff (1990), and Bautista and Wallender (1993) developed multiple examples of the hydrodynamic model.

2.4.2 Zero-Inertia Model

Elliott et al. (1982) were the first researchers who used the concept of zero inertia in their model to simulate the advance phase in the irrigation in furrow irrigation. Then, a number of the surface irrigation models were developed based on the concept of zero inertia (Ross, 1986; Schwankl and Wallender 1988; Strelkoff, 1991). In this model, because the flow rate in surface irrigation is low, the acceleration and inertia components were assumed to be zero by Katopodes and Strelkoff (1977), the equation of motion became as follows:

$$\frac{\partial y}{\partial x} = S_0 - S_f. \tag{2.19}$$

2.4.3 Kinematic Wave Model

This model was developed for hydrological works (Lai and Pandya, 1972; Walker and Humpherys, 1983). With the modifications made to the model, it is also used for the furrow irrigation studies (Walker and Skogerboe, 1987). In this model, it is assumed that the flow is uniform and equations such as Manning and Chezy formula and Darcy–Weisbach equation are used instead of the momentum equation. By considering some assumptions in this model, the momentum equation is simplified to the following equation (Ebrahimian and Liaghat, 2011):

$$S_0 - S_f. \tag{2.20}$$

It is also assumed that there is a uniform relationship between the flow discharge and the flow cross-section. This model is called the normal depth models, and also uniform flow models.

Oweis (1984) performed the sensitivity analysis of the zero-inertia model in simulating the surge flow in furrows on the parameters of the inflow rate, slope, and coefficients of the infiltration equation. The results of this study indicated the sensitivity of the advance phase to the inflow rate, the increased sensitivity of the model to the reduction of the slope, and the effect of the coefficients of infiltration equation on the advance and depletion phases. By analyzing the sensitivity of the nonlinear zero-inertia model for the surge flow in furrow irrigation, Wallender and Rayej (1985) concluded that the sensitivity of this nonlinear model was consistent with that of the linear model provided by Oweis (1984) despite the different sensitivity of the model to the Manning roughness coefficient and the geometric parameters of the cross-section. Izuno and Podmore (1985) expanded the kinematic surge for the water flow in furrows through a surge method and analyzed the sensitivity of the proposed model. The results showed that the farm slope, the Manning roughness coefficient, and the geometric parameters of the furrow slightly affected the results of the model in forecasting, although the effects of infiltration coefficients and surge irrigation management factors were significant in prediction and performance of the model. Oweis and Walker (1990) developed the zero-inertia model for the surge flow irrigation in furrows. The sensitivity analysis of this model to the parameters of inflow rate, furrow slope, and infiltration coefficients showed that the model is sensitive to the inflow rate in both surge and continuous flows. Moreover, the sensitivity of this model to the farm was high, especially in the depletion phase of the model. However, Oweis and Walker (1990) stated that the sensitivity of the model to these parameters would not be problematic, due to the possibility of accurately determining the slope value, as well as the significant effects of the infiltration coefficients on the prediction accuracy of the model.

2.5 Conclusions

Water infiltration rate is regarded as one of the important parameters in designing and managing irrigation systems, hydrological studies, water resources management, and soil conservation, as well as designing and implementing drainage and soil erosion control projects in watersheds. The selection of irrigation systems in each area depends on the characteristics of the water infiltration in the study area. On the other hand, the management of irrigation systems and irrigation planning in these systems is performed based on the water infiltration rate. Further, soil characteristics such as salinity and SAR can affect the infiltration of water into the soil. So far, comparing and evaluating the performance of infiltration equations have led to the different results and some contradictory cases, due to the changing nature of the infiltration process of water to soil, so that even the performance of an equation can be different in two relatively similar soils. Depending on the infiltration measurement method and the initial and boundary conditions of the process (including water height on the soil surface, soil moisture before infiltration, etc.), the flow of water in the soil will be different and lead to the different results. In many parts of the world, most Faryab lands are irrigated by the surface methods. Surface irrigation is one of the most common and oldest irrigation methods, which is performed in different ways, depending on the condition of the soil, water, and plants. Generally, surface irrigation methods include tape irrigation, basin irrigation, furrow irrigation, and flood irrigation.

The major problem with surface irrigation methods is the low efficiency of irrigation water, which is mainly caused by poor management of irrigation. In surface irrigation methods, furrow irrigation is used more than the other surface irrigation methods and is more compatible with the development of mechanized agriculture. Further, a small volume of water moves in the width unit of the farm relative to the basin and tape irrigations, although it fails to provide a yield equal to their yield. In this regard, deep infiltration and runoff are the main problems in surface irrigation. It should be noted that managing surface irrigation allows increasing the efficiency of irrigation and even minimizing the pollution from agricultural pollutants. By changing the physical and chemical properties of the soil, irrigation management can affect soil's dispersion and aggregates breakdown and cause the formation of surface seals and changes in permeability functions. Considering the temporal and spatial changes in soil characteristics, it is possible to obtain the high efficiency in surface irrigation through the appropriate application of irrigation management.

References

Abedi-Koupai, J., Mostafazadeh-Fard, B., Afyuni, M. and Bagheri, M.R., 2006. Effect of treated wastewater on soil chemical and physical properties in an arid region. *Plant Soil and Environment*, 52(8), 335.

Abu-Sharar, T.M., Bingham, F.T. and Rhoades, J.D., 1987. Reduction in hydraulic conductivity in relation to clay dispersion and disaggregation 1. *Soil Science Society of America Journal*, 51(2), 342–346.

Akan, A.O., 1992. Horton infiltration equation revisited. Journal of Irrigation and Drainage Engineering, 118(5), 828–830.

Amer, K.H., 2010. Corn crop response under managing different irrigation and salinity levels. *Agricultural Water Management*, 97(10), 1553–1563.

Ayars, J.E., Fulton, A. and Taylor, B., 2015. Subsurface drip irrigation in California—Here to stay. *Agricultural Water Management*, 157, 39–47.

Ayars, J.E., Phene, C.J., Hutmacher, R.B., Davis, K.R., Schoneman, R.A., Vail, S.S. and Mead, R.M., 1999. Subsurface drip irrigation of row crops: A review of 15 years of research at the water management research laboratory. *Agricultural Water Management*, 42(1), 1–27.

Ayers, R.S. and Westcot, D.W., 1985. *Water Quality for Agriculture*, 29, Food and Agriculture Organization of the United Nations, Rome, Italy.

Bautista, E. and Wallender, W.W., 1993. Identification of furrow intake parameters from advance times and rates. *Journal of Irrigation and Drainage Engineering*, 119(2), 295–311.

Bishop, A.A., 1980. Surge flow: The most efficient irrigation system. *Crops Soils*, 33(2), 135–147.

Brainard, D.C. and Bellinder, R.R., 2004. Weed suppression in a broccoli–winter rye intercropping system. *Weed Science*, 52(2), 281–290.

Brosz, D.D. and Wiersma, J.L., 1974. Comparing trickle, subsurface and sprinkler irrigation systems. *Journal of the Irrigation and Drainage Division*, 100(3), 321–338.

Chemura, A., Kutywayo, D., Chagwesha, T.M. and Chidoko, P., 2014. An assessment of irrigation water quality and selected soil parameters at Mutema irrigation scheme, Zimbabwe. *Journal of Water Resource and Protection*, 14(6), 132–140.

Childs, J., Wallender, W.W. and Hopmans, J.W., 1993. Spatial and seasonal variation of furrow infiltration. *Journal of Irrigation and Drainage Engineering*, 119(1), 74–90.

de Lima, R.S.N., de Assis, F.A.M.M., Martins, A.O., de Deus, B.C.D.S., Ferraz, T.M., de Assis Gomes, M.D.M., de Sousa, E.F., Glenn, D.M. and Campostrini, E., 2015. Partial root zone drying (PRD) and regulated deficit irrigation (RDI) effects on stomatal conductance, growth, photosynthetic capacity, and water-use efficiency of papaya. *Scientia Horticulturae*, 183, 13–22.

Deng, X.P., Shan, L., Zhang, H. and Turner, N.C., 2006. Improving agricultural water use efficiency in arid and semiarid areas of China. *Agricultural Water Management*, 80(1–3), 23–40.

Dixon, R.M., 1976. Comment on 'Derivation of an equation of infiltration' by HJ Morel-Seytoux and J. Khanji. *Water Resources Research*, 12(1), 116–118.

Doorenbos, J. and Kassam, A.H., 1979. Yield response to water. Irrigation and *drainage paper*, 33(1), 257–325.

Douh, B., Mguidiche, A., Bhouri-Khila, S., Mansour, M., Harrabi, R. and Boujlben, A., 2013. Yield and water use efficiency of cucumber (*Cucumis sativus* L.) conducted under subsurface drip irrigation system in a Mediterranean climate. *Journal of Environmental Science, Toxicology and Food Technology*, 2(4), 46–51.

Dry, P.R. and Loveys, B.R., 1998. Factors influencing grapevine vigour and the potential for control with partial rootzone drying. *Australian Journal of Grape and Wine Research*, 4(3), 140–148.

Duan, R., Fedler, C.B. and Borrelli, J., 2011. Field evaluation of infiltration models in lawn soils. *Irrigation Science*, 29(5), 379–389.

Ebrahimian, H. and Liaghat, A., 2011. Field evaluation of various mathematical models for furrow and border irrigation systems. *Soil and Water Research*, 6(2), 91–101.

Ebrahimian, H., 2014. Soil infiltration characteristics in alternate and conventional furrow irrigation using different estimation methods. *KSCE Journal of Civil Engineering*, 18(6), 1904–1911.

Eching, S., 2000. *CIMIS Agricultural Resource Book*. State of California, The Resources Agency, Department of Water Resources, USA.

Elliott, R.L., Walker, W.R. and Skogerboe, G.V., 1982. Zero-inertia modeling of furrow irrigation advance. *Journal of the Irrigation and Drainage Division*, 108(3), 179–195.

English, M., 1990. Deficit irrigation. I: Analytical framework. *Journal of Irrigation and Drainage Engineering*, 116(3), 399–412.

FAO. 2002. *Crops and Drops: Making the Best Use of Water for Agriculture*, 28, FAO. Information Brochure, Rome, Italy.

Fernández-Gálvez, J., Simmonds, L.P. and Barahona, E., 2006. Estimating detailed soil water profile records from point measurements. *European Journal of Soil Science*, 57(5), 708–718.

Gardner, W.R. and Fireman, M., 1958. Laboratory studies of evaporation from soil columns in the presence of a water table. *Soil Science*, 85(5), 244–249.

Gowdish, L. and Muñoz-Carpena, R., 2009. An improved Green–Ampt infiltration and redistribution method for uneven multistorm series. *Vadose Zone Journal*, 8(2), 470–479.

Green, W.H. and Ampt, G.A., 1911. Studies on soil physics. *The Journal of Agricultural Science*, 4(1), 1–24.

Haie, N., 1985. *Hydrodynamic Simulation of Continuous and Surged Surface Flow*, Unpublished PhD. Thesis. Utah State University, Logan.

Hanson, B., Grattan, S.R. and Fulton, A., 1999. *Agricultural Salinity and Drainage*. University of California Irrigation Program, University of California, Davis.

Hillel, D., 2003. *Introduction to Environmental Soil Physics*. Elsevier Academic Press, Amsterdam.

Horton, R.E., 1940. An approach towards the physical interpretation of infiltration-capacity. *Soil Science Society of America Proceedings*, 5, 325–333.

Horton, R.E., 1941. An approach toward a physical interpretation of infiltration-capacity 1. *Soil Science Society of America Journal*, 5(C), 399–417.

Hoyos, E.M. and Cavalcante, A.L.B., 2015. Sensitivity analysis of one-dimensional infiltration models. *The Electronic Journal of Geotechnical Engineering*, 20(10), 4313–4324.

Irmak, S., Burgert, M.J., Yang, H.S., Cassman, K.G., Walters, D.T., Rathje, W.R., Payero, J.O., Grassini, P., Kuzila, M.S., Brunkhorst, K.J. and Eisenhauer, D.E., 2012. Large-scale on-farm implementation of soil moisture-based irrigation management strategies for increasing maize water productivity. *Transactions of the ASABE*, 55(3), 881–894.

Israelsen, O.W. and Hansen, V.E., 1962. *Irrigation Principles and Practices*, John Wiley and Sons Inc, New York, USA; London, UK.

Izuno, F.T. and Podmore, T.H., 1985. Kinematic wave model for surge irrigation research in furrows. *Transactions of the ASAE*, 28(4), 1145–1150.

Jalali, M., Merikhpour, H., Kaledhonkar, M.J. and Van Der Zee, S.E.A.T.M., 2008. Effects of wastewater irrigation on soil sodicity and nutrient leaching in calcareous soils. *Agricultural Water Management*, 95(2), 143–153.

Johnson, R. and Cody, B.A., 2015. *California Agricultural Production and Irrigated Water Use*. Sacramento: Congressional Research Service, California, USA.

Kang, S. and Zhang, J., 2004. Controlled alternate partial root-zone irrigation: Its physiological consequences and impact on water use efficiency. *Journal of Experimental Botany*, 55(407), 2437–2446.

Kang, S., Liang, Z., Pan, Y., Shi, P. and Zhang, J., 2000a. Alternate furrow irrigation for maize production in an arid area. *Agricultural Water Management*, 45(3), 267–274.

Kang, S.Z., Shi, P., Pan, Y.H., Liang, Z.S., Hu, X.T. and Zhang, J., 2000b. Soil water distribution, uniformity and water-use efficiency under alternate furrow irrigation in arid areas. *Irrigation Science*, 19(4), 181–190.

Kang, Y., Chen, M. and Wan, S., 2010. Effects of drip irrigation with saline water on waxy maize (*Zea mays* L. var. ceratina Kulesh) in North China Plain. *Agricultural Water Management*, 97(9), 1303–1309.

Katopodes, N.D. and Strelkoff, T., 1977. Hydrodynamics of border irrigation—Complete model. *Journal of the Irrigation and Drainage Division*, 103(3), 309–324.

Kemper, W.D., Trout, T.J. and Kincaid, D.C., 1987. Cablegation: Automate supply for surface irrigation. *Advances in Irrigation*, 4, 1–66.

Kemper, W.D., Trout, T.J., Humpherys, A.S. and Bullock, M.S., 1988. Mechanisms by which surge irrigation reduces furrow infiltration rates in a silty loam soil. *Transactions of the ASAE*, 31(3), 821–0829.

Kijne, J.W., Barker, R. and Molden, D.J. eds., 2003. *Water Productivity in Agriculture: Limits and Opportunities for Improvement*, 1, CABI, Oxfordshire.

Kirkham, D. and Powers, W.L., 1972. *Advanced Soil Physics*, Wiley, New York, USA.

Kostiakov, A.N., 1932. On the dynamics of the coefficient of water percolation in soils and the necessity of studying it from the dynamic point of view for the purposes of amelioration. *Society of Soil Science*, 14, 17–21.

Lai, R. and Pandya, A.C., 1972. Volume balance method for computing infiltration rates in surface irrigation. *Transactions of the ASAE*, 15(1), 69–0072.

Lesch, S.M. and Suarez, D.L., 2009. A short note on calculating the adjusted SAR index. *American Society of Agricultural and Biological Engineers*, 52 (2), 493–496.

Liu, T., Bruins, R., J. and Heberling, M.T, 2018. Factors influencing farmers' adoption of best management practices: A review and synthesis. *Sustainability*, 10(2), 432–453.

Loáiciga, H.A. and Huang, A., 2007. Ponding analysis with Green-and-Ampt infiltration. *Journal of Hydrologic Engineering*, 12(1), 109–112.

Machiwal, D., Jha, M.K. and Mal, B.C., 2006. Modelling infiltration and quantifying spatial soil variability in a wasteland of Kharagpur, India. *Biosystems Engineering*, 95(4), 569–582.

Miller, E.E. and Miller, R.D., 1956. Physical theory for capillary flow phenomena. *Journal of Applied Physics*, 27(4), 324–332.

Morison, J.I.L., Baker, N.R., Mullineaux, P.M. and Davies, W.J., 2008. Improving water use in crop production. *Philosophical Transactions of the Royal Society B: Biological Sciences*, 363(1491), 639–658.

Müller, T., Bouleau, C.R. and Perona, P., 2016. Optimizing drip irrigation for eggplant crops in semi-arid zones using evolving thresholds. *Agricultural Water Management*, 177, 54–65.

Nolz, R. and Loiskandl, W., 2017. Evaluating soil water content data monitored at different locations in a vineyard with regard to irrigation control. *Soil and Water Research*, 12(3), 152–160.

Oster, J.D., Shainberg, I. and Abrol, I.P., 1999. Reclamation of salt-affected soils. *Agricultural Drainage*, 38, 659–691.

Oweis, T.Y. and Walker, W.R., 1990. Zero-inertia model for surge flow furrow irrigation. *Irrigation Science*, 11(3), 131–136.

Oweis, T.Y., 1984. *Surge Flow Furrow Irrigation Hydraulics with Zero Inertia*. Ph.D. Dissertation, College of Engineering, Dep. of Ag. And Irrigation Engineering, Utah State University, Logan, Utah, USA.

Oyonarte, N.A., Mateos, L. and Palomo, M.J., 2002. Infiltration variability in furrow irrigation. *Journal of Irrigation and Drainage Engineering*, 128(1), 26–33.

Pal, R., Singh, S. and Poonia, S.R., 1980. Effect of water quality on the water transmission parameters of unsaturated soils. *Journal of the Indian Society of Soil Science*, 28(1), 1–9.

Philip, J.R., 1957. The theory of infiltration: 1. The infiltration equation and its solution. *Soil Science*, 83(5), 345–358.

Posadas, A., 2008. *Partial Root-Zone Drying: An Alternative Irrigation Management to Improve the Water Use Efficiency of Potato Crops*. International Potato Center, Peru.

Qin, W., Hu, C. and Oenema, O., 2015. Soil mulching significantly enhances yields and water and nitrogen use efficiencies of maize and wheat: A meta-analysis. *Scientific Reports*, 5, 16210.

Ragab, A.A.M., Hellal, F.A. and Abd El-Hady, M., 2008. Water salinity impacts on some soil properties and nutrients uptake by wheat plants in sandy and calcareous soil. *Australian Journal of Basic and Applied Sciences*, 2(2), 225–233.

Rahil, M.H. and Antonopoulos, V.Z., 2007. Simulating soil water flow and nitrogen dynamics in a sunflower field irrigated with reclaimed wastewater. *Agricultural Water Management*, 92(3), 142–150.

Ramon, C., Martin, J.T. and Powell, M.R., 1987. Low-level, magnetic-field-induced growth modification of *Bacillus subtilis*. *Bioelectromagnetics*, 8(3), 275–282.

Rasoulzadeh, A. and Sepaskhah, A.R., 2003. Scaled infiltration equations for furrow irrigation. *Biosystems Engineering*, 86(3), 375–383.

Rhoades, J.D., Oster, J.D., Ingvalson, R.D., Tucker, J.M. and Clark, M., 1974. Minimizing the salt burdens of irrigation drainage waters. *Journal of Environmental Quality*, 3(4), 311–316.

Richards, L.A., 1931. Capillary conduction of liquids through porous mediums. *Physics*, 1(5), 318–333.

Robbins, C.W., 1984. Sodium adsorption ratio-exchangeable sodium percentage relationships in a high potassium saline-sodic soil. *Irrigation Science*, 5(3), 173–179.

Romero, P., Muñoz, R.G., Fernández-Fernández, J.I., del Amor, F.M., Martínez-Cutillas, A. and García-García, J., 2015. Improvement of yield and grape and wine composition in field-grown Monastrell grapevines by partial root zone irrigation, in comparison with regulated deficit irrigation. *Agricultural Water Management*, 149, 55–73.

Ross, P., 1986. Zero-inertia and kinematic wave models for furrow and border irrigation. Unpublished notes.

Sadeghi, M., Ghahraman, B. and Davari, K., 2008. Scaling and prediction of soil moisture profile during redistribution phase. *Journal of Water and Soil*, 22(2), 417–431.

Sande, L. and Chu, X., 2012. Laboratory experiments on the effect of microtopography on soil-water movement: Spatial variability in wetting front movement. *Applied and Environmental Soil Science*, 147, 1–8.

Schwankl, L.J. and Wallender, W.W., 1988. Zero inertia furrow modeling with variable infiltration and hydraulic characteristics. *Transactions of the ASAE*, 31(5), 1470–1475.

Sepaskhah, A.R. and Afshar-Chamanabad, H., 2002. SW—Soil and water: Determination of infiltration rate for every-other furrow irrigation. *Biosystems Engineering*, 82(4), 479–484.

Sepaskhah, A.R. and Ahmadi, S.H., 2012. A review on partial root-zone drying irrigation. *International Journal of Plant Production*, 4(4), 241–258.

Shahnazari, A., Liu, F., Andersen, M.N., Jacobsen, S.E. and Jensen, C.R., 2007. Effects of partial root-zone drying on yield, tuber size and water use efficiency in potato under field conditions. *Field Crops Research*, 100(1), 117–124.

Sharma, M.L., Gander, G.A. and Hunt, C.G., 1980. Spatial variability of infiltration in a watershed. *Journal of Hydrology*, 45(1–2), 101–122.

Shukla, M.K., Lal, R., Owens, L.B. and Unkefer, P., 2003. Land use and management impacts on structure and infiltration characteristics of soils in the North Appalachian region of Ohio. *Soil Science*, 168(3), 167–177.

Stieber, T.D. and Shock, C.C., 1995. Placement of soil moisture sensors in sprinkler irrigated potatoes. *American Potato Journal*, 72(9), 533–543.

Stoll, M., Loveys, B. and Dry, P., 2000. Hormonal changes induced by partial rootzone drying of irrigated grapevine. *Journal of Experimental Botany*, 51(350), 1627–1634.

Strelkoff, T., 1990. *SRFR: A Computer Program for Simulating Flow in Surface Irrigation: Furrows-Basins-Borders*. Water Conservation Laboratory, Agricultural Research Service,Department of Agriculture, Washington, DC.

Strelkoff, T., 1991. SRFR—A model of surface irrigation—Version 20. *Irrigation and Drainage*, 52(3), 676–682,

Tagar, A.A., Siyal, A.A., Brohi, A.D. and Mehmood, F., 2010. Comparison of continuous and intermittent leaching methods for the reclamation of a saline soil. *Pakistan Journal of Agriculture, Agricultural Engineering and Veterinary Sciences*, 26, 36–47.

Taylor, S.A. and Ashcroft, G.L., 1972. *Physical Edaphology—The Physics of Irrigated and Nonirrigated Soils*. W.H.Freeman, San Francisco, CA.

Tuli, A., Kosugi, K. and Hopmans, J.W., 2001. Simultaneous scaling of soil water retention and unsaturated hydraulic conductivity functions assuming lognormal pore-size distribution. *Advances in Water Resources*, 24(6), 677–688.

Turner, E.R., 2006. *Comparison of Infiltration Equations and Their Field Validation with Rainfall Simulation*, Doctoral Dissertation, University of Maryland, College Park.

US Department of Agriculture, Natural Resources and Conservation Service. 1974. *National Engineering Handbook*. Section 15. Border Irrigation. National Technical Information Service, Chapter 4. Washington, DC, USA.

Vahabi, J. and Mahdian, M.H., 2008. Rainfall simulation for the study of the effects of efficient factors on run-off rate. *Current Science*, 95(10): 1439–1445.

van Schilfgaarde, J., Bernstein, L., Rhoades, J.D. and Rawlins, S.L., 1974. Irrigation management for salt control. *Journal of the Irrigation and Drainage Division*, 100(3), 321–338.

Walker, W.R. and Humpherys, A.S., 1983. Kinematic-wave furrow irrigation model. *Journal of Irrigation and Drainage Engineering*, 109(4), 377–392.

Walker, W.R. and Skogerboe, G.V., 1987. *Surface Irrigation. Theory and Practice*. Prentice-Hall, Englewood Cliffs, NJ.

Wallender, W.W. and Rayej, M., 1985. Zero-inertia surge model with wet-dry advance. *Transactions of the ASAE*, 28(5), 1530–1534.

Wang, D., Kang, Y. and Wan, S., 2007. Effect of soil matric potential on tomato yield and water use under drip irrigation condition. *Agricultural Water Management*, 87(2), 180–186.

Wang, X., Yang, J., Liu, G., Yao, R. and Yu, S., 2015. Impact of irrigation volume and water salinity on winter wheat productivity and soil salinity distribution. *Agricultural Water Management*, 149, 44–54.

Warrick, A.W., 1980. Spatial variability of soil physical properties in the field. *Application of Soil Physics*, 32, 319–344.

Woodrow, J.E., Seiber, J.N., LeNoir, J.S. and Krieger, R.I., 2008. Determination of methyl isothiocyanate in air downwind of fields treated with metam-sodium by subsurface drip irrigation. *Journal of Agricultural and Food Chemistry*, 56(16), 7373–7378.

Woolhiser, D.A. and Brakensiek, D.L., 1982. Hydrologic modeling of small watersheds. *Hydrologic Modelling of Small Watersheds*, 65(5): 3–16.

Wösten, J.H.M., Pachepsky, Y.A. and Rawls, W.J., 2001. Pedotransfer functions: Bridging the gap between available basic soil data and missing soil hydraulic characteristics. *Journal of Hydrology*, 251(3–4), 123–150.

WRI. (2005), *World Resources Institute: Freshwater Resources 2005*, http://earthtrends.wri.org/pdf_library/data_tables/wat2_2005.pdf.

Yactayo, W., Ramírez, D.A., Gutiérrez, R., Mares, V., Posadas, A. and Quiroz, R., 2013. Effect of partial root-zone drying irrigation timing on potato tuber yield and water use efficiency. *Agricultural Water Management*, 123, 65–70.

Yazar, A., Gökçel, F. and Sezen, M.S., 2009. Corn yield response to partial rootzone drying and deficit irrigation strategies applied with drip system. *Plant, Soil and Environment*, 55(11), 494–503.

Zegbe, J.A. and Serna-Pérez, A., 2011. Partial rootzone drying maintains fruit quality of 'Golden Delicious' apples at harvest and postharvest. *Scientia Horticulturae*, 127(3), 455–459.

Zhang, Y., Wu, P., Zhao, X. and Li, P., 2012. Evaluation and modelling of furrow infiltration for uncropped ridge–furrow tillage in Loess Plateau soils. *Soil Research*, 50(5), 360–370.

Zheng, S., Ni, K., Ji, L., Zhao, C. Chai, H., Yi, X., He, W. and Ruan, J., 2021. Estimation of evapotranspiration and crop coefficient of rain-fed tea plants under a subtropical climate. *Agronomy*, 11(11), 23–39.

Zolfaghari, A.A., Mirzaee, S. and Gorji, M., 2012. Comparison of different models for estimating cumulative infiltration. *International Journal of Soil Science*, 7(3), 108–115.

3

Movement of Water in Soil

Sushant Mehan
Colorado State University

Saeid Eslamian
*Isfahan University
of Technology*

3.1 Introduction

Water is like a free object that moves from a state of higher energy to lower energy. Many factors result in the difference in energy that enables water to move, including the force of gravity, the interaction of water molecules with each other and the different particles of soil, and external pumping or forced flooding. Other than that, the soil consists of particles that make its texture (a proportion of sand, silt, and clay). Different textures of soil will have different porosity or pore space. Water moves through such pore spaces as it can bind with the different water molecules and walls of the medium it flows through, e.g., soil. More water flows in sand-dominated soil where pore space is more extensive than clayey soil. The soil particles in the clayey soils are close enough, and pore spaces are relatively small. Besides flowing through the ground, some part of the water gets stored in the soil and is used by plants for its growth, and the other becomes part of the surface and subsurface water bodies. The climate, topography, and soil properties along with the use and recharge of subsurface water decide on the water stored in the soil, which is dynamic and can be explained with a mass balance equation as shown below:

DOI: 10.1201/9780429290114-4

$$\text{Soil moisture storage} = \text{Input water} \left(\text{precipitation, irrigation} \right)$$

$$- \text{output water} \left(\text{ET, deep percolation, surface runoff, lateral flow} \right) \quad (3.1)$$

where, ET is evapotranspiration. The hydrologic processes in soil influence the sustainability of water resources, and henceforth, most of the terrestrial and aquatic life forms (O'Geen et al., 2010). Water in soil is a principal abiotic determinant factor that controls the plant growth, photosynthesis, and movement of organic and inorganic molecules within the plant. For every carbon dioxide molecule absorbed for photosynthesis, there is a loss of an average of 400 water molecules (McElrone et al., 2013). The process requires a considerable water supply within the soil. It is, therefore, becomes more important to study the soil–water relation. This chapter will discuss the structure and properties of water; what kind of forces are responsible for the movement of water in different media; mass, energy, and volume relationships; and the factors affecting available water. Figure 3.1 demonstrates the critical pathways of water through land and the atmosphere.

3.2　Structure of Water and Its Properties

Water is a polar compound, and each molecule of water is composed of oxygen and two hydrogen atoms (Figure 3.2). Water forms a dipole as one end of the water molecule is positively charged due to the protons of the hydrogen, and the other end of the water is charged negative because of the two lone pairs of electrons. Each hydrogen atom can share a couple of electrons with a single oxygen atom via a covalent bond forming an angle of 103–106 degrees, with an oxygen atom as an apex of the angle (Figure 3.2). Hence, the geometry of a water molecule is asymmetrical (V-Shaped). Opposite charges on the end of water molecules govern the interaction within the different water molecules, alter many physio-chemical reactions, and, thus, drive the soil–plant–water relationship. The polar nature of water increases the affinity of ionic components toward the water. In other words, the polarity of water is responsible for the interaction of water with cations and negatively charged clay particles. The list of selected properties of water is provided in Table 3.1.

Two forces, hydrogen bonding and the van der Waals–London force, bind the water molecules together with each other and with the medium (soil) through which it flows.

3.2.1　Hydrogen Bonding

The hydrogen atom within a water molecule forms a point of electrostatic attraction that links the two water molecules with a force of about 1.3–4.5 kcal/mol in water (Kramer, 1983). The force of attraction between the negative lone-pair electrons of one water molecule and a positive proton (H) of another water molecule forms a hydrogen bond. It leads to the polymerization and lattice structure of water. Therefore, the high boiling point, specific heat, and viscosity of water compared with the compounds having similar molecular weight.

3.2.2　Van der Waals–London Forces

The Dutch physicist van der Waals (1837–1923) described the van der Waals–London forces. Unlike hydrogen bonding, such forces are independent of a net electrical charge and occur between neutral non-polar molecules. It is a weak force of attraction between two water molecules. Water is a life-sustaining element that makes earth hospitable for human living by its wide range of physical and chemical properties, some of which are as follows.

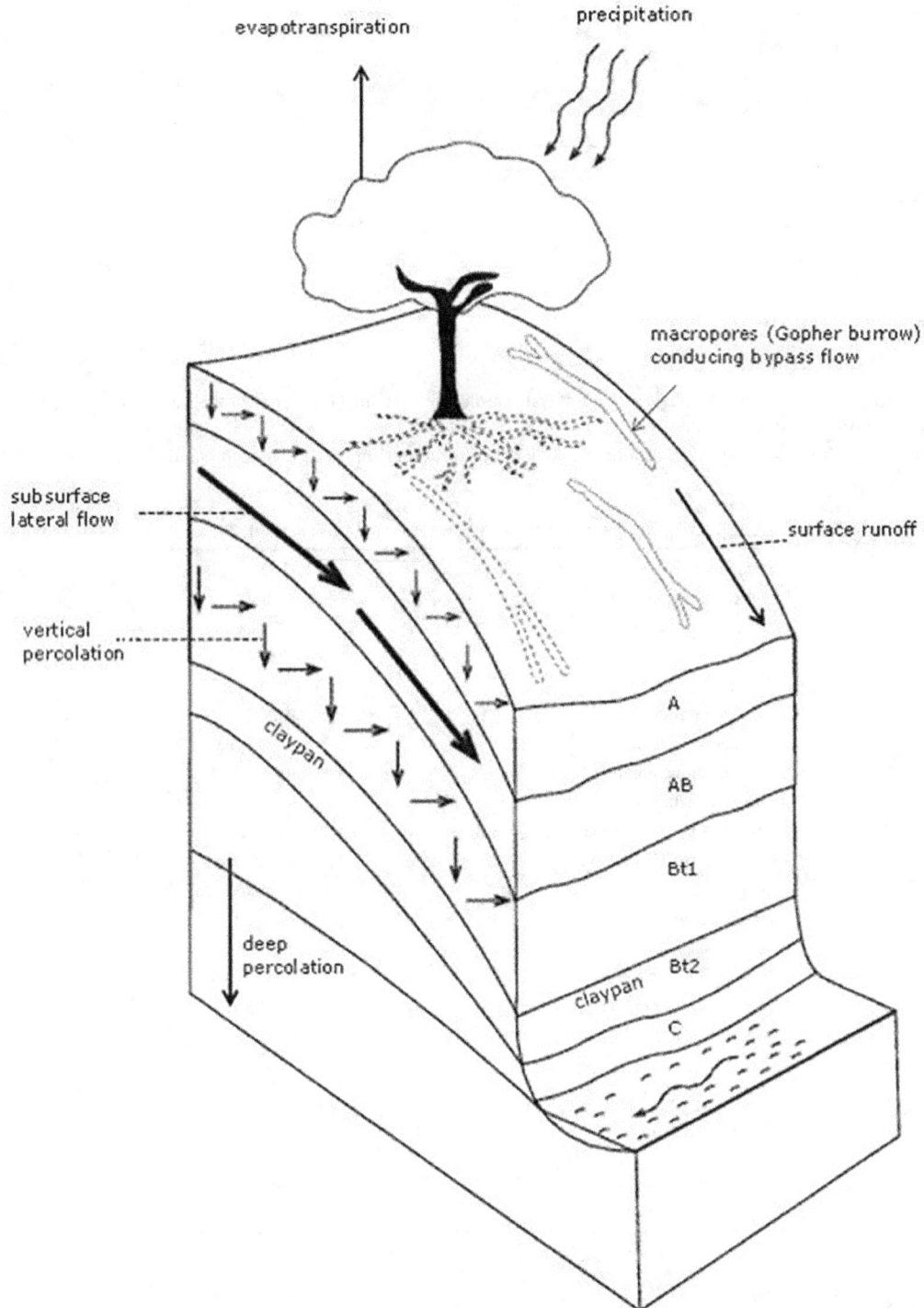

FIGURE 3.1 Conceptual diagram of a soil profile illustrating the critical paths of water through land and atmosphere. © 2013 Nature Education Modified from O'Geen et al. 2010. All rights reserved.

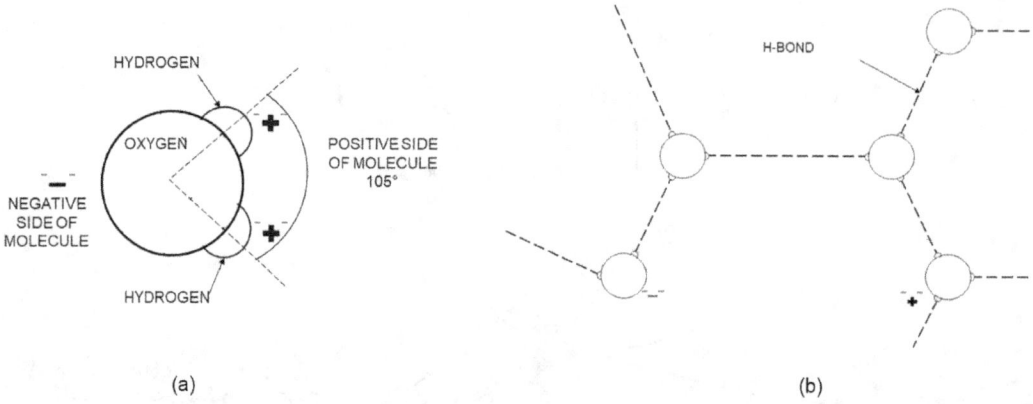

FIGURE 3.2   Two-dimensional structure of water molecule showing (a) polarity of the water molecule and (b) hydrogen bonding of water to itself.
Source: http://www.soilmanagementindia.com/soil-water/soil-waterimportance-concepts-and-classification/1790.

TABLE 3.1   Selected Properties of Water (Horvath, 1975; Perry, 1985)

Chemical formula	H_2O
Molecular weight	18.0148
Critical temperature	373.91°C
Critical pressure	22.05 MPa
Critical density	315.0 kg/m³
Triple point temperature	0.01°C
Triple point pressure	615.066 Pa
Normal boiling point	100.0°C
Normal freezing point	0.0°C
Density of ice at the usual melting point	918.0 kg/m³
Maximum density, 3.98°C	999.973 kg/m³
Viscosity, 25°C	0.889 mN s/m²
Surface tension, 25°C	72 mN/m
Heat capacity, 25°C	4.1796 kJ/kg K
Enthalpy of vaporization, 100°C	2,257.7 kJ/kg
Enthalpy of fusion, 0°C	333.8 kJ/kg
Velocity of sound, 0°C	1.403 km/s
Dielectric constant, 25°C	78.40
Electrical conductivity, 25°C	8 µS/m
Refractive index, 25°C	1.333
Liquid compressibility, 10°C	480×10^{-12} m²/N
Coefficient of thermal expansion, 25°C	256.32×10^{-6} K⁻¹
Thermal conductivity, 25°C	0.608 W/m K

3.2.3 Specific Heat

Specific heat is the amount of heat energy required to raise a per unit mass of a substance by per unit increase in temperature. Dimensionally, it is $[L^2T^{-2}K^{-1}]$. After liquid ammonia, water has the highest specific heat. The specific heat of water is at its lowest value of 4.1779 J/g°C at 35°C.

3.2.4 Heat of Vaporization

The heat of vaporization is the amount of heat required to transform per unit mass of liquid into vapor without any change in the temperature of the liquid. Because of the high heat of vaporization in water leads to cooling when the water evaporates and warming when the water condenses. The heat of vaporization for water is 40.65 kJ/mol or 540 cal/g at 100°C, water's boiling temperature. The latent heat of vaporization is temperature-dependent given by Harrison (1963):

$$L_v = 2.501 - 2.361T \times 10^{-3}, \tag{3.2}$$

where

L_v is in MJ/kg, and
T is in °C.

3.2.5 Heat of Fusion

The heat of fusion is the amount of energy required to transform per unit mass of solid to a liquid without any change in temperature. The heat of fusion for water is approximately 334 J (80 cal/g) at 0°C. The phenomenon of the heat of fusion is crucial in areas where crops receive irrigation water during frost times.

3.2.6 Heat Conduction

Heat conduction is the quantity of heat in calories, which is transmitted per second through a plate one centimeter thick across an area of one square centimeter when the temperature difference is 1°C. At 20°C, water has a thermal conductivity of 0.00144 cal/scm°C. Copper, a metal, at 18°C, has a thermal conductivity of 0.918 cal/scm°C, 1,000 times greater than water.

3.2.7 Transparency to Visible Radiation and Opaqueness to Infrared Radiation

White light can penetrate through water, and because of its transparent nature for visible radiation, life forms such as algae can survive at different depths of the sea. Thus, water acts as an excellent heat absorber because it is opaque to the longer wavelengths.

3.2.8 Other Properties of Water

- Water has a high internal intramolecular cohesive force and surface tension, providing enough tensile strength to rise in plants.
- Because water has its maximum density at 4°C instead of at its freezing point, water expands on freezing.
- Water can be ionized and has a high dielectric constant, which makes it an excellent insulator. Saltwater is different from pure water, which has a low dielectric constant.
- Water can form bonds and act as a suitable solvent for electrolytes because of its ionization.
- Water forms a strong bond, and surfaces like clay adsorb water.
- Also, water has a high-viscosity, internal friction. It is because of the viscosity of water that liquid flows through capillaries. Like surface tension, viscosity is also a function of temperature.

3.3 Terminology Relating to Soil and Soil–Water Relationships

Soil, from an agricultural perspective, serves three primary functions: (1) soil acts as a moisture reservoir, (2) soil also acts as a nutrient reservoir, and (3) soil provides mechanical stability for plants.

3.3.1 Mass and Volume Relationships

The soil itself, along with the pores and voids within the soil, comprises solids, liquids, and gases. M_a is the mass of air, which is considered negligible and assumed to be zero, M_w is the mass of water, and M_s is the mass of soil solids. The corresponding weights (i.e., the product of mass and acceleration due to gravity) are often considered instead of the masses. The volumes of the three components are represented by V_a for soil air, V_w for soil water and V_s for soil solids. If V_f represents the total pore space; and M_t, the total mass; and V_t, the total volume, then the following relationships holds.

$$V_f = V_a + V_w \tag{3.3}$$

$$M_t = M_a + M_w + M_s = M_w + M_s \tag{3.4}$$

and

$$V_t = V_a + V_w + V_s = V_f + V_s \tag{3.5}$$

3.3.2 Soil Texture

In a soil texture triangle (Figure 3.3), the proportions of sand, silt, and clay particles determine the soil texture. It is an important property because tighter soils like clay soils are heavier textured soils, which corresponds to the slow movement of water, unlike sandy soil. However, clay soils will generally hold more water than other soils. Additionally, nutrient holding capability is highest in clay.

3.3.3 Soil Structure

The individual soil constituents or soil particles hold together and determine the soil structure (Figure 3.4). Sand particles are not held together tightly and, thus, they form a single grain structure, but clay forms a blocky structure, having granules with very few spaces making it difficult for water to move through it. Even plant roots don't grow very well under such circumstances. A soil structure can be manipulated in different ways: growing crops and leaving the residue in the ground, enhancing soil's physical properties with more organic matter.

3.3.4 Particle Density

Particle density, P_d, is defined as the ratio of the total mass of the solid particles to the total volume of the solid particles, and expressed as:

$$P_d = \frac{M_s}{V_s}. \tag{3.6}$$

Heat-capacity calculations, sedimentation analysis, or estimating volumes or mass of soil requires the information on particle density. Table 3.2 presents the particle density of the different soil textural classes.

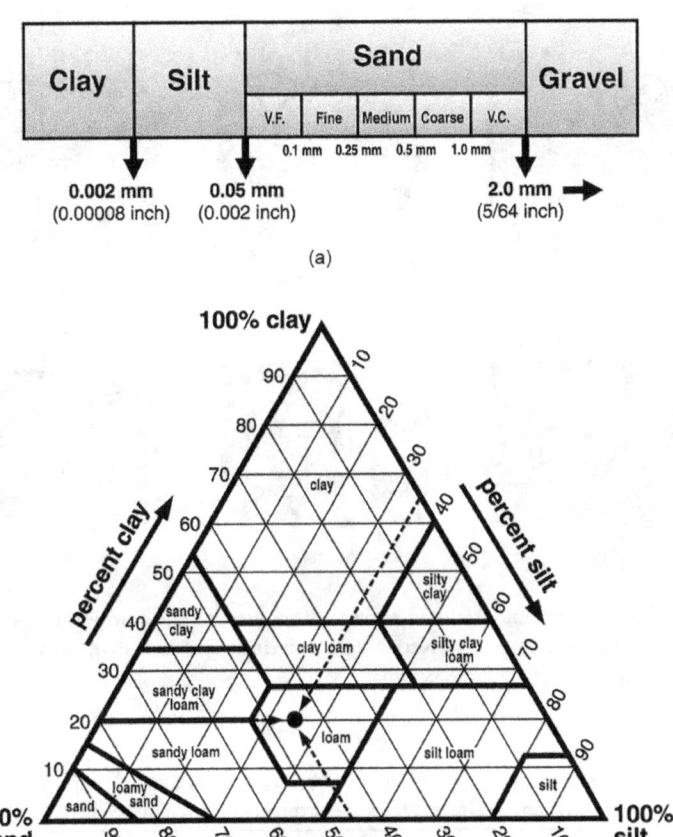

(a)

(b)

FIGURE 3.3 (a) Textural classification by the size of the primary soil particles (U.S. Department of Agriculture Soil Classification System). V.F. in the diagram refers to very fine and V.C. to very coarse. (b) The Soil Texture Triangle (U.S. Department of Agriculture (USDA), 1987, https://www.nrcs.usda.gov/Internet/FSE_DOCUMENTS/stelprdb1044818.pdf). The point depicts a loam soil with 45% sand, 35% silt, and 20% clay content.

3.3.5 Bulk Density

Bulk density, B_d, is defined as the dry weight of a given volume of soil in its natural condition. The bulk density of soil, B_d, is given by:

$$B_d = \frac{M_s}{V_t} = \frac{M_s}{V_s + V_w + V_a}. \tag{3.7}$$

An increase in the soil bulk density improves strength, limiting soil porosity further, leading to soil compaction. Table 3.3 presents the values of the bulk density of different soil textural classes.

3.3.6 Wet Bulk Density

Wet bulk density, B_{dw}, is defined as the ratio of the total mass of the moist soil to the total volume of the soil, and expressed as:

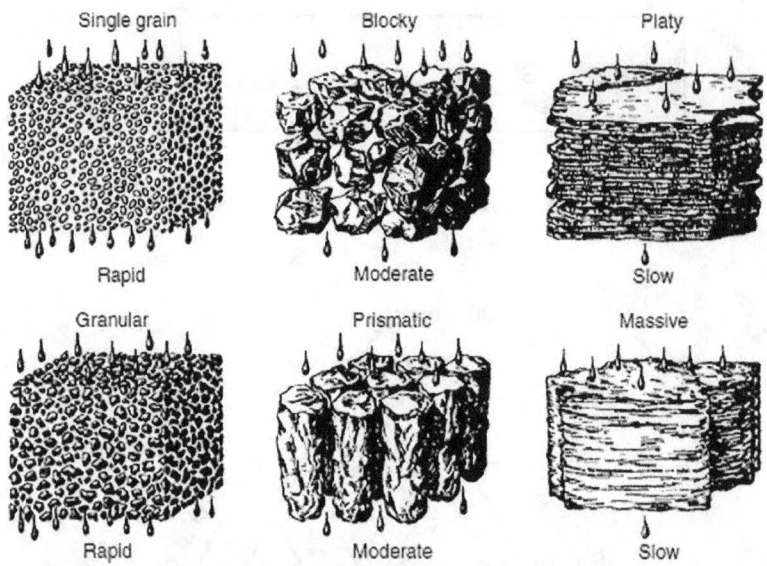

FIGURE 3.4 Some of most common soil structures with an effect on the downward movement (infiltration) of water. Adapted from NRCS, Section 15 of the National Engineering Handbook, 1991, https://directives.sc.egov.usda.gov/OpenNonWebContent.aspx?content=18350.wba.

TABLE 3.2 Particle Diameter (mm) of Sand, Silt, and Clay Based on the United States Department of Agriculture System

Texture Size Class	Particle Diameter (mm)
Very coarse sand (vcos)	1.0–2.0
Coarse sand (cos)	0.5–1.0
Medium sand (s)	0.25–0.5
Fine sand (fs)	0.1–0.25
Very fine sand (vfs)	0.05–0.1
Silt (si)	0.002–0.05
Clay (c)	<0.002

Source: NEH 15-1, 1991, https://directives.sc.egov.usda.gov/OpenNonWebContent.aspx?content=18350.wba)

$$B_{dw} = \frac{M_s + M_w}{V_t} = \frac{M_s + M_w}{V_s + V_w + V_a}. \tag{3.8}$$

3.3.7 Specific Volume

The specific volume of the soil is the volume per unit mass of the dry soil. It characterizes the level of compaction or looseness index of the soil. In other words, it is the reciprocal of the dry bulk density of soil.

$$V_b = \frac{V_t}{M_s} \tag{3.9}$$

TABLE 3.3 Relationship between Bulk Density and Root Growth Based on Soil Texture

Textural Class	Ideal Bulk Density (g/cm³)	Bulk Density that May Affect Root Growth (g/cm³)	Bulk Density that Restricts Root Growth (g/cm³)
Sands, loamy sands	<1.60	1.69	>1.80
Sandy loams, loams	<1.40	1.63	>1.80
Sandy clay loams, clay loams	<1.40	1.60	>1.75
Silts, silt loams	<1.30	1.60	>1.75
Silt loams, silty clay loams	<1.40	1.55	>1.65
Sandy clays, silty clays, some clay loams (35%–45% clay)	<1.10	1.49	>1.58
Clays (>45% clay)	<1.10	1.39	>1.47

Source: Adapted and modified from Schoonover and Crim (2015); *Source:* Arshad et al. (1996).

3.3.8 Porosity

It refers to the percent of soil's volume occupied by pore space. Porosity, f is expressed as:

$$f = \frac{V_f}{V_t} = \frac{V_a + V_w}{V_s + V_w + V_a}. \tag{3.10}$$

The values of particle density and bulk density can help in calculating the total pore space or porosity of the soil. The particle density of soil is generally assumed to be 2.65 g/cm³. Knowing the bulk density and particle density, the porosity can be calculated by:

$$f = 1 - \frac{B_d}{P_d}. \tag{3.11}$$

3.3.9 Void Ratio

The Void ratio of the soil is the ratio of the total pores in the soil to the volume of soil solids. It is unlike porosity, which is related to the total soil volume. This property is termed as 'Void Ratio' and written as:

$$e = \frac{V_a + V_w}{V_s} = \frac{V_f}{V_s}. \tag{3.12}$$

3.3.10 Mass Wetness

The mass wetness of the soil is the amount of water held per unit mass of the dry soil. It is also referred to as the soil moisture expressed on a weight basis and conventionally denoted by w. The mass wetness of soil can be represented as a fraction or as a percentage. Analytically, it is:

$$w = \frac{M_w}{M_s}. \tag{3.13}$$

In terms of percentage:

$$w\left(\% \text{ by weight}\right)\left(\frac{g}{g}\right) = w \times 100 \tag{3.14}$$

3.3.11 Volume Wetness

Volume wetness or volumetric water content is the volume of water held per unit volume of oven-dried soil. Volume wetness is a dimensionless fraction or a percentage, as shown below. θ is generally used to represent the volume wetness, which is equal to:

$$\theta = \frac{V_w}{V_t} = \frac{V_w}{V_s + V_f}.$$

(3.15)

$$\theta\left(\% \text{ by volume}\right) = \theta \times 100.$$

(3.16)

The conversion of gravimetric water content (water content expressed on a weight basis) to volumetric water content is simple and given by:

$$\theta = \frac{w \times B_d}{\rho_w},$$

(3.17)

where ρ_w is the density of water.

3.3.12 Degree of Saturation

The degree of saturation, θ_e, is defined as the ratio of the volume of water present in the soil to the total pore space (Eslamian et al. 2018).

$$\theta_e = \frac{V_w}{V_f} = \frac{V_w}{V_a + V_w}$$

(3.18)

3.3.13 Air-Filled Porosity

This is a property used to characterize soil aeration and defined as the ratio of the volume of pore spaces filled with air to the total soil volume, i.e.,

$$f_a = \frac{V_a}{V_t} = \frac{V_a}{V_s + V_w + V_a}.$$

(3.19)

3.3.14 Soil Moisture Tension

Tension or soil moisture tension is the tightness with which the soil will hold water at a specific moisture content. It's a function of the moisture content. If there is dry soil, the water in the soil, at that moisture content, is held quite tightly by the soil, it's not readily available for the plants to use. In a moist environment but not too wet, plants will use more water if the water is readily available. Soil moisture tension can be measured using a tensiometer.

3.3.15 Field Capacity

During irrigation scheduling and planning, the water applied to the field is usually at the field capacity. Saturated soil creates the anaerobic condition for plants, thus inhibiting the plant growth in the absence

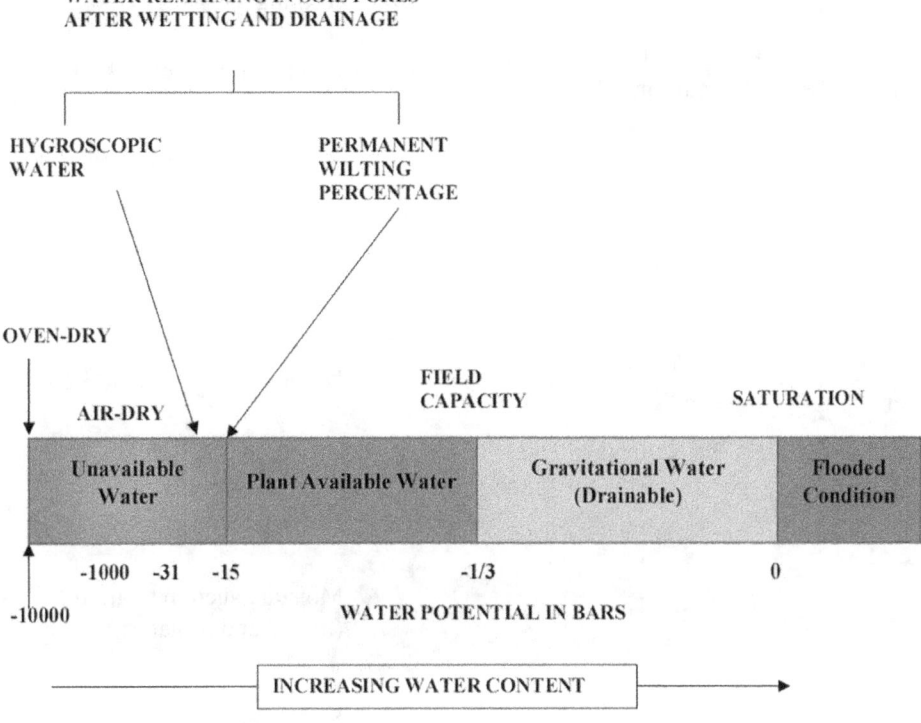

FIGURE 3.5 Different types of soil water at different water content and soil conditions at respective water content.

of oxygen for roots to respire (Figure 3.5). The soil will drain water after the irrigation process, and there's water left in the soil when that drainage stops, the water content at the specific point of time is called field capacity (Figure 3.5).

3.3.16 Permanent Wilting Point

When plants use water, some of the water evaporates off. If the soil gets quite dry, so dry that the plants now would irreversibly wilt because they can't get water fast enough, and they would die at that moisture content – and the threshold is called the permanent wilting point (Figure 3.5).

Plant available water is the difference in water content between field capacity and permanent wilting point and depends on the number of factors as presented in the line diagram in Figure 3.6. The amount of water that drains from macropores by gravity between saturation and field capacity (in the absence of evaporation) is called the drainable porosity (Figure 3.5)

3.3.17 pF

pF is used to express the negative pressure potential of soil water in terms of equivalent hydraulic head of magnitudes of 10,000–100,000 cm of water. It is the negative logarithm of either the tension or suction head of water, e.g., a pF of 3 is the tension head of 1,000 cm of water. Classification of soil based on the relationship between pF values and height of water columns (in cm) and pressure (bar) is presented in Table 3.4.

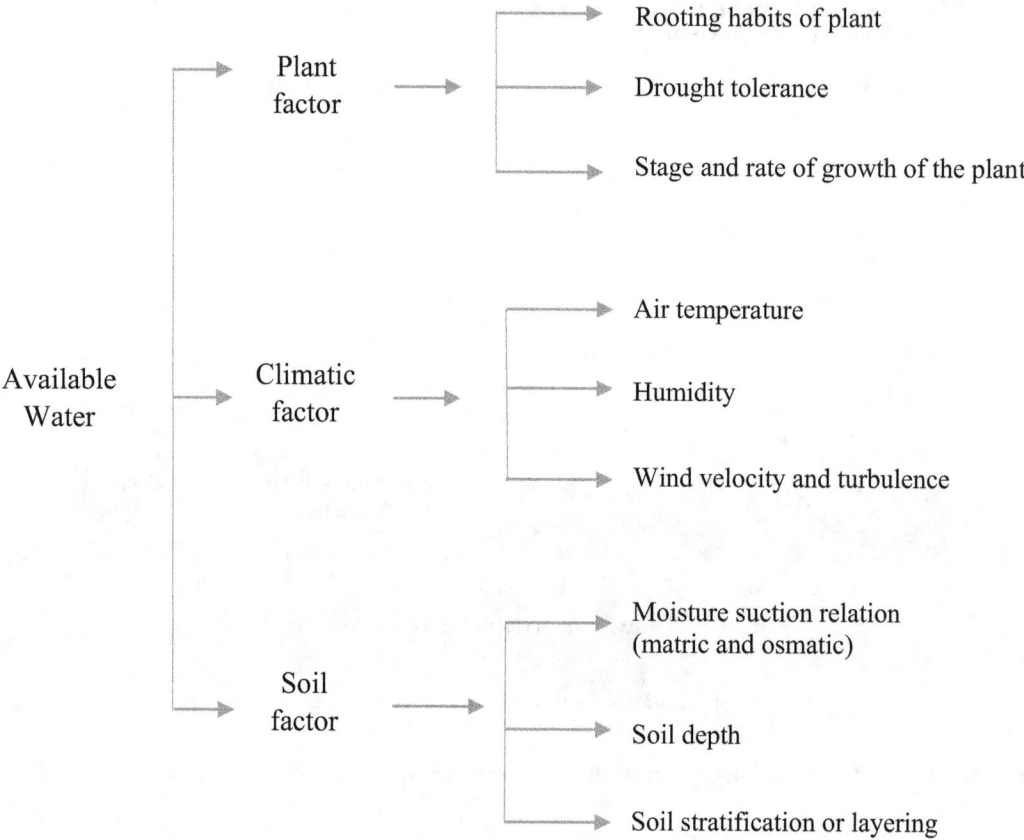

FIGURE 3.6 Line diagram presenting different factors affecting plant available water.

TABLE 3.4 The Relationship among pF Values, Height of Water Column, and Pressure or Atmosphere

Heights of Water Columns (cm)	Pressure (Atmosphere or bars)	pF Values
1	0.001	0 Saturated soil
10	0.01	1
10^2	0.1	2
346	1/3	2.53 Field capacity
10^3	1.0	3
10^4	10.0	4
15.849	15.0	4.18 Wilting point
31.623	31.0	4.50 Hygroscopic
10^5	100.0	5
10^6	1000.0	6
10^7	10000.0	7 Oven dry soil

3.3.18 Soil Water Potential

Different kinds of energy between the water molecules and soil water determine the retention and movement of water in the soil, its translocation within plants, and, eventually, its loss into the atmosphere.

The energy which drives solute-solvent movement within the soil can occur in different forms, including potential, kinetic, and electrical.

The forces that exist within soil water include matric force, gravitational force, osmotic force, and water pressure. The sum of matric, gravitational, and osmotic forces total the water potential. All such forces consider pure or free water at a specific height as the reference point for these potential forces. Unlike pure free water, soil water potential is usually negative because of forces that include adsorption, absorption, and cohesion, which hold water molecules together with each other and with the soil.

Water rising in a small cylindrical medium is possible because of matric potential or capillary potential. The matric potential can be measured using a tensiometer. The matric potential is usually negative and numerically equivalent to the soil moisture tension.

The potential energy associated with water at a specific elevation is gravitational potential energy or gravitational potential. The gravitational potential defines the direction of water movement rising or infiltrating based on a particular reference level. If the reference level is below the reference point, there is a need for work on the water, and gravity potential is positive and vice versa.

The osmotic potential is significant when determining the water movement within the saline soil. It is numerically equivalent to the solute potential attributable to the attraction of solutes for water. The osmotic pressure is measured using an osmometer.

The pressure potential is the potential energy due to the weight of the water at any reference position. A piezometer or a manometer measures the pressure potential. Another term synonymous with the pressure potential is air pressure potential under unsaturated conditions and hydrostatic pressure potential under saturated conditions.

ISSS (1963) define the total potential of soil water as "the amount of work that must be done per unit quantity of pure water to transport reversibly and isothermally an infinitesimal quantity of water from a pool of pure water at a specified elevation and atmospheric pressure to the soil water (at the point under consideration)." At any given point in space and time, the potential can be either hydrostatic pressure, vapor pressure, or elevation.

Analytically, the total potential of soil water is the sum of the matric, osmotic, gravitational, and pressure potential and is written as follows:

$$\varphi_t = \varphi_m + \varphi_o + \varphi_g + \varphi_p, \tag{3.20}$$

where φ_t is the total potential, φ_m is the matric potential, φ_o is the osmotic potential, φ_g is the gravitational potential, and φ_p is the pressure potential (Figure 3.7). Table 3.5 explains the energy levels presented in Figure 3.5 during different soil conditions at different pressure potentials.

3.3.19 Soil Water Content

The water content and the soil moisture describe the state of water in the soil. The amount of water in the ground, either by weight or volume, determines the water content. Gravimetrically, when a constant mass of soil loses water at 105°C, the water loss per unit mass of the dry soil or units of volume of water lost per unit bulk volume of the soil is water content expressed in kg/kg or m^3/m^3, respectively. Table 3.6 represents the amount of available soil moisture for various soil textures, and Table 3.7 presents the different methods to measure the soil water content, strengths, and limitations. The conservation of soil moisture is possible by using protected and precision farming (Mehan and Singh, 2015).

There exist a relation between soil water content and soil water potential at different states of soil, water, and atmospheric conditions, as shown in Figure 3.8.

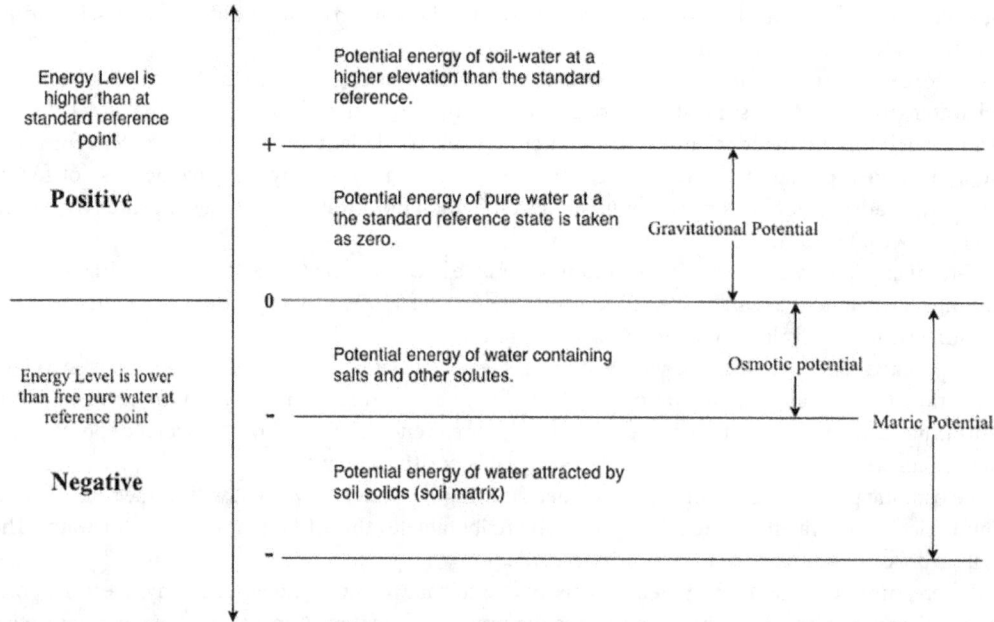

FIGURE 3.7   Different soil water potential and energy levels with reference to pure water and relationship with soil.

TABLE 3.5   Relative Energy Potential for Different Soil, Plant, and Atmospheric Conditions

Soil, Plant, and Atmospheric Conditions	Bars and Atmospheres (bars and atm)	Pounds Per Square Inch (psi)	Kilo Pascals (kPa)	Relative Energy Potential
Saturated soil	0	0	0	High
Field capacity	−0.33	−5	−33	Medium
Available water	−0.33 to −15	−5 to −225	−33 to −1,500	Medium
PWP (Permanent Wilting Point)	At or below −15	At or below −225	At or below −1,500	Low
Air dried soil	−31	−465	−3,100	Low
Oven-dried soil	Below −31	Below −465	Below −3,100	Low
Root tissue	−3 to −20	−45 to −300	−300 to −200	Low
Leaf tissue	−15 to −30	−225 to −450	−1,500 to −3,000	Low
Atmosphere	−100 to −500	−1,500 to 7,500	−10,000 to −50,000	Very low

TABLE 3.6   Available Soil Moisture-Holding Capacity for Various Soil Textures

Soil Texture	Available Soil Moisture	
	inches/inch	inches/foot
Coarse sand and gravel	0.02–0.06	0.2–0.7
Sands	0.04–0.09	0.5–1.1
Loamy sands	0.06–0.12	0.7–1.4
Sandy loams	0.11–0.15	1.3–1.8
Fine sandy loams	0.14–0.18	1.7–2.2
Loams and silt loams	0.17–0.23	2.0–2.8
Clay loams and silty clay loams	0.14–0.21	1.7–2.5
Silty clays and clays	0.13–0.18	1.6–2.2

TABLE 3.7 Various Methods, Strengths, and Weaknesses with Response Time to Measure In-Situ Soil Moisture or Moisture Conditions over a Large Area

Method	Measured Parameter	Strengths	Limitations	Response Time
Overview of Methods Used for In-Situ Soil Measurement				
Gravimetric	Mass of water content (percent of dry vs. wet soil weight)	1. High level of accuracy	1. Destructive 2. Repeated sampling is not possible 3. Time and labor-intensive 4. High level of potential uncertainty	24 hours
Neutron scattering	Volumetric water content (percentage of volume)	1. Straightforward implementation 2. Nondestructive 3. Enables measurement at several depths 4. High level of accuracy	1. Radiation hazard 2. Time-intensive 3. Low sensitivity in the upper 20 cm of the soil profile	1–2 minutes
Gamma attenuation	Volumetric water content	1. Nondestructive 2. The operation can be automated	1. Low accuracy 2. Requires a high level of expertise 3. Restricted to a soil depth of 2.5 cm	Instantaneous
TDR	Volumetric water content aided by the propagation of electromagnetic wave measurement	1. Nondestructive 2. Easy to use 3. The operation can be automated at multiple points 4. High level of accuracy	1. Environment-sensitive, especially in saline soils	Instantaneous
FDR		1. Nondestructive	1. Environment-sensitive	Instantaneous
Tensiometer techniques	Soil water potential (capillary potential)	1. Nondestructive 2. High level of accuracy 3. Easy to install, operate and maintain for extended periods	1. Only allow indirect estimation of SMC (Soil Moisture Content) 2. Fragile automated operation impractical	2–3 hours
Resistive sensor (gypsum)	Soil moisture tension	1. Measurement of soil moisture in the same location in the field over an extended period	1. Each block requires individual calibration 2. Calibration changes with time 3. Provides inaccurate measurements 4. Life of device limited	2–3 hours
Hygrometric techniques	Soil water potential	1. Automated measurements	1. Sensing element deteriorates through interaction with soil components	<3 minutes
Overview of Methods Used for Large Areas Using Remote Sensing				
Optical remote sensing	Reflectance	1. High spatial resolution 2. Multiple satellites available 3. Large Area Soil Moisture estimation, simple to use	1. Change of color of soil can decrease the accuracy 2. Poor penetration capability 3. Restricted its use for soil moisture underneath crop cover	
Thermal remote sensing	Thermal inertia	1. Good spatial resolution 2. Multiple satellites available 3. Large area soil moisture estimation	1. Poor temporal resolution 2. Weak relationship to soil moisture when there is high amount of vegetation cover	

(Continued)

TABLE 3.7 (*Continued*)   Various Methods, Strengths, and Weaknesses with Response Time to Measure In-Situ Soil Moisture or Moisture Conditions over a Large Area

Method	Measured Parameter	Strengths	Limitations	Response Time
Active microwave remote sensing	Dielectric constant	1. Large area soil moisture estimation 2. Direct sensitivity toward the soil moisture 3. High spatial resolution 4. Cost-effective	1. Surface roughness and vegetation cover and noise parameters	
Passive microwave remote sensing	Brightness temperature	1. Regional- and global-scale soil moisture map 2. Cost-effective	1. Very coarse spatial resolution of the order of 25–50 km	

Source: Adapted from Sharma et al. (2018).

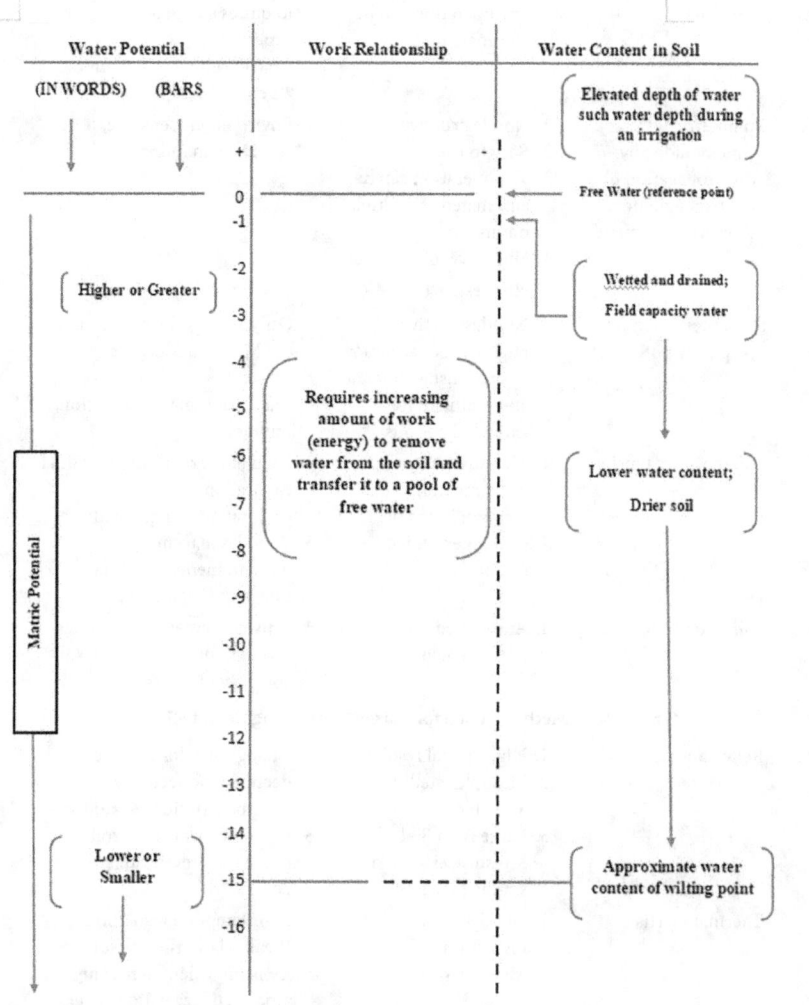

FIGURE 3.8 Line diagram explaining the relationship between soil water potential and soil water content at different soil conditions.

3.3.20 Soil Moisture Characteristic Curve and Hysteresis

The characterization of the soil using a soil moisture characteristic curve determines how and what happens to the soil–water relationship and explains the retention of soil water within the soil. Figure 3.9 shows a soil moisture characteristic curve, where the horizontal axis represents the water content. At a pressure of zero bar, the soil is saturated, which means that there's no tension pulling water out of the soil, and the vertical axis (unlike the horizontal axis) represents negative pressure. As soil goes from zero to more suction or negative pressure, drainage slowly pulls the water out. Soil dries, and there is an occurrence of a residual saturation point where there is no more draw on water, pulling it out. It means the water films around the soil particle breaks down. At this point, the pressure is water entry pressure or air entry pressure; that's when the water had been thoroughly filling the pores, and essentially the largest pores start releasing their water as the air is entering, and this is called the drainage curve. As draining goes this way, soil can take more water. So, two curves present the same pressure at very different moisture contents, called hysteresis. Hysteresis is significant because as the plants apply pressure, they pull water out of the soil. If the soil was saturated, the plant can quickly get the water out, whereas if the soil were dry, that same water content would be at a much lower value, and the plant may die because of lack of water. Hysteresis effect is a fundamental piece of the soil characteristic curve that describes plant–water relations, infiltration processes, or hydrology of the soil.

The various reasons leading to the hysteresis effect are:

a. The non-uniformity in the geometry of the individual pores.
b. The difference in contact angle and radius of curvature during advancing meniscus (greater) than receding meniscus (smaller), resulting in water content that tends to exhibit higher suction in desorption (drying) than in sorption (wetting).

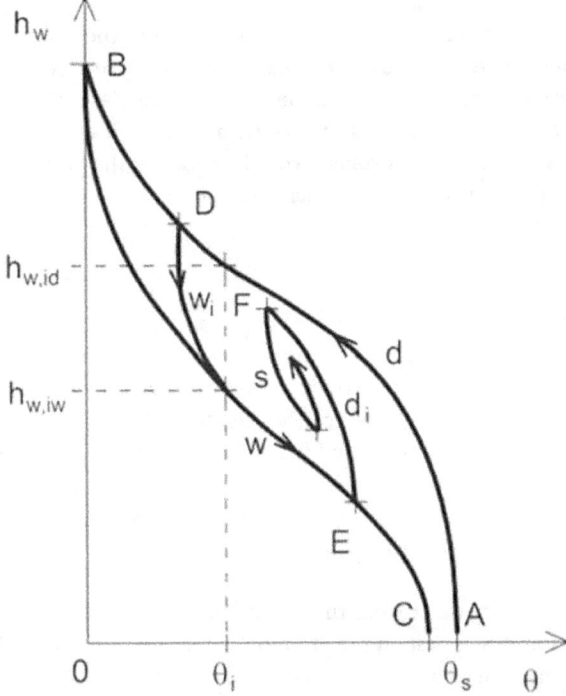

FIGURE 3.9 Soil-water matric potential h_w and volumetric soil-water content Θ. Pairs of possible values of (Θ_i, h_{wi}) are confined by main drying (*d*) and main wetting (*w*) branches of the soil-water retention curve. Adapted from Novák and Hlaváčiková, 2019.

c. Decreased water content due to entrapped air, which further reduces the water content of the newly wetted soil.

d. Alternate soil wetting and soil heating can alter the soil structure, resulting in swelling, shrinking, or aging. These physical processes impact the soil, air, and water relationship, affecting the suction wetness in wetting and drying systems.

3.4 Different Hydrologic Processes Involved When Water Contacts Soil

The various processes involved when water meets the soil surface include infiltration, percolation, and drainage.

3.4.1 Infiltration

Soil surface characteristics and present soil moisture conditions determine the downward movement of water. Such downward movement of water is infiltration. Barren land usually has less infiltration than the ground covered with vegetation. Due to more considerable temperature differences, dry or warm soil absorbs more water than wet or colder soil. Soil texture also influences the infiltration process. Soils with coarser, granular, and high organic matter observe higher infiltration rates. Different factors affecting the infiltration process include the presence or absence of clay minerals, soil texture and structure, antecedent moisture content, vegetative cover, and topography.

3.4.2 Percolation

Percolation is the process when water moves through a column of the soil profile. It is an important hydrological process that regulates groundwater recharge, which can be used later for irrigation through wells or pumping. The water percolating through the plant roots carries certain plant nutrients, and the process is known as *leaching*. If a land area is covered with vegetation and has a higher water table, the percolation losses are less. Drier the atmospheric conditions lower the percolation rate while the high precipitation results in higher percolation. Coarser soils, sand, has a higher percolation rate than the finer soils, clay.

3.4.3 Permeability

The property of the soil that reflects the ease with which water can move within the soil is known as permeability or soil permeability. It is an intrinsic property of the porous material itself without considering the effects of the type of fluid. However, the readiness of transmissivity of fluids through the soil or medium is known as hydraulic conductivity. Table 3.8 depicts the rate of permeability for various soil textures.

3.4.4 Drainage

The process of removal of excess water from the ground or the root zone is called drainage. In many areas with excess water, some part of it on the ground makes its way to the outlet, including swamps, lakes, and rivers, termed natural drainage. Such a drainage method is insufficient to drain the subsurface water and, at times, even the surface water. Artificial drainage using mole and tile drains, helps under such conditions and is gaining importance. Soils are into different drainage classes based on their structure and texture, and those are: (1) very poorly drained, (2) poorly drained, (3) imperfectly drained, (4) moderately well-drained, (5) well-drained, (6) somewhat excessive, and (7) excessive. Identification

TABLE 3.8   Classification of Permeability Rates for Different Soil Textural Classes

Permeability Class	Permeability (cm/hr.)	Textural Class
Very slow	<0.13	Clay
Slow	0.13–0.5	Sandy clay, silty clay
Moderately slow	0.5–2.0	Clay loam, sandy clay loam, silty clay loam
Moderate	2.0–6.3	Very fine sandy loam, loam, silt loam, silty clay loam, silt
Moderately rapid	6.3–12.7	Sandy loam, fine sandy loam
Rapid	12.7–25.4	Sand, loamy sand
Very Rapid	>25.4	Coarse sand

Source: Based on the National Cooperative Soil Survey recommendation:
https://www.nrcs.usda.gov/wps/portal/nrcs/detail/national/home/?cid=nrcs142p2_053573.

of appropriate drainage class is significant as it drives the different hydrologic processes, especially in tile-drained watersheds (Mehan et al., 2019).

3.4.5 Types of Water Flow in Soil

The water movement in the soil can either be a saturated flow, or an unsaturated flow, or a flow of water in vapor form within the soil.

3.4.5.1 Saturated Flow

When water fills all of the soil pores, micropores, and macropores, the water flow through the soil is called the saturated flow. Water potential larger than −33 kPa occurs under such conditions. The conditions where the saturated flow occurs or is noticed include water-bearing sediments and rock layers, flooded soil, and lower horizons of soil with limited drainage. Pore size is a significant factor and is related as the fourth power to the rate of saturated flow. Poiseuille law forms the basis under saturated flow conditions. Usually, the gravitation force is the main driver behind the existence of such flow in nature.

3.4.5.2 Unsaturated Flow

While air fills the larger pores, macropores, of the soil, and small pores hold and transmit water, such condition is known as unsaturated flow. Water potential recorded under such circumstances is lower than −1/3 bar. The matric potential will move the water toward the region of lower potential. In the case of uniform soil, water moves to the drier fronts from wetter areas. Adhesion and capillarity are prime forces driving such movement.

3.4.5.3 Water Vapor Movement

The forces of diffusion and convection regulate the flow of water in the vapor form through the soil medium. Water vapor movements are the internal movements, where water in soil pores covert to vapor, and external movement, where vapors contacting the land surface lose into the atmosphere. As in unsaturated and saturated flow conditions, water moves from the region of high-water potential (wet) to a low potential area (dry), a vapor pressure gradient, the difference in vapor pressure of two points separated by a unit distance, drives the movement of vapor. The higher the difference is, the more rapid diffusion will be; hence, more vapor quantity transfers at the instant of time. In other words, like water in liquid form, vapor's movement takes place from moist soil having high vapor pressure to dry soil (low vapor pressure). In dry conditions, water movement occurs in the vapor form supply water to the drought-resistant plants. Soil moisture and soil temperature affect the water vapor movement. Other properties like soil organic matter content, vegetative cover over the soil, and soil color also affect the flow of water vapor.

TABLE 3.9   Different Flow Laws in Context to Soil–Plant–Water Relationships

	Linear Flow Laws			
Law	Quantity Transported per second Across Flow Region	Transport Coefficient of Flow Region	Difference in Driving Potential Across System	Form Factor F for System (e.g., Rectangular Box)[b]
Ohm $I = \dfrac{\sigma \Delta V A^a}{L} = \dfrac{V}{R^b}$	$I =$ Amperes or coulombs per second	$\sigma =$ Specific electrical conductivity (S/cm)	$\Delta V =$ Difference in electrical potential between input and output (V)	$F = \dfrac{A}{L}$
Darcy $Q = \dfrac{k \Delta h A}{L}$	$Q = \text{cm}^3\big/\text{second}$	$k =$ Hydraulic conductivity (cm / second)	$\Delta h =$ Difference in the head of water column between input and output (cm)	$F = \dfrac{A}{L}$
Fick $Q = \dfrac{D \Delta C A}{L}$	$Q =$ g/second	$D =$ Diffusion coefficient (cm^2/second)	$\Delta C =$ Difference in the concentration of gas at input and output (g/cm^3)	$F = \dfrac{A}{L}$
Fourier $Q = \dfrac{K \Delta T A}{L}$	$Q =$ cal/second	$K =$ Thermal conductivity (cal/second/cm^2 for a thickness of 1 cm and temperature difference of 1°C)	$\Delta T =$ Difference in temperature between input and output (°C)	$F = \dfrac{A}{L}$

[a] $A =$ Cross-sectional area of the system (box) perpendicular to the direction of flow; $L =$ Length of the system (box) through which the flow occurs. The form factor is for two-dimensional flow problems; it is the same for equal geometries of the flow region.

[b] R in this equation = resistance in Ohms, Ω (1S = 1/Ω = mho).

3.4.5.4 Rate of Water Movement

The rate of water movement is the product of a driving force causing water to move and the hydraulic conductivity of the soil. Henry Darcy in 1856 formalized this as:

$$q = -K\frac{dTH}{dx},\tag{3.21}$$

where K is the saturated hydraulic conductivity of the soil, q is the rate of water movement, expressed as the volume of water flowing through a unit cross-sectional area of soil per unit time, x is the position coordinate in the direction of flow, and TH is the total hydraulic head.

This equation is known as Darcy's Law. For uniformly saturated soils, the equation is:

$$q = K\frac{TH_A - TH_B}{L_{AB}}\tag{3.22}$$

where hydraulic conductivity, K, represents the ease with which water flows through the soil.

Table 3.9 represents different linear flow laws in context to soil–plant–water relationships (Kirkham, 2015)

3.5 Soil Water Movement in Plants

Knowing that gravity is trying to pull the water down, the trees like Redwood trees and the other plants can pull it up. Plants or such trees can do pull water up by the process called capillarity or capillary action. It's a force that opposes gravity. Because this force pulls water up, the force of capillarity

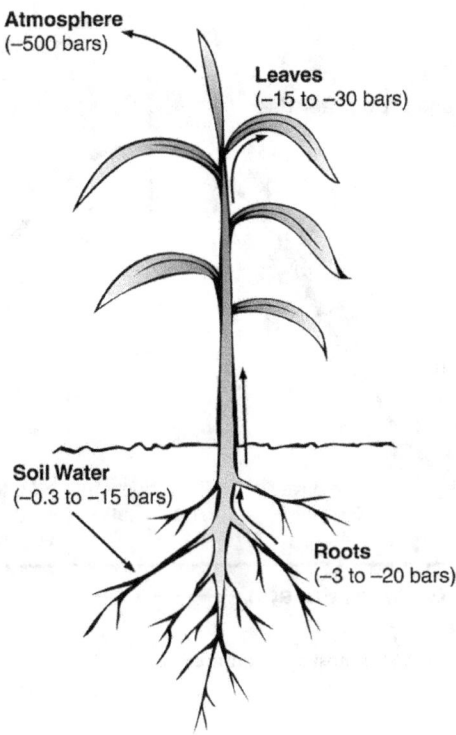

Atmosphere
(–500 bars)

Leaves
(–15 to –30 bars)

Soil Water
(–0.3 to –15 bars)

Roots
(–3 to –20 bars)

FIGURE 3.10 Illustration of the energy differentials that drive the water movement from the soil into the roots, up the stalk, into the leaves, and out into the atmosphere. The water moves from a less negative soil moisture tension to a more negative tension in the atmosphere.

only occurs when very narrow spaces build pressure enough for water to move up. In a plant, there are microscopic cells with areas facilitating water movement. Similarly, in soil, the size of the pores defines whether the spaces are narrow enough for the water to move up. The smaller the pores, the higher up the water will go.

Figure 3.10 demonstrates the amount of work required by the plant to transport the soil water from roots to the top of the plant, and Figure 3.11 illustrates the water use curves based on plant growth.

3.5.1 Other Important Processes Relating to Soil Water Movement

3.5.1.1 Salinity and Salinization

After evaporation of water containing salts is applied for irrigation; a salt crust appears on the soil; the process is called salinization and the amount of dissolved salts in the soil solution is soil salinity. Saline soil has an electrical conductivity of less than 4 mmho/cm. Such soils have a pH of less than 8.5. The common ions present in saline soils include chlorides, potassium, sodium sulfate, and magnesium. Table 3.10 illustrates the classification of soil into saline, sodic, and saline-sodic based on the electrical conductivity, pH, and sodium absorption ratio.

Water containing dissolved salts of carbonate, sodium, and potassium seeps through unlined canal banks and raises the water table. Such water makes soil saline after evaporation. Alluvial soils contain a large number of salts. Such salts in water build the water table, and when water evaporates, it leaves the salt behind, making soil saline. Soil salinity destroys soil structure and hence affects the soil water movement. Plant growth becomes poor or almost stunted in saline soil. The high salt content in the saline soil containing Ca^{++} and Mg^{++} prevents seed germination.

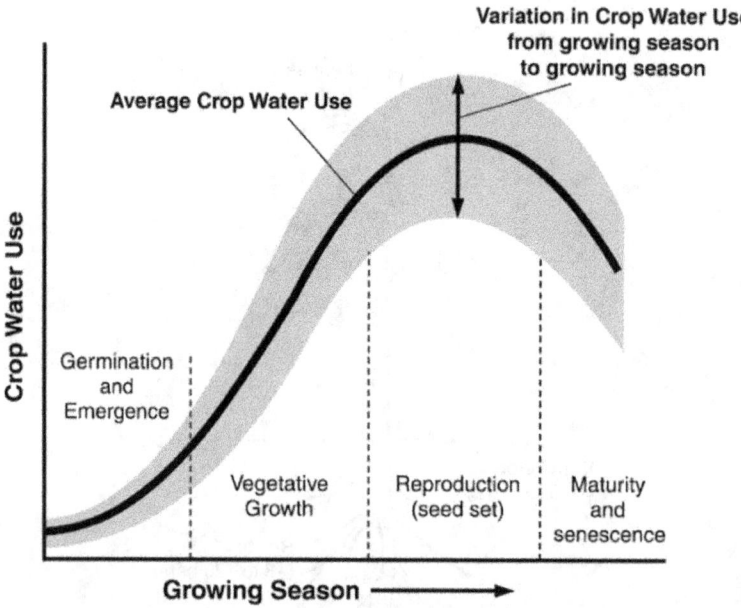

FIGURE 3.11 Typical water use curve for most agronomic crops.
Source: https://www.ag.ndsu.edu/publications/crops/soil-water-and-plant-characteristics-important-to-irrigation.

TABLE 3.10 Soil Chemistry Measurements Used to Classify Saline, Sodic and Saline-Sodic Soils

	Electrical Conductivity[a] (mmho/cm)	pH	Sodium Adsorption Ratio[a] (SAR)
Saline soil	Greater than 4	Less than 8.5	Less than 13
Sodic soil	Less than 4	8.5–10	Greater than 13
Saline-sodic soil	Greater than 4	Less than 8.5	Greater than 13

[a] Measured from a saturated soil extract.

Water can help in leaching excessive salts from the soil. The amendments like sulfuric acid, iron sulfate, and gypsum also help in mitigating salinity. Tube wells lower the water table by pumping out water, thus beneficial in saline soil reclamation. Using manure at the rate of 10–20 tons/acre improves the organic matter and, hence, soil structure and soil porosity, reducing the occurrence of salt crust on the soil. Also, adding manure improves soil nutrition, thus alleviating the salinity issue in the soil. Some crops such as soybean, cabbage, cauliflower, and sweet potato can be grown in moderate saline soil and help to reduce the excessive salts in the soil. Bio-drainage also helps in getting rid of salinity. Some salt-tolerant grasses like *Sporobolus arabicus*, *Chloris barbata*, and herbs like *Portulaca orelacea*, can be grown in saline soil. Some plant species like *Suaeda monoica* accumulate salts in crystalline form in their stems and the cells of their leaves, thus controlling soil salinity.

3.5.1.2 Waterlogging

Waterlogging is a phenomenon when soil becomes wet enough that it leads to anaerobic conditions in soil pores. Under such circumstances, plant roots are unable to respire. Besides the lack of oxygen, gases like carbon dioxide and ethylene, detrimental to the plant root zone, accumulate. Though there is no reference for oxygen amount to identify the waterlogging conditions, every plant differs in its oxygen demand. However, there are different advantages and disadvantages to prevent waterlogging situations, as described in Table 3.11.

TABLE 3.11 Summary of Advantages and Disadvantages of Different Soil and Crop Management Practices

Soil and Crop Management Practices	Advantages (In Addition to Reducing Waterlogging)	Disadvantages	Reference
Surface drainage	Both installation and maintenance are simplest and cheapest	Open drains with less cropping area; needs periodic maintenance	Food and Agriculture Organization [FAO] (2002), Ritzema et al. (2008), Ayars and Evans (2015), Palla et al. (2018)
Raised bed system	Improvements in soil structure	Efficiency depends on height of water table; poorer weed control in furrows; cost of modifying machinery; less cropping area	Bakker et al. (2005b, 2007), Roth et al. (2005), Zhang (2005), Acuña et al. (2011), Gibson (2014)
Pipe drains	Well-tested method for severe waterlogging	Needs outfall and periodic maintenance; cost of installation is high	Tanji (1990), Food and Agriculture Organization [FAO] (2002), Filipović et al. (2014), Teixeira et al. (2018)
Vertical drainage	Well-tested method for severe waterlogging	Maintenance and operational costs are higher than for horizontal pipe drainage systems	Christen et al. (2001), Food and Agriculture Organization [FAO] (2002), Kijne (2006), Prathapar et al. (2018)
Mole drains	Well-tested method; cheaper than other underground drainage	Needs periodic maintenance; will not maintain integrity in dispersive soils	Tuohy et al. (2016, 2018), Dhakad et al. (2018)
Controlled traffic farming	Reduced soil compaction, erosion, tillage costs, water and nutrient losses	Variable results with different conditions, such as different crops, soil types, and tillage	Zhang (2005), Chamen et al. (2006), Guenette and Hernandez-Ramirez (2018), Thomsen et al. (2018), Bennett et al. (2019)
Strategic deep tillage and subsoil manuring	Decreases soil strength resulting in deeper and denser rooting	SDT with no added amendment is often short-term nature, less effective in hostile subsoils, such as acidity, sodicity, or subsoil salinity	Gajri et al. (1994), Bakker et al. (2007), Roper et al. (2015)
Early sowing and vigorous crop	Use of existing soil water provides a buffer; avoids terminal waterlogging events	Minor benefit with severe waterlogging	Stapper and Harris (1989), Setter and Waters (2003), Bassu et al. (2009), Ploschuk et al. (2018), Sundgren et al. (2018), Wollmer et al. (2018)
Bio-drainage	Tried and tested at many locations with success	Needs proper plantation techniques, expertise, thinning, pruning, and harvesting	Kapoor (2000), Food and Agriculture Organization [FAO], (2002), Heuperman and Kapoor (2003), Dash et al. (2005), Lin et al. (2011), Lerch et al. (2017), Muñoz-Carpena et al. (2018), Sarkar et al. (2018), Singh and Lal (2018)

(Continued)

TABLE 3.11 (*Continued*)   Summary of Advantages and Disadvantages of Different Soil and Crop Management Practices

Soil and Crop Management Practices	Advantages (In Addition to Reducing Waterlogging)	Disadvantages	Reference
Nutrient application, in particular, N	Improving plant growth and development	Appropriate methods, nutrient types, timing, and rate should be considered for large-scale application	Rao et al. (2002), Pang et al. (2007b), Guo et al. (2010), Ashraf et al. (2011), Habibzadeh et al. (2012), Wu et al. (2012), Li et al. (2013), Najeeb et al. (2015), Kaur et al. (2017, 2018), Pereira et al. (2017), Zheng et al. (2017)
Plant growth regulators	Promote stomatal conductance and photosynthetic capacity of waterlogged plants	Appropriate methods, timing, and rate should be considered for large-scale application; unproven in broad-scale agriculture	Drew et al. (1979), Lin et al. (2006), Habibzadeh et al. (2013), Ren et al. (2016, 2018)
Use of anti-ethylene agents	Increase both photosynthesis and fruit retention; diminish crop loss induced by ethylene accumulation	Untested in broad-scale agriculture	Kawakami et al. (2010), Shabala (2011), Najeeb et al. (2018)
Pretreatment with hydrogen peroxide	Protect crops from oxidative damage caused by waterlogging	Untested in broad-scale agriculture	Gechev et al. (2002), Ishibashi et al. (2011), Rajaeian and Ehsanpour (2015), Savvides et al. (2016), Andrade et al. (2018)

Source: Adapted from Manik et al. (2019).

3.6 Discussions

With the increasing frequency of extreme events around the world, it has become necessary to understand the movement of water in the soil to mitigate the harmful effects and preserve the ecosystem. Pore size is the most fundamental soil property affecting water movement through the soil. Larger soil pores conduct water more rapidly, such as in sand, than in clay with smaller pores. The texture of the soil and its composition determine the rate at which water passes through the soil.

In soils, capillary and gravitational forces act together while water moves through the soil. Small pores have higher capillary forces than large pores. Adhesion and cohesion together constitute a capillary action. The attraction of liquid to solid surfaces is adhesion, while cohesion is the attraction of water molecules to each other. Sandy soils contain larger pores than clay soils but

do not provide as much total pore space as clay soil. Clay soils retain a higher water per unit volume of soil than sandy soils. In soil layers with tiny pores such as clay, capillary forces are stronger than in soil with the large pores such as sand. Gravity pulls water down when there is no capillary action, as evident in saturated soils (Figure 3.12).

Soil texture, soil structure, organic matter, and bulk density are the factors that affect the soil pore size and shape that influence water movement. Farm operations and activities, including tillage, compaction, residue decay, and wormholes, also affect the soil water movement. The composition of different soil layers affects the rate and direction of water moving through soils. The water movement is affected by the abrupt changes in the soil profile. As clay soil lies on top of the sand, the downward movement of water may temporarily stop at the sand–clay interface until the soil above the interface is almost saturated. The rate of water movement is slower in clay soil than in sand. When a coarse-textured soil such as sand receives water, the water moves downward. Once the wetting front contacts the clay soil, a long-term buildup of a perched water table is formed above the sand–clay interface if water continues to infiltrate the soil surface.

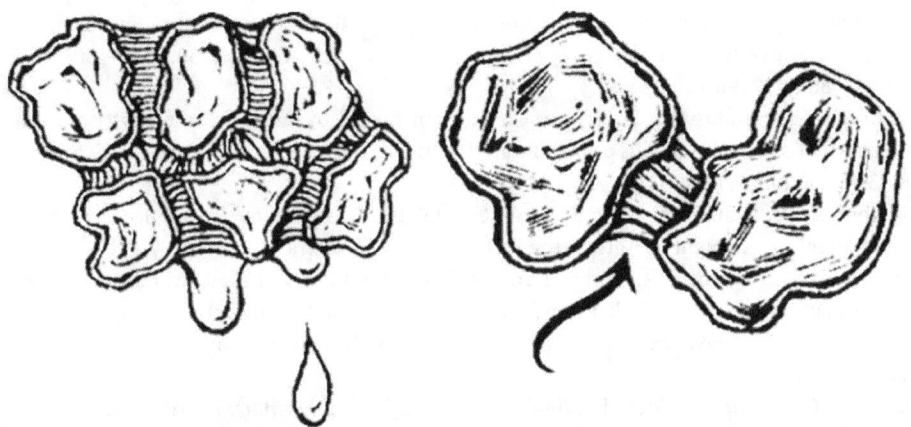

Gravitational water. The pore spaces are filled with water in excess of their capillary capacity, and the excess, or gravitational water, drains downward.

Capillary water is held in the pore space against the force of gravity.

FIGURE 3.12 Illustration to demonstrate the gravitational and capillary water.

3.7 Conclusions

To summarize, the natural physical processes resulting in soil formation, including translocation, transformation, additions, and removal of soil constituents, depending on the water storage dynamics of the environment. The type of soil and its chemical, morphological, and physical properties determine the movement of water through soil and, thus, the different hydrologic processes. Though soil structure can be modified, it is not feasible to change the soil texture to improve soil water movement. Soil acts as a reservoir or a sponge holding water essential for maintaining the plant–water relations under the face of changing climate and land use and cover. It is, therefore, pivotal for better and planned decision-making to improve the soil–plant–water relationships.

References

Acuña, T. B., Dean, G., and Riffkin, P. (2011). Constraints to achieving high potential yield of wheat in a temperate, high-rainfall environment in southeastern Australia. *Crop Pasture Sci.* 62, 125–136. doi: 10.1071/CP10271.

Andrade, C. A., de Souza, K. R. D., de Oliveira Santos, M., da Silva, D. M., and Alves, J. D. (2018). Hydrogen peroxide promotes the tolerance of soybeans to waterlogging. *Sci. Hortic.* 232, 40–45. doi: 10.1016/j.scienta.2017.12.048.

Arshad, M. A., B. Lowery, and B. Grossman. 1996. Physical tests for monitoring soil quality. In: *Methods for assessing soil quality*, J.W. Doran and A.J. Jones (Eds.). Soil Science Society of America Special Publication 49, SSSA, Madison, WI, USA, pp. 123–142.

Ashraf, M. A., Ahmad, M. S. A., Ashraf, M., Al-Qurainy, F., and Ashraf, M. Y. (2011). Alleviation of waterlogging stress in upland cotton (*Gossypium hirsutum* L.) by exogenous application of potassium in soil and as a foliar spray. *Crop Pasture Sci.* 62, 25–38. doi: 10.1071/CP09225.

Ayars, J. E., and Evans, R. G. (2015). Subsurface drainage—What's next? *Irrig. Drain.* 64, 378–392. doi: 10.1002/ird.1893.

Bassu, S., Asseng, S., Motzo, R., and Giunta, F. (2009). Optimising sowing date of durum wheat in a variable Mediterranean environment. *Field Crops Res.* 111, 109–118. doi: 10.1016/j.fcr.2008.11.002.

Bennett, J. M., Roberton, S. D., Marchuk, S., Woodhouse, N. P., Antille, D. L., Jensen, T. A., et al. (2019). The soil structural cost of traffic from heavy machinery in Vertisols. *Soil Till. Res.* 185, 85–93. doi: 10.1016/j.still.2018.09.007.

Blann, K. L., Anderson, J. L., Sands, G. R., and Vondracek, B. (2009). Effects of agricultural drainage on aquatic ecosystems: a review. *Crit. Rev. Environ. Sci. Technol.* 39, 909–1001. doi: 10.1080/10643380801977966.

Bakker, D., Hamilton, G., Houlbrooke, D., Spann, C., and Van Burgel, A. (2007). Productivity of crops grown on raised beds on duplex soils prone to waterlogging in Western Australia. *Aust. J. Exp. Agric.* 47, 1368–1376. doi: 10.1071/EA06273.

Bakker, D., Houlbrooke, D., Hamilton, G., and Spann, C. (2005b). *A Manual for Raised Bed Farming in Western Australia*. Department of Agriculture and Food, Perth, Australia.

Blann, K.L., Anderson, J.L., Sands, G.R., and Vondracek, B. (2009). Effects of agricultural drainage on aquatic ecosystems: a review. *Critic. Rev. Environ. Sci. Technol.* 39(11), 909–1001.

Brady, N. C. and Weil, R. R., 1999, *The Nature and Properties of Soils*, 12th ed. Upper Saddle River, NJ: Prentice-Hall.

Chamen, T., Cottage, C. C., and Maulden, B. (2006). *'Controlled Traffic' Farming: Literature Review and Appraisal of Potential Use in the UK, HGCA Research Review No. 59*. Maulden, UK: The Home-Grown Cereals Authority.

Christen, E., and Skehan, D. (2001). Design and management of subsurface horizontal drainage to reduce salt loads. *J. Irrig. Drain. Eng.* 127, 148–155. doi: 10.1061/(ASCE)0733-9437(2001)127:3(148)

Christen, E. W., Ayars, J. E., and Hornbuckle, J. W. (2001). Subsurface drainage design and management in irrigated areas of Australia. *Irrig. Sci.* 21, 35–43.

Darcy, H. (1856). *The Public Fountains of the City of Dijon*. Dalmont, Paris, 647.

Dash, C., Sarangi, A., Singh, A., and Dahiya, S. (2005). Bio-drainage: an alternate drainage technique to control waterlogging and salinity. *J. Soil Water Conserv. India* 4, 149–155.

Davies, M., and Hillman, G. (1988). Effects of soil flooding on growth and grain yield of populations of tetraploid and hexaploid species of wheat. *Ann. Bot.* 62, 597–604. doi: 10.1093/oxfordjournals.aob.a087699

Dhakad, S., Ambawatia, G., Verma, G., Patel, S., Rao, K. R., and Verma, S. (2018). Performance of Mole drain system for soybean (glycine max)-wheat (*Triticum aestivum*) cropping system of Madhya Pradesh. *Int. J. Curr. Microbiol. Appl. Sci.* 7, 2107–2112. doi: 10.20546/ijcmas.2018.702.251.

Drew, M., Sisworo, E., and Saker, L. (1979). Alleviation of waterlogging damage to young barley plants by application of nitrate and a synthetic cytokinin, and comparison between the effects of waterlogging, nitrogen deficiency and root excision. *New Phytol.* 82, 315–329. doi: 10.1111/j.1469-8137.1979.tb02657.x.

Eslamian, S., Sayahi, M., Ostad-Ali-Askari, K., Basirat, S., Ghane, M., Matouq, M. (2018). Saturation, In: Bobrowsky P., Marker B. (eds), *Encyclopedia of Engineering Geology, Encyclopedia of Earth Sciences Series*, Springer.

Filipovic, V., Mallmann, F. J. K., Coquet, Y., and Šimůnek, J. (2014). Numerical´ simulation of water flow in tile and mole drainage systems. *Agric. Water Manag.* 146, 105–114. doi: 10.1016/j.agwat.2014.07.020.

Food and Agriculture Organization [FAO] (2002). *Food and Agriculture Organization of the United Nations*. Available at: http://www.fao.org/3/abc600e.pdf.

Gajri, P. R., Arora, V. K., and Chaudhary, M. R. (1994). Maize growth responses to deep tillage, straw mulching and farmyard manure in coarse textured soils of N.W. India. *Soil Use Manage.* 10, 15–19. doi: 10.1111/j.1475-2743.1994.tb00451.x.

Gardner, B., Nielsen, D., and Shock, C. (1992). Infrared thermometry and the crop water stress index. I. History, theory, and baselines. *J. Prod. Agric.* 5, 462–466. doi: 10.2134/jpa1992.0462.

Gechev, T., Gadjev, I., Van Breusegem, F., Inzé, D., Dukiandjiev, S., Toneva, V., et al. (2002). Hydrogen peroxide protects tobacco from oxidative stress by inducing a set of antioxidant enzymes. *Cell. Mol. Life Sci.* 59, 708–714. doi: 10.1007/s00018-002-8459-x.

Gibson, G. (2014). *Utilising Innovative Management Techniques to Reduce Waterlogging*. Moama, NSW: Nuffield Australia Farming Scholars.

Gill, M. B., Zeng, F., Shabala, L., Böhm, J., Zhang, G., Zhou, M., et al. (2018). The ability to regulate voltage-gated K+-permeable channels in the mature root epidermis is essential for waterlogging tolerance in barley. *J. Exp. Bot.* 69, 667–680. doi: 10.1093/jxb/erx429.

Guenette, K. G., and Hernandez-Ramirez, G. (2018). Tracking the influence of controlled traffic regimes on field scale soil variability and geospatial modeling techniques. *Geoderma* 328, 66–78. doi: 10.1016/j.geoderma.2018.04.026.

Guggenheim, E.A. (1945). The principle of corresponding states. *J. Chem. Phys.* 13(7), 253–261.

Guo, W. Q., Chen, B. L., Liu, R. X., and Zhou, Z. G. (2010). Effects of nitrogen application rate on cotton leaf antioxidant enzyme activities and endogenous hormone contents under short-term waterlogging at flowering and boll-forming stage. *Yingyong Shengtai Xuebao*, 21, 53–60.

Habibzadeh, F., Sorooshzadeh, A., Pirdashti, H., and Sanavy, S. (2012). Effect of nitrogen compounds and tricyclazole on some biochemical and morphological characteristics of waterlogged-canola. *Int. Res. J. Appl. Basic Sci.* 3, 77–84.

Habibzadeh, F., Sorooshzadeh, A., Pirdashti, H., and Modarres-Sanavy, S. A. M. (2013). Alleviation of waterlogging damage by foliar application of nitrogen compounds and tricyclazole in canola. *Aust. J. Crop Sci.* 7, 401–406.

Harrison, L.P. (1965). *Fundamental Concepts and Definitions Relating to Humidity*. Reinhold Publishing Company, New York.

Heuperman, A., and Kapoor, A. (2003). *Biodrainage Status in India and Other countries*. New Delhi: Indian National Committee on Irrigation and Drainage, 147.

Hillel, D. (2013). *Introduction to Soil Physics*. Academic Press.

Horvath, A. L. (1975). *Physical Properties of Inorganic Compounds*. Arnold.

Huang, X., Fan, Y., Shabala, L., Rengel, Z., Shabala, S., and Zhou, M. (2018). A major QTL controlling the tolerance to manganese toxicity in barley (*Hordeum vulgare* L.). *Mol. Breed.* 38, 16.

Ishibashi, Y., Yamaguchi, H., Yuasa, T., Iwaya-Inoue, M., Arima, S., and Zheng, S.-H. (2011). Hydrogen peroxide spraying alleviates drought stress in soybean plants. *J. Plant Physiol.* 168, 1562–1567. doi: 10.1016/j.jplph.2011.02.003.

ISSS International Society of Soil Science. (1963). *Soil Phys. Terminol. Bull.* 23(7).

Jongepier, R., et al. (2015). Management options for water-repellent soils in Australian dryland agriculture. *Soil Res.* 53, 786–806. doi: 10.1071/SR14330.

Kapoor, A. (2000). Bio-drainage feasibility and principles of planning and design. In *Proceedings of the Eighth ICID International Drainage Workshop*, New Delhi: International Commission on Irrigation and Drainage, 17–32.

Kawakami, E. M., Oosterhuis, D. M., and Snider, J. L. (2010). Physiological effects of 1-methylcyclopropene on well-watered and water-stressed cotton plants. *J. Plant Growth Regul.* 29, 280–288. doi: 10.1007/s00344-009-9134-3.

Kaur, G., Nelson, K., and Motavalli, P. (2018). Early-season soil waterlogging and N fertilizer sources impacts on corn N uptake and apparent N recovery efficiency. *Agronomy* 8, 102. doi: 10.3390/agronomy8070102.

Kaur, G., Zurweller, B. A., Nelson, K. A., Motavalli, P. P., and Dudenhoeffer, C. J. (2017). Soil waterlogging and nitrogen fertilizer management effects on corn and soybean yields. *Agron. J.* 109, 97–106. doi: 10.2134/agronj2016.07.0411.

Kijne, J. W. (2006). Abiotic stress and water scarcity: identifying and resolving conflicts from plant level to global level. *Field Crops Res.* 97, 3–18. doi: 10.1016/ j.fcr.2005.08.011.

Kirkham, M. B. (2014). *Principles of Soil and Plant Water Relations*. Academic Press, 96.

Li, M.-F., Zhu, J.-Q., and Jiang, Z.-H. (2013). Plant growth regulators and nutrition applied to cotton after waterlogging, In *Proceedings of the Intelligent System Design and Engineering Applications (ISDEA), Third International Conference on*, Piscataway, NJ: IEEE, 1045–1048. doi: 10.1109/ISDEA.2012.246.

Lin, C. H., Lerch, R. N., Goyne, K. W., and Garrett, H. E. (2011). Reducing herbicides and veterinary antibiotics losses from agroecoystems using vegetative buffers. *J. Environ. Qual.* 40, 791–799. doi: 10.2134/jeq2010.0141.

Lin, K.-H. R., Tsou, C.-C., Hwang, S.-Y., Chen, L.-F. O., and Lo, H.-F. (2006). Paclobutrazol pre-treatment enhanced flooding tolerance of sweet potato. *J. Plant Physiol.* 163, 750–760. doi: 10.1016/j.jplph.2005.07.008.

Lerch, R. N., Lin, C. H., Goyne, K. W., Kremer, R. J., and Anderson, S. H. (2017). Vegetative buffer strips for reducing herbicide transport in runoff: effects of buffer width, vegetation, and season. *J. Am. Water Resour. Assoc.* 53, 667–683. doi: 10.1111/1752-1688.12526.

Manik, S. N., Pengilley, G., Dean, G., Field, B., Shabala, S. and Zhou, M. (2019). Soil and crop management practices to minimize the impact of waterlogging on crop productivity. *Front. Plant Sci.* 10, 140.

Marshall, T.J., Holmes, J.W., and Rose, C.W. (1996). *Soil Physics*. Cambridge University Press, Cambridge.

McElrone, A. J., Choat, B., Gambetta, G. A. and Brodersen, C. R. (2013). Water uptake and transport in vascular plants. *Nat. Educ. Knowled.* 4(5), 6.

Mehan, S., Aggarwal, R., Gitau, M. W., Flanagan, D. C., Wallace, C. W., and Frankenberger, J. R. (2019). Assessment of hydrology and nutrient losses in a changing climate in a subsurface-drained watershed. *Sci. Total Environ.* 688, 1236–1251.

Mehan, S. and Singh, K.G. (2015). Use of mulches in soil moisture conservation: A review. *Best Manag. Pract. Drip Irrig. Crops* 2, 283.

Muñoz-Carpena, R., Fox, G. A., and Sabbagh, G. J. (2010). Parameter importance and uncertainty in predicting runoff pesticide reduction with filter strips. *J. Environ. Qual.* 39, 630–641. doi: 10.2134/jeq2009.0300.

Najeeb, U., Bange, M. P., Tan, D. K., and Atwell, B. J. (2015). Consequences of waterlogging in cotton and opportunities for mitigation of yield losses. *AoB Plants* 7, lv080. doi: 10.1093/aobpla/plv080.

Najeeb, U., Tan, D. K., Bange, M. P., and Atwell, B. J. (2018). Protecting cotton crops under elevated CO_2 from waterlogging by managing ethylene. *Funct. Plant Biol.* 45, 340–349. doi: 10.1071/FP17184.

Novák, V., Hlaváčiková, H. (2019). Soil-water retention curve. In: *Applied Soil Hydrology. Theory and Applications of Transport in Porous Media*, Vol 32. Springer, Cham, 77–96.

O'Geen, A., Dahlgren, R., Swarowsky, A., Tate, K., Lewis, D., & Singer, M., 2010. Research connects soil hydrology and stream water chemistry in California oak woodlands. *California Agric.* 64(2), 78–84.

Palla, A., Colli, M., Candela, A., Aronica, G., and Lanza, L. (2018). Pluvial flooding in urban areas: the role of surface drainage efficiency. *J. Flood Risk Manag.* 11, S663–S676. doi: 10.1111/jfr3.12246.

Pang, J., Ross, J., Zhou, M., Mendham, N., and Shabala, S. (2007b). Amelioration of detrimental effects of waterlogging by foliar nutrient sprays in barley. *Funct. Plant Biol.* 34, 221–227. doi: 10.1071/FP06158.

Pereira, E. I., Nogueira, A. A. R., Cruz, C. C., Guimaraþes, G. G., Foschini, M. M., Bernardi, A. C., et al. (2017). Controlled urea release employing nanocomposites increases the efficiency of nitrogen use by forage. *ACS Sustain. Chem. Eng.* 5, 9993–10001. doi: 10.1021/acssuschemeng.7b01919.

Perry, R. H., and Green, D. ed. (1985). *Perry's Chemical Engineers' Handbook*, 6th ed., McGraw-Hill.

Prathapar, S., Rajmohan, N., Sharma, B., and Aggarwal, P. (2018). Vertical drains to minimize duration of seasonal waterlogging in Eastern Ganges Basin flood plains: a field experiment. *Nat. Hazards*, 92, 1–17. doi: 10.1007/s11069-018-3188-0.

Rajaeian, S., and Ehsanpour, A. (2015). Physiological responses of tobacco plants (*Nicotiana rustica*) pretreated with ethanolamine to salt stress. *Russ. J. Plant Physiol.* 62, 246–252. doi: 10.1134/S1021443715020156.

Rao, R., Li, Y., Bryan, H. H., Reed, S. T., and D'ambrosio, F. (2002). Assessment of foliar sprays to alleviate flooding injury in corn (*Zea mays* L.). In *Proceedings of the Florida State Horticultural Society*, Lake Alfred, FL: Florida State Horticultural Society, 208–211.

Ren, B., Zhang, J., Dong, S., Liu, P., and Zhao, B. (2018). Exogenous 6 benzyladenine improves antioxidative system and carbon metabolism of summer maize waterlogged in the field. *J. Agron. Crop Sci.* 204, 175–184. doi: 10.1111/jac.12253.

Ren, B., Zhu, Y., Zhang, J., Dong, S., Liu, P., and Zhao, B. (2016). Effects of spraying exogenous hormone 6-benzyladenine (6-BA) after waterlogging on grain yield and growth of summer maize. *Field Crops Res.* 188, 96–104. doi: 10.1016/j.fcr.2015.10.016.

Risk Management Agency: RMA. (2020). https://www.rma.usda.gov/SummaryOfBusiness/CauseOfLoss. Accessed on 02/22/2020.

Ritzema, H., Satyanarayana, T., Raman, S., and Boonstra, J. (2008). Subsurface drainage to combat waterlogging and salinity in irrigated lands in India: Lessons learned in farmers' fields. *Agric. Water Manag.* 95, 179–189. doi: 10.1016/j.agwat.2007.09.012.

Roper, M. M., Davies, S. L., Blackwell, P. S., Hall, D. J. M., Bakker, D. M., Ploschuk, R. A., Miralles, D. J., Colmer, T. D., Ploschuk, E. L., and Striker, G. G. (2018). Waterlogging of winter crops at early and late stages: impacts on leaf physiology, growth and yield. *Front. Plant Sci.* 9, 1863. doi: 10.3389/fpls.2018 01863.

Roth, C. H., Fischer, R. A., Piggin, C., and Meyer, W. (2005). *Evaluation and Performance of Permanent Raised Bed Systems in Asia, Mexico and Australia: A Synopsis* Griffith, NSW, Australia and Mexico: 200–208.

Sarkar, A., Banik, M., Ray, R., and Patra, S. (2018). Soil moisture and groundwater dynamics under biodrainage vegetation in a waterlogged land. *Int. J. Pure Appl. Biosci.* 6, 1225–1233. doi: 10.18782/2320-7051.6052.

Savvides, A., Ali, S., Tester, M., and Fotopoulos, V. (2016). Chemical priming of plants against multiple abiotic stresses: mission possible? *Trends Plant Sci.* 21, 329–340. doi: 10.1016/j.tplants.2015.11.003.

Schoonover, J.E. and Crim, J.F., 2015. An introduction to soil concepts and the role of soils in watershed management. *J. Contemp. Water Res. Educ.* 154(1), 21–47.

Setter, T., and Waters, I. (2003). Review of prospects for germplasm improvement for waterlogging tolerance in wheat, barley and oats. *Plant Soil* 253, 1–34. doi: 10.1023/A:1024573305997.

Shabala, S. (2011). Physiological and cellular aspects of phytotoxicity tolerance in plants: the role of membrane transporters and implications for crop breeding for waterlogging tolerance. *N. Phytol.* 190, 289–298. doi: 10.1111/j.1469-8137.2010.03575.x.

Sharma, P. K., Kumar, D., Srivastava, H. S., and Patel, P. (2018). Assessment of different methods for soil moisture estimation: a review. *J. Remote Sens. GIS.* 29, 9(1), 57–73.

Singh, G., and Lal, K. (2018). Review and case studies on biodrainage: An alternative drainage system to manage waterlogging and salinity. *Irrig. Drain.* 67, 51–64. doi: 10.1002/ird.2252.

Stapper, M., and Harris, H. (1989). Assessing the productivity of wheat genotypes in a Mediterranean climate, using a crop-simulation model. *Field Crops Res.* 20, 129–152. doi: 10.1016/0378-4290(89)90057-9.

Sundgren, T. K., Uhlen, A. K., Lillemo, M., Briese, C., and Wojciechowski, T. (2018). Rapid seedling establishment and a narrow root stele promotes waterlogging tolerance in spring wheat. *J. Plant Physiol.* 227, 45–55. doi: 10. 1016/j.jplph.2018.04.010.

Tanji, K. K. (1990). Nature and extent of agricultural salinity. agricultural salinity assessment and management, In *ASCE Manuals and Reports on Engineering Practice No. 671*, ed. K. K. Tanji, New York, NY: American Society of Civil Engineers, 619.

Teixeira, D. L., de Matos, A. T., de Matos, M. P., Miranda, S. T., and Vieira, D. P. (2018). Evaluation of the effects of drainage and different rest periods as techniques for unclogging the porous medium in horizontal subsurface flow constructed wetlands. *Ecol. Eng.* 120, 104–108. doi: 10.1016/j.ecoleng.2018.05.042.

Thomsen, M. N., Tamirat, T. W., Pedersen, S. M., Lind, K. M., Pedersen, H. H., de Bruin, S., et al. (2018). Farmers' perception of Controlled Traffic Farming (CTF) and associated technologies, No. 2018/12. In *IFRO Working Paper*, Copenhagen: University of Copenhagen, Denmark.

Tuohy, P., Humphreys, J., Holden, N., and Fenton, O. (2016). Runoff and subsurface drain response from mole and gravel mole drainage across episodic rainfall events. *Agric. Water Manag.* 169, 129–139. doi: 10.1016/j.agwat.2016.02.020.

Tuohy, P., O'Loughlin, J., and Fenton, O. (2018). Modeling performance of a tile drainage system incorporating mole drainage. *Trans. ASABE*, 61, 169–178. doi: 10.13031/trans.12203.

Whitaker, S. (1986). Flow in porous media I: A theoretical derivation of Darcy's law. *Transp. Porous Media*, 1(1), 3–25.

Wollmer, A. C., Pitann, B., and Mühling, K. H. (2018). Nutrient deficiencies do not contribute to yield loss after waterlogging events in winter wheat (*Triticum aestivum*). *Ann. Appl. Biol.* 173, 141–153. doi: 10.1111/aab.12449.

Wu, Q.-X., Zhu, J.-Q., Liu, K.-W., and Chen, L.-G. (2012). Effects of fertilization on growth and yield of cotton after surface waterlogging elimination. *Adv. J. Food Sci. Technol.* 4, 398–403.

Zheng, W., Liu, Z., Zhang, M., Shi, Y., Zhu, Q., Sun, Y., et al. (2017). Improving crop yields, nitrogen use efficiencies, and profits by using mixtures of coated controlled-released and uncoated urea in a wheat-maize system. *Field Crops Res.* 205, 106–115. doi: 10.1016/j.fcr.2017.02.009.

Zhang, S. (2005). *Soil Hydraulic Properties and Water Balance Under Various Soil Management Regimes on the Loess Plateau, China*. Ph.D. thesis, Swedish University of Agricultural Sciences, Umeå, Sweden.

Zhou, M., Li, H., and Mendham, N. (2007). Combining ability of waterlogging tolerance in barley. *Crop Sci.* 47, 278–284. doi: 10.2135/cropsci2006.02.0065.

II

Evapotranspiration and Water Requirements

4

An Introduction to Soil–Water–Plant Relationship

Mahbub Hasan
Alabama A&M University

Saeid Eslamian
*Isfahan University
of Technology*

DOI: 10.1201/9780429290114-6

4.1 Introduction

This chapter explains the concept of the interrelationship among soil, water, and plants. Details of soil structure and texture, water infiltration, and the volume and rate of infiltrated water, how much of this infiltrated water could be stored in the crop root zone have been described. The individual components of soil, water, and plants are thoroughly considered to determine and schedule the crop water requirement. In any plant production environment, soil acts as the medium in which the seed germinates, grows up, and becomes matured after passing through different growth stages and finally producing the desired crop. Soil and plant relate the properties of soil and the nurturing of the plants in a medium, which is soil. There are, naturally, innumerable spaces in between the smallest soil particles. They are called voids. These voids are empty spaces that accommodate air and water. Volume of these void spaces is dependent on the soil type. Soil provides the room for holding the moisture inside the voids. The plants are nourished with this moisture, which is mixed with nutrients and passes through the roots of the plant. Therefore, the soil can be considered as the medium and water is the carrier of nutrients for the plants. It is not always true that the total amount of rainwater is enough for nourishing the plants or crops. The amount of rainwater may be insufficient for the total plant growth requirement. Hence, an additional amount of water from other sources, either from surface water or groundwater resources, is required to satisfy the total demand of water for the plants in the form of *irrigation*. Water intake rate into the soil surface, its retention, movement, and availability to the plants are all parts of the physical phenomena. Therefore, it is not only important to study the physical properties of soil but also the activities of water in the root zone of the plants and how this water is contributing with its potentiality and economically.

4.2 Physical Properties of Soil

Soil has three phases. The solid phase is made of minerals, organic matters, various chemical compounds, and solid soil particles. The liquid phase is termed *soil moisture*, and the gaseous phase is called the *soil air*. The two phases of moisture and soil air are held suspended inside the soil's voids, which are formed within the soil's inter particle spaces. The solid soil particles are of different sizes and shapes, and they form the pore spaces of different geometric shapes. The pore spaces are filled with water and air in different proportions depending on the soil's moisture volume. A typical silt loam soil contains 50% solid soil particles, 30% water, and 20% air. In addition to these three basic components of soil, it may contain living organisms, such as bacteria, fungi, algae, protozoa, insects, and small animals, which directly or indirectly affect *soil structure* and its *infiltration* and *water-holding capacity*. There are other soil properties such as soil structure, soil texture, depth of water table, capillary conductivity, and soil profile conditions.

4.3 Mechanical Composition of Soil

The composition of soil that refers to its solid phase composed of mineral fraction is termed as the mechanical composition of soil. It is very important to study the soil's physical, chemical, and biological composition, as the mechanical composition of soil gives an insight into the characteristics of the soil [1]. This is important to have the information needed for the plant growth and irrigation studies. The organic soils are most stable. They comprise rock particles developed in situ by a process called weathering, which leads the soil to be deposited by water or wind. The mineral particles are the principal components of most soils. Soil particles and minerals are classified according their sizes. Minerals between 2 cm and 2 mm in diameter are called gravel, minerals smaller than 2 mm in diameter are called fine earth. Components of fine earth are further classified into sand, silt, and clay. Classifications of sand, silt, and clay have been established by the various agencies, and national and international organizations. United States Department of Agriculture (USDA) suggested the method, which is commonly used

TABLE 4.1 Soil Classification by International Soil Science Society (ISSS) and United States Department of (USDA)

Fraction	Particle Diameter	
	USDA (mm)	ISSS (mm)
Gravel	>2 mm	>2 mm
Very coarse sand	1–2	-
Coarse sand	0.5–1	0.2–2
Medium sand	0.25–0.5	-
Fine sand	0.1–0.25	0.02–0.2
Very fine sand	0.05–0.1	-
Silt	0.002–0.05	0.002–0.02
Clay	<0.002	<0.002

for classifying the soil [2]. The classification of soil as suggested by International Soil Science Society (ISSS) and USDA is mostly accepted and used globally. This classification is shown in Table 4.1.

The finest soil particles, clay, are further divided into coarse clay with a diameter of 0.002–0.0002 mm, whereas soil particles less than 0.0002 mm in diameter are termed as colloidal clay. The colloidal soil particles are microscopically visible.

Sand and silt particles are spherical or cubical in shape. Actually, the soil that is formed in a certain place and not transported from other places is sharp edged, whereas the soil transported by water or air is smooth, as the edges are smoothed off due to transportation. On the contrary, clay particles are plate or lathe shaped. The shape again depends on the structure of the crystal of a specific type of clay.

4.3.1 Soil Texture

The comparative amount of sand, silt, and clay forming a soil determines the soil texture. This is represented by the names of the major size portions and the word *loam* whenever all three major size portions occur in sizable proportions. For example, the name *silty clay* designates a soil predominated by clay, and which also encompasses a significant amount of silt. In general, coarse soils are called sand, the medium textured soils are called silt, and the fine-textured soils are called clay.

Figure 4.1 represents the textural classification chart as originally described by USDA for 12 main textural classes.

This figure shows that the least textural group is sand, which comprises less than 15% silt and clay. The characteristic of this soil is having a simple capillary with a bigger volume of noncapillary pore spaces ensuring good drainage and aeration. Sandy soil possesses the following typical characteristics:

1. Chemically inert,
2. Loose and non-cohesive,
3. Lower water-holding capacity, and
4. Higher infiltration rate.

On the other hand, the exact opposite nature is associated with clay soil. It has the following characteristics:

1. Complex particle structure and cohesive,
2. Higher water-holding capacity, and
3. Lower infiltration rate.

Clay contains 40% or more of clay particles and 45% sand or silt particles. Clay particles are aggregated to form a granule and complex structure. The surface area of clay soil particle is larger compared to sand and silt because of the plate-like the shape of clay particles.

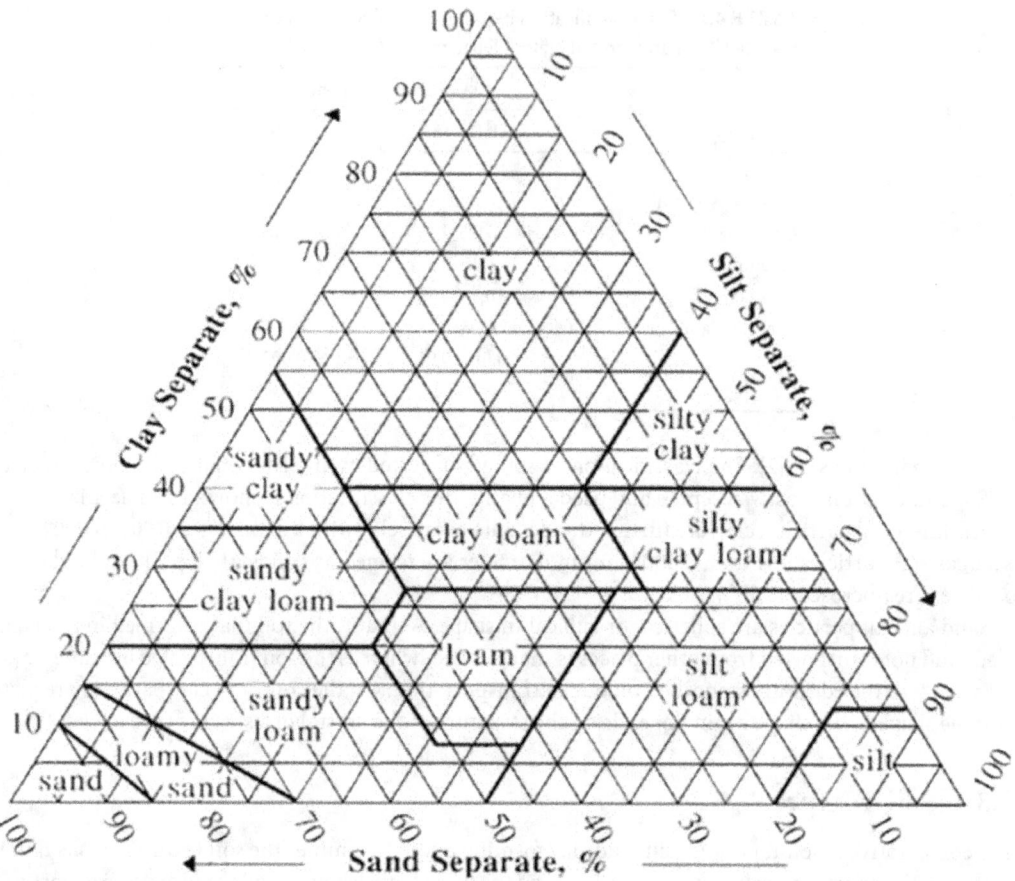

FIGURE 4.1 USDA soil textural classification chart [3].

Loam soil consists of almost equal amounts of sand, silt, and clay. The loam soil has properties that lie between sand and clay. This is the most suitable soil for crop cultivation. This soil can hold water and serve the plants if required. On the other hand, the sand portion of loam soil can make the root zone environment favorable for aeration.

Soil texture and classification represent the characteristic as an approximate relationship that describes the behavior of soil as the plant growth medium. Soil texture can be modified by adding or removing organic matter. Soil texture constitutes the basic matrix, and the geometry of the voids created in this soil matrix is dependent on the class of soil texture [1]. Therefore, the soil texture impacts and controls the other phases of the soil (air and water) confined in the same spaces in the soil matrix.

4.3.2 Soil Structure

Soil structure refers to the pattern and arrangement of individual soil particles. The structural units are formed due to the grouping of the soil particles. It is worth mentioning that the size and shape of the structural units, bonding due to the strength of the soil particles, and the proportion of the soil particles vary significantly among soils. There are two kinds of phases documented in the development of soil aggregates. One is the development of interparticle bonding that leads to soil aggregate stability and the other is the disintegration of the soil particles from each other due to the size and shape of the soil aggregates. Clay content of soil is the main source and is considered to assess soil stability. Also, the stability of the soil aggregate determines the extent of flocculation, the organic and inorganic relation, the microbial cohesiveness of the soil aggregate, and the presence of aluminum oxides and iron oxides (known as mineral cementing materials).

Soil structure varies due to the shape of the soil aggregates, and they are classified as follows:

1. Simple structure – shows indistinct or total absence of natural cleavage planes.
2. Compound structure – shows natural cleavage planes prominently.

Simple structure may be a single *grain structure* or *massive structure*. This is observed normally only in sand or silt with low organic matter content. The soil with single grain structure is advantageous for plant growth due to enough aeration and unobstructed capillary movement of soil moisture. Soils with single grain structure, like in sand, facilitates capillary movement of water, nutrients, and aeration. Massive structure soils are coherent-type soils but like the single grain structure soils.

Size of aggregate is the criteria that identify the standards and criteria of soil structure. It is well-established fact that the soil aggregate consisting of sand-sized soil particles are better for plant growth than the soil aggregate consisting of exceedingly small or exceptionally large soil particles.

Soil pore spaces are important for better aeration, capillary water movement, and favorable soil infiltration rate. Soil structure frequently demonstrates the arrangement of small, medium, and large soil pores into a structural pattern. One of the primary effects of soil pattern is its *porosity*. Soil structure has a distinct effect on such soil properties as follows:

1. Erodibility,
2. Hydraulic conductivity,
3. Porosity,
4. Water-holding capacity, and
5. Infiltration.

Soil structure quality is expressed in various characteristics of soil particles in an aggregate, such as:

1. Porosity,
2. Aggregation, and
3. Cohesiveness or permeability for water or air.

Among the above characteristics of soil, porosity plays a key role because the biological and chemical processes occur in the spaces between the soil particles, which are the pore spaces. Effects of pore spaces and their sizes are dependent on:

1. Crop response (depends on crop species),
2. Climatic conditions, and
3. Depth of water table.

Scientists opined that the pore spaces filled with either irrigation or rainwater are not that important, whereas the pore spaces suitable for aeration are better medium for plant growth. Aeration pores are essential for optimum plant growth.

Soil aggregate must be stable to withstand the impact of rainfall and sustain against submergence due to excess water due to rain or drainage water accumulation. Cohesiveness of soil is associated with the moisture content. Soil aids better plant growth when soil particles are not too loose. A soil aggregate with sticky particles that makes a compact soil structure retards the normal penetration, extension, and spreading of roots into the soil. It also restricts the healthier aeration opportunity for optimum plant growth. Therefore, the soil structure should possess a large infiltration capacity, medium percolation capacity, and enough aeration capacity without being excessive.

4.3.3 Soil Structure Measurement

Soil structure is classified by four properties, and they can be measured and quantified. These properties are:

1. Porosity,
2. Aggregation,
3. Cohesiveness, and
4. Permeability.

The weight by volume of the soil samples whose real density is already known allows for the calculation of the whole pore spaces. The aggregate distribution is determined by sieve analysis with oven-dried soil samples. Soil particles' size distribution and stability are determined by wet sieve analysis.

4.3.4 Soil Structure Management

Soil structure needs good management with a target to obtain the maximum yield under the natural conditions of the following factors:

1. Climate,
2. Moisture availability, and
3. Nutrient supply.

The other objectives of soil structure management are:

1. Prevention of soil particles' detachability and transportability by water and air (soil particles protection against erosion), and
2. To increase the infiltration and percolation capacities so that there is less runoff and to minimize the erosion if the infiltration is less.

The common methods of soil structure management are:

1. Appropriate land uses,
2. Suitable tillage practices at an optimum moisture content,
3. Subsoiling depending on the necessity,
4. Adding organic matter,
5. Selecting the best crop rotation practices,
6. Optimum fertilizer-level applications,
7. Appropriate time and depth of mulching,
8. Ensuring efficient drainage of unnecessary volume of water,
9. Ensuring controlled irrigation,
10. Soil and water conservation practices,
11. Avoid and protect wet soils against compaction, and
12. Occasional application of soil conditioners.

4.4 Volume and Mass Relationships of Soil Elements

It has already been mentioned that the soil aggregate consists of three constituents, namely:

1. Air,
2. Water, and
3. Solid soil.

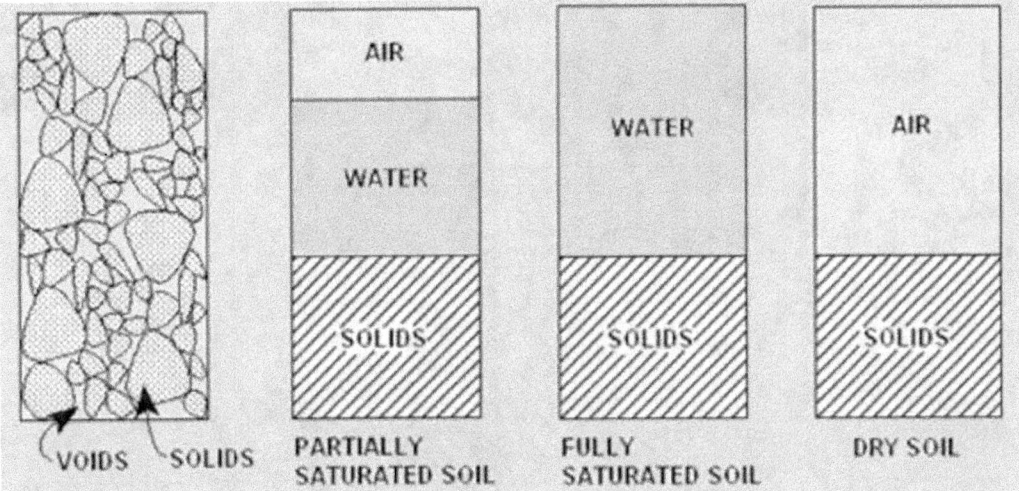

FIGURE 4.2 A soil continuum showing soil sample in saturated, partially saturated, and dry condition [4].

Figure 4.2(a) shows the soil particles and voids formed within the inter particles spaces. This soil profile may contain water and air, which are naturally available in normal atmospheric conditions. In this case, water and air share the void spaces. Sharing of void space is dependent on:

1. Capillary water movement,
2. Water-holding capacity of the soil, and
3. Infiltration capacity of the soil.

Soil with air and water is called partially saturated soil as represented in Figure 4.2b. The total void space may be occupied by water in a fully saturated condition. This situation may occur due to rainfall and flooding for a long time. This is shown in Figure 4.2c. In some cases, soil has no water content at all. Deserts with higher temperature may have this type of soil as shown in Figure 4.2d. This condition may occur due to:

1. Higher percolation or drainage that happens in sandy soils, and
2. Higher evaporation rate

Figure 4.3 shows a schematic diagram to define mass and volume relationship of three soil phases. In this diagram, air was placed at the top-most place of the column, considering that air is the lightest weight material among all three phases. Water has an intermediate weight and is placed in the middle from the top or bottom of this soil column. Soil is the heaviest phase and is shown at the bottom of the column.

Figure 4.3 shows both volumes and masses of distinct phases at the left and the right of the soil column, respectively.

In this figure:

M_a = Mass of air (approximately zero)
M_w = Mass of water
M_s = Mass of solid soil particles
M_t = Total mass ($M_a + M_w + M_s$)
V_a = Volume of air
V_w = Volume of water
V_v = Volume of void ($V_a + V_w$)
V_s = Volume of soil
V_t = Total volume ($V_a + V_w + V_s$) = ($V_v + V_s$)

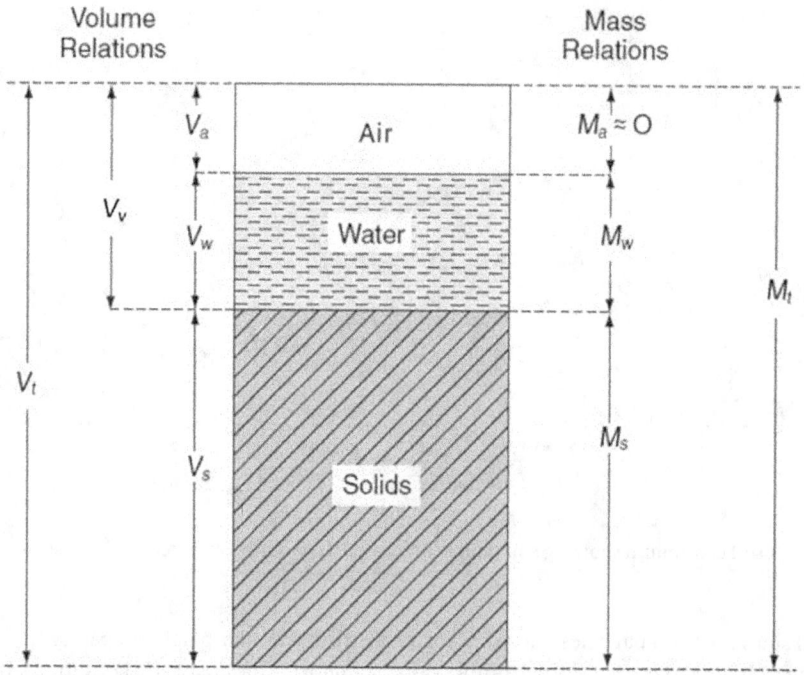

FIGURE 4.3   A schematic diagram to show three phases system of soil [4].

4.4.1 Definitions

1. Density of soil (ρ_s) is defined as the ratio of soil solid to its volume. It can be shown as:

$$\rho_s = \frac{M_s}{V_s \rho_w}$$ (4.1)

   ρ_w = density of water at 4°C
   Density of soil is 2.65 g/cc and the specific gravity is about 2.65. However, the density of organic matter ranges between 1.3 and 1.5 g/cc.

2. Dry bulk density (ρ_b) is defined as the ratio of mass of dried soil solid particles to the total volume of soil (including particles and pores).

$$\rho_b = \frac{M_s}{V_t} = \frac{M_s}{V_a + V_w + V_s} = \frac{M_s}{V_v + V_s}$$ (4.2)

   Bulk density has the unit of g/cc. Apparent specific gravity is used synonymously, and it has no unit.

3. Total bulk density is defined as the ratio of mass of wet soil to the total volume. Because the mass of soil is considered as wet soil, it is sometimes called the wet bulk density:

$$\rho_t = \frac{M_t}{V_t} = \frac{M_s + M_w}{V_a + V_w + V_s} = \frac{M_s + M_w}{V_v + V_s}$$ (4.3)

4. Porosity (η) is defined as the ratio of the volume of pores (or it is called volume of voids) to the total soil volume:

$$\eta = \frac{V_v}{V_t} = \frac{V_a + V_w}{V_a + V_W + V_S} = \frac{V_a + V_w}{V_v + V_s} \qquad (4.4)$$

Porosity is a measure that indicates the volume of pores. It is dependent on textural and structural characteristics of soil. The porosity of sandy soil usually varies from 35% to 50%, while porosity of clay soil varies from 40% to 60%. Finer particles of soil show the higher porosity [1].

5. Void ratio is the number expressing the ratio of pore spaces to solid soil volume. In other words, it can be termed as relative porosity.

$$e = \frac{V_f}{V_s} = \frac{V_a + V_w}{V_s} \qquad (4.5)$$

Void ratio (e) has some advantage over porosity (η). In the case of void ratio, the total soil volume changes with the change of voids, but in the case of porosity, the volume of pores may change without the change of soil.

The relationships are true between porosity and void ratio to apparent and true specific gravity:

$$\rho_b = \rho_s \left(1 - \frac{\eta}{100} \right) \qquad (4.6)$$

$$\rho_s = \rho_b (1 + e) \qquad (4.7)$$

4.4.2 Soil Wetness

Soil wetness refers to the understanding as to how much water or moisture there is with respect to the solid soil particles.

$$\text{Mass wetness} = \frac{M_w}{M_s} \times 100\% \qquad (4.8)$$

This is very often called the *soil moisture content*. This is sometimes expressed in terms of decimal or percentage.

Volume wetness or soil wetness is defined as the ratio of volume of soil to total volume of soil and water (pores). Mathematically, it can be written as:

$$\text{Volume wetness} = \frac{V_s}{V_t} = \frac{V_s}{V_v + V_s} \qquad (4.9)$$

This is also called the *volumetric water content*.

Degree of saturation is another way to express the relative water content with respect to the pore spaces available in the soil. This is the ratio of volume of water to the volume of voids (pore spaces).

$$\text{Degree of saturation} = \frac{V_w}{V_v} = \frac{V_w}{V_a + V_w} \qquad (4.10)$$

4.4.3 Soil Bulk Density Determination

The traditional method of determination of bulk density or apparent specific gravity is to find an incompact soil sample of known volume [1]. A core sampler is used to collect undisturbed soil samples. The cylinder of the core sampler is designed to have a cutting edge. The sampler is driven into the soil and an incompact core of soil is obtained within the tube. The sample is carefully trimmed at both ends of the core cylinders. These soil samples are kept inside an electric oven for 24 hours at a temperature of 105°C. The weight of each soil core is taken before and after drying. The differences in weights of the collected soil sample and the oven dried soil sample is the weight of soil moisture. The volume of the soil core is the same as the inside volume of the core cylinder. The oven-dried soil weight in grams is divided by the volume of the soil core in cc. of soil. This is the value of bulk density or apparent specific gravity. Most clay soils have bulk density almost equal to 1.0 g/cc. The sandy soil has a bulk density of about 1.8 g/cc.

4.4.4 Capillary and Noncapillary Pores

There are two major types of soil pores. They are:

1. Capillary pores, and
2. Noncapillary or aeration pores.

Capillary pores hold water after the soil's free drainage is completed. Capillary porosity is the percentage of pore spaces that may be occupied by capillary water. Noncapillary or large pore spaces cannot hold water. Non-capillary porosity is the percentage of pore spaces saturated with air after soil drains to field capacity. An ideal soil should have almost equal capillary and noncapillary pore spaces because this type of soil has both aeration and water-holding capacity that is suitable for plant growth.

4.4.5 Kinds of Soil Water

If dry soil gets water either from irrigation or rainfall, it is distributed in the soil particles and is held by adhesive or cohesive forces. This water displaces air from the pore spaces and fills the pore spaces and finally this water captures the pore spaces. When the soil pore spaces are filled with water, it is called saturated soil at its maximum retention capacity. There are three main classes of soil water:

1. Hygroscopic water – where water is held with strong bonding with the soil particle surfaces by adsorption forces,
2. Capillary water – where water is held by forces due to the surface tension and exists as a continuous film around the soil particles, and
3. Gravitational water – water that moves through the pore spaces of soil with gravitational forces and drains out of the soil without any obstruction.

4.5 Infiltration: Movement of Water into Soils

Movement of water into the soil from soil surface is called *infiltration*. Infiltration is the key factor that influences irrigation water application, irrigation efficiency, pump operation time, when and how much of water is needed to apply into the field, etc. [5].

Infiltration rate determines the maximum rate at which water enters the soil under some specific conditions.

Infiltration rate is rapid in the initial stage of water infiltration and gradually it gets slower and infiltration rate reaches a constant value, which is developed after some time has elapsed from the start of irrigation. This is called the *basic infiltration rate*.

Accumulated infiltration is the quantity of water that has entered the soil in each time. This is also called cumulative infiltration. Infiltration rate and accumulated infiltration are the two major components than govern irrigation and its efficiency. The empirical formula for estimating accumulated infiltration is:

$$y = at^{\alpha} + b \qquad\qquad (4.11)$$

where

    y = Accumulated infiltration in time t, centimeter
    t = Elapsed time or infiltration opportunity time, minute
    a, α, b = Characteristic constants

4.5.1 Factors Affecting Infiltration

Infiltration capacity is dependent on:

1. Soil texture,
2. Soil structure,
3. Soil cover,
4. Existing soil moisture content,
5. Soil hydraulic conductivity,
6. Soil porosity,
7. Existing soil swelling colloids and organic matter,
8. Irrigation or rainfall duration, and
9. Viscosity of water.

For irrigating the crops, the quantity of water requirement should be determined and applied considering infiltration, runoff, water application efficiency and water use efficiency [5]. Infiltration has the unit of velocity, like centimeter/hour. From this definition, it is well understood that the distance or height of entrance of water per unit time is dependent on soil characteristics. Compactness of soil, porosity, and soil type play a vital role in allowing the water to enter and flow downward. Moreover, if the antecedent moisture content is already abundant in the soil profile, there will be less opportunity for the incoming water to get into the soil profile. Water will infiltrate until the water has enough space in the soil profile or if the soil becomes saturated; the excess water will start filling up the soil depressions and finally runoff will take place following the surface gradient into the nearest reservoir.

4.5.2 Infiltration Measurement

There are three methods for determining the infiltration characteristics for any irrigation system's design and water management practices. They are: (1) cylindrical infiltrometer method, (2) accumulation infiltration estimation from waterfront advance data, and (3) depletion of free water surface measurement in a large basin. Out of the above-mentioned methods, cylindrical infiltrometer method is most commonly used. Cylindrical infiltrometer method offers advantages over the other two methods. It also helps avoiding the cumbersome procedure in collecting dependable data from the field while estimating from waterfront advance data. It is also necessary to consider the evaporation loss due to atmospheric influences on the large basin while measurement of water depletion is considered [5]. Hence, the cylindrical infiltrometer is comparatively reliable for measuring the infiltration rate and accumulated infiltration. Infiltration characteristics can be measured by using a metal, cylindrical, round-shaped hollow drum driven to a certain length into the soil surface and then ponding this cylinder with water and simultaneously recording the time required to deplete water and enter into the soil surface. In early days, only one cylinder was used to measure the height of water lowered in the cylinder. That procedure resulted in several drawbacks and a higher degree of variability due to the uncontrollable movement of lateral seepage and movement of water to and from the cylinder. This lateral movement of water has been well controlled by another concentric cylinder similarly with ponded water as the inner cylinder.

Figure 4.4 shows the dimensions of a common infiltrometer. Two cylinders of diameters 30 cm and the 60 cm, respectively, are driven concentrically into the soil surface about 10 cm and the total height of these two ring cylinders is 25 cm. The material used for making these cylinders is rolled steel of 2 mm.

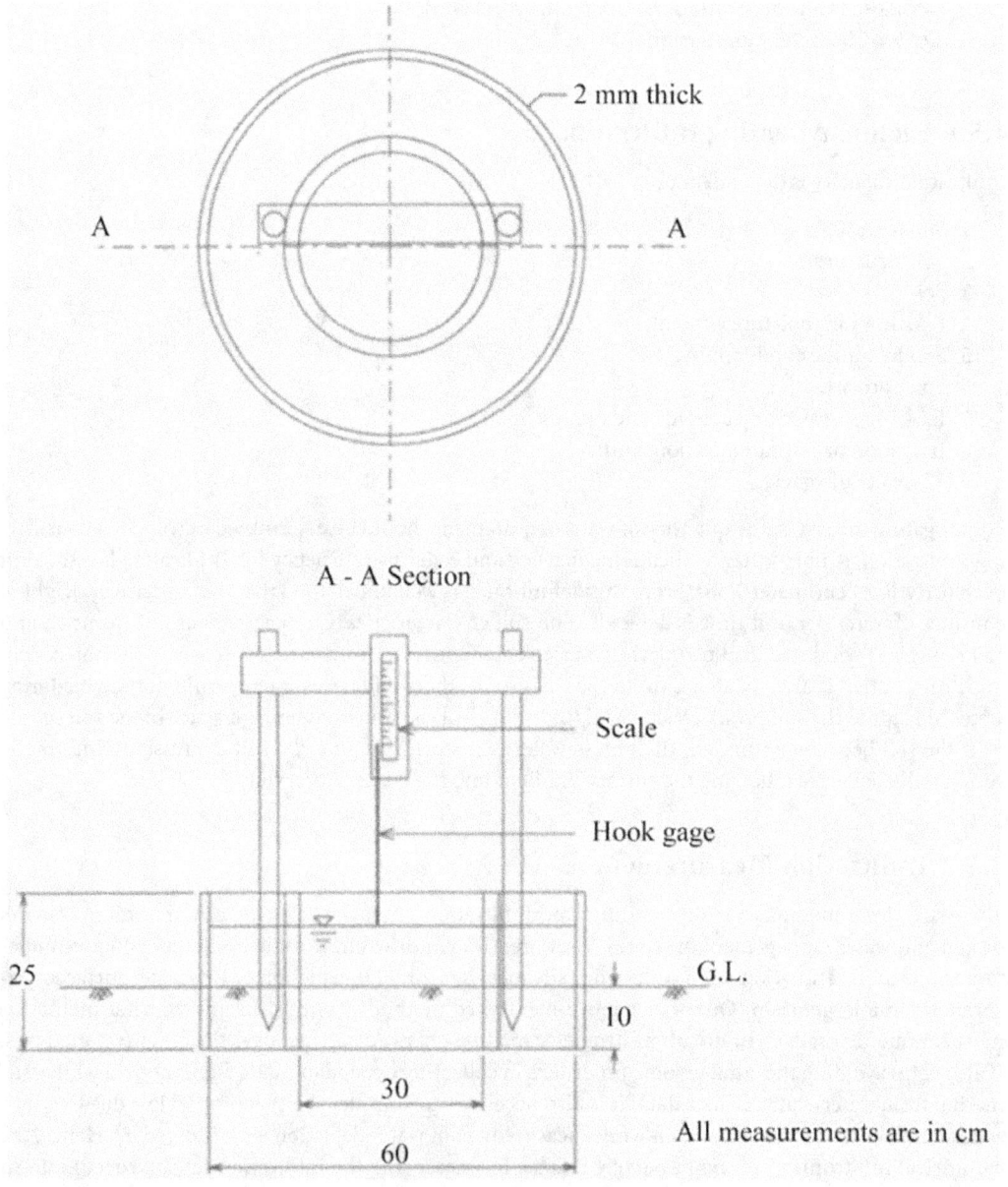

FIGURE 4.4 Plan and cross-sectional views of an infiltrometer.

The inner and the outer cylinders are both ponded with water. The outer cylinder is used as a buffer pond to avoid the lateral movement of water from and to the inner cylinder. Care should be taken against beveling of the cylinder bottoms. The cylinders are driven into the soil by a falling weight hammer striking on top of a wooden plank placed on top of the cylinders to avoid any damage to the edge of the cylinders. The main objective of this study was to develop a model for studying the infiltration characteristic by the modified Kostiakov method and to calculate the accumulated infiltration and infiltration rate with specific focus on: (1) deriving the constant values of the modified Kostiakov method for the soil under consideration, (2) judging the applicability of the model using the field data, and (3) finding the percentage of error between the actual values and the values calculated by the model.

4.5.3 Infiltration Measurement Methodology

After the infiltrometer is set up, as shown in Figure 4.4, water is poured into both inner and outer rings, and water levels of the inner cylinder were read by a needle-type pointed hook gauge whose sharp and pointed headend was only touching the water level for initial water height reading and the tail end was set with a scale to read the difference after a predetermined time when there was a depletion in water height.

The pointed head is adjusted again to touch the depleted water surface. The difference between the initial and the final readings is the height of water that infiltrated during the predetermined time. This height divided by time is the one defined as the infiltration rate. Here, time was recorded as minutes, but they were converted into hours for calculation purposes and the infiltration rate was expressed as centimeter/hour. After some time, there is no more depletion of water taking place and the curve of accumulated infiltration (ordinate) vs. time (abscissa) is the constant infiltration rate; the characteristic of this point and hereafter is called asymptote. At the initial stages, time vs. depletion of water recordings were taken frequently, and refill of water was done as quickly as possible so that the pace of infiltration could be kept constant. Water levels in the inner and the outer cylinders were kept approximately the same to keep the water pressure the same between the inner and outer cylinders and to avoid the lateral water movement due to dissimilar height between the inner and the outer cylinders. The average value of accumulated infiltration y and average infiltration rate have been plotted against time t and shown in Figure 4.5.

The modified Kostiakov method [1] can be tested as to whether it suits the local soil condition and if it can be used to assess the accumulated infiltration and infiltration rate. The relationship of accumulated

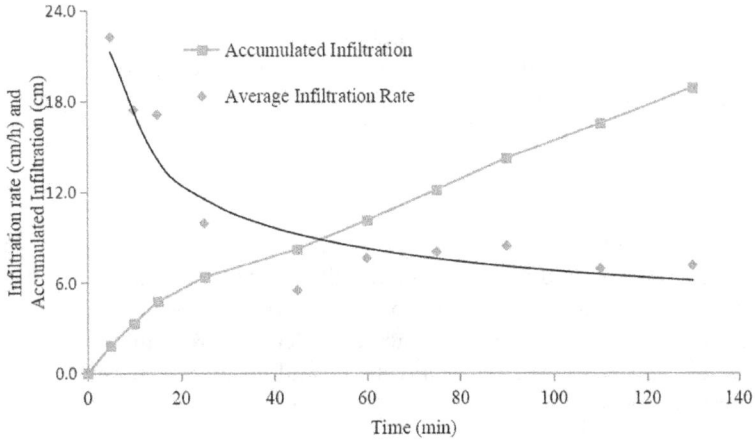

FIGURE 4.5 Average infiltration rate (centimeter/hour) and accumulated infiltration (centimeter) vs. time (minute).

infiltration y with respect to time t can be mathematically defined by the equation (4.11), known as the modified Kostiakov method. The values of a, α, and b are three characteristic constants, and they can be calculated by the method suggested by Davis [4]. The values of a, α, and b are usually less than 1 [1]. The steps are described below:

1. Plot the values of y and t;
2. Select a pair of points (t_1, y_1) and (t_2, y_2) from this plot; points selected on the plotted line should be near the extremities of the line to cover a wide range of interpolated values;
3. Calculate a third value for time t_3 using the values of t_1 and t_2 from the procedure followed in the previous step. The equation for calculating t_3 is as follows:

$$t_3 = t_1 \times t_2 \tag{4.12}$$

The values of t_1 and t_2 are available from the plotted line (described in step 2). The value of b can be calculated by the formula used in regression analysis. The formula is as follows:

$$b = \frac{y_1 \times y_2 - y_3^2}{y_1 + y_2 - 2y_3} \tag{4.13}$$

Equation (4.11) can be rearranged and written as follows:

$$y - b = at^\alpha \tag{4.14}$$

Taking log on both sides:

$$\log(y - b) = \log(a) + \alpha \log(t) \tag{4.15}$$

Data collected from the field can be arranged to get accumulated infiltrations with respect to various times, and the diagram is plotted as shown in Figure 4.5. Data from the different accumulated infiltration curves can be plotted and plugged in Eq. (4.15). Steps to calculate the values for a and α are:

1. value of b to be plugged in from Eq. (4.13), and
2. values of a and α can be available by solving two equations formed as Eq. (4.15). At least two equations are needed to solve for a and α.

4.5.4 Soil Moisture Tension

Water is being held in the pore spaces with some force, and the same amount of force is to be exerted to remove moisture called the soil moisture tension. Tenacity is measured in terms of potential energy of water in the soil, and it is measured with respect to free water that has the unit called atmospheres. This is a measure that represents the average air pressure at sea level. There is another unit to measure this soil moisture tension. This is expressed in centimeter or millimeter of mercury. The relation between atmosphere and centimeter is:

1 atmosphere = 1,036 cm (about 33.99 ft) of water or 76.39 cm (about 2.51 ft) of mercury. This is also represented in terms of bar, which can be shown as:
1 bar = 10^6 dynes/cm^2 = 1,023 cm (about 33.56 ft) of water column and 1 mill bar = 1/1,000 bar.

Soil tension has been brought to the smallest possible dimension by surface tension, which is associated with capillary tension. This has been defined as the *capillary potential* [3] representing the energy with which the water is held by soil. However, this term is not applicable over the entire moisture range. If the capillary column is filled with water, it is called the *hygroscopic potential*. In the intermediate range, the term "capillary potential" is well suited and in the dry range, the term "hygroscopic potential" is suitable. It may be stated here that the terms "soil moisture potential", "soil moisture suction," and "soil moisture tension" are often synonymously used to cover the entire range of moisture [6].

4.5.5 Soils Power Factor *pF*

The logarithmic expression of soil moisture tension expressed as *pF* is an exponential expression of a free-energy difference (based on the height of a water column above free water level in centimeters). This definition means the negative pressure of the soil moisture expressed in centimeters of water. It can be shown as:

$$pF = \log_{10} h \tag{4.16}$$

where h = Soil moisture tension in centimeters of water.

4.5.6 Total Soil–Water Potential

The total soil–water potential can be defined as the amount of work that must be done per unit quantity of pure water to transport reversely and isothermally an infinitesimal quantity of water from a pool of pure water at a specified elevation at atmospheric pressure to soil water at the point under consideration [1].

The total potential includes chemical potential and the potential of external force fields. When expressed per unit amount of water (ψ_k) in the force field, or

$$\psi_k = -\int F_k dl \tag{4.17}$$

where

F_k = force per unit mass due to field "*K*"

The total potential can be obtained by calculating the algebraic sum of component potentials corresponding to the various force fields. It can be written as:

$$\psi = \int_k \psi_k = -\sum_k \int F_k \cdot dl \tag{4.18}$$

where

ψ = Total potential

Naturally, soil is under the influence of various force fields. Therefore, the total potential is calculated as the sum of the individual force potential as follows:

$$\psi = \psi_g + \psi_{p(m)} + \psi_o \tag{4.19}$$

where

    ψ = Total potential
    ψ_g = Gravitational potential
    $\psi_{p(m)}$ = Pressure (or matric) potential
    ψ_o = Osmotic potential

Gravitational potential $\left(\psi_g\right)$ is the potential that is related to the gravitational force, which is dependent on the height or elevation. This is the amount of work that a quantity of water in an equilibrium soil–water (or plant–water) system at an arbitrary level is capable of doing when it moves to another equilibrium identical in all respects other than that is at a reference level [1]. As the gravitational potential of soil water at each point is determined by the elevation of the point relative to some arbitrary reference level, it is customary to set the reference level at the elevation of a pertinent point within the soil or below the soil profile being considered, so that this potential can always be taken as zero or positive.

Considering a point at height Z above the reference point, the gravitational potential energy is defined as:

$$E_g = MgZ = \rho_\omega gVZ \tag{4.20}$$

where

    M = Mass of water in gram
    g = Acceleration due to gravity in gram/second^2
    ρ_ω = Density of water in gram/centimeter^3
    V = Volume of water in cm^3

Gravitational potential can be defined as:

$$\psi_g = \left(\text{mass}\right) \text{ per unit mass} = \frac{E_g}{M} = \frac{MgZ}{M} = gZ \tag{4.21}$$

$$\psi_g = \left(\text{volume}\right) \text{ per unit volume} = \frac{E_g}{V} = \frac{\rho_\omega gZ}{V} = \rho_\omega gZ \tag{4.22}$$

$$\psi_g = \left(\text{weight}\right) \text{ per unit weight} = \frac{\rho_\omega VgZ}{\rho_\omega Vg} = Z \tag{4.23}$$

Pressure potential ψ_p is defined as the amount of work that a unit quantity of water in an equilibrium soil–water (or plant–water) system is capable of doing when it moves to another equilibrium system identical in all respects except that it is at a reference pressure. In the field, the pressure potential at a point is a direct result of the overlying water, which is where the air pressure on a soil–water system is determined by the depth h of the point below the water table. This is also referred to as the submerged potential. The hydrostatic pressure P of water with a reference to atmospheric pressure is given by:

$$P = \rho_\omega gh \tag{4.24}$$

The potential energy of the water is Pdv, where dv is the infinitesimal volume of water. The pressure potential can be defined as:

$$\psi_p = (\text{mass}) \text{ per unit mass} = \frac{\rho_\omega gh}{\rho_\omega} = gh \tag{4.25}$$

$$\psi_p = (\text{volume}) \text{ per unit volume} = \frac{Pdv}{dv} = P \tag{4.26}$$

$$\psi_p = (\text{weight}) \text{ per unit weight} = \frac{\rho_\omega gh}{\rho_\omega g} = h \tag{4.27}$$

Matric potential (ψ_m) is defined as the amount of work that a unit quantity of water in an equilibrium soil–water (or plant–water) system is capable of doing when it moves to another equilibrium system identical in all respects except that there is no matrix present. It is the portion of the total water potential that is attributable to the solid colloidal matrix of the soil system. This is the negative pressure potential that results from the capillary and adsorptive forces.

Considering dv as an infinitesimal of total volume of water, with a pressure deficit of P, the matric potential can be defined as:

$$\psi_m = (\text{volume}) \text{ per unit volume} = \frac{Pdv}{dv} = P = \rho_\omega gh \tag{4.28}$$

$$\psi_m = (\text{mass}) \text{ per unit mass} = \frac{\rho_\omega gh}{\rho_\omega} = gh \tag{4.29}$$

$$\psi_m = (\text{weight}) \text{ per unit weight} = \frac{\rho_\omega gh}{\rho_\omega g} = h \tag{4.30}$$

Osmotic potential (ψ_o) is defined as the amount of work that a unit quantity of water in an equilibrium soil–water system is capable of doing when it moves to another equilibrium system identical in all respects except that there are no solutions. Osmotic potential, also termed the solute potential, is the portion of water potential resulting from the solute present in the soil system.

$$\psi_o = -\pi \tag{4.31}$$

where

$\pi =$ Osmotic pressure due to dissolved salts.

4.5.7 Soil–Moisture Characteristic

Soil moisture content does not represent the soil's moisture content nor the amount of moisture available for plant growth at any specific tension or water requirement for specific plant growth stage. Soil moisture characteristics are dependent on soil texture, structure, and other characteristics of soil determined separately for any specific type of soil.

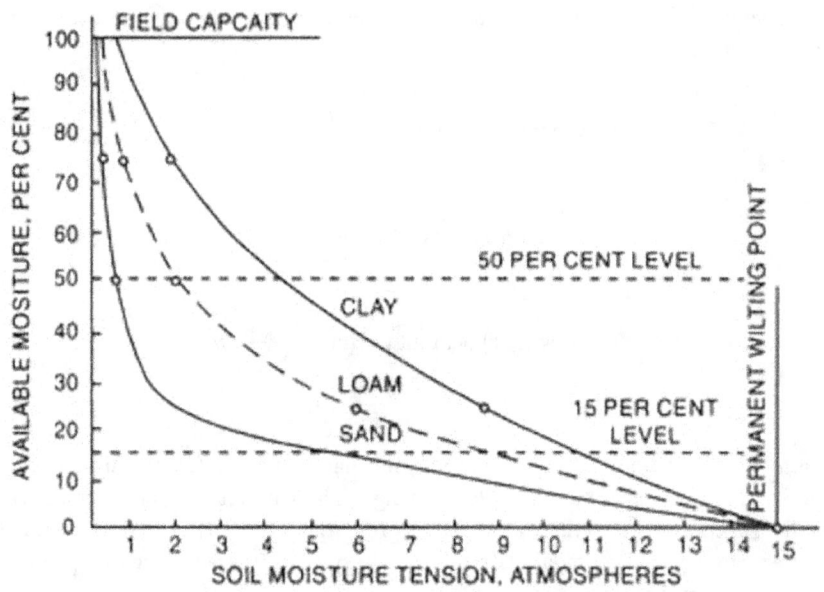

FIGURE 4.6　Typical moisture characteristic curves of sand, silt, and clay [7].

The general characteristic of sandy soil is that it drains the total amount of moisture at a lower tension, whereas the fine-textured clay soil still holds considerable amount of moisture even at such high tension that the plant that is growing may wilt. Moisture extraction curves or moisture characteristic curves are shown in Figure 4.6. This figure shows the amount of moisture in a given soil that holds at various tensions. An understanding of how much water is held by the soil is necessary to determine how much water is available for the plants. This idea further helps to decide how long (if the flow size is known) or how much water should be applied to the crop field once it is quantified keeping in mind the amount of percolation. Seepage and percolation are again two important factors to consider determining the optimum amount of water that can be applied in the root zone without losses (not being held in the soil or utilized by the roots).

4.5.8 Soil–Moisture Stress

Salt and minerals in soil water increase the force that needs to be exerted to extract water and affect the amount of water available to plants. The amount of force increases tension by minerals and salts due to osmotic pressure. If two solutions differing in concentration are separated by an impermeable membrane, such as a cell membrane in a plant root, water moves from the solution of lower concentration to higher concentration. This force with which water moves across such a membrane is called *osmotic pressure*, and it is measured in *atmospheres*.

Soil moisture stress is the sum of soil moisture tension and osmotic pressure of soil solution. Plant growth is related to soil moisture stress. Controlling of soil moisture in the root zone is a critical issue for better plant growth against higher osmotic pressure due to the existence of excessive salt in soil moisture. The osmotic pressure developed by the soil solution retards the uptake of water by plant through its root system. For example, if a soil has a tension of 1 atmosphere, it may extract the required quantity of water. But if the soil has an osmotic pressure like 7 atmospheres, it cannot extract enough water for its normal growth. Thus, the saline soil must be treated by allowing leaching as much as possible and lowering the soil moisture tension.

4.5.9 Various Soil–Moisture Constants

There are two driving forces that cause the movement of soil moisture. These driving forces are:

1. Pressure gradient, and
2. Vapor pressure difference.

Therefore, soil moisture cannot be said to be constant at a specific pressure. However, certain moisture contents described herewith are important from an agricultural point of view. These are called the soil moisture *constants*.

Saturation capacity: If the soil pores are filled with water, the soil is saturated. Saturation capacity: If soil pores are filled with water, the soil is called saturated. It is also called the maximum *water-holding capacity*. Obviously, the tension is almost zero. This is equal to free water surface.

Field capacity: If a soil has drained out all of the gravitational water and when the moisture content is almost stable and there is insignificant movement of water, it is termed as *field capacity*. In field capacity condition, the larger pores are filled with air and the finer pores are filled with water. This is the upper range or limit of available water for plant growth. But soil moisture tension during field capacity ranges from 1/10th to 1/3rd atmospheric pressures.

Moisture equivalent: This is defined as the amount of water retained by a sample of initially saturated soil material after being subjected to a centrifugal force of 1,000 times that of gravity for about half an hour. To determine the moisture equivalent, a small sample of soil is whirled in a centrifuge with a centrifugal force of 1,000 times that of gravity. The moisture remaining in the sample is determined. This moisture content is expressed as percentage on oven-dry basis, and it gives the value of *moisture equivalent*.

Permanent wilting point: This is the soil moisture content at which plants can no longer extract moisture to meet the requirement to transpire. They remain wilted unless water is added to the soil. At the verge of permanent wilting, the film of water around the soil particles is held so firmly that roots in contact with the soil cannot remove the water at a sufficiently rapid rate to prevent wilting of the plant leaves. The soil moisture tension at the permanent wilting point ranges from 7 to 32 atmospheres, and this value depends on the soil texture, structure, type and condition of the plants, amount of soluble salts in the solution, and to some extent the climatic conditions. Generally, a value of 15 atmospheres is considered as the *permanent wilting point*.

Available water: This is the water within the range between field capacity and permanent wilting point, called readily available moisture (RAM). Table 4.2 presents the different soil textural groups and their respective water-holding capacities. The total water-holding capacity is calculated based on the root zone depth of mature plants to be grown for designing the irrigation system.

TABLE 4.2 Range of Available Water-Holding Capacity of Different Soils [1]

| | Percent of Moisture Based on | | Depth of Available Water |
| Soil | Dry weight of soil | | Per unit of soil |
Type	Field capacity	Permanent wilting point	Centimeter per meter depth of soil
Fine sand	3–5	1–3	2–4
Sandy loam	5–15	3–8	4–11
Silt loam	12–18	6–10	6–13
Clay loam	15–30	7–16	10–18
Clay	25–40	12–20	16–30

4.5.10 Movement of Water within Soils

Movement of water is particularly important to analyze from the perspectives of various aspects like:

Rate and accumulated infiltration within the root zone not only indicates the amount infiltrated into the soil surface but also informs the amount of water that is supplied to plant roots and the rate of underground flow to streams and recharging the groundwater. Water movement takes place through the pores under the action of gravity. If the soil is unsaturated, the thin film of water movement takes place by surface tension. Water diffusion takes place as vapor through the air-filled pore spaces along pressure gradients from higher to lower vapor pressure.

4.5.11 Terminology

Water intake: The movement of water into and through the soil from the soil surface is termed as *water intake*. It represents several factors, like infiltration, percolation, seepage, etc.

Percolation: When a part of water moves downward through saturated or nearly saturated soil due to gravitational force, it is termed as *percolation*. It occurs at a lower tension, about ½ atmosphere. Percolation is also sometimes called *infiltration*.

Interflow: This is the flow or movement of seepage water in the lateral direction in between relatively pervious and impervious or less pervious layers of soil. It may reappear at the surface of soil at a lower elevation.

Seepage: This is the infiltration of water into the soil. It moves vertically downward and laterally towards substrata of soil from a water source like lakes, ponds, or irrigation canals. Seepage water may reappear on the surface as wet spots, or seeps, or may percolate to contribute to the groundwater storage or reservoir. Sometimes, the seepage water may find its way toward the streams or springs, which take place as a subsurface flow. Seepage rate is related to the wetted perimeter of the reservoirs or the canals and the soil's capacity to transport water horizontally and vertically.

Permeability: This is the characteristic of soil that permits water to flow though it (porous media). This is the specific property of the soil that governs the rate with which a porous media transmits fluid under standard conditions. The term *intrinsic permeability* represents the permeability factor independent of the fluid. It must be remembered that the factors which tend to change the permeability of the soil matrix to water, will influence this value and prevent its use unless they can be measured or evaluated separately.

Hydraulic conductivity: This is the proportionality factor k in Darcy's law ($v=ki$, in which v is the velocity of flow of water or fluid though porous media and i is the hydraulic gradient). The value of k depends on the properties of fluid and the soil as well. They represent the characteristic of interaction between soil and fluid such as swelling or shrinkage of a soil. A soil with higher and coarse open texture will have a higher hydraulic conductivity than a soil with higher porosity but finer soil texture because of the resistance of flow of fluid through finer soil particles. Furthermore, in fine-textured soils, hydraulic conductivity depends mostly on structural pores. In some soils, particles are cemented together to form an impervious layer, called *hardpan*.

Hydraulic head: This is the elevation with respect to a standard datum at which water stands in a riser pipe (manometer) connected to the point in question in the soil. This elevation is the rise of water due to the sum of *velocity head*, *pressure head*, and *elevation head*. The velocity head is negligible for a non-turbulent flow of water in soil. Hydraulic head has a dimension of length (L).

Hydraulic gradient: This is the rate of change of hydraulic heads at two points regarding the distance between them.

Hydraulic equilibrium of water in soil: This is the condition for zero flow rate of liquid or film water in the soil. This condition is satisfied when the pressure gradient force is only equal and opposite to the gravity force.

4.5.12 Movement of moisture under Saturated Condition

Soil is saturated when all its pores are filled with water. Therefore, there is no tension. Moisture movement in this condition is explained by *Poiseuille's* law, which expresses the flow of water in a narrow tube as:

$$q = \frac{P\pi r^4}{8l\mu} \tag{4.32}$$

where

q = Discharge or volume of flow per unit time (cm³/s)
P = Pressure difference between two ends of the tube of length l (dynes/cm³)
r = Radius of tube (cm)
l = Length of tube (cm)
μ = Viscosity of liquid (dynes/s/cm²)

Poiseuille's law was the basic law to develop different equations to determine hydraulic conductivity of soil considering and using the knowledge of pore size distribution. Equation (4.32) indicates that pore size plays a significant role as its fourth power is directly proportional to the flow size. This further shows that saturated flow under otherwise identical conditions decreases as the pore size decreases. The sequence of flow size with respect to the various soil textures is shown as:

Sand > loam > clay

Darcy's law: This is very frequently used to calculate the hydraulic conductivity of soil. It states that a quantity of water passing through a porous media (soil) is directly proportional to the gradient of the hydraulic head and area of cross section of the porous media. The mathematical expression is:

$$q = kia \tag{4.33}$$

where

q = Volume of flow of water per unit time (cm³/s)
k = Hydraulic conductivity (cm/s)
a = Area of cross section (cm²)
i = Hydraulic gradient (dimensionless) = $\dfrac{h_1 - h_2}{l}$

Here, h_1 and h_2 are the hydraulic heads at points of measurement 1 and 2, respectively, and l is the distance between points 1 and 2. Equation (4.33) can be simplified further as:

$$\frac{q}{a} = ki = v \tag{4.34}$$

Value of $\dfrac{q}{a}$ is also called the velocity flux.

4.5.13 Measurement of Soil Moisture

Measurement of soil moisture varies in importance based on the purpose of the soil to be served. In the case of irrigation and water management, it is very important to know the soil moisture to plan for the amount of water needed to be applied to meet the water requirement of the crop. Measurement of soil moisture with time is important to estimate evapotranspiration (ET). There are also many experimental situations where careful measurements and control of soil moisture is necessary if the results of investigations on soil–plant–water relationships are to be interpreted properly.

The traditional methods of expressing the soil moisture are:

1. by the amount of water in a given amount of soil
2. the stress or tension under which the water is held by the soil

The amount of soil moisture held by a certain volume of soil can be expressed as weight or volume percent. Soil moisture on weight basis is expressed with respect to dry weight of the sample as follows:

$$MC_w = \frac{w_w - w_d}{w_d} \times 100 \text{ \%}$$
(4.35)

where

MC_w = Moisture content per unit weight (%)
w_w = Weight of moist soil (g)
w_d = Weight of oven-dried soil (g)

Expression of soil moisture content as a percent with respect to dry weight will not indicate the amount of water available to the plants unless a soil moisture characteristic curve or the field capacity and permanent wilting points are known. In some cases, it is necessary to convert moisture per unit weight into moisture per unit volume. This conversion can be done by:

$$MC_v = MC_v \times BD$$
(4.36)

where

MC_v = Moisture content per unit volume (%)
MC_w = Moisture content per unit weight (%)
BD = Bulk density of soil (g/cm³)

The percentage of moisture by volume obtained by the above equation is numerically equal to the moisture available as centimeter of water per meter depth of soil. For example, if the moisture content (percent by volume) of soil is 12 and the bulk density is 1.6 g/cm³. Then, the moisture is 1.6×12=19.2 cm (about 7.56 in)/m depth of soil.

Tensiometer: This is a device that directly measures the water that is held by soil particles within their pore spaces. This is expressed as the matric or capillary potential. Figure 4.7 below shows a typical tensiometer used to determine soil water availability.

The tensiometer consists of a porous cup made of ceramic, which is buried into the soil up to a desired depth. This cup is connected to a water filled tube to a manometer (mercury manometer). The scales are calibrated in either 100ths of an atmosphere (atm) or centimeters of water. Mercury tensiometer is preferred because they yield precision results with higher accuracy.

Tensiometer reading does not yield the moisture directly other than indicating the tension due to the lack of moisture in the soil. Tensiometer indicates the soil condition and to decide *when to irrigate*. It does not offer any information on *how much of water to apply*. Hence, a special moisture characteristic curve for the soil is needed to convert the moisture tension measurements versus soil moisture available in percentage must be prepared through appropriate calibration.

Figure 4.8 shows calibration results of a tensiometer. This figure can be used very easily to find the moisture content in percent once the soil tension is observed in the tensiometer. This calibration results and graph must be available for a specific soil to infer the correct moisture content.

Figure 4.8 was developed to show the soil moisture's characteristic with respect to the tension. This is a quite easy way to determine the soil moisture content. It should be mentioned here that the tensiometer does not measure and yield the moisture content of the entire range of soil types satisfactorily, but

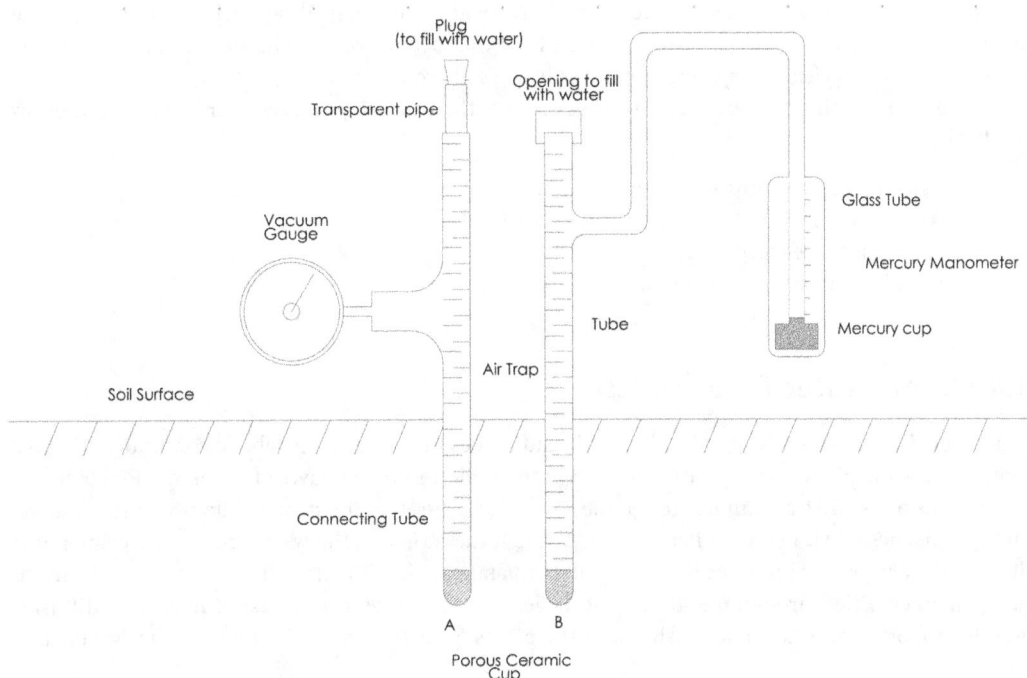

FIGURE 4.7 Tensiometers installed in the field to measure the soil moisture content.

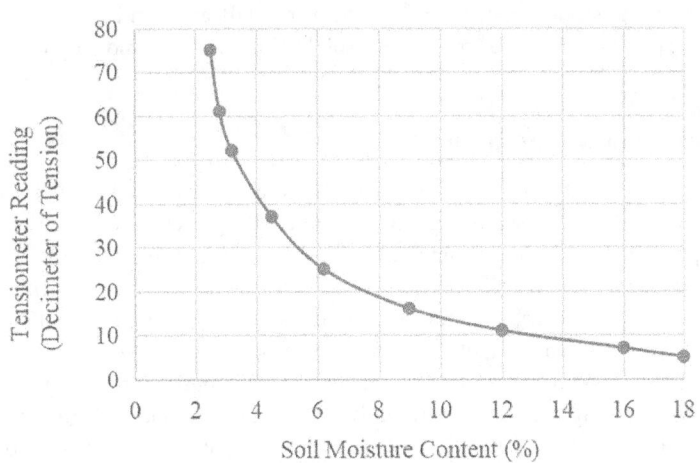

FIGURE 4.8 Calibration curve of a tensiometer.

this is considered the best way to determine the moisture condition in the wet range. They are best suited in sandy soils since in this soil a large part of the moisture available to the plants is held at a tension less than 1 atmosphere. Tensiometers are less suited for the sine textured soil in which they only have a small part of the available moisture and has a tension of less than 1 atmosphere. The usual practice is to use tensiometers at different depths to identify the root zone depth and to irrigate when the tensiometers show that the matric suction has reached some prescribed value. Use of tensiometers at different depths can indicate the amount of water needed to reach the requirement of field capacity, and it is possible by using the calibration curve as shown in Figure 4.8 for the specific soil type in the field under consideration. They can also yield the *hydraulic gradient* in the soil profile.

There are some other methods for directly or indirectly measuring soil moisture contents. They are as follows:

1. Pressure membrane pressure plate technique,
2. Measurement of solute (Osmotic) potential method,
3. Electric resistance method, and
4. Neutron moisture meter method.

4.6 Plant–Water Relationships

Water constitutes 80%–90% of most plant cells and tissues under active metabolic conditions [1]. It is a continuous liquid phase starting from the roots to the epidermis of the leaves of the plants. Plants absorb water by the roots and transmit it through the conducting vessels of the stem and finally up to the leaves through the mesophyll tissues. After moving through these tissues, the water reaches the evaporation sites, which are primarily the walls of the sub-stomatal cavities. The final transfer of water from the sub-stomatal cavities through the stomata of the leaves takes place in a process of molecular diffusion of water vapor. Plant-water relationship and the plants' growth and yields responses are grouped as follows:

1. *Soil factors:* texture, structure, soil moisture content, salinity, density, aeration, fertility, drainage, and temperature.
2. *Plant factors:* type and variety of crop, depth of root zone and their density, aerodynamic roughness of the crop, varietal effects, and drought tolerance.
3. *Weather factors:* sunshine, temperature, humidity, rainfall, and wind.
4. *Other factors:* plant spacing and soil volume, soil fertility, and soil and crop management.

4.6.1 Water as a Plant Component

Metabolic activity of the plant cells and plant itself is directly related and dependent on water content. Plant growth is controlled by the rate of cell divisions and enlargement and by the supply of inorganic and organic compounds essential for the synthesis of new protoplasm and cell walls. A decreasing water content is accompanied by a loss of cell turgidity and wilting, cessation of cell enlargement, closure of stomata reduction of photosynthesis and interference with many basic metabolic processes. This is very important for plant growth. Lack of water causes wilting, ending of cell amplification, shutting of stomata lessening in photosynthesis, and interference with many basic metabolic processes. Eventually, if dehydration means shortage of water continually taking place, it causes disorganization of the protoplasm and death of most organisms. The relation of water content during physiological activities increases manifold as the water content increases.

The total quantity of water necessary for physiological activities in a plant is usually less than 5% of all the water absorbed [1]. Most of the water absorbed by the root system is lost through *transpiration* and contributes little to its growth. However, failure to respond to the water lost by transpiration causes the loss of turgidity, cessation of plant body growth, and eventually death of the plant from dehydration.

The following are the main areas of water–plant relationship:

1. Water absorption,
2. Water conduction and translocation, and
3. Water loss or transpiration.

4.6.2 Amount of Water in Plant

The amount of water varies in various parts of the plant. The apex of the root and stem contains about 90% or more water. Young fruits and leaves are also the parts of the plant where water content is higher. With the maturity of the plants, water content decreases. The wood portion of the stem contains about 50%–60% of water. The stems of wheat, barley, and sorghum contain 60%–70% of water, and it reduces to 5%–10% at the harvesting time. The freshly harvested crops may contain 10%–15% of moisture and it will determine their storage life, viability, and germinability.

4.6.3 Movement of Water within the Soil–Plant–Atmosphere System

The total path of water movement starting from soil to the leaves through plant stem systems may be analyzed by evaluating the potential difference between the soil and atmosphere in contact with root and leaf, respectively. The path of water movement can be sequentially shown as follows:

1. Supply of water to the root zone,
2. Entry of the water into the root, and
3. Movement of water vapor through and out of the leaves.

The rate of the water movement is directly proportional to the potential gradient and inversely proportional to the resistance to flow.

Table 4.3 shows the approximate values of water potential in the soil–plant–atmosphere system. It is clearly understood from the data shown in this table that the difference in total water potential in the soil–atmosphere system could generate a dynamic force for water movement from the soil through the plant to atmosphere. If this continuum is broken, the dynamic force will automatically disappear.

4.6.4 Absorption of Water in Soil–Plant–Atmosphere System

Water extraction from the soil takes place through the plant's roots. The root system varies across plants. The variability is there due to:

1. Rooting depth,
2. Root length,
3. Root distribution horizontally or laterally.

These are further influenced by environmental and genetic factors.

Scientists have found that the roots of cereal crops extend to 200–4,000 cm/cm^2 of soil surface area as against 15–200 cm/cm^2 for nongerminomatous plants. It is wise to consider the water absorption in the total soil–plant–atmosphere system instead of considering the root zone alone. The flow rate of water in the soil–plant–atmosphere system may be shown as follows:

$$\text{Flow rate} = \frac{\Psi_{\text{soil}} - \Psi_{\text{root surface}}}{r_{\text{soil}}} = \frac{\Psi_{\text{root surface}} - \Psi_{\text{xylem}}}{r_{\text{root}}} = \frac{\Psi_{\text{xylem}} - \Psi_{\text{leaf}}}{r_{\text{xylem}} + r_{\text{leaf}}} = \frac{\Psi_{\text{leaf}} - \Psi_{\text{air}}}{r_{\text{leaf}} + r_{\text{air}}} \tag{4.36}$$

TABLE 4.3 Water Potential Magnitude and Soil–Plant–Atmosphere System [1]

Component	Water Potential (bars)
Soil	−0.10 to −20
Leaf	−0.50 to −50
Atmosphere	−1,000 to −2,000

Here, ψ represents the water potential at various points of the system and r is the corresponding resistance.

Water absorption by roots is dependent on the supply of water at the root surface. Two main phenomena directly related with water absorption are:

1. Movement of water to the root surface, and
2. Growth of roots into the soil.

Under the condition where the water extracted by roots is not frequently replaced by rain or irrigation, it is important that the root system must be expanded continuously or else have already occupied a large enough volume of soil to provide the plant with sufficient water to replace the transpiration losses. Hence, all the factors that affect root growth or the occupation by roots of a large enough soil volume will also affect the absorption of water by plants [1].

The actual entry of water into the roots is dependent on:

1. Extent of the absorbing zone of the roots,
2. Permeability of the root cortex to water movement, and
3. Water potential at the root surface.

The movement of water through the root and conducting elements of the leaves' xylem to the leaves is initiated and mostly controlled by the transportation from the leaves in response to the water potential gradient extending from the soil water, through the plant to the atmosphere. The water moves from the xylem strands of the leaf across the mesophyll tissue and through the cell walls bordering the sub-stomatal cavities where the liquid vaporizes and diffuses out of the leaves through the stomatal openings. Transpiration through an energy-controlled process is modified by the soil, plant, and atmospheric factors, which govern the potential gradients in the various parts of the water path to the leaf surface.

4.6.5 Plant Response against Moisture Stress

Plant and water relations are dependent on various interrelated and interdependent processes. Therefore, the internal water balance or degree of turgidity of a plant depends on the relative rates of water absorption and water loss, and is affected by the complex of atmospheric, soil, and plant factors that modify the rates of absorption and transpiration [1].

Water movement takes place due to the potential gradient. A base level of turgor or plant–water potential is reached when the plant roots are in equilibrium with soil water potential and the water potential gradient is near or equal to zero. Due to the absence of sunlight during night and the early morning period, the value of water potential remains at or near zero. An increase in the rate of transpiration, in conjunction with evaporation during daytime, causes a decrease in the turgor pressure of the upper leaves and the development of water potential gradients through the plant from the evaporating surface of the leaves to the absorbing surface of the roots. Therefore, there is a possibility that the rate of water loss is greater than the rate of water absorption, causing an internal water deficit to develop in the plant. This condition may occur along with many physiological processes in the plant, which are directly accountable for the growth and yield of the crop if the current condition is prevailing.

The yield of a crop is an integrated function of various physiological processes. Lack of water may cause water stress to the plant and ultimately affect the biological processes of *photosynthesis* and *transpiration*. This condition further restricts growth and reproduction. Ultimately, water deficit causes the cessation of leaves' growth and leaves' area, minimizes the size of cells, and collapses the intercellular volumes. Water stress produces tremendous changes in *carbohydrate* and *nitrogen* metabolism in plants. There are some critical plant growth stages greatly affected by water deficits and injuries compared to other stages. As an example, water application during crown root initiation enhances the yield in wheat crop.

4.6.6 Drought Tolerance of Plants

There are various means for the plants to survive against drought. Short duration crops that avoid extensive drought periods are better drought resistant compared to other varieties. Plants show drought tolerance due to either of the following reasons:

1. Because of the factors that affect their intake or loss of water, or
2. The plants can survive tissue desiccation.

Drought-resistant plants usually have smaller cells than those living in watery habitats. The smaller cells get a much smaller proportion of reduction in volume than larger cells. Usually, higher osmotic values are characteristic of the drought-resistant plants. The higher values of osmotic pressure increase the ability of cells to retain water and may increase the resistance of protoplasm to dehydration. A widespread and deeper root system can save the plants against drought as seen in sorghum. Tuber roots, potatoes, and onions suffer from drought very easily because they have very sparse and widely spreading rooting systems.

4.6.7 Water and Nutrients' Movement in Plants

Water and solutes are always in movement in the cells, cell to cell, and tissue to tissue. There are two groups of forces believed to be contributing to this movement. They are as follows:

1. Passive movement by diffusion and mass flow, and
2. Active transport, depending on the expenditure of metabolic energy.

Movement of materials by mass flow occurs when force is exerted on the moving substance by an external agent so that the molecules tend to move in the same direction in mass, whereas the diffusion results from random movement of individual molecules or iron by their own *kinetic energy*. Absorption of salts by plants involves movement of ions from soil to root surfaces; ions get accumulated in the root cells, movement of ions to root surface to xylem and finally translocation from roots to shoots. The amount and kinds of ions by the root cells vary widely among different species of plants, with the metabolic activities of roots, the concentration of ions in the soil, temperature of soil, and aeration. Mobile ions, such as potassium and nitrate, move in the direction of roots in the soil solution but immobile ions like phosphates are available only from the soil in the closest vicinity of roots. Therefore, extension of roots is an important factor in ion absorption. Ions which reach the xylem sap of the roots are usually carried to the shoots in the transpiration stream. However, many salts are removed from the xylem sap by the adjacent living cells, and the salts' concentration of the xylem decreases from roots of shoots.

4.6.8 Effective Root Zone

There are several factors that determine the amount of available soil moisture for the plants. They are as follows:

1. Moisture characteristics of soil,
2. Depth to which the plant roots extend, and
3. Propagation or density of the roots.

Soil moisture characteristic of the soil is again a function of *field capacity* and *wilting point*, which are again dependent on soil texture and organic matter. A little change may bring these limits to any greater extent. Promising possibilities may be changing the characteristic of the plant allowing for a deeper root zone system and enhancing the root zone area to become larger. The root proliferation and root density are very important. Water moves very slow, and it is only a few centimeters in an unsaturated condition. To utilize water effectively in the soil profile, roots need to be highly dense and proliferated within the

TABLE 4.4 Some Common Plants with Their Root Zone Depths [8]

Shallow Rooted	Moderately Deep Rooted	Deep Rooted	Very Deep Rooted
		Root Zone Depth	
60 cm	90 cm	120 cm	180 cm
Rice	Wheat	Maize	Sugarcane
Potato	Tobacco	Cotton	Citrus
Cauliflower	Castor	Sorghum	Coffee
Cabbage	Groundnut	Pearl millet	Apple
Lettuce	Muskmelon	Soybean	Grapevine
Onion	Carrots	Sugar beat	Safflower
	Pea	Tomato	Lucerne
	Bean		
	Chilli		

unexploited root zones throughout the total plant's growth cycle. Roots, sometimes during the moisture stress period, can elongate rapidly and try to extract and maintain the plant–water requirement. If a good root system is developed during favorable growing periods, a plant can draw its moisture supply from deeper soil layers.

It has been found that vegetable crops such as onions and potatoes have a sparse rooting system and they are not that capable of extracting all the soil moisture within the root zone. Forage grasses, sorghum, maize, and such other crops have very fibrous and dense roots. Another factor that affects the moisture relation with crop is whether the plant is annual or perennial. An annual plant must extend its root system favorable for potential uses out of the total moisture available. However, a perennial plant has already an established root system. Except it just needs extra branching to utilize the maximum level of water available in the soil.

Development of the root system into the soil surface may be affected by several reasons. They are as follows:

1. Higher groundwater water table,
2. Shallow soils, and
3. Any impermeable soil layer near the ground surface that restricts the depth of rooting.

Salt status and fertility also play an important role and influence penetration of the rooting depth. Studies have proved that the rooting pattern of common plants vary widely from soil to soil. For example, maize crop has been found to extend its roots as deep as 1.5 m in medium to coarse-textured soil while in a fine-textured soil, there is a shallower root system.

The effective root zone is the depth from which the roots of an average mature plant are capable of reducing soil moisture to the extent that it should be replaced by irrigation.

Table 4.4 shows the depth of the root zone of some common plants. Root development of any one crop varies widely with the type of soil and other factors mentioned earlier.

4.6.9 Moisture Extraction Pattern within Root Zone

If root length is divided into different depths, all of them are not equally absorbing moisture from the soil. It gives an idea about relative amounts of moisture extracted from different depths within the crop root zone. Figure 4.9 shows the soil moisture extraction pattern of common plants growing in deep uniform soils. It is usually observed that 40% of moisture is used in the first quarter, 30% in the second quarter, 20% in the third quarter, and 10% in the last quarter of root zone. Hence, to have a reliable estimate of the soil moisture status, at least two depths should be selected within the root zone depth.

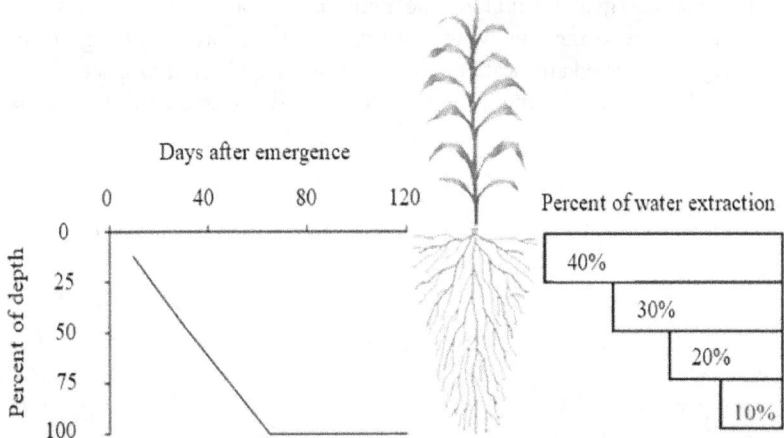

FIGURE 4.9 Soil moisture extra pattern from surface to the root zone depth [3].

4.7 Evaporation and Transpiration

Evaporation and *Transpiration* are the two main factors that govern estimating irrigation requirement and planning irrigation systems.

4.7.1 Terminology

Evaporation: This is the process that transforms water into the gaseous state. This is a fundamental and important component of the hydrological cycle by which water changes into vapor in heat energy. This is the form of moisture that transfers from land to the atmosphere.

Evaporation from natural surfaces, such as open water, bare soil, or vegetative cover is a diffusive process by which water in the form of vapor is transferred from the underlying surface to the atmosphere. Evaporation requirements are as follows:

1. Source of heat that vaporizes the liquid water, and
2. Presence of a gradient of concentration of the liquid water vapor between the evaporating surface and the surrounding air.

The source of energy for evaporation may be solar energy, the air blowing over the surface or the underlying surface itself. The energy requirement for evaporation, regardless of the surface where evaporation is taking place, is 590 cal/g of water evaporated at 20°C [1]. Evaporation can, however, occur only when vapor concentration at the evaporating surface exceeds that in the overlying air.

Dalton in 1882 initially described the fundamentals and principles of evaporation from a free surface of water. He said evaporation is a function of the difference between water's vapor pressure and air's vapor pressure. The empirical equation of Dalton's theory can be shown as:

$$E = \left(e_s - e_d\right) f\left(u\right) \tag{4.37}$$

where

E = Evaporation
e_s = Saturation vapor pressure at the temperature of the evaporating surface in mm Hg
e_d = Saturation vapor pressure at the dew point temperature of the atmosphere in mm Hg
$f\left(u\right)$ = Wind velocity function

The requirement of solar energy that vaporizes the liquid in the evaporation process provides the basis of energy balance approach to the study and research about evaporation. A significant fraction of incoming solar energy is reflected and scattered by the atmosphere back into space. This reflected portion of energy never plays any role in energy balance at the earth's surface. The thermal balance can be expressed as follows:

$$R_s - R_r - R_{lw} = H_a + H_s + H_e \tag{4.38}$$

where

R_s = Incoming solar radiation,
R_r = Reflected solar radiation,
R_{lw} = Net outgoing longwave radiation
H_a = Sensible heat flow into the air
H_s = Heat flow into the soil
H_e = Evaporative heat or latent heat flow into the air.

Evaporation from land surface: Evaporation from land surface is affected mainly by

1. The degree of saturation of soil surface,
2. Soil surface and air temperature,
3. Humidity, and
4. Wind velocity.

Many of the factors stated above are considerably influenced by the vegetative cover. Density of vegetative cover plays an important role in the process of evaporation. Due to the soil textural classes and structure, soil moisture characteristics are different. Hence, it is difficult to generalize on the amounts of evaporation from the land surfaces. Literally, at the highest soil moisture content, the evaporation is almost equal to the amount that occurs from free water surfaces, whereas, at lower moisture levels, evaporation may decrease in proportion to the content of water remaining in the soil. Evaporation from land surfaces is usually confined to the shallow depths and the evaporation rate reduces rapidly with the increase in depth from the soil surface. For reducing the evaporation, mulches can make it very effective after a few days of rain. They restrict air movement, maintain a higher vapor pressure near the soil surface, and shield the soil from solar energy.

Transpiration: Transpiration is the process by which moisture leaves the living plant body and mixes with the atmosphere.

Transpiration involves a continuous process of movement of water from the soil to the roots, through the stem, and out through the leaves into the atmosphere. This process again involves either *cuticular transpiration* or direct evaporation into the atmosphere from moist membranes through the cuticle, and *stomatal transpiration* or outward diffusion into the atmosphere through the stomata and lenticels, of water vapor previously evaporated from imbedded membranes into intercellular spaces within the plant.

Transpiration is the same as the evaporation process from the free surface of water, but it takes place through the plant structure as a modified process of evaporation. Transpiration is the dominant factor in plant–water relations because evaporation of water yields the energy gradient that causes the movement of water into and through the plants. The rate of transpiration depends on the supply of energy to vaporize water, the water vapor pressure or concentration gradient in the atmosphere, which constitutes the driving force, and the resistance to diffusion in the vapor pathway. The transpiration process is influenced by climate, soil, and the plant factors. The light intensity, atmospheric vapor pressure, temperature, and wind play an effective role while considering climatic factors. The soil factors are those controlling water supply to the roots, and plant factors include the extent and efficiency of root system in

TABLE 4.5 Soil Moisture Used by Crops under Different Climatic Conditions [9]

Climatic Conditions	Peak Rate of Soil Moisture Removal (mm/day)
Cool, humid	3
Cool, dry	4
Moderate, humid	4
Moderate, dry	5
Hot, humid	5
Hot, dry	8

moisture absorption, the leaf area, leaf arrangement and structure, and stomatal behavior. The interval over which evaporation is considered and evaporation from a free surface of water is of great importance. Vegetation ceases transpiration process at night while the stomata are closed, but the evaporation process continues to take place at a slower rate. Hence, evaporation from a free water surface may be greater than the transpiration of a crop over a period (or a month). In other words, in a few hours during the day, the crop may evaporate more than a free water surface.

Evapotranspiration (ET): ET means the addition of two processes of evaporation and transpiration together. This is also called the *consumptive use*. It defines the quantity of water transpired by plants during their growth, or retained in the plant tissue, in addition to the quantity of moisture evaporated from the surface of the soil and vegetation.

In designing water use or requirement by crops, evaporation and transpiration are combined together and considered in the determination of water requirement and water application into the crop field process. The combination of evaporation and transpiration is called *evapotranspiration* or *consumptive use* of water. It is not indistinguishable to determine individually either evaporation or transpiration for determining crop water requirement. Since the metabolic activities of the plant take place with water along with the process of transpiration process, the quantity of water is very insignificant (less than 1% of ET) compared to evapotranspiration. The term consumptive use is generally equivalent to ET. It thus includes all the water used by the plants plus the water evaporated from the bare land and water surfaces in the area occupied by the crop.

Table 4.5 shows some typical values of the peak rate of soil moisture removed by crops under different climatic conditions.

Wind is important in removing water vapor from the cropped area and the prevailing temperature and humidity conditions result from the interaction of two processes. Usually, a close relationship exists between net incoming solar radiation and evapotranspiration. There is another factor that needs to be given emphasis in the context of crop water requirement and soil–plant–water relationship. That is, the influence of wind and crop growth stages influence considerably in consumptive use rate. This is specifically true for annual crops that generally have three distinct crop growth stages. They are as follows:

1. Emergence and development of complete vegetative cover during which time consumptive use rate increases rapidly from a low value and approaches the maximum,
2. Period of maximum vegetative cover during which time the consumptive use rate is maximum if plentiful soil moisture is available, and
3. Crop maturation stage during which, for most crops, the consumptive use rate decreases.

There are four principal methods to measure evaporation. They are as follows:

1. Lysimeter method,
2. Field experimental plot method,
3. Soil moisture depletion method,
4. Water balance method.

The above methods of determination of evaporation are laborious, costly, and time-consuming.

Seasonal consumptive use: The total amount of water used as evapotranspiration by a cropped area during the entire crop growing period is called seasonal consumptive use. This is expressed as the depth of water in centimeter or volume as hectare-centimeter per hectare. This is a good parameter to evaluate and determine seasonal water applied into the crop fields as irrigation.

Potential evapotranspiration (PET): This is the evaporation that takes place from a land surface covered with large vegetation with adequate moisture available at all times. It was defined by Thornthwaite in 1948. He thought that since the moisture supply was not restricted, the PET was dependent totally on available energy. Penman in 1947 defined PET as the ET from an actively growing short green vegetation completely shading the ground and never facing any shortage of moisture availability.

Apparently, the understanding of PET can be visualized as the integral effect of all of the climatic factors that govern the evapotranspiration process. It may be defined as the ET that takes place when the ground is entirely covered by the actively growing vegetation and where there is no restriction in the soil moisture. It may be considered to be the upper limit of evapotranspiration for a crop in a given climatic condition.

4.7.2 Estimation of Evapotranspiration (ET) from Evaporation Data

There is a very close relationship between the rate of consumptive use by crops and the rate of evaporation from a properly located evaporation pan. The US Weather Bureau Class A open pan evaporimeter or sunken screen open pan evaporimeter, as described by Sharma and Dastane in 1968, may be used for measuring evapotranspiration. If the data for evapotranspiration or evaporation pan are not available, it can be estimated by climatological data. Relationship between evapotranspiration and evaporation is obtained by crop factor.

The value of crop factor for any crop depends on foliage characteristics, growth stage, environmental and geographical location. Consumptive use value is low during the early growth stage and increases as the plants mature. It shows a general declining characteristic of consumptive use at the later stage periods.

Sunken screen pan evaporimeter: Figure 4.10 shows the sunken screen evaporimeter, which was developed by Sharma and Dastane in 1968 [8].

The sunken screen evaporimeter shown in Figure 4.10 provides a simple device to estimate the reasonable value for consumptive use. Ratio between evaporation and evapotranspiration, in sunken screen evaporimeter, is reported to be 0.95 to –1.05. It has been claimed that the evaporation value obtained by using this equipment closely approximated the evapotranspiration. However, this equipment still needs research and studies and to be tested in varying climatic conditions and with different crops.

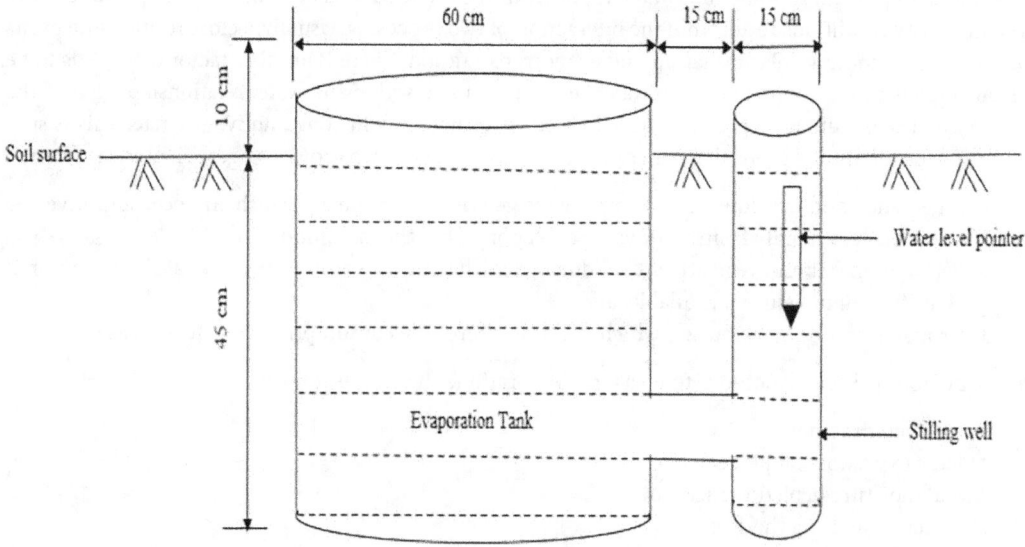

FIGURE 4.10 Sunken screen pan evaporimeter [10].

4.7.3 Estimation of Evapotranspiration (ET) from Climatological Data

There are many unavoidable reasons that make it not a reliable and dependable value of direct measurement of pan evaporation under the field conditions other than only an indication and idea. Evaporation is often predicted using climatological data. Using climatological data, modeling, and expressing by mathematical equation give a better result than direct field measurement of evaporation under field conditions.

Different scientists have worked using the concept of using climatological factors like:

1. Temperature day length,
2. Humidity,
3. Wind,
4. Sunshine, etc.

Among a number of methods, the commonly used formulae in estimating evapotranspiration are:

1. *Blaney–Criddle method:*

    Blaney–Criddle in 1950 developed the equation:

$$U = KF = \Sigma kf = \Sigma u = \Sigma \frac{ktp}{100} \tag{4.39}$$

where

U = Seasonal consumptive use of water by the crop for a given period, inches
u = Monthly consumptive use, inches
K = Empirical seasonal consumptive use coefficient for growing season
F = Sum of monthly consumptive use factors (f) for the growing season
k = Empirical consumptive use crop coefficient for the month, u/f
$f = t \times p/100$
t = Mean monthly temperature, °F
p = Monthly daylight hours expressed as percent of daylight hours of the year.

2. *Thornthwaite method:*

    Thornthwaite in 1948 assumed that an exponential relationship existed between mean monthly temperature and mean monthly consumptive use. The relationship was based on the climatic conditions of central and eastern United States. No considerations were taken for crops of varying land uses and different crops. The formula was developed for the purpose of a rational classification of the broad climatic patterns of the world. Suitable coefficients should, therefore, be developed locally for reliable estimations of evapotranspiration (ET) values. The proposed formula was:

$$e = 1.6 \left(\frac{10t}{I} \right)^a \tag{4.40}$$

where

e = Unadjusted evapotranspiration, centimeter per month (month of 30 days each and 12 hours day time).
t = Mean air temperature, °C
I = Annual seasonal heat index, the summation of 12 values of monthly heat indices (i) when
    $i = \left(\frac{t}{5} \right)^{1.514}$
a = An empirical exponent computed by the equation
    $a = 0.000000675 I^3 - 0.0000771 I^2 + 0.01792 I + 0.49239$

The unadjusted values of e are corrected for actual day light hours and days in a month. The correction factors of e need to be incorporated in the calculation process. In North America, where the development of this equation took place, a reasonable value was found because of the higher correlativeness between temperature and radiation. An additional assumption should be taken for calculating consumptive use in other parts of the world. The drawbacks of this formula has been studied and reported by Chang in 1968. He described it as follows:

1. Temperature alone is not a good indication for the energy available for evapotranspiration,
2. Air temperature of a place lags behind radiation,
3. According to the formula, ET is caused when the mean temperature is below 0°C which by no means is true, although the amount of evaporation will be very small,
4. The formula doesn't take into account the wind effect, which might be an important factor in some areas, and
5. It also doesn't consider the effect of warm and cool air on the temperature of a place.

3. *Penman method*: Penman proposed an equation to calculate the evaporation from open water surface in 1948 in combination with energy balance and sink strength, which can be shown below as [11]:

$$E_o = \frac{\Delta Q_n + \gamma E_a}{\Delta + \gamma} \tag{4.41}$$

where

E_o = Open water surface evaporation, millimeter/day,

Δ = Slope of saturated vapor pressure versus temperature curve (de_a/dT) at the mean air temperature T_a, mm Hg per °C (refer Figure 4.11),

Figure 4.11 shows the relationship between slope of saturated vapor pressure, Δ in mm Hg/°C, and temperature in °C.

e_a = Saturation vapor pressure of the evaporating surface (e_s) in mm Hg at mean air temperature T_a [refer Figure 4.12. Here (e_s) is considered equal to (e_a) by assuming zero temperature gradient between surface (s) and air temperatures].

T_a = Mean air temperature in °K = 273 + °C,

Q_n = Net radiation in millimeter of water

$$= Q_A(1-r)\left(0.18 + 0.55\frac{n}{N}\right) - \sigma T_a^4\left(0.55 - 0.092\sqrt{e_d}\right)\left(0.10 + 0.90\frac{n}{N}\right)$$

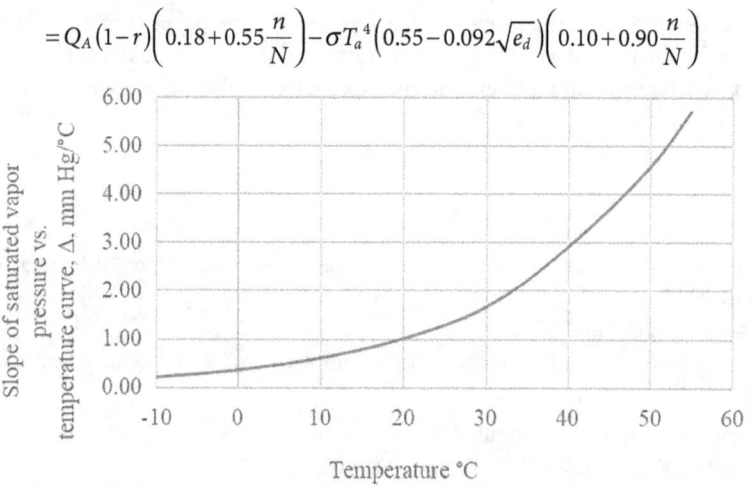

FIGURE 4.11 Slope of saturated vapor pressure vs. temperature curve.

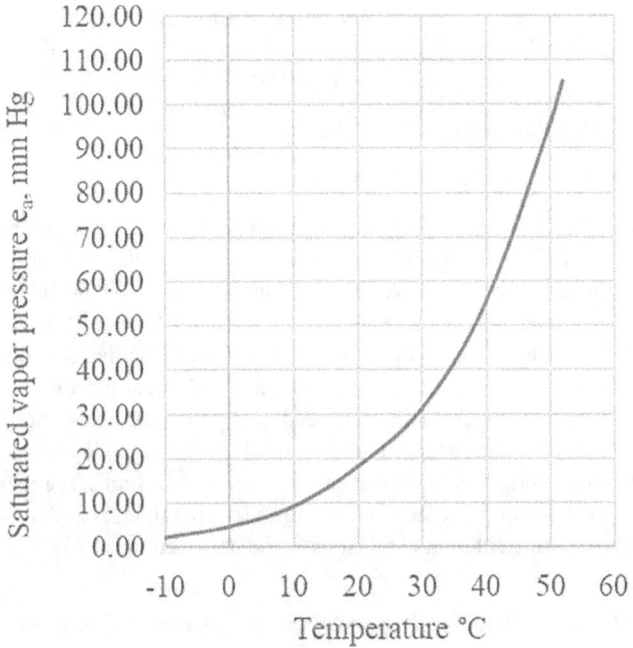

FIGURE 4.12 Saturated vapor pressure vs. temperature.

TABLE 4.6 Values of σT_a^4 for Various Temperature When Computing Evaporation by the Penman Method [12]

Temperature (°K)	σT_a^4 (mm water/day)
270	10.73
275	11.51
280	12.40
285	13.20
290	14.26
295	15.30
300	16.34
305	17.46
310	18.60
315	19.85
320	21.15
325	22.50

r = Reflection coefficient of evaporating surface.
    This is considered 0.06 for open water surface,

Q_A = Angot's value of mean monthly extraterrestrial radiation, millimeter of water/day,
$\dfrac{n}{N}$ = Ratio of actual and possible sunshine hours of bright sunshine,
σ = Stefan–Boltzmann constant (refer Table 4.6 for σT_a^4),
e_d = Saturation vapor pressure of the atmosphere, mm Hg at dew point temperature
  = $\dfrac{RH_{mean}}{100} \times e_a$, here RH is the mean relative humidity,

γ = Psychometric constant or the ratio of specific heat of air to the latent heat of evaporation of water (0.49 for 0°C and mm Hg),

E_a = An aerodynamic component = $0.35(e_a - e_d)(1 + 0.0098u_2)$,

Here, u_2 = Wind speed, miles per day at 2 m height = $u_1\left(\dfrac{\log 6.6}{\log h}\right)$

u_1 = Wind speed, miles per day at any other height h in feet.

Since the Penman equation holds good for calculating evaporation, it needs some modification to calculate evapotranspiration (ET) for crops. This is obtained by multiplying evaporation with a crop coefficient (K). Values of K for short grasses were suggested by Penman as 0.6 from November to February, 0.7 for March, April, September, and October, and 0.8 for May and August [1]. To obtain a more realistic and precise value of K, it should be locally determined. PET can be calculated using appropriate value of reflection coefficient, (r) for green vegetation, which is taken as 0.25 for most crops.

Several other constants in Penman's equation need evaluation with reference to the specific place because of different climatic conditions. Penman's method is working well and satisfactory in arid and humid regions. This is also better applicable in calm weather conditions. The main drawbacks of the Penman's method are the handling of a substantial number of parameters and mind-numbing calculations using numerous data with different conditions of the parameters.

4.7.4 Modified Penman Method for the Estimation of Evapotranspiration

Many intensive studies were done on the prediction of ET. Doorenbos and Pruitt (1975) proposed the modified Penman's method for estimating a reliable and accurate reference crop ET and suggested several tables for necessary computations. They derived the equation as follows:

$$\text{ET}_c^* = W \cdot R_n + (1 - w) \cdot f(u) \cdot (e_a - e_d) \tag{4.42}$$

where

ET_c^* = Reference crop evapotranspiration, millimeter/day (unadjusted)

e_a = Saturation vapor pressure, mbar at the mean air temperature in degree Celsius (Values of e_a with respect to mean air temperature, are available in a different Hydrology book or [5])

e_d = Mean actual vapor pressure of air, mbar = $e_a \times \dfrac{\text{RH}_{\text{mean}}}{100}$, where RH is relative humidity. This can be determined using dry and wet bulb temperature too.

$f(u)$ = Wind function (values available in a different Hydrology book or [5])

$(1 - w)$ = Weighting factor related to temperature and elevation for the effect of wind and humidity on ET_c, and

R_n = Net radiation (same as $Q_n = R_{ns} - R_{nl}$)

in which,

R_{ns} = Net incoming shortwave solar radiation = $R_A(1 - \alpha)(0.25 + 0.50n/N)$; $R_A = Q_A$ is the extraterrestrial radiation in millimeter/day, α is the reflection coefficient usually taken as 0.25 for most of the crops.

R_{nl} = Net longwave radiation = $f(t) \cdot f(e_d) \cdot f(n/N)$

Values of $f(t)$, $f(e_d)$, and $f(n/N)$ are available in various Hydrology books or [5].

4.7.5 Crop Coefficient for Estimating ET$_{(crop)}$

Above, different common methods of calculation of potential evapotranspiration PET have been discussed. PET needs to be adjusted for actual crop ET. Conversion of PET into ET, a crop coefficient (K_c) has to be incorporated. Value of coefficient (K_c) depends on the following:

1. Crops and the varieties,
2. Soil types, textures, and structures,
3. Climatic conditions, and
4. Different crop growing stages.

The equation is;

$$ET_{(crop)} = K_c \cdot PET \tag{4.43}$$

The coefficient (K_c) represents the evapotranspiration of a crop grown under optimum conditions, producing yields. Each of the methods described above predict PET, and only one set of (K_c) values are required. Ratio of ET$_{(crop)}$/PET usually ranges from 0.2 at the establishment stage to 1.0 at the time when the crop develops maximum canopy and the root system. There is a sudden drop off in the ratio of ET$_{(crop)}$/PET as the crop reaches its maturity stage. The value of this ratio may exceed 1.0 in areas prone to advection since the same is not reflected in PET estimated by most of the formulae. The correct estimation of ET from predicted value of PET, therefore, depends on accuracy of K_c values that are going to be adopted. The K_c values should be verified from the direct method available for ET determination.

4.8 Conclusions

The most important part of irrigation and water management is irrigation scheduling. It means (1) when the plants need to be irrigated, and (2) how much water needs to be applied. These two steps need the overall knowledge about soil, plants, and water. This chapter will help to gather and apply the knowledge to perform the irrigation scheduling and finally to maximize crop production. A little more information on different water-related efficiencies (water conveyance, water application, water use, etc.) and water resources availability for using in the crop field are needed to quantify the amount of water more precisely. This will allow the crop to be irrigated with the exact amount of water based on crop water requirement. Hence, this chapter can be regarded as the application and ultimate goal for best management practices based on the information and knowledge of soil–water–plant continuum.

References

1. Hasan, M., Chowdhury, T., Drabo, M., Kassu, A., and Chance, G. (2015). Modeling of infiltration characteristics by modified Kostiakov method. *J. Water Resour. Protect.*, 7, 1309–1317.
2. Blaney, H.F. and Criddle, W.D. (1950). Determination of water requirements in irrigated areas from climatological and irrigation data. In: H.H. Bennett (ed), *SCS-TP 96*, Chief, US Department of Agriculture, Soil Conservation Services, Washington 25, D.C., USA.
3. https://www.google.com/search?q=Sunken+screen+pan+evaporimeter&tbm=isch&source=iu& ictx=1&fir=EbsN9THoXt9yeM%253A%252CSHix925_M1SlpM%252C_&vet=1&usg=AI4_-kS-aDLK-zUQA0EF41LP9xkY93GmnA&sa=X&ved=2ahUKEwjO1sb09NDhAhVG7qwKHZNaAVg Q9QEwAnoECAkQBg#imgrc=EbsN9THoXt9yeM accessed on 04.14.2019.
4. Davis, D.S. (1943). *Empirical Equations and Monography.* McGraw Hill Book Co., New York, 200.
5. Doorenbos, J. and Pruitt, W.C. (1975). Guideline for predicting crop water requirements. In: Edouarda Saouma (ed), *Irrigation and Drainage Paper 24.* Director, Land & Water Development Division, FAO, Rome, Italy, 197p.

6. Gargouri-Ellouze, E., Eslamian, S., Ostad-Ali-Askari, K., Chérif, R., Bouteffeha, M., Slama, F. (2017). Infiltration, In: Bobrowsky P., Marker B. (eds), *Encyclopedia of Engineering Geology, Encyclopedia of Earth Sciences Series.* Springer, New York City.

7. http://ecoursesonline.iasri.res.in/mod/page/view.php?id=1558 accessed on 04.11.2019.

8. https://upload.wikimedia.org/wikipedia/commons/e/e6/SoilComposition.png accessed on 3.19.2019.

9. https://www.researchgate.net/figure/Textural-triangle-diagram-according-to-ISSS-system-of-classification-of-soil-particles_fig1_297737054 accessed on 04.08.2019.

10. https://www.google.com/search?rlz=1C1SQJL_enUS820US820&q=soil+moisture+extraction+pattern&tbm=isch&source=univ&sa=X&ved=2ahUKEwiqv6bal8nhAhURPK0KHejsAaoQsAR6BAgJEAE&biw=1536&bih=754 accessed on 04.11.2019.

11. Michael, A.M. (1997). *Irrigation Theory and Practices.* Vikash Publishing House Pvt. Ltd., New York City

12. https://www.google.com/search?q=mass+and+volume+relations+of+soil+constituents+diagram&tbm=isch&source=univ&client=firefox-b-1-d&sa=X&ved=2ahUKEwjkx5z8nKHhAhXKz4MKHSaHA9YQsAR6BAgJEAE&biw=1280&bih=607 accessed on 9.26.2019.

5

Plant Evapotranspiration: Concepts and Problems

Parth J. Kapupara
Parul University

Saeid Eslamian
*Isfahan University
of Technology*

Mousa Maleki
*Illinois Institute
of Technology*

DOI: 10.1201/9780429290114-7

5.1 Introduction

Evapotranspiration (ET), or consumptive use, denotes the quantity of water transpired by plants during their growth, or retained in the plant tissue, plus the moisture evaporated from the surface of the soil and the vegetation. The concept of potential ET, PET, was defined by Thornthwaite (1948), as the ET from a large vegetation-covered land surface with adequate moisture at all times. He felt that since the moisture supply was not restricted, the PET depended solely on available energy. Penman (1948) defined PET as the ET from an actively growing short green vegetation completely shading the ground and never short of moisture availability. Though Penman's definition specifies the important characteristic of reference vegetation, it does not specify the name of the vegetation. Jensen (1968) assumed PET as the upper limit of ET that would occur with a well-watered agricultural crop having an aerodynamically rough surface such as Lucerne with 30–50 cm of top growth. In designating water use by crops, evaporation and transpiration are combined into one term "evapotranspiration (ET)," as it is difficult to separate these two losses in the cropped fields.

The term "consumptive use" is used to designate the losses due to ET and the water that is used by the plant for its metabolic activities since the water used in the actual metabolic processes is insignificant (less than 1% of ET), the term consumptive use is generally taken equivalent to ET. It thus includes all of the water consumed by plants plus the water evaporated from the bare land and water surfaces in the area occupied by the crop. The relative amounts of direct evaporation from land, water surfaces, and transpiration depend usually on the amount of ground cover. For the most crops covering the soil surface, only a very small amount of water is lost from the ground surface. Under field conditions, incoming solar radiation supplies the energy for the ET process. The movement of air is crucial for eliminating water vapor from agricultural regions, and the prevailing temperature and humidity conditions arise from the interplay between these two phenomena.

Other factors being equal, the stage of growth of a crop has a considerable influence on its consumptive use rate. This is particularly true for annual crops which generally have three rather distinct stages of growth. These are: (1) emergence and development of complete vegetative cover, during which time consumptive use rate increases rapidly from a low value and approaches its maximum; (2) the period of maximum vegetative cover during which time the consumptive use rate may be maximum if abundant soil moisture is available; and (3) crop maturation stage, when for most crops, the consumptive use rate begins to decrease.

5.2 Evapotranspiration Process

ET process includes the following important steps:

- Evaporation and transpiration occur simultaneously and there is no easy way of distinguishing between the two processes.
- This fraction decreases over the growing period as the crop develops and the crop canopy shades more and more of the ground area.
- When the crop is small, water is predominately lost by soil evaporation, but once the crop is well developed and completely covers the soil, transpiration becomes the main process.
- At sowing nearly 100% of ET comes from evaporation, while at full crop cover more than 90% of ET comes from transpiration.

5.3 Computation of Evaporation

Evaporation is the process during which a liquid change into a gas. The process of evaporation of water in nature is one of the fundamental components of the hydrological cycle by which water changes to vapour through the absorption of heat energy. This is the moisture that transfers from the lands and oceans into atmosphere.

Estimation of evaporation is of utmost importance in many hydrologic problems associated with planning and operation of reservoirs and irrigation systems (Hasanzadeh Saray et al., 2020). In arid zones, this estimation is particularly important to conserve the scarce water resources. However, the exact measurement of evaporation from a large body of water is indeed one of the most difficult tasks. The amount of water evaporated from a water surface is estimated by the following methods:

- Evaporimeter data usage
- Empirical evaporation equations
- Analytical methods.

Evaporation is directly proportional to the following parameters.

5.3.1 Vapour Pressure

The rate of evaporation is proportional to the difference between the saturation vapour pressure at the water temperature, e_w and the actual vapour pressure in the air e_a.

$$E = C(e_w - e_a),$$ (5.1)

where:

E is the rate of evaporation (mm/day),
C is the a constant, and
e_W and e_a are in mm of mercury.

The above equation is known as Dalton's law of evaporation.

Evaporation takes place till $e_w > e_a$; condensation occurs, if $e_w < e_a$.

5.3.2 Atmospheric Temperature

The rate of evaporation increases if the water temperature is increased. The rate of evaporation also increases with the air temperature.

5.3.3 Wind Speed

As the wind speed increases, evaporation increases.

5.3.4 Heat Storage in the Water Body

Deep bodies can store more heat energy than the shallow water bodies, which causes more evaporation in winter than in the summer.

5.4 Types of Evaporation

Types of evaporation are as follows:

Soil evaporation: Evaporation from the water stored in the pores of the soil.
Canopy evaporation: Evaporation from the plant/tree canopy.
Total evaporation: Summation of both soil and canopy evaporation.

5.5 Measurement of Evaporation

For measuring evaporation, the following methods could be used.

5.5.1 Piche Evaporimeter

It is usually kept suspended in a Stevenson Screen. It consists of a disc of filter paper kept constantly saturated with water from a graduated glass tube (Figure 5.1). The loss of water from the tube over a known period gives the average rate of evaporation. Though it is a simple instrument, the observations obtained are often more erratic than those from the standard pans.

5.5.2 Types of Pan Evaporimeters

Evaporimeters are water-containing pans, which are exposed to the atmosphere and the loss of water by evaporation measured in them at regular intervals (Figure 5.2). Meteorological data, such as humidity, wind movement, air and water temperatures and precipitation, are also noted along with the evaporation measurement. Many types of evaporimeters are in use and a few commonly used pans are described here.

5.5.2.1 USDA Class A Evaporation Pan

It is a standard pan of 1,210 mm diameter and 255 mm depth used by the US Weather Bureau and is known as Class A Pan. The depth of water is maintained between 18 and 20 cm. The pan is normally

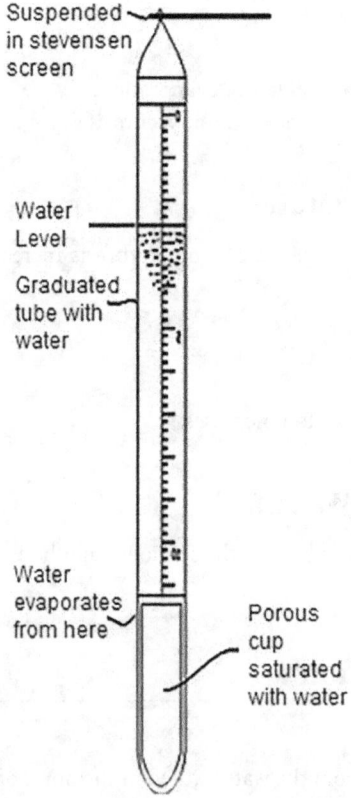

FIGURE 5.1   Piche evaporimeter.

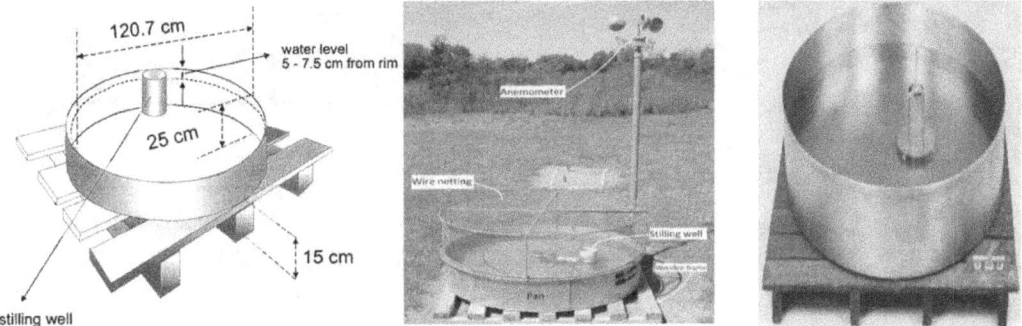

FIGURE 5.2 Pan evaporimeter.

made of the unpainted galvanized iron sheet. Monel metal is used where corrosion is a problem. The pan is placed on a wooden platform of 15 cm height above the ground to allow the free circulation of air below the pan. Evaporation measurements are made by measuring the depth of water with a hook gauge in a stilling well.

5.5.2.2 ISI Standard Pan

This pan evaporimeter specified by IS: 5973-1970, also known as modified Class A Pan, which is shown in Figure 5.3, consists of a pan 1,220 mm in diameter with 255 mm of depth. The pan is made of copper sheet of 0.9 mm thickness, tinned inside and painted white outside. A fixed-point gauge indicates the level of water. A calibrated cylindrical measure is used to add or remove water maintaining the water level in the pan to a fixed mark. The top of the pan is covered fully with a hexagonal wire netting of galvanized iron to protect the water in the pan from birds. Further, the presence of a wire mesh makes the water temperature more uniform during day and night. The evaporation from this pan is found to be less by about 14% compared to that from unscreened pan. The pan is placed over a square wooden platform of 1,225 mm width and 100 mm height to enable the circulation of air below the pan.

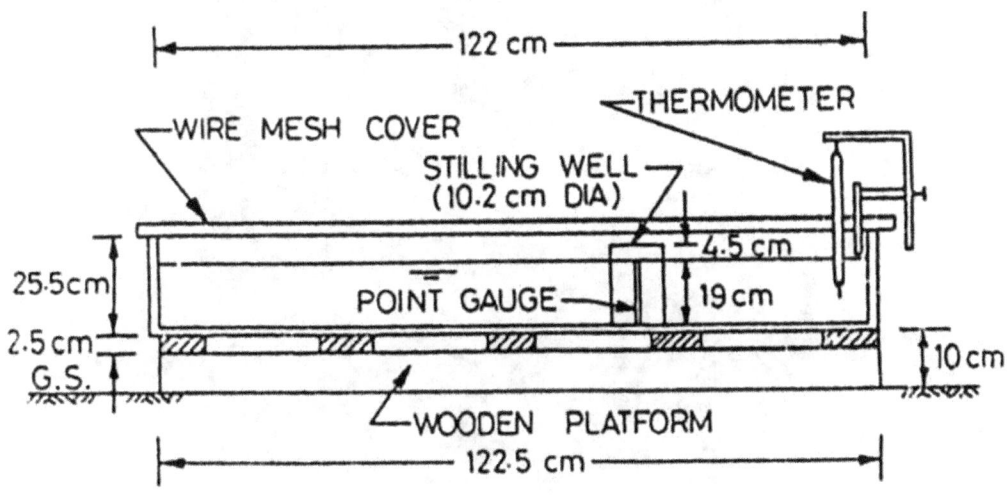

FIGURE 5.3 ISI standard pan.

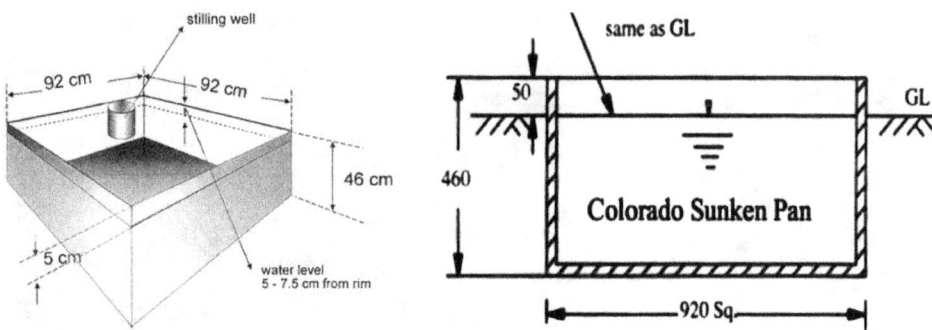

FIGURE 5.4   Colorado Sunken Pan.

5.5.2.3 Colorado Sunken Pan

This pan, 920 mm square and 460 mm deep, is made up of unpainted galvanized iron sheet and buried into the ground within 100 mm of the top which can be seen in Figure 5.4. The chief advantage of the sunken pan is that radiation and aerodynamic characteristics are similar to those of a lake. However, it has the following disadvantages:

- Difficult to detect leaks.
- Extra care is needed to keep the surrounding area free from tall grass, dust, etc.
- Expensive to install.

5.5.2.4 US Geological Survey Floating Pan

With a view to simulate the characteristics of a large body of water, this square pan (900 mm side and 450 mm depth), as depicted in Figure 5.5, supported by drum floats in the middle of a raft (4.25 m × 4.87 m), is set afloat in a lake. The water level in the pan is kept at the same level as the lake, leaving a

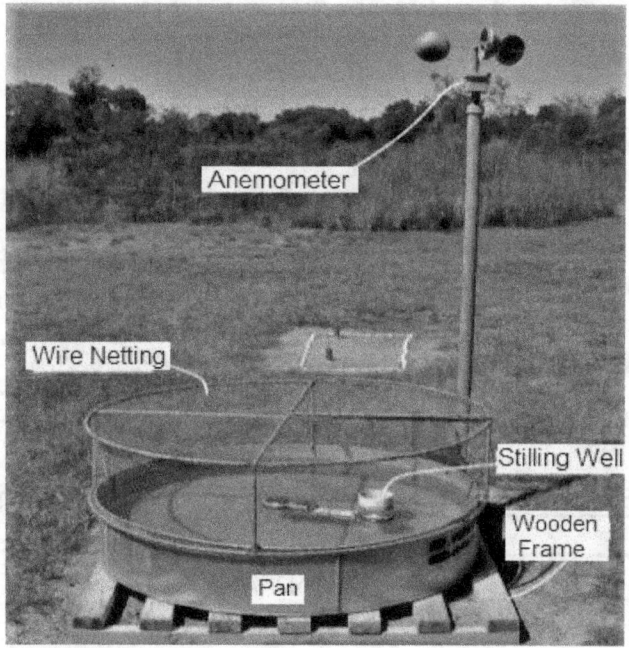

FIGURE 5.5   Floating pan.

rim of 75 mm. Diagonal baffles provided in the pan reduce the surging in the pan due to wave action. Its high cost of installation and maintenance together with the difficulty involved in performing measurements are its main disadvantages.

5.5.2.5 Pan Coefficient C_p

Evaporation pans are not the exact models of large reservoirs and have the following major drawbacks:

- They differ in the heat-storing capacity and heat transfer from the sides and bottom. The sunken pan and floating pan aim to reduce this deficiency. As a result of this factor, the evaporation from a pan depends to a certain extent on its size. While a pan of 3 m diameter is known to give a value which is about the same as from a neighbouring large lake, a pan of size 1.0 m diameter indicates about 20% excess evaporation than that of the 3 m diameter pan (Table 5.1).
- The height of the rim in an evaporation pan affects the wind action over the surface. Also, it casts a shadow of variable magnitude over the water surface.
- The heat transfer characteristics of the pan material are different from that of the reservoir.
- In view of the above, the evaporation observed from a pan must be corrected to get the evaporation under similar climatic and exposure conditions. Thus, a coefficient is introduced as:

$$\text{Evaporation(mm)} = C_p(\text{pan coefficient})\text{pan evaporation(mm)}.$$

TABLE 5.1 The Values of C_p in Use for Different Pans (Masoner et al., 2008)

Type of Pan	Ave. Value	Range
Class A land pan	0.70	0.60–0.80
ISI pan (modified Class A)	0.80	0.65–1.10
Colorado Sunken Pan	0.78	0.75–0.86
USGS floating pan	0.80	0.70–0.8

5.6 Evaporation Stations

It is usual to install the evaporation pans in such locations where other meteorological data are also simultaneously collected. The World Meteorological Organization recommends the minimum network of evaporimeter stations as below:

- Arid zones: One station for every 30,000 km^2,
- Humid temperate climates: one station for every 50,000 km^2, and
- Cold regions: One station for every 100,000 km^2.

The evaporation tank should be installed in the soil with the top of the tank being 76 mm above the surrounding. When siting the tank or pan, a care should be taken to avoid the errors that may result from:

- Out-splashing of water from the tank in high wind.
- In-splashing of rain from the surrounding ground. In-splashing can be kept to a minimum by growing grass on the tank grounds. The top of the pan should be left to be at least 0.5051 m above the tops of the blades of grass.
- Lack of proper exposure. Exposure should be like one for the rain gauge so that the amount of rain caught by the tank is accurately expressed as a represented by the amount caught by the rain gauge. Exposure must be such that it represents the surroundings. A choice of a sheltered position as for the rain gauge is necessary.
- A still-water pond of 0.102 m^2 and 0.305 m is provided inside the tank to avoid the ripples on the surface of the water during measurement. The still-water pond has a round hole in the center of the base.

5.7 Empirical Equation to Estimate Evaporation

The empirical equations to estimate evaporation are as follows:

5.7.1 Water Budget Method

The water budget equation for estimating evaporation (Horton, 1943) can be written as Eq. 5.2:

$$E = I + P - O - O_s + \Delta S, \tag{5.2}$$

where:

E is the evaporation,
I is the inflow,
P is the precipitation,
O is the outflow,
O_s is the seepage, and
ΔS is the change in storage.

Here, inflow, outflow, precipitation, and change in storage can be measured reasonably accurate. Seepage, O_s, cannot be measured or evaluated directly and accurately, and the extent to which this quantity is accurate, will affect the true value of evaporation.

The water budget method of determining long-term evaporation can be used as a standard for comparing the other methods. This method is not perfect, but it is satisfactory for the practical purposes.

5.7.2 Energy Budget Method

Energy budget method is based on the below equation:

$$R_n = E + A + S + C, \tag{5.3}$$

where:

R_n is the net radiation available energy from sun,
E is the energy used for evaporation,
A is the energy used in heating air,
S is the energy used in heating water, and
C is the energy used in heating the surrounding of water.

5.7.3 Mass Transfer Method

The mass transfer method is one of the oldest methods and is still an attractive method for estimating the free water surface evaporation because of its simplicity and reasonable accuracy. The mass transfer methods are based on the Dalton Equation, which for free water surface can be written as:

$$E_0 = C\,[e_s - e_a], \tag{5.4}$$

where:

E_0 is the free water surface evaporation,
e_s is the saturation vapour pressure at the temperature of the water surface,
e_a is the vapour pressure in the air = the saturation vapour pressure at dew point temperature,
C is the aerodynamic conductance = $1/r_a$, and
r_a is the aerodynamic resistance.

The parameter C can also be interpreted as the amount of water the evaporated from unit vapour pressure deficit. Although C depends on the horizontal wind speed, surface roughness and thermally induced turbulence, it is normally assumed to be dependent on wind speed, u.

Therefore, above equation may be expressed as:

$$E_0 = f(u)[e_s - e_a],$$ (5.5)

where:

$f(u)$ is the wind function.

This function depends on, among other factors, the observational heights of the wind speed and vapour pressure measurements.

5.7.4 Combination Method (Penman)

Evaporation can be computed by the aerodynamic method when energy supply is not a limiting factor and energy method when vapour transport is not a limiting factor. When both factors are limiting, a combination of above methods should be used.

Deviation of weighting factors is based on the physical processes:

D = vapour pressure deficit = $4098es/(237.3 + T)^2$
γ = psychometric constant = ~66.8 Pa/°C

Combination method is the most accurate and most commonly used method if meteorological information is available. Particularly, it is good for the small and well-monitored areas.

The requirements are net radiation, air temperature, humidity and wind speed.

5.8 Evaporation Equations

Evaporation equations include the following equations:

5.8.1 Meyer's Equation (1942)

$$E = K_m(e_w - e_a)(1 + (u_9/16)),$$ (5.6)

where:

E is the evaporation in mm/day,
e_w is the saturated vapour pressure at the water surface temperature,
e_a is the actual vapour pressure of over lying air at a specified height,
u_9 is the monthly mean wind velocity in km/hour at about 9 m above the ground, and
K_m is the coefficient, 0.36 for large deep waters and 0.50 for small shallow waters.

5.8.2 Lake Hefner Equation

This equation (Marciano and Harbeck, 1954) was developed for a water-supply reservoir (Lake Hefner) located near Oklahoma City, USA:

$$E = 0.00177u(e_0 - e_a),\qquad(5.7)$$

where:

> E is the evaporation rate (in/day),
> e_0 is the saturation vapour pressure (mb) at the water temperature,
> e_a is the vapour pressure (mb) of the air, and
> u is the mean daily wind speed (mph).

5.8.3 Rohwer Equation

Rohwer (1931) provided an equation:

$$E = 0.771(1.465 - 0.0186P)(0.44 + 0.118u)(e_s - e_a),\qquad(5.8)$$

where:

> E is the evaporation rate (in/day),
> e_s is the saturation vapour pressure (mb) at the water temperature,
> e_a is the vapour pressure (mb) of the air,
> u is the mean daily wind speed (mph), and
> P is the air pressure in Pa.

5.8.4 Ryan and Harleman Equation

This equation was developed to estimate the evaporation from heated water bodies. In that case, both forced (wind driven) convection and free (buoyancy driven) convection effectively control the evaporation rates, while for natural water bodies, the forced convection is the dominant factor. The evaporative heat flux can be converted into the equivalent evaporation (mm/day).

$$Q_e = (2.7\Delta Q_v^{1/3} + 3.1u)(e_s - e_z),\qquad(5.9)$$

where:

> Q_e is the evaporative heat flux (W/m²),
> ΔQ_v is the virtual temperature difference (°C),
> u is the wind speed measured at 2 m (m/s), and
> e_s and e_z are the saturation vapour pressure at the water surface and vapour pressure of the air (mb), respectively.

5.8.5 Morton Equation

Morton formula is as follows:

$$E = (300 + 50u)(e_s - e_a),$$

where:

E is the evaporation in inches/month,

u is the wind speed in mph, and

e_s and e_a are the saturation vapour pressure at the water surface and vapour pressure of the air (mb), respectively.

5.8.6 Zeykov Equation

Zeykov equation is as follows:

$$E = (0.15 + 0.1u)(e_s - e_a), \tag{5.10}$$

where:

E is the evaporation in inches/month,

u is the wind speed in mph, and

e_s and e_a are the saturation vapour pressure at the water surface and vapour pressure of the air (mb), respectively.

5.8.7 Kohler Equation

Kohler equation is given as follows:

$$E = 0.00304u(e_s - e_a), \tag{5.11}$$

where:

E is the evaporation in inches/month,

u is the wind speed in mph, and

e_s and e_a are the saturation vapour pressure at the water surface and vapour pressure of the air (mb), respectively.

5.8.8 Penman Equation

Penman equation is described as follows:

$$E_{\text{penOW}} = \frac{\Delta}{\Delta + \gamma} \frac{R_{nw}}{\lambda} + \frac{\gamma}{\Delta + \gamma} E_a \tag{5.12}$$

where:

E_{penOW} is the daily potential evaporation (mm/day = kg/m²/day) from a saturated surface,

R_n is the net daily radiation to the evaporating surface (MJ/m²/day),

E_r (mm/day) is a function of the average daily wind speed (m/s), saturation vapour pressure (kPa) and average vapour pressure (kPa),

Δ is the slope of the vapour pressure curve (kPa/°C) at air temperature,

γ is the psychrometric constant (kPa/°C),

λ is the latent heat of vaporization (MJ/kg), and

E_a is the aerodynamic component in the Penman equation.

5.8.9 Priestley–Taylor Equation

The Priestley–Taylor equation (Priestley and Taylor, 1972) allows the potential evaporation to be computed in terms of the energy fluxes without an aerodynamic component as follows:

$$E_{PT} = \alpha_{PT}\left[\frac{\Delta}{\Delta+\gamma}\frac{R_n}{\lambda}-\frac{G}{\lambda}\right], \tag{5.13}$$

where:

E_{PT} is the Priestley–Taylor potential evaporation (mm/day=kg/m²/day),
R_n is the net daily radiation at the evaporating surface (MJ/m²/day),
G is the soil flux into the ground (MJ/m²/day),
Δ is the slope of the vapour pressure curve (kPa/°C) at air temperature,
γ is the psychrometric constant (kPa/°C),
λ is the latent heat of vaporization (MJ/kg), and
α_{PT} is the Priestley–Taylor constant=1.26.

5.8.10 Aerodynamic Method

It is based on the Dalton-type approach in which evaporation is the product of a wind function and the vapour pressure deficit between the evaporating surface and the overlying atmosphere.

McJannet et al. (2012) proposed the following empirical relationship to estimate the open-surface water evaporation.

$$EL_{area} = (2.36+1.67u_2)A^{-0.05}(V_s-V_a), \tag{5.14}$$

where:

EL_{area} is an estimate of open-surface water evaporation (mm/day), which is as a function of evaporating area, A (m²),
u_2 is the wind speed (m/s) over land at 2 m height,
V_s is the saturated vapour pressure (kPa) at the water surface, and
V_a is the vapour pressure (kPa) at air temperature.

5.6.11 The PenPan Model for Evaporation

It was developed by Linacre (1994) for the open surface water evaporation:

$$E_{penpan} = \frac{\Delta}{\Delta+a_p\gamma}\frac{R_{Npan}}{\lambda}+\frac{a_p\gamma}{\Delta+a_p\gamma}f_{pan}(u)(e_s-e_a), \tag{5.15}$$

where:

E_{penpan} is the modelled Class A (unscreened) pan evaporation (mm/day=kg/m²/day),
R_{Npan} is the net daily radiation at the pan (MJ/m²/day),
Δ is the slope of the vapour pressure curve (kPa/°C) at air temperature,
γ is the Psychrometric constant (kPa/°C),

λ is the latent heat of vaporization (MJ/kg),
a_p is a constant $= 2.4$,
$e_s - e_a$ is the vapour pressure deficit (kPa),
$f_{pan}(u)$ is defined as $f_{pan}(u) = 1.202 + 1.62u_2$, and
u_2 is the average daily wind speed at 2 m height (m/s).

5.8.12 Cummings and Richardson Equation for Evaporation

This formula is independent of the wind but contains the terms that are affected by the altitude.

$$E = \frac{(H - S - C)}{L(1 + R)}, \tag{5.16}$$

where:

E is the evaporation,
H is the net radiation,
S is the heat stored in a column of water of unit cross section,
C is a correction for interchange of heat through the walls surrounding water,
L is the latent heat of water, and
R is the Bowen's ratio.

5.8.13 Horton's Equation (1917)

This formula is as follows:

$$E = C(\Psi e_s - e_d), \tag{5.17}$$

where:

C is a coefficient, and
ψ is a factor that takes care of the wind.

5.8.14 Bigelow Equation

The most complete series of the evaporation experiments ever made was conducted by Bigelow (1907–1910) for the United States Weather Bureau.

Bigelow made his observations at Reno, Nev. and at the Saltón Sea in California. He collected a great mass of evaporation data and developed the formula:

$$E = 0.138 \frac{e_s}{e_d} \frac{d_e}{d_s} (1 + 0.07W), \tag{5.18}$$

where:

E is the evaporation (cm/day),
e_s and e_d are the vapour pressures (mm),
W is the wind velocity (km/hour), and
d_e/d_s is the rate of change in the maximum vapour pressure with temperature.

5.8.15 Stelling Equation

Based on the pan experiments in Russia (1875–1882), is in metric units, this formula gives the results that are too high.

$$E = (0.8424 + 0.01056W)(e_s - e_a),$$

where:

  E is the evaporation,
  $e_s - e_a$ is the difference in vapour pressure (mm), and
  W is the wind velocity (m/s).

5.8.16 Fitzgerald Equation

He worked in Boston (1876–1887) and made a very careful and the complete series of observations on evaporation both under controlled conditions in the laboratory and under natural conditions outside. He proposed the formula:

$$E = (0.40 + 0.199W)(e_s - e_a). \tag{5.19}$$

5.8.17 Carpenter Equation

He carried on the experiments in 1887 on a sunken tank 3 feet square at the Colorado Agricultural College to determine the constants for the Fitzgerald formula applicable to the western conditions. From these experiments, he derived the formula which does not differ materially from Fitzgerald's formula.

$$E = (0.39 + 0.187W)(e_s - e_a) \tag{5.20}$$

5.8.18 Russell Equation

From observations on the evaporation from Piche evaporimeter at 18 Weather Bureau stations distributed over the United States, Russell (1888) derived the evaporation formula:

$$E = \frac{(1.96e_w + 43.88)(e_w - e_d)}{B}, \tag{5.21}$$

where:

  e_w is the vapour pressure, mercury inch, corresponding to the mean wet-bulb temperature;
  B is the mean barometer reading, in inches of mercury at 32°F; and
  e_s and e_d = vapour pressures are in inches of mercury.

5.8.19 Ritchie Equation

To evaluate the net radiation at the soil surface, Ritchie assumed that it decreases with an increase in leaf area index as shown below. This relationship was obtained from experimental data gathered for grain sorghum, cotton, corn, soybeans and snap beans. By using the experimental data, Ritchie proposed the following relationship for the potential soil evaporation.

$$E_p = \left(\frac{\Delta}{\Delta + \gamma c}\right) Rn \exp(-0.398 LAI) \tag{5.22}$$

5.9 Computation of Transpiration

5.9.1 Transpiration Definition

Transpiration is the process by which water vapour leaves the living plant body and enters the atmosphere.

5.9.2 Weighing Method

A small lightweight potted plant can be weighed before and after the end of a certain period of time. The soil surface and the pot should be fully covered to prevent the evaporation from the surfaces other than the plant. The loss in weight by the plant during that time is due to the loss of water by transpiration.

An improvement in the weighing method can be made by using a glass bottle with a graduated side tube, filled with water and a tube fitted into it as shown in Figure 5.6. This would indicate the volume of water loss that can be compared with the loss in weight with the help of a weighing machine (*B*) or by converting cc into grams (1 cc water weighs 1 g).

Another weighing experiment can be made by using a test tube filled with water and inserting a leafy shoot (no roots) in it and pouring some oil on the surface to prevent loss of water by evaporation. The test tube could be rested in a small beaker and weighted them together. The intact test tube should be removed and kept it straight in the test tube stand for a few hours. It should be weighted again by keeping it in the same beaker. Any difference in weight will indicate loss of water by the shoot (due to transpiration).

5.9.3 Potometer Method

Potometer (potos: drink, meter: measure) means a device that measures the water taken in by a plant.
Apparatus required: Capillary tube, tap, scale, sharp razor, sample plant, container.

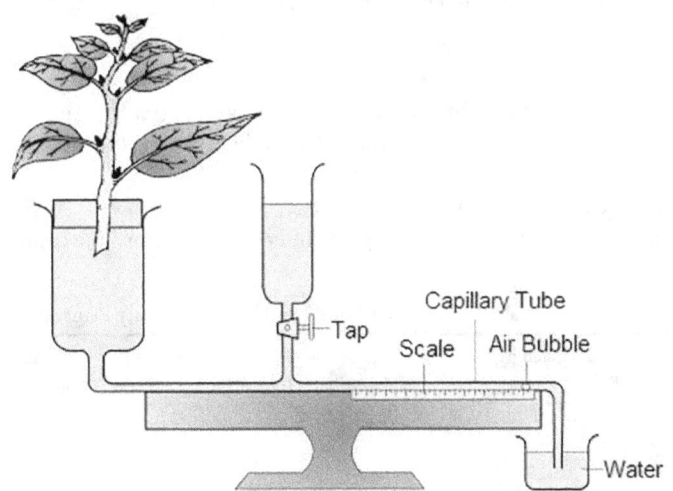

FIGURE 5.6 Potometer method.

Principle: A potometer is a device used for measuring the rate of water uptake of a leafy plant shoot. The main reason for water uptake by a cut shoot is transpiration (evaporation in plants) and is affected by the transpiration stream.

5.9.3.1 Potometer Procedure

1. Use a sharp razor blade to cut a leafy shoot under water.
2. Insert the leafy shoot through the hole of the stopper provided with the potometer.
3. Fill the potometer with water and fit the stopper holding the leafy shoot to the apparatus.
4. Use vaseline to seal all of the connections of the apparatus.
5. Trap an air bubble in the capillary tube by the following procedure:

   - dip the end of the capillary tube into a beaker of water,
   - close the tap of the reservoir,
   - take away the beaker of water and allow the plant to transpire for a while, and
   - re-immerse the capillary tube into the beaker of water again.

6. Wait for about 5 minutes for the plant to equilibrate.
7. Estimate the rate of transpiration by measuring the distance moved by the air bubble per unit time. Take another measurement and average the two readings.
8. Find out the total leaf surface area of the experimental plant using a graph paper.
9. Transpiration rate can be expressed in terms of water transpired per unit time per unit area of leaf surface.
10. If time permits, estimate the rate under different environmental conditions, e.g. under direct incandescent light illumination, under low light condition, in rapid air movement (provided by a fan) or in still air.

The following observations could be done, and their values are tabulated in Table 5.2.

- Distance moved by air bubble (d), cm
- Cross-sectional area of capillary tube (A), cm^2
- Volume of water transpired (V), m^3.

Calculation:

$$\text{Volume}(V) = \text{Area}(A)\,\text{Distance}(d)$$

Application:

- Set up the conditions of the experiment. Alterations to lighting (placing the plant in bright light or shadow), wind (directing a fan at the plant), and humidity (placing the plant in a humid chamber) are typical.
- Allow the bubble reach a "zero" point in the tube.
- Measure the movement of the bubble at regular intervals and record the results.

TABLE 5.2   Observations for Calculating the Rate of Water Uptake of a Leafy Plant Shoot

S. No.	Distance Moved by Air Bubble d (cm)	Cross-Sectional Area of Capillary Tube A (cm²)	Volume of Water Transpired $V = A \times d$ (cm³)
(1)	(2)	(3)	(4) = (2) × (3)
1			
2			
3			

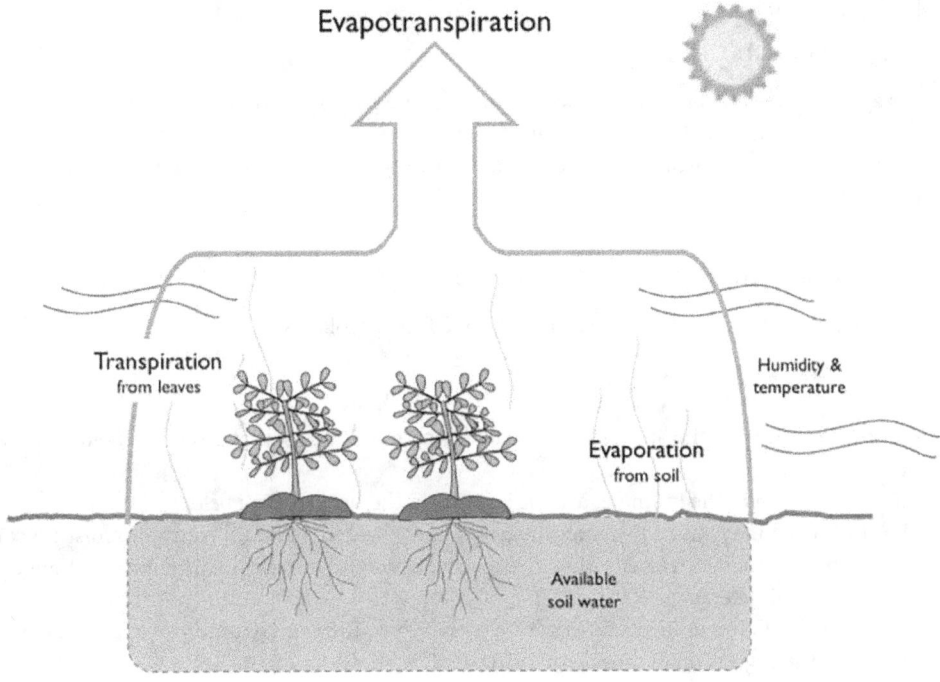

FIGURE 5.7   Phytometer.

5.9.3.2 Ganong's Potometer

A twig of some suitable plant (e.g. coleus) cut with a sharp knife is fixed in an apparatus as shown in Figure 5.7. The entire apparatus is filled with water so that no air spaces are present. An air bubble is introduced into the horizontal graduated capillary tube which is dipping into the beaker containing water. (This is done by lifting the bent capillary tube above the coloured water so that air will be sucked in due to suction pull and is again dipped into the water.) As the transpiration proceeds, i.e. as the water is lost from the twig, a suction force is set up which pulls the water from the beaker and the bubble in the capillary tube moves along. The readings on the capillary tube would give the volume of water lost in a given time. The air bubble can be brought back to its original position by releasing some more water into the capillary tube by opening the stopcock.

5.9.3.3 Darwin's Potometer

Darwin is the simplest type of potometer used to measure the rate of transpiration. It consists of a glass tube with a side tube. A twig (fresh) is cut obliquely under water and is inserted in the side tube through a single hole cork. The upper end of the straight tube is also corked.

At the lower end a single tube cork with a capillary tube is fitted. A scale remains fitted on the capillary tube, whose lower part is dipped in a beaker containing water. An air bubble is introduced in the capillary tube through which water is absorbed the rate of movement of bubble over a fixed distance is noted. The comparison of bubble movement under different conditions gives the rate of transpiration.

Potometer does not measure the water lost during transpiration but measures the water uptake by the cut shoot (L. potos: drink). Some of the water is used by the cells to carry out other processes, e.g. manufacture of food (photosynthesis). The potometer should be made completely watertight and the twig should be cut obliquely (to allow larger surface for the water intake) and under water to avoid suction of an air bubble into the twig which will stop the absorption of water into the xylem.

5.9.3.4 Phytometer

Phytometer is explained as follows:

- It is a large vessel filled with soil in which one or more plants are rooted.
- The soil surface is sealed to prevent evaporation.
- The only escape of moisture is by transpiration which can be determined by weighing the plant and container at desired intervals.

5.9.4 Measurement of Evapotranspiration

The principal methods for the direct measurement of ET are as follows:

5.9.4.1 Lysimeter Experiments

- A lysimeter is a special watertight tank containing a block of soil and set in a field of growing plants.
- The plants grown in the lysimeter are the same as in the surrounding field.
- ET is estimated in terms of the amount of water required to maintain constant moisture conditions within the tank measured either volumetrically or gravimetrically through an arrangement made in the lysimeter.
- Lysimeters should be designed to accurately reproduce the soil conditions, water content, type and size of the vegetation of the surrounding area, as shown in Figure 5.8.
- A lysimeter is a measuring device which can be used to measure the amount of actual ET which is released by plants, usually crops or trees.
- By recording the amount of precipitation that an area receives, and the amount lost through the soil, the amount of water lost to ET can be calculated.
- In general, a lysimeter consists of the soil-filled inner container and retaining walls or an outer container, as well as special devices for measuring percolation and changes in the soil moisture content.
- There is no universal international standard lysimeter for measuring ET.
- The surface area of lysimeters in use varies from 0.05 to 100 m^2 and their depth varies from 0.1 to 5 m.
- Monolithic weighable lysimeters are a tool for water balance studies and solute transport determination.
- Lysimeter studies are time-consuming and expensive.
- Lysimeters are of two types: (1) weighing and (2) non-weighing.

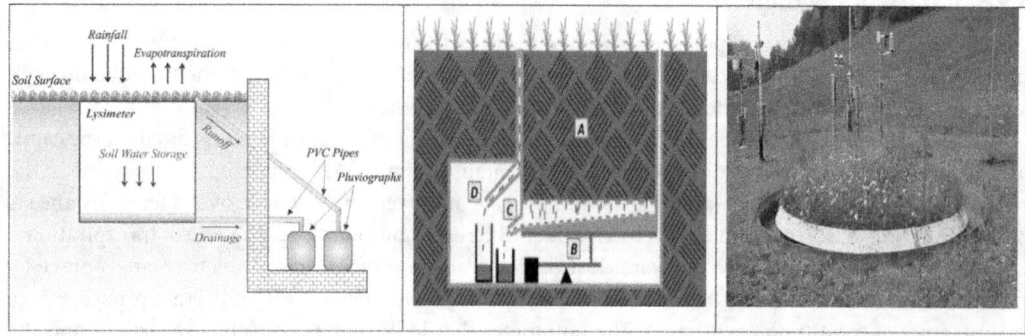

FIGURE 5.8 Lysimeters mechanism.

5.9.4.2 Field Experimental Plots

Measurement of water supplies to the field and changes in soil moisture content of the field plots are sometime more dependable for computing seasonal water requirement of crops than measurement with lysimeters which do not simulate field conditions.

The seasonal water requirements are computed by adding measured quantities of irrigation water, the effective rainfall received during the season and the contribution of moisture from the soil.

Field water balance may be expressed by the following relationship:

$$\text{WR} = \text{IR} + \text{ER} + \sum_{i=1}^{n} \frac{M_{bi} - M_{ei}}{100} \cdot A_i \cdot D_i, \tag{5.23}$$

where:

WR is the seasonal water requirement, mm;
IR is the total irrigation water applied, mm;
ER is the seasonal effective rainfall, mm;
M_{bi} is the moisture percentage at beginning of season in the ith layer of soil;
M_{ei} is the moisture percentage at end of season in the ith layer of soil;
A_i is the apparent specific gravity of the ith layer of the soil;
D_i is the depth of the ith layer of the soil within the root zone, mm; and
n is the number of soil layers in the root zone D.

The method requires that the amount of water applied to a field is measured accurately. This method, though satisfactory for computing seasonal water requirements, does not provide the information on intermediate soil moisture conditions, short-term use, profile use, deep percolation losses and peak use rate of the crop.

5.9.4.3 Soil Moisture Depletion Studies

The soil moisture depletion method is usually employed to determine the consumptive use of irrigated field crops grown on uniform soils when the depth to the ground water is such that it will not influence the soil moisture fluctuation within the root zone. These studies involve measurement of soil moisture from various depths at a number of times throughout the growth period. The greater the number of measurements, the more the information obtained from such studies. Consumptive use (Cu) is calculated from the change in soil water content in successive samples from the following relationship:

$$u = \sum_{i=1}^{n} \frac{M_{1i} - M_{2i}}{100} \cdot A_i \cdot D_i, \tag{5.24}$$

where:

u is the water used from root zone between sampling, mm;
M_{1i} is the moisture percentage at first sampling in the ith layer of the soil;
M_{2i} is the moisture percentage at second sampling in the ith layer of the soil;
A_i is the apparent specific gravity of the ith layer of the soil;
D_i is the depth of the ith layer of the soil within the root zone, mm; and
n is the number of soil layers in the root zone.

Seasonal consumptive use (Cu = Σu) is calculated by summing the consumptive use values of each sampling interval. A correction is made by adding the PET values for accelerated water loss for the intervals just after irrigations and before soil moisture sampling.

5.9.4.4 Water Balance Method

The water balance method, also called the inflow-outflow method, is suitable for large areas (watersheds) over long periods. It may be represented by the following hydrological equation:

$$P = RO + S + DL + CWU.$$

Total rainfall = surface runoff + soil moisture storage + deep percolation losses + crop water use.

This method necessitates adequate measurement of all factors, except ET; the value of ET is computed from the measured data.

5.9.4.4.1 Water Balance Study for Maize Crop

As an example, the data for maize crop are as follows:

Irrigation water applied $(I) = 75$ mm
Total rainfall $(P) = 916.0$ mm
Total surface runoff $(RO) = 254.6$ mm
Soil moisture storage $(S) = 36.8$ mm
Deep percolation losses $(DL) = 276.0$ mm
Crop water use $(CWU) = 423.6$ mm

$$P + I = RO + S + DL + CWU \tag{5.25}$$

The crop field is shown in Figure 5.9.

5.9.4.4.2 Water Balance Study for Wheat (Lok-1)

As the second example, the data for wheat crop are as follows:

Irrigation water applied $(I) = 245$ mm
Total rainfall $(P) = 116.9$ mm

FIGURE 5.9 Water balance study for maize crop.

FIGURE 5.10 Water balance study for wheat.

Total surface runoff (RO)=0.0 mm
Soil moisture storage (S)=26.5 mm
Deep percolation losses (DL)=0.0 mm
Crop water use (CWU)=335.3 mm

$$P + I = RO + S + DL + CWU \qquad (5.26)$$

The field is shown in Figure 5.10.

5.9.5 Estimation of Evapotranspiration

5.9.5.1 Blaney–Criddle Method

This method requires the use of only two factors, temperature and information of daylight hours which is a factor based purely on the latitude of the place. Using Blaney–Criddle approach (Sammis et al. 1982) PET can be expressed as follows.

In metric unit:

$$PET = 0.46\,p(T + 17.8)$$
$$= p(0.46T + 8.188) \qquad (5.27)$$
$$= p(0.46T + 8),$$

where:

PET is the potential ET (mm/month) (mean value of month),
p is the monthly percent of total day time hours of the year=monthly day time (hour)×100/total annual day time (hour), and
T is the mean monthly temperature (°C) (Avg. of daily max. and min. values).

In English units:

$$PET = T_p / 100, \qquad (5.28)$$

where:

PET is the monthly potential ET (inches/month) (mean value over the month),
p is the mean daily percentage of maximum possible annual daylight hours=monthly day time (hour)×100/total annual day time (hour), and
T is the mean monthly temperature (°F).

Conversion of PET to actual ET of consumptive use can be given by

$$U = KF = \sum kf = \sum u,$$ (5.29)

where:

U is the seasonal consumptive use (mm),
u is the monthly consumptive use (mm),
K is the empirical seasonal consumptive use coefficient for the growing season,
F is the sum of the monthly consumptive use factor (f) for the growing season,
k is the empirical consumptive use crop coefficient for the month (u/f), and
f is the value of monthly PET (mm).

Blaney–Criddle method assumes that the CU of water is dependent only on temperature and day length which is not true.

Doorenbos and Pruitt (1977) did not agree to with the use of crop coefficient (K_c), developed by Blaney–Criddle, because:

- The original crop coefficient (K_c) is dependent on local conditions.
- The relationship between Blaney–Criddle f-values and PET can be adequately described for a wide range of temperatures for areas having minor variation in RH, sunshine and wind velocity.
- Once PET has been determined by any standard method one set of crop factors (K_c) can be used to determine crop ET.

Example 5.1: Calculation of the Consumptive Use of Water for Sugarcane Crop at Delhi (28°N) by Using Blaney–Criddle Formulae

The solving steps are summarized in the below table:

Month	Mean Monthly Temperature (T) (°C)	Monthly Crop Coefficient (k)	% Daylight Hours (p)	Monthly Consumptive Use $u = k_p(0.46T + 8.188)$ (mm/month)
January	13.17	0.75	7.38	581.93
February	15.78	0.80	7.02	608.98
March	20.72	0.85	8.39	1,060.20
April	27.00	0.85	8.69	1,322.80
May	31.11	0.90	9.48	1,819.76
June	33.50	0.95	9.41	1,985.08
July	30.61	1.00	9.60	2,052.27
August	29.00	1.00	8.18	1,440.49
September	28.22	0.95	8.33	1,395.46
October	24.78	0.90	8.01	1,131.02
November	13.72	0.85	7.25	647.80
December	13.17	0.75	7.24	560.06
				$U = \sum u = 14{,}605.86$

5.9.5.2 Thornthwaite Method

Thornthwaite method assumes of an exponential relationship between the mean monthly temperature and mean monthly consumptive use.

$$e = 1.6 \left(10t/I\right)^{\alpha}, \tag{5.30}$$

where:

e is the unadjusted PET (cm per month),
t is the mean air temperature (°C), and
I is the annual or seasonal heat index, the summation of 12 values of monthly heat indices (i).

when $i = (t/5)1.514$

$$\alpha = 0.0000006751I^{3} - 0.0000771I^{2} + 0.01792I + 0.49239$$

The drawbacks suggested by Chang (1968) are as follows:

- Temperature alone is not a good indication for the energy available for ET.
- Air temperature of a place lags radiation.
- According to this formulae ET will cease when mean temperature is below 0°C which by no mean is true, although the amount of evaporation will be very small.
- The formulae do not consider the wind effect which might be an important factor in some areas.
- It does not consider the effect of warm and cool air on the temperature of a place.

Example 5.2: Calculation of Potential Evapotranspiration for Delhi (28° 4′ N) by Using Thornthwaite Formulae

The solving steps are summarized in the below table:

Month	Mean Monthly Temperature (T) (°C)	Monthly Heat Index (i)	Unadjusted PET (cm/month) $e=1.6 (10t/I)\alpha$	Correction Factor for "e"	Corrected Values of PET (cm/month)
January	12.6	4.05	1.36	0.91	1.24
February	15.8	5.70	2.72	0.87	2.36
March	20.7	8.59	6.39	1.03	6.58
April	27.0	12.85	14.27	1.07	15.26
May	31.1	16.04	22.04	1.17	25.78
June	33.5	17.73	27.64	1.16	32.06
July	30.6	15.53	20.91	1.19	24.88
August	29.0	14.32	17.74	1.13	20.04
September	28.2	13.65	16.30	1.03	16.78
October	24.8	11.30	10.19	0.98	9.98
November	18.9	7.49	4.72	0.90	4.24
December	13.7	4.60	1.75	0.89	1.55
		Total=131.85 (=132.00=I)			

$$\alpha = 0.0000006751I^{3} - 0.0000771I^{2} + 0.01792I + 0.49239$$

$$\alpha = 0.0000006751(132)^{3} - 0.0000771(132)^{2} + 0.01792(132) + 0.49239$$

$$\alpha = 3.07$$

5.9.5.3 Hargreaves' Method

Hargreaves, based on his work on data from grass lysimeter, proposed the following relationship to estimate ET:

$$PET = 0.0135(t + 17.78)R_s, \tag{5.31}$$

where:

PET is the reference crop potential CU, well-watered grass, langleys/day;
t is the mean daily temperature (°C); and
R_s is the incident solar radiation in langleys/day=0.10 $R_{\text{reference}}$ grass $(S)1/2$.

Conversion:

mm/day=langleys/day $\times$ 10/latent heat of water (λ)

5.9.5.4 Hargreaves' Radiation Formula

Hargreave's radiation formula, adjusted and tested at a large number of weather stations under varying climate conditions, is expressed as follows (FAO, 1998):

$$R_s = R_a k_{rs} \sqrt{(T_{\max} - T_{\min})}, \tag{5.32}$$

where:

R_a is the extra-terrestrial radiation (MJ/m²/day),
$T_{\max}$ is the maximum air temperature (°C),
$T_{\min}$ is the minimum air temperature (°C),
k_{rs} is the adjustment coefficient (t),
k_{rs}=0.16 for the interior regions where air masses not influenced by water body, and
k_{rs}=0.19 for the coastal regions where air masses influenced by a water body.

$$ET_0 = 0.0023(T_{\text{mean}} + 17.8)(T_{\max} - T_{\min})0.5R_a$$

5.9.5.5 Radiation Method (Makkink, 1957)

It is recommended for areas where the weather data are not sufficient to use the Penman formula. The essential climate data include air temperature and sunshine or cloudiness or radiation. Only general levels of wind and humidity, which could be estimated from local weather stations, are adequate. ET_0 is estimated by (FAO, 1998):

$$ET_0 = c(W - R_s)(\text{mm} / \text{day}), \tag{5.33}$$

where:

ET_0 is the reference crop evaporation in mm/day;
R_s is the solar radiation at ground level, equivalent evaporation in mm/day;
W is the weighing factor, which is a function of the temperature and altitude; and
c is the adjustment factor, depends on mean humidity and daytime wind conditions.

The radiation method is considered superior to Blaney–Criddle method, particularly in equatorial regions, on small islands and at high altitudes, even if recorded data on sunshine or cloudiness are not available. The radiation method has proved valuable, particularly in humid region.

5.9.5.6 Penman Formula

Penman (1948) proposed an equation for evaporation from open surface water, based on a combination of energy balance and sink strength which is given below with changes in certain symbols in view of the recent trends:

$$E_0 = \frac{\Delta Q_n + \gamma E_a}{\Delta + \gamma}, \tag{5.34}$$

$$Q_n = Q_A(1-r)\left(0.18 + 0.55\frac{n}{N}\right) - \sigma T_a^4\left(0.55 - 0.092\sqrt{e_d}\right)\left(0.10 + 0.90\frac{n}{N}\right), \tag{5.35}$$

where:

E_0 is the evaporation from open surface water (mm/day),
Δ is the slope of saturation vapour pressure vs temperature curve (de_a/dT) at the mean air temperature T_a mm Hg per °C,
Q_n is the net radiation (mm of water),
e_a is the saturation vapour pressure of evaporating surface (e_s) in mm Hg at T_a,
T_a is the mean air temperature in °K=273+°C,
r is the reflection coefficient of evaporating surface, 0.06 for open surface water,
Q_A is the Angot's value of mean monthly extra-terrestrial (mm of water/day),
n/N is the ratio between actual and possible hours of bright sunshine,
σ is the Stefan–Boltzmann constant,
e_d is saturation vapour pressure of atmosphere, in mm Hg,=(RH$_{mean}$/100) e_a,
γ is the psychromatic constant (0.49 for 0°C and mm Hg),
E_a is an aerodynamic component=0.35 $(e_a - e_d)$ $(1 + 0.0098 \, u_2)$,
u_2 is wind speed in miles/day at 2 m height=u_1 (log 6.6/log h), and
u_1 is the wind speed in miles per day at any other height h in feet.

Example 5.3: Calculation of Evaporation from Open Surface Water (E_0) in mm/day for Delhi (28° 4′ N), Using Penman's Formula for the Period of August 1–10, 2015

• Mean air temperature, T_a	28.7°C
• Mean relative humidity	88.5%
• Mean sunshine hours, n	3.43
• Possible sunshine hours, N	13.1
• Value of ratio n/N	0.26
• Crop coefficient, k	0.80
• Wind speed at 3 m height, u_1	90.02 miles/hour
• Wind speed at 2 m height, u_2=(log 6.6/log 3)	83.7 miles/hour
• Extra-terrestrial radiation, Q_A	15.7 mm/day
• Reflection coefficient, r	0.06
• σT_a^4	16.72
• Δ	1.65

- Y 0.49PET
- Vapour pressure, $e_a =$ 29.2
- $e_d =$ 25.5
- $\sqrt{e_d}$ 5.04

$$Q_n = Q_A(1-r)\left(0.18+0.55\frac{n}{N}\right) - \sigma T_a^4\left(0.55-0.092\sqrt{e_d}\right)\left(0.10+0.90\frac{n}{N}\right)$$

$$Q_n = 15.7(1-0.06)(0.18+0.55(0.26)) - 16.72\left(0.55-0.092(5.04)\right)(0.10+0.90(0.26))$$

$$Q_n = 4.72 - 0.38 = 4.34$$

- $E_a = 0.35\ (e_a - e_d)(1+0.0098\ u2) = 0.35\ (29.2-25.5)\ (1+0.0098(83.7)) = 2.37$

$$E_0 = \frac{\Delta Q_n + \gamma E_a}{\Delta + \gamma} = \frac{1.65 \times 4.34 + 0.49 \times 2.37}{1.65 + 0.49} = 3.89\,\text{mm/day}$$

- $\text{PET} = k \times E_0 = 0.80 \times 3.89 = 3.11$ mm/day

5.9.5.7 Modified Penman Formula

The original Penman equation (1948) predicts the evaporation losses from an open surface water (E_0). The Penman equation consists of two terms: the radiation term and aerodynamic (wind and humidity) term.

Doorenbos (1975) and Doorenbos and Pruitt (1977) proposed modified Penman methods for estimating accurately the reference crop ET and gave the tables and graphs to facilitate the necessary computations. The modified Penman equation involves a revised wind function term. The methods use mean daily climate data. Since day and nighttime weather conditions affect the level of ET, and adjustment for this is included.

The formula contains components which are to be derived from measured climate data; when no direct measurement of net radiation is available, this can be obtain from measured solar radiation, sunshine duration or clouding along with measured humidity and temperature.

The modified Penman equation is stated as follows (FAO 24, 1977):

$$\text{ET}_0 = c\left[W \cdot R_n + (I-W) \cdot f(u) \cdot (e_a - e_d)\right], \tag{5.36}$$

where:

ET_0 is the reference crop evapotranspiration (mm/day),
e_a is the saturation vapour pressure in mbar at the mean air temp in °C,
e_d is the mean actual vapour pressure of the air in mbar $= (\text{RH}_{\text{mean}}/100)\ e_a$,
$f(u)$ is a wind related function $= 0.27(1+u_2/100)$,
u_2 is the wind speed in miles/day at 2 m height $= u_1\ (\log 6.6/\log h)$,
u_1 is the wind speed in miles per day at any other height h in feet,
$(1-W)$ is a temperature and elevation related weighing factor for the effect of wind and humidity on ET_0,

W is the temperature and elevation related weighing factor for effect of radiation on ET_0,
R_n is the net radiation (same as $Q_n = R_{ns} - R_{nl}$),
R_{ns} is the net incoming shortwave solar radiation $= RA(1-\alpha)\ (0.25 + 0.50\ n/N)$,
RA $=$ QA $=$ extra-terrestrial radiation, equivalent evaporation in mm/day,
$n/N =$ ratio between actual and possible hours of bright sunshine,
$\alpha = r =$ reflection coefficient $= 0.25$ for most crops,
$R_{nl} =$ the net long wave radiation $= f(t)\ f(e_d)\ f(n/N)$, and
$c =$ adjustment factor to compensate for the effect of day and night weather condition (approximate values) for different areas.

Example 5.4: Calculation of Evaporation from Open Surface Water (E_0) in mm/day for Delhi (28° 4′ N), Using Penman's Formula for the Period of August 1–10, 2015

- Mean air temperature, T_a 28.7°C
- Mean relative humidity   88.5%
- Mean sunshine hours, n 3.43
- Possible sunshine hours, N 13.1
- Value of ratio n/N 0.26
- Adjustment factor, c 1.09
- Wind speed at 3 m height, u_1 145.2 km/hour
- Wind speed at 2 m height, $u_2 = (\log 6.6/\log 3)$ 135.0 km/hour
- Extra-terrestrial radiation, $Q_A = R_a$ 15.7 mm/day
- Reflection coefficient, $\alpha = r$ 0.25
- Weighing factor, W 0.78
- Weighing factor, $(1 - W)$ 0.22
- $f(u)$ 0.635
- $f(t)$ 16.5
- $f(e_d)$ 0.08
- $f(n/N)$ 0.34
- Vapour pressure, $e_a =$ 39.4 mbar
- $ed = (RH_{mean}/100)\ e_a = 0.885 \times 39.4$ 34.9 mbar
- $e_a - e_d = 39.4 - 34.9$ 4.5 mbar
- $R_{ns} = RA\ (1-\alpha)\ (0.25 + 0.50\ n/N) = 15.7\ (1 - 0.25)\ (0.25 + 0.50(0.26)) = 4.46$
- $R_{nl} = f(t)\ f(e_d)\ f(n/N) = 16.5 \times 0.08 \times 0.34 = 0.45$
- $R_n = R_{ns} - R_{nl} = 4.46 - 0.45 = 4.01$
- ET_0 (unadjusted) $= [W\ R_n + (I - W)\ f(u)\ (e_a - e_d)] = (0.78)\ (4.01) + (0.22)\ (0.635)\ (4.5) = 3.76$ mm/day
- ET_0 (adjusted) $=$ adjustment factor $(c) \times$ ET0 (unadjusted) $= 1.09 \times 3.76 = 4.10$ mm/day

---

5.9.5.8 FAO Penman–Montieth Formula

The combination formula recommended by the Expert Consultation on FAO Methodologies for Crop Water Requirements (UNFAO, 1990) derived by combining the equation of aerodynamic resistance and surface resistance is stated below:

$$
\mathrm{ET_0} = \frac{0.408\Delta(R_n - G) + \gamma \dfrac{900}{T+273} u_2(e_s - e_a)}{\Delta + \gamma(1 + 0.34u_2)}, \tag{5.37}
$$

where:

$\mathrm{ET_0}$ is the reference evapotranspiration (mm/day),
R_n is the net radiation at the crop surface (MJ/m²/day),
G is the soil heat flux density (MJ/m²/day),
T is the mean daily air temperature at 2 m height (°C), and
u_2 is the wind speed at 2 m height (m/s).

$$
u_2 = u_h \frac{4.87}{\ln(67.8h - 5.42)},
$$

u_h is the measured wind speed at h m above ground surface (m/s),
h is the height of measurement above ground surface (m), and
e_s is the saturation vapour pressure (kPa).

$$
e_s = \frac{e^0(T_{max}) + e^0(T_{min})}{2}
$$

$$
e^0(T) = 0.6108 \exp\left[\frac{17.27T}{T + 237.3}\right],
$$

e_a is the actual vapour pressure (kPa).

1. For $\mathrm{RH_{max}}$ and $\mathrm{RH_{min}}$,

$$
e_a = \frac{e^0(T_{min})\dfrac{\mathrm{RH_{max}}}{100} + e^0(T_{max})\dfrac{\mathrm{RH_{min}}}{100}}{2},
$$

$e_0(T_{min})$ is the saturation vapour pressure at daily minimum temperature (kPa),
$e_0(T_{max})$ is the saturation vapour pressure at daily max. temperature (kPa),
$\mathrm{RH_{max}}$ is the maximum relative humidity (%), and
$\mathrm{RH_{min}}$ is the minimum relative humidity (%).

$$
\text{For } \mathrm{RH_{max}}\, e_a = e^0(T_{min})\frac{\mathrm{RH_{max}}}{100},
$$

$e_s - e_a$ is the saturation vapour pressure deficit (kPa), and
Δ is the slope of vapour pressure curve (kPa/°C).

$$\Delta = \frac{4{,}098\left[0.6108\exp\left(\dfrac{17.27T}{T+237.3}\right)\right]}{(T+237.3)^2},$$

γ is the psychometric constant (kPa/°C).

$$\gamma = \frac{c_p P}{\varepsilon\lambda} = 0.665 \times 10^{-3}P,$$

c_p is the specific heat at constant pressure=1.013×10⁻³ (MJ/kg/°C),
ε is the ratio of molecular weight of water vapour/dry air=0.622,
λ is the latent heat of vaporization=2.45 (MJ/kg), and
P is the atmospheric pressure (kPa).

$$P=101.3\left(\frac{293-0.0065\ z}{293}\right)^{5.26},$$

z is the elevation above mean sea level (m), and
900=a conversion factor.

When no measured radiation data are available, the net radiation can be estimated as follows:

$$R_n = R_{ns} - R_{nl}, \tag{5.38}$$

where:

R_n is the net radiation (MJ/m²/day),
R_{ns} is the net sort wave radiation (MJ/m²/day)=$(1-\alpha)\ R_s$,
α is the albedo or canopy reflection factor=0.23 for grass reference crop,
R_s=$(0.25+0.50\ n/N)\ R_a$,
N is the daylight hour=$(24/\Pi)\ \omega s$,
R_s is the incoming solar radiation (MJ/m²/day), and
R_a is the extra-terrestrial radiation (MJ/m²/day).

$$R_a = \frac{24(60)}{\pi}G_{sc}\ d_r\ \left[\omega_s\sin(\varphi)\sin(\delta)+\cos(\varphi)\cos(\delta)\sin(\omega_s)\right],$$

G_{sc} is the solar constant=0.0820 MJ/m²/day, and
d_r is the inverse relative distance Earth-Sun.

$$d_r =1+0.33\cos\left(\frac{2\pi}{365}J\right),$$

J is the number of day in a year=365 or 366, and
ω_s is the sunset hour angle (rad).

$$\omega_s = \arccos\left[-\tan(\varphi)\tan(\delta)\right],$$

φ is the latitude (rad), and

δ is the solar declination (rad).

$$\delta = 0.409 \sin\left(\frac{2\pi}{365} J - 1.39 \right),$$

n/N is the relative sunshine fraction, and

R_{nl} is the net long wave radiation (MJ/m²/day).

$$R_{nl} = 2.45 \times 10^{-9} \left(0.9 \frac{n}{N} + 0.1 \right)\left(0.34 - 0.14 \sqrt{e_a} \right)\left(T_{kx}^4 + T_{kn}^4 \right),$$

T_{kx} is the maximum temperature (K) = T_{max} (°C) + 273.16,

T_{kn} is the minimum temperature (K) = T_{main} (°C) + 273.16,

E_a is the actual vapour pressure (kPa),

$G = 0.14 (T_{month\ n} - T_{month\ n-1})$,

G is the soil heat flux (MJ/m²/day),

T_{meani} is the mean temperature of the month (°C), and

$T_{mean\ i-1}$ is the mean temperature of the previous months (°C).

Example 5.5

Estimate the value of ET$_0$ for the month of March based on the following monthly average climate data for the 25-year period (1957–1981) at the Regional Research Station of Kerala Agricultural University at Pattambi, District Palakkad, Kerala, India. The station is located at 10°48′N latitude and 76°12′E longitude. The altitude of the station is 25 m above mean sea level. The values of daily maximum and minimum temperatures for each month during the period are available at the weather station attached to the research station

- Daily (average) maximum temperature (T_{max})   36.470°C
- Daily (average) minimum temperature (T_{min})   23.220°C
- Maximum relative humidity (RH_{max})   87.480%
- Minimum relative humidity (RH_{min})   39.420%
- Wind speed measured at 3 m height   4.520 km/hour = 1.256 m/s
- Actual hours of sunshine (n)   9.830 hours

Solution:

$$T_{mean} = \frac{T_{max} + T_{min}}{2} = \frac{36.470 + 23.220}{2} = 29.845 \text{ °C}$$

$$\Delta = \frac{4{,}098\left[0.6108 \exp\left(\frac{17.27 T}{T + 237.3} \right) \right]}{(T + 237.3)^2} = \frac{4{,}098\left[0.6108 \exp\left(\frac{17.27 \times 29.845}{29.845 + 237.3} \right) \right]}{(29.845 + 237.3)^2} = 0.241 \text{ kPa/°C}$$

$$P = 101.3 \left(\frac{293 - 0.0065 \, z}{293} \right)^{5.26} = 101.3 \left(\frac{293 - 0.0065 \times 25}{293} \right)^{5.26} = 101.005 \, \text{kPa}$$

$$\gamma = \frac{c_p P}{\varepsilon \lambda} = 0.665 \times 10^{-3} P = 0.665 \times 10^{-3} \times 101.005 = 0.067 \, \text{kPa/°C}$$

$$u_2 = u_h \frac{4.87}{\ln(67.8h - 5.42)} = 1.256 \frac{4.87}{\ln((67.8 \times 3) - 5.42)} = 1.157 \, \text{m/s}$$

$$e^0(T_{max}) = 0.6108 \exp\left[\frac{17.27 T_{max}}{T_{max} + 237.3} \right] = 0.6108 \exp\left[\frac{17.27 \times 36.470}{36.470 + 237.3} \right] = 6.096 \, \text{kPa}$$

$$e^0(T_{min}) = 0.6108 \exp\left[\frac{17.27 T_{min}}{T_{min} + 237.3} \right] = 0.6108 \exp\left[\frac{17.27 \times 23.220}{23.220 + 237.3} \right] = 2.847 \, \text{kPa}$$

$$e_s = \frac{e^0(T_{max}) + e^0(T_{min})}{2} = \frac{6.096 + 2.847}{2} = 4.472 \, \text{kPa}$$

$$e_a = \frac{e^0(T_{min}) \dfrac{RH_{max}}{100} + e^0(T_{max}) \dfrac{RH_{min}}{100}}{2} = \frac{2.847 \times \dfrac{87.480}{100} + 6.096 \times \dfrac{39.420}{100}}{2} = 2.447 \, \text{kPa}$$

Vapour pressure deficit $= e_s - e_a = 4.472 - 2.447 = 2.025 \, \text{kPa}$

$$R_a = \frac{24(60)}{\pi} G_{sc} d_r \left[\omega_s \sin(\varphi)\sin(\delta) + \cos(\varphi)\cos(\delta)\sin(\omega_s) \right]$$

$G_{sc} = $ solar constant $= 0.0820 \, \text{MJ/m}^2/\text{day}$

J for 15th March $= 74$

$$d_r = 1 + 0.33\cos\left(\frac{2\pi}{365} J \right) = 1 + 0.33\cos\left(\frac{2\pi}{365} \times 74 \right) = 1.010 \, \text{rad}$$

$$\delta = 0.409\sin\left(\frac{2\pi}{365} J - 1.39 \right) = 0.409\sin\left(\frac{2\pi}{365} \times 74 - 1.39 \right) = -0.047 \, \text{rad}$$

$\varphi = $ latitude $= 10°48' = 10.80 \; 3.14/180 = 0.188 \, \text{rad}$

$$\omega_s = \arccos\left[-\tan(\varphi)\tan(\delta) \right] = \cos^{-1}\left[-\tan(0.188)\tan(-0.047) \right] = 1.562 \, \text{rad}$$

$$R_a = \frac{24(60)}{\pi} G_{sc} d_r \left[\omega_s \sin(\varphi)\sin(\delta) + \cos(\varphi)\cos(\delta)\sin(\omega_s) \right]$$

$$R_a = \frac{24(60)}{\pi} 0.082 \times 1.01 \left[1.562 \times \sin(0.188)\sin(-0.047) + \cos(0.188)\cos(-0.047)\sin(1.562) \right]$$

$$R_a = 36.706 \, \text{MJ/m}^2/\text{day}$$

$$N = \frac{24}{\pi} \omega_s = \frac{24}{\pi} 1.562 = 11.933 \, \text{hours}$$

$$R_s = (0.25 + 0.50 \, n/N) R_a = (0.25 + 0.50(9.830/11.933))36.706 = 24.295 \left(\text{MJ/m}^2/\text{day} \right)$$

$$R_{ns} = \text{net solar radiation} = (1 - \alpha) R_s = (1 - 0.23)24.295 = 18.707 \left(\text{MJ/m}^2/\text{day} \right)$$

$$R_{so} = (0.75 + 2 \times 10^{-5} \times z) R_a = (0.75 + 2 \times 10^{-5} \, 25)36.706 = 27.548 \left(\text{MJ/m}^2/\text{day} \right)$$

$$R_{nl} = 2.45 \times 10^{-9} \left(0.9 \frac{n}{N} + 0.1 \right) \left(0.34 - 0.14\sqrt{e_a} \right) \left(T_{kx}^4 + T_{kn}^4 \right)$$

$$R_{nl} = 2.45 \times 10^{-9} \left(0.9 \frac{9.83}{11.933} + 0.1 \right) \left(0.34 - 0.14\sqrt{2.447} \right) \left((36.470 + 273.16)^4 + (23.220 + 273.16)^4 \right)$$

$$R_{nl} = 2.45 \, (0.84) \, (0.12) \, (16.90) = 4.218 \, (\text{MJ/m}^2/\text{day})$$

$$R_n = R_{ns} - R_{nl} = 18.707 - 4.218 = 17.789 \left(\text{MJ/m}^2/\text{day} \right)$$

$T_{\text{mean } i-1}$ is the mean temperature of the previous months (i.e. February) = 28°C.

$$G = 0.14 \left(T_{\text{month } n} - T_{\text{month } n-1} \right) = 0.14(29.845 - 28.205) = 0.230 \left(\text{MJ/m}^2/\text{day} \right)$$

Grass reference ET:

$$\text{ET}_0 = \frac{0.408\Delta(R_n - G) + \gamma \dfrac{900}{T + 273} u_2(e_s - e_a)}{\Delta + \gamma(1 + 0.34 u_2)} = (1) + (2)$$

$$(1) = \frac{0.408\Delta (R_n - G)}{\Delta + \gamma (1 + 0.34 u_2)} = \frac{0.408 \times 0.241 (14.489 - 0.23)}{0.241 + 0.067(1 + 0.34 \times 1.157)} = 4.195$$

$$(2) = \frac{\gamma \dfrac{900}{T + 273} u_2(e_s - e_a)}{\Delta + \gamma(1 + 0.34 u_2)} = \frac{0.067 \dfrac{900}{29.845 + 273} 1.157 \times 2.025}{0.241 + 0.067 (1 + 0.34 \times 1.157)} = 1.392$$

$$\text{ET}_0 = (1) + (2) = 4.195 + 1.392 = 5.587 \, \text{mm/day}$$

Table 5.3 summarizes equations of different ET methods. Table 5.4 lists the equations and symbols used in ET_0 using FAO 56 PM model. Table 5.5 gives the advantages, limitations and application timestep of different ET estimation models.

TABLE 5.3 Summary of Equations of Different Evapotranspiration Methods

No.	Methods	Equations
1	Blaney–Criddle (1942)	$PET = 0.46p(T + 17.8)$ $U = KF = \sum kf = \sum u$
2	Thornthwaite (1948)	$e = 1.6(10t / I)^{\alpha}$
3	Hargreaves (1956) (°F) Hargreaves and Samani (1985) (°C)	$ET_0 = 0.0075 R_s TF$ $ET_0 = 0.0023 R_s(T_a + 17.8) (T_{max} - T_{min})0.5$
4	Christiansen (1968)	$E = K\, R_a\, C$
5	Penman-Related Equations	
5.1	*Original Penman Equation (1948)*	$$ET = \frac{\left(\Delta(R_n - G) + k_w(e_s - e_a) f(u)\gamma\right)}{\lambda(\Delta + \gamma)}$$ $$f(u) = a_w + b_w u_2$$
5.2	*CIMIS Penman Method*	$a_w = 0.29, b_w = 0.53$ for $R_n > 0$ $a_w = 1.14, b_w = 0.40$ for $R_n \leq 0$
5.3	*Penman–Monteith (1965)*	$$\lambda E = \frac{0.408\Delta(R_n - G) + \gamma \dfrac{900}{T_a + 273} u_2(e_s - e_a)}{\Delta + \gamma(1 + 0.34 u_2)}$$
6	Priestley–Taylor (1972)	$\lambda E = \alpha\, \Delta\, (R_n - G)/(\Delta + \gamma)$ where $\alpha = 1.26$ for water surfaces with minimum advection.
7	Fixed Surface Resistance Approach	
7.1	*FAO 56 PM (1998)*	$$ET_0 = \frac{0.408\Delta(R_n - G) + \gamma \dfrac{900}{T + 273} u_2(e_s - e_a)}{\Delta + \gamma(1 + 0.34 u_2)}$$
7.2	*ASCE-EWRI Standardized PM Equation (2005)*	$$ET_{sz} = \frac{0.408\Delta(R_n - G) + \gamma C_n u_2 \dfrac{(e_s - e_a)}{T + 273}}{\Delta + \gamma(1 + C_d u_2)}$$
7.3	*Valiantzas Model (2006, 2013)*	$$ET_0 = 0.0393 R_s \sqrt{T_a + 9.5} - 0.19 R_s^{0.6}\, \varnothing^{0.15} + 0.148(T_a + 20)\left(1 - \frac{RH}{100}\right) u^{0.7}$$
8	Variable Surface Resistance Approach	
8.1	*Katerji–Perrier Model (1983)*	$$\frac{r_c}{r_a} = a\frac{r^*}{r_a} + b \qquad r^* = \frac{\Delta + \gamma}{\Delta\gamma}\frac{\rho\, c_p}{R_n - G} D$$
8.2	*Todorovic Model (1999)*	$$t = \frac{\gamma}{\Delta}\frac{D}{(\Delta + \gamma)}$$
8.3	*Li et al. Model (2009)*	$$t = \frac{\gamma}{\Delta}\frac{DC}{(\Delta + \gamma)} \qquad C = \frac{\left(\dfrac{\Delta}{\gamma}\right)\left(\dfrac{1}{r_i}\right)\dfrac{1}{r_a}}{\left(1 + \dfrac{\gamma}{\Delta}\right)\left(\dfrac{1}{r_c}\right) + \left(\dfrac{\gamma}{\Delta}\right)\left(\dfrac{1}{r_a}\right)}$$
8.4	*Shuttleworth Model (2006, 2009)*	$$K_c = \left[\frac{\dfrac{R_c^{50}}{u_3} + \dfrac{D_{50}}{D_2} r_{clim}}{\dfrac{302}{u_3} + \dfrac{D_{50}}{D_2} r_{clim}}\right]\left[\frac{(\Delta + \gamma)\dfrac{302}{u_2} + 70\gamma}{(\Delta + \gamma)\dfrac{R_c^{50}}{u_2} + \gamma(r_s)_c}\right]$$

(Continued)

TABLE 5.3 (*Continued*)   Summary of Equations of Different Evapotranspiration Methods

No.	Methods	Equations

where:

Δ: slope of saturation vapour pressure with air temperature, Pa/°C.

γ: psychometric constant, kPa/°C.

φ: latitude of site, rad.

ET: evapotranspiration rate, mm/hour or mm/day or inches/day.

PET: potential ET rate, mm/hour or mm/day or inches/day.

ET_0: reference evapotranspiration rate from a grass surface, mm/hour or mm/day or inches/day.

ET_{sz}: reference evapotranspiration rate from a standardized surface, mm/hour or mm/day.

ET_c: crop evapotranspiration, mm/hour or mm/day.

u: monthly consumptive use (ET), inches/month or mm/month.

U: seasonal consumptive use (ET), inches/season or mm/season.

K_c: crop coefficient developed by FAO 56 method.

k: empirical crop coefficient for monthly period.

K: empirical crop coefficient for irrigation season or growing period.

T_a: mean monthly/daily/hourly air temperature, °C.

T_F: mean monthly/daily/hourly air temperature, °F.

t: difference between actual canopy temperature and canopy temperature in wet conditions, °C.

u_2: wind speed at 2 m height, m/s.

r_c: canopy surface resistance, s/m.

r_a: aerodynamic resistance, s/m.

r_l: daily average stomatal resistance: s/m.

r_i: climatological resistance, s/m.

r^*: climatic resistance, s/m.

RH: relative humidity, %.

R_n: net radiation, MJ/m²/day or MJ/m²/h.

R_s: incoming solar radiation, MJ/m²/day or MJ/m²/hour.

R_a: extra-terrestrial radiation, MJ/m²/day or MJ/m²/hour.

e_a: actual vapour pressure, kPa.

e_s: saturation vapour pressure, kPa.

D: vapour pressure deficit, kPa.

p: monthly percentage of daytime hours of the year, %.

f: monthly consumptive use (ET) factor.

F: sum of monthly consumptive use (ET) factors for the period.

i: heat index.

I: sum of the 12 monthly heat index i.

S: measured sunshine hours times 100 divided by the number of possible sunshine hours.

K_{RS}: calibration coefficient.

TD: mean maximum − mean minimum temperature, °C.

K: dimensionless constant developed empirically from data analysis.

C: dimensionless coefficient related to climatic parameters.

G: soil heat flux, MJ/m²/day or MJ/m²/hour.

$f(u)$: wind speed function.

J: Julian day of the year.

λ: latent heat of evaporation, MJ/kg.

ρ: air density, kg/m³.

C_p: specific heat capacity of air at constant pressure, J/kg/K.

D: zero plane displacement height, m.

h_c: crop height, m.

z_m: height of wind measurements, m.

z_h: height of humidity measurements, m.

z_{om}: roughness length governing momentum transfer, m.

z_{oh}: roughness length governing heat transfer.

k: von-Korman's constant (0.41).

U_z: wind speed at height z, m.

LAI: leaf area index, m²/m².

C_n: numerator constant that changes with reference type and calculation timestep, K mm s³ M/g/day.

C_d: denominator constant that changes with reference type and calculation timestep, s/m.

W_{aero}: empirical weighted factor.

R_c^{50}: aerodynamic coefficient for a crop of height h_c at blending height of 50 m.

D_{50}: vapour pressure deficit at blending height of 50 m (kPa).

D_2: vapour pressure deficit at 2 m height (kPa).

r_{clim}: climatological resistance (s/m).

$(r_s)_c$: surface resistance of crop (s/m)

TABLE 5.4 List of Equations and Symbols Used in ET_0 Using FAO 56 PM Model

No.	Equation	Quantity	Unit
1	$P = 101.3\left(\dfrac{293-0.0065\,z}{293}\right)^{5.26}$	Atmospheric pressure	kPa
2	$\gamma = \dfrac{c_p P}{\varepsilon\lambda} = 0.665\times10^{-3}\,P$	Psychometric constant	kPa/°C
3	$\Delta = \dfrac{4{,}098\left[0.6108\exp\left(\dfrac{17.27T}{T+237.3}\right)\right]}{(T+237.3)^2}$	Slope of saturation vapour pressure curve	kPa/°C
4	$e_s = \dfrac{e^0\left(T_{\max}\right)+e^0\left(T_{\min}\right)}{2}$	Mean of saturation vapour pressure	kPa
5	$e^0(T)=0.6108\exp\left[\dfrac{17.27T}{T+237.3}\right]$	Saturation vapour pressure at maximum or minimum air temperature	kPa
6	$e_a = \dfrac{e^0\left(T_{\min}\right)\dfrac{RH_{\max}}{100} + e^0\left(T_{\max}\right)\dfrac{RH_{\min}}{100}}{2}$	Actual vapour pressure	kPa
7	$e_s - e_a$	Vapour pressure deficit	kPa
8	$R_a = \dfrac{24(60)}{\pi}G_{sc}d_r\left[\omega_s\sin(\varphi)\sin(\delta)+\cos(\varphi)\cos(\delta)\sin(\omega_s)\right]$	Extra-terrestrial radiation	MJ/m²/day
9	$d_r = 1+0.33\cos\left(\dfrac{2\pi}{365}J\right)$	Inverse relative distance Earth-Sun	Radian
10	$\delta = 0.409\sin\left(\dfrac{2\pi}{365}J-1.39\right)$	Solar declination	Radian
11	$\omega_s = \arccos\left[-\tan(\varphi)\tan(\delta)\right]$	Sun hour angle	Radian
12	$N = \dfrac{24}{\pi}\omega_s$	Possible daylight hour	Hour
13	$R_s = (0.25+0.50\,n/N)\,R_a$	Solar radiation	MJ/m²/day
14	$R_{so} = (0.75+2\times10^{-5}\times z)\,R_a$	Clear sky solar radiation	MJ/m²/day
15	$R_{ns} = (1-\alpha)\,R_s$	Net shortwave radiation	MJ/m²/day
16	$R_{nl} = 2.45\times10^{-9}\left(0.9\dfrac{n}{N}+0.1\right)\left(0.34-0.14\sqrt{e_a}\right)\left(T_{kx}^4 + T_{kn}^4\right)$	Net long wave radiation	MJ/m²/day
17	$R_n = R_{ns} - R_{nl}$	Net radiation	MJ/m²/day
18	$u_2 = u_h\dfrac{4.87}{\ln\,(67.8h-5.42)}$	Wind speed at height h	m/second
19	$ET_0 = \dfrac{0.408\Delta\left(R_n-G\right)+\gamma\dfrac{900}{T+273}u_2(e_s-e_a)}{\Delta+\gamma\,(1+0.34u_2)}$	Reference evapotranspiration	mm/day

TABLE 5.5 Advantages, Limitations and Application Timestep of Different ET Estimation Models

No.	Methods	Variables Used	Advantages	Limitations	Application Timestep
1	Blaney–Criddle Method (1942)	T_a, T_F, p, k, K	Simplicity	ET underestimation, in general.	Monthly
2	Thornthwaite Method (1948)	T_a	Simplicity	ET underestimation in advective condition.	Monthly
3	Hargreaves and Samani (1985)	R_s, T_p, T_a, R_a	Simplicity	Problems of over and under estimation of ET.	Weekly
4	Christiansen Method (1968)	K, R_a, C	More or less accurate to predict ET for monthly timestep	Not accurate to calculate ET for daily or shorter timesteps.	Monthly
5	Penman-Related Equations				
5.1	*Original Penman Equation (1948)*	Δ, R_n, G, e_s, e_a, γ, $f(u)$, λ	Physical equation based on the combination of surface energy balance equation and aerodynamic equation	Wind speed function is difficult to obtain. The equation was mainly developed for evaporation from free water surfaces.	Daily, hourly
5.2	*CIMIS Penman Method*	Δ, R_n, G, e_s, e_a, γ, $f(u)$, λ	a_w and b_w coefficients used in $f(u)$ are easy to obtain. Also applicable for hourly timesteps	The coefficients used in this equation were developed for Californian condition, hence may not be applicable elsewhere.	Hourly
5.3	*Penman–Monteith equation (1965)*	Δ, R_n, G, e_s, e_a, γ, ρ, c_p, r_a, r_c	Physical equation with the inclusion of r_c	It is difficult to directly implement this equation to calculate actual crop ET, as r_c is difficult to obtain.	Daily, hourly
6	Priestley–Taylor Method (1972)	α, Δ, γ, R_n, G	Relatively simple	ET underestimation mainly in advective condition.	Daily
7	Fixed Surface Resistance Approach				
7.1	*FAO 56 PM Equation (1998)*	Δ, R_n, G, e_s, e_a, γ, T_a, u_2	Considered very accurate to calculate grass reference ET on daily basis	May not be applicable to apply for hourly timestep.	Daily
7.2	*ASCE-EWRI Standardized PM Equation (2005)*	Δ, R_n, G, e_s, e_a, γ, T_a, u_2	It can calculate both grass and alfalfa reference crop ET on both hourly and daily timesteps	K_c needs to be developed also for alfalfa reference surfaces. The use of fixed r_c for entire day may induce some errors in estimating reference ET.	Daily, hourly
7.3	*Valiantzas Model (2006, 2013)*	T, R_s, φ, RH, u_2	Relatively simple, can be used when some parameters like wind speed is missing	It is semi-empirical, so may not be accurate enough as PM equation.	Daily
8	Variable Surface Resistance Approach				
8.1	*Jarvis Model (1976)*		New concept to calculate stomatal resistance	It is not easy to obtain canopy resistance (r_c) from stomatal resistance.	
8.2	*Katerji–Perrier Model (1983)*	Δ, R_n, G, e_s, e_a, γ, T_a, u_2, a, b, r_a	Relatively simple to calculate actual crop ET in one step process	This is empirical method. The coefficients "a" and "b" need to be tested for different species and also for different climatic conditions.	Daily, hourly
8.3	*Todorovic Model (1999)*	Δ, R_n, G, e_s, e_a, γ, T_a, u_2, r_a	Mechanistic equation to calculate r_c	Some flaws in the procedure as shown by Li et al. (2009).	Hourly
8.4	*Li et al. Model (2009)*	Δ, R_n, G, e_s, e_a, γ, T_a, u_2, r_a	Relatively simple to implement	Only applicable for winter wheat crop in North China Plain.	Hourly
8.5	*Shuttleworth Model (2006, 2009)*	Δ, R_n, G, e_s, e_a, γ, T_a, u_2, r_a, K_c	Provides one step ET for daily timestep based by calculating r_c based on K_c	Complicated to use, r_c is a function of FAO 56 K_c values; in other words, r_c depends on the accuracy of K_c.	Daily
8.6	*Irmak and Mutibwaa Models (2008, 2009, 2010, 2013)*	PPFD, LAI	Already implemented models successfully for corn and soybean	Complicated to use. Needs many variables to calculate r_c including PPFD (photosynthetic photon flux density).	Hourly

5.10 Conclusions

ET is the main parameter to compute the water use of the plant, which includes quantity of water transpired by plants during their growth, or retained in the plant tissue, plus the moisture evaporated from the surface of the soil and vegetation. Estimation of evaporation is of extreme importance in many hydrologic problems associated with planning and operation of reservoirs and irrigation systems. The above-mentioned are various analytical and empirical methods to determine the plant ET. Many scientists have developed various models to determine the plant ET by using available meteorological data. Among the above-mentioned methods, Penman–Monteith method is widely accepted (Allen and Pruitt, 1991; Córdova et al., 2015). As Penman–Monteith method requires all weather data, viz. maximum and minimum temperature, maximum and minimum relative humidity, wind speed, etc. (Allen et al., 1998), meanwhile Hargreaves' method can be used even with mean daily temperature data only. This chapter can be referred by students, academicians and young researchers.

References

Allen, R.G., Pereira, L.S., Raes, D. and Smith, M., 1998. Crop evapotranspiration-Guidelines for computing crop water requirements. *FAO Irrigation and Drainage Paper 56. FAO, Rome, Italy, 300*(9), D05109.

Allen, R.G. and Pruitt, W.O., 1991. FAO-24 reference evapotranspiration factors. *Journal of Irrigation and Drainage Engineering, 117*(5), 758–773.

Christiansen, J.E., 1968. Pan evaporation and evapotranspiration from climatic data. *Journal of the Irrigation and Drainage Division, 94*(2), 243–266.

Córdova, M., Carrillo-Rojas, G., Crespo, P., Wilcox, B. and Célleri, R., 2015. Evaluation of the Penman-Monteith (FAO 56 PM) method for calculating reference evapotranspiration using limited data. *Mountain Research and Development, 35*(3), 230–239.

Doorenbos, J., 1977. Guidelines for predicting crop water requirements. *Food and Agriculture FAO Irrigation and Drainage Paper, Italy, 24*, 1–179.

Doorenbos, J. and Pruitt, W.O., 1977. Guidelines for prediction of crop water requirements. *FAO Irrigation and Drainage Paper, Italy, 24*, 1–179.

Hargreaves, G.H., 1956. Irrigation requirements based on climatic data. *Journal of the Irrigation and Drainage Division, 82*(3), 1–11.

Hargreaves, G.H. and Samani, Z.A., 1985. Reference crop evapotranspiration from temperature. *Applied Engineering in Agriculture, 1*(2), 96–99.

Hasanzadeh Saray, M., Eslamian, S. S., Klöve, B., Gohari, A., 2020. Regionalization of potential evapotranspiration using a modified region of influence. *Theoretical and Applied Climatology, 140*, 115–127.

Horton, R.E., 1917. A new evaporation formula developed. *Engineering News-Record, 78*, 196–199.

Horton, R.E., 1943. Hydrologic interrelations between lands and oceans. *Eos, Transactions American Geophysical Union, 24*(2), 753–764.

Jensen, M.E., 1968. Water consumption by agricultural plants, Chapter 1, In: Kozlowski, T.T., (ed.) *Plant Water Consumption and Response. Water Deficits and Plant Growth*, Vol. II. 1–22. Academic Press, New York, USA.

Katerji, N. and Perrier, A., 1983. A model of actual evapotranspiration (ETR) for a field of lucerne: The role of a crop coefficient. *Agronomie, 3*(6), 513–521.

Linacre, E.T., 1994. Estimating US Class A pan evaporation from few climate data. *Water International, 19*(1), 5–14.

Makkink, G.F., 1957. Testing the Penman formula by means of lysimeters. *Journal of the Institution of Water Engineers, 11*, 277–288.

Marciano, J.J. and Harbeck, G.E., 1954. Mass transfer studies, water loss investigations: Lake Hefner studies technical report. *Geological Survey Professional Paper*, U.S. Department of the Interior, USA, 269.

Masoner, J.R., Stannard, D.I. and Christenson, S.C., 2008. Differences in evaporation between a floating pan and class a pan on land 1. *JAWRA Journal of the American Water Resources Association, 44*(3), 552–561.

McJannet, D.L., Webster, I.T. and Cook, F.J., 2012. An area-dependent wind function for estimating open water evaporation using land-based meteorological data. *Environmental Modelling and Software, 31*, 76–83.

Meyer, A.F., 1942. *Evaporation from Lakes and Reservoirs: A Study Based on Fifty Years' Weather Bureau Records*. Minnesota Resources Commission, USA.

Michael, A. M., 2014. *Irrigation: Theory and Practice. Second Edition*. Vikas Publishing House Pvt. Ltd., Delhi. ISBN: 978-81-259-1867-7.

Penman, H.L., 1948. Natural evaporation from open water, bare soil and grass. *Proceedings of the Royal Society of London. Series A. Mathematical and Physical Sciences, 193*(1032), 120–145.

Priestley, C.H.B. and Taylor, R.J., 1972. On the assessment of surface heat flux and evaporation using large-scale parameters. *Monthly Weather Review, 100*(2), 81–92.

Rohwer, C., 1931. *Evaporation from Free Water Surfaces (No. 271)*. US Department of Agriculture, USA.

Sammis, T.W., Gregory, E.J. and Kallsen, C.E., 1982. Estimating evapotranspiration with water-production functions or the Blaney-Criddle method. *Transactions of the ASAE, 25*(6), 1656–1661.

Shuttleworth, W.J., 2006. Towards one-step estimation of crop water requirements. *Transactions of the ASABE, 49*(4), 925–935.

Suresh, R., 2010. *Micro Irrigation: Theory and Practice*. Standard Publishers Distributors, Delhi. ISBN: 978-81-8014-150-8.

Thornthwaite, C.W., 1948. An approach toward a rational classification of climate. *Geographical Review, 38*(1), 55–94.

Todorovic, M., 1999. Single-layer evapotranspiration model with variable canopy resistance. *Journal of Irrigation and Drainage Engineering, 125*(5), 235–245.

Valiantzas, J.D., 2006. Simplified versions for the Penman evaporation equation using routine weather data. *Journal of Hydrology, 331*(3–4), 690–702.

Valiantzas, J.D., 2013. Simplified forms for the standardized FAO-56 Penman–Monteith reference evapotranspiration using limited weather data. *Journal of Hydrology, 505*, 13–23.

6

Plant Water Requirements and Evapotranspiration

Mobushir Khan
Charles Sturt University

Monzur A. Imteaz
Swinburne University
of Technology

6.1 Introduction

Global population is estimated to reach 9.8 billion in 2050 with a continued increasing trend till 2100 with a population of 11.2 billion people living on Earth (United Nations, 2017). Population growth is putting high pressure on our land and water resources to meet its demands. Since the last century, water use has grown twice the rate of population increase. Food and Agriculture Organization of the United Nations (FAO) estimates that about 60% more food will be needed by 2050 to meet the food requirements of a growing global population, which shows that water demand will increase manifolds, whereas current estimates forecast only 10% more water can be withdrawn for agricultural purposes (FAO, 2017). The world contains an estimated 1,400 million km^3 of water with only 45,000 km^3 as freshwater usable for drinking, hygiene, agriculture and industry (Hoogeveen et al., 2015). Climate change exacerbates the challenge to meet demand for food especially in the developing countries (MacAlister and Subramanyam 2018). Climate change has a significant impact on the water cycle through factors such as shifts in rainfall patterns and increased temperature, hence affecting the availability and quality of water (Kløve et al., 2014).

The increasing demand for water resources combined with limited availability a crucial problem that requires sustainable water resources management. Globally, agriculture accounts for 70% of all water consumption, 20% for industry and 10% for domestic use (Bates et al., 2008). This demands for scientific research on water resources to quantify the water needed and how this requirement varies spatially in order to optimally manage water resources in sustainable manner. Since irrigation water is limited, its efficient utilization is focus of the agricultural and water resources policy makers. Thus, better tools and technologies are required not only to irrigate the fields but also to estimate water requirements of crops at the first place (Eslamian et al., 2012). This will ensure that water is supplied where it is needed and as per requirements of the food production areas. Accurate and timely estimates of crop water needs will support the global endeavors of water conservation.

DOI: 10.1201/9780429290114-8

It has been observed that most hydrologic studies focus on the supply side of the water problem as shifts in rainfall patterns, glacier/snow melt, quality and quantity of groundwater, whereas the demand side has largely been ignored, i.e. evapotranspiration (ET). ET is one of the main components of the water cycle, and its accurate estimation is needed on large spatial scales for water management at different decision levels. However, this is difficult to achieve in practice because actual ET (ETA) or crop ET (ET_{crop} or ET_c) cannot be measured directly and varies considerably in time and space with changing land cover/land use. ET is the major element of water cycle after precipitation for most of the global land surface. Therefore, the assessment of spatial and temporal inconsistency of ET is fundamental for studies in various disciplines, i.e. water resources planning, hydrologic budgeting, agricultural irrigation, crop water requirement, drought monitoring and climate change studies (Liu et al., 2010). ETA is the quantity of water removed from surface of earth by evaporation and transpiration. Temporal and spatial variability in ETA depends on various factors such as vegetation, soil, topography and meteorological conditions (Bastiaanssen et al., 1998; Mu et al., 2007). Water productivity analysis, irrigation scheduling and planning can be addressed appropriately if timely and accurate information about water consumption by crops is available (Kite and Droogers, 2000).

The amount of water that a crop needs comprises the water that is transpired by the plant and the water that is stored in the tissue of the plant as a result of photosynthesis. The water stored in the plant's tissue is negligible proportion (less than 5%) of the total amount of water used by the plant. So, the water use of a crop is equal to transpiration and/or evaporation by the plant. Since transpiration constitutes most of the water used by the crop, we measure the water use of a crop as the rate of ET, which is the process by which liquid water moves from the soil or plants to vapor form in the atmosphere. ET is comprised of two processes, evaporation of water from soil and transpiration of water from leaves. ET is an important part of the hydrologic cycle as it is the pathway by which water moves from the earth's surface into the atmosphere.

Through estimations of ET using available methods and comparing those with actual measured ET from two regions, this chapter presents a modified equation for estimating net irrigation requirement for the crop.

6.2 Factors Affecting Plant Water Requirements

Evaporation is the conversion of water into water vapor and hence the removal of water from a surface, such as a water bodies, soil or wet vegetation, into the air. Evaporation rates are affected by solar radiation (sunshine hours), temperature (minimum and maximum), relative humidity and the wind (speed and direction). Since ET includes evaporation from soils and transpiration from plants, ET rate is also affected by the weather variables. This tells us that the crop water use will depend upon solar radiation, temperature, relative humidity and the wind. More water evaporates from plants and soils in conditions of higher air temperature, less humidity, higher solar radiation (longer sunshine hours) and high speed of the wind (Allen et al., 1998).

Transpiration is the removal of water from the surface of leaf cells in actively growing plants. The soil moisture is carried from roots to small pores of leaves (located at the undersides) and from here it evaporates in the atmosphere. The transpiration portion of ET depends upon structure, age and health of crop, as well as other plant factors, can also affect the rate of transpiration. For example, desert plants are adapted to transpire at slower rates (through their leave structures) than plants belonging to more humid regions. Some desert plants have the capability to reduce transpiration through keeping their stomata closed during the day. Plant adaptations to conserve moisture also include wilting, pointing leave structures' small leaves, silvery and hairy leaves to reduce transpiration through controlled evaporation.

The amount of water needed by a crop can be expressed as a relation of two main factors, potential ET and the types of crops. Potential ET depends upon the weather conditions of an area. The relationship of five major weather variables is described later in the chapter to estimate potential ET. Potential ET or the reference ET is the ET from a reference surface which is grass. The crop coefficient (K_c) is the ratio of the crop ET (ET_c) to the reference ET (ET_0). K_c represents an integration of the effects of primary characteristics that distinguish the crop from reference grass. These characteristics include crop height, reflectance, resistance of crop canopy and soil evaporation. In a nutshell, the amount of water that a crop needs is measured by the ET rate of a crop. The ET rate includes water that is transpired or evaporated through the plants. Further, ET rate varies depending on climatic conditions, the plant characteristics and the soil conditions. The following equation describes the relation between ET and plant characteristics:

$$ET_{crop} = ET * K_c,$$ (6.1)

where:

ET_{crop} = crop evapotranspiration or the actual evapotranspiration
ET = evapotranspiration or the reference evapotranspiration (it is also expressed as ET_0)
K_c = crop coefficient

6.3 Calculating Evapotranspiration

There are a variety of methods to measure and estimate ET directly through instrumentation or indirectly by means of relationship among parameters (weather, crops and soil). The estimation of physical parameters can be expressed via model. These methods can be divided into various categories, namely hydrological approaches—soil-water balance (indirect method) and weighing lysimeter (direct method); plant physiology approaches—Penman–Monteith model, sap flow method, chambers system, soil-water balance modeling approach and crop coefficient approach; and micrometeorological approaches—Bowen ratio and energy balance (indirect method), eddy covariance (direct method) and aerodynamic method.

Installation of lysimeter is costly and it is difficult to maintain. So other direct and indirect methods are used to measure ET . Eddy correlation and Bowen ratio instruments are used to measure various weather parameters from the field to calculate ET . The ET data from these methods are used to calibrate and validate models for long-term ET estimation (Acharya and Yang 2015). Most of the ET estimation models require measurement of meteorological variables which influence evaporation and ET from a reference surface. Some of these models make use of detailed meteorological variables while others use a few to predict ET values. The spatial and temporal scale of measuring and estimating ET of each method differs. To measure local and regional ET, a wide range of numerical models were introduced. The applications of these models are confined to those areas where the requisite input data (vegetation, soil and weather) is available since long (Nouri et al., 2013).

A number of empirical methods have been used by many research studies to estimate ET such as Pan Evaporation method, Blaney–Criddle Method, Hargreaves method, Jensen–Haise method and Penman–Monteith method using different meteorological variables (McMahon et al., 2013).

Standard form of evaporation pan is stainless steel with dimensions of Class A, 54 mm height with 1,206 mm diameter. This pan is installed by using wooden support, where we avoid from bushes, trees and other natural obstacles around pan. Results are measured on daily basis with still well, with measuring range being 100 mm and accuracy up to 0.02 mm. Evaporation is based on temperature, wind speed and humidity and other conditions. Level sensor can also be used for automatic calculations. Blaney–Criddle method is the simplest method with low accuracy, which provides rough estimate (Xystrakis and Matzarakis, 2011). Only daily average temperature and daily percentage of annual daytime hours are needed to calculate ET_0 using the following formula:

$$ET = p * (0.46 * T_{mean} + 8),$$ (6.2)

where

ET = Reference crop evapotranspiration (mm/day) as an average for a period of one month
T_{mean} = mean daily temperature (°C)
p = mean daily percentage of annual daytime hours

The FAO recently recommended Penman–Monteith method as standard method. Climatic data used for Penman–Monteith method are minimum air temperature, maximum air temperature, relative humidity, sunshine duration and wind speed. The formula for this method is (*Source*: http://www.fao.org/docrep/x0490e/x0490e06.htm):

$$\text{ET}_0 = \frac{0.408\Delta(R_n - G) + \gamma\dfrac{900}{T + 273}u_2(e_s - e_a)}{\Delta + \gamma(1 + 0.34u_2)},$$

(6.3)

where

ET_0 = reference evapotranspiration (mm/day)
R_n = net radiation at the crop surface (MJ/m²/day)
G = soil heat flux density (MJ/m²/day)
T = mean daily air temperature at 2 m height (°C)
u_2 = wind speed at 2 m height (m/s)
e_s = saturation vapor pressure (kPa)
e_a = actual vapor pressure (kPa)
$e_s - e_a$ = saturation vapor pressure deficit (kPa)
Δ = slope vapor pressure curve (kPa/°C)
γ = psychrometric constant (kPa/°C)

The analysis of the performance of the various algorithms revealed the need for formulating a standard method for the computation of the reference crop ET. For this reason, the FAO Penman–Monteith method has been recommended as a standard (Allen et al., 1998).

During calculation of ET grass is used as reference, other crops may not use same quantity of water due to difference in crop root depth, growth, height, ground cover and variety of crop. Therefore, for estimating precise ET, it is important to adjust the reference ET (ET_0). Crop coefficient (K_c) is used to adjust of ET value for a specific crop; several K_c can be used for single crop depending upon crop stage and type. Crop coefficient (K_c)-based estimation of crop ET is one of the most commonly used methods for irrigation water management. However, uncertainties of the generalized dual crop coefficient (K_c) method of the FAO Irrigation and Drainage Paper No. 56 can contribute to crop ET estimates that are substantially different from actual crop ET. Determination of ETA has traditionally been done based on a two-step approach. The ET of a reference standard crop (ET_0) is first estimated on the basis of site meteorological variables. A semi-empirical coefficient (crop coefficient, K_c) is then applied to take into account all the other crop and environmental factors (Magliulo et al., 2003). To estimate crop ET the following equation is used:

$$\text{ETA} = \text{ET}_0 * K_c.$$

(6.4)

Using standard method for ET_0 calculation, one can calculate ETA by multiplying values of ET_0 with K_c values, K_c values calculated by International Water Management Institute (Ullah et al., 2001). Temporally dynamic values of crop coefficients have been found highly correlated with vegetation indices (VIs) derived from satellite data. Several studies have shown potential for modeling crop coefficient as a function of the VIs. This became feasible from a technical and economical point of view by remote sensing technology. High spatial resolution such as WorldView-2 imagery and high temporal resolution such as MODIS and SPOT Vgt have been used for mapping Normalized Difference Vegetation Index (NDVI) and its relationship to temporal urban landscape ET factors (Irmak et al., 2011) For this purpose, the estimation of ETA at regional scale has been widely studied in recent years by combining conventional meteorological ground measurements with remotely sensed data. Several methods for assessing ET have been developed at various spatial and temporal scales. These methods vary in complexity from statistical/semi-empirical direct approaches to more analytical approaches with a physical base, and finally to numerical models simulating the heat and water flux through the soil, the vegetation and the atmosphere (Kustas and Norman, 1996).

6.3.1 Estimation of ET_0

Firstly, weather data were used to generate ET or reference ET . Secondly, crop area maps were generated using crop statistics and remote sensing data. Thirdly, K_c values were identified using the generated maps and K_c values published in FAO Irrigation and Drainage Paper No. 24. Lastly, ET was multiplied with the K_c values to obtain the maps of crop ET .

ET values measured or calculated at different locations or in different seasons are comparable as they refer to the ET_0 from the same reference surface. The only factors affecting ET_0 are weather parameters which were used to calculate ET_0 using Penman–Monteith formula (Eq. 6.3).

6.3.2 Interpolation and Validation of ET_0

After obtaining ET_0 values for each weather station, interpolation was performed in Geographic Information System environment. Krigging interpolation technique was used to predict the ET_0 values for the unmeasured area to provide evapotranspiration throughout the study area. While executing interpolation, 90% of the values were used, whereas remaining 10% were used as validation data set. The predicted values for the corresponding validation data set were obtained by overlay function in ArcGIS and scatter plots were made to check the accuracy of the interpolation results for validation purpose.

6.3.3 Crop Mapping

The next requirement for calculation of monthly ETA or ET crop is the values of crop coefficients (K_c) as per Eq. 6.3. This requires the spatial knowledge of various crops grown in the field so that their K_c values could be obtained from the literature. FAO has provided a list of K_c values of various crops all over the world on monthly basis in its volume 56. For this purpose, available land-use land-cover maps can be used but those may be outdated. Here, remote sensing provides options to map various crops in space and time. VIs which are ratios of surface reflectance at two or more wavelengths have been widely used to map different types of vegetations through various digital image processing techniques. Selection of a specific vegetation index and a particular method depends upon the purpose of mapping. In case of mapping crop areas, NDVI has been widely used (Khan et al., 2010; de Bie et al., 2011). VIs computed from the satellite data such as MODIS NDVI (MOD13Q1) data available after every 16 days with 250 m resolutions can be acquired free of cost to prepare spatially explicit crop maps. Other satellite data include SPOT VGT s10 and Copernicus Sentinels can be used to prepare updated crop area maps. The method used to classify should be able to capture the crop calendar of a particular area so that it can be able to differentiate various crops grown over an area.

6.3.4 Case Study 1

The following case study, which was conducted by the authors for Australian Center for International Agricultural Research (ACIAR) in South East Asia, used the methodology based on Eq. 6.1.

6.3.5 Estimation of ET_0

Monthly weather data were acquired from Meteorological Department for five weather stations in the study region (Figure 6.1). The weather parameters included minimum temperature, maximum temperature, relative humidity, wind speed and sunshine hours.

The results of interpolated ET_0 calculation are provided in Figures 6.2 and 6.3 (only two seasons, winter and summer).

6.3.6 Crop Mapping for Estimation of Crop Coefficients

Clustering approach was used to generate specific crop maps. Unsupervised classification was implemented using ISODATA clustering method to identify existing cropping patterns in study area. ISODATA clustering has minimum user inputs, which are number of iterations (set to 50) and convergence threshold (set to 0.95) for every classification run (Khan et al., 2010). A total number of 20 classification runs were executed starting from 5 classes till 100 classes with an increment of five classes in each run. The output of each classification run was a classified image and its corresponding signature. Optimal classification was selected by using divergence statistics (Khan et al., 2010).

After selection of optimal classification, the corresponding signature was used, and matched practiced crop calendar of the study area and each class was related with the crop calendar. Individual crop areas were then identified both in winter and summer seasons by linking the results of image processing

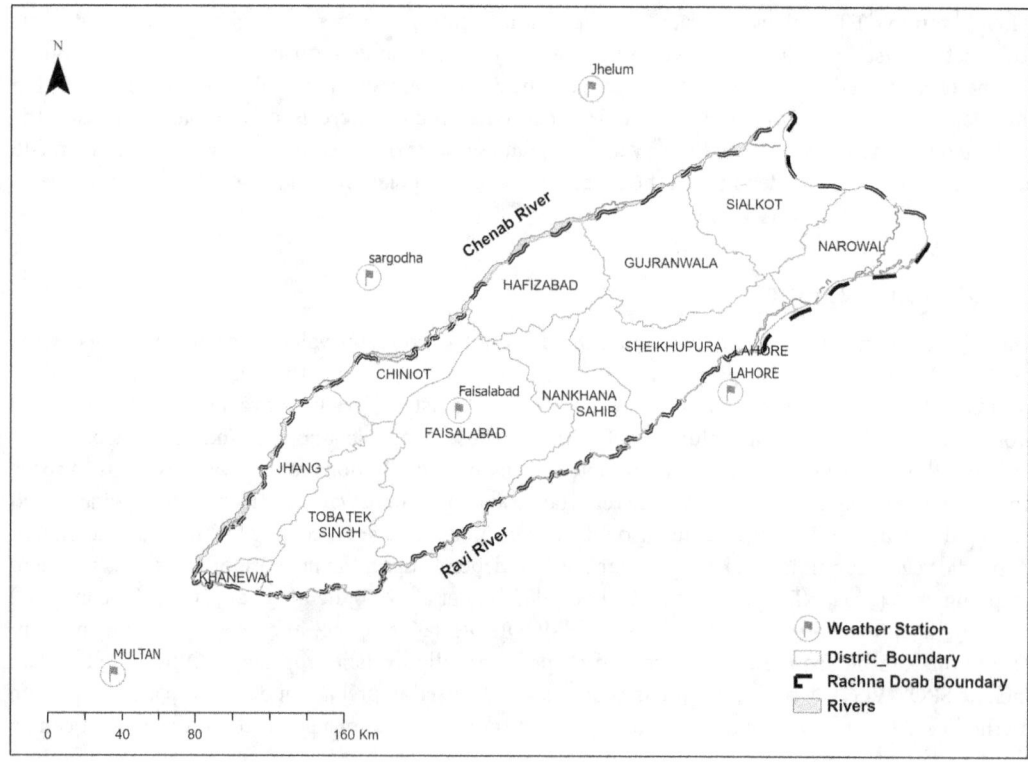

FIGURE 6.1   Study area and the location of weather stations.

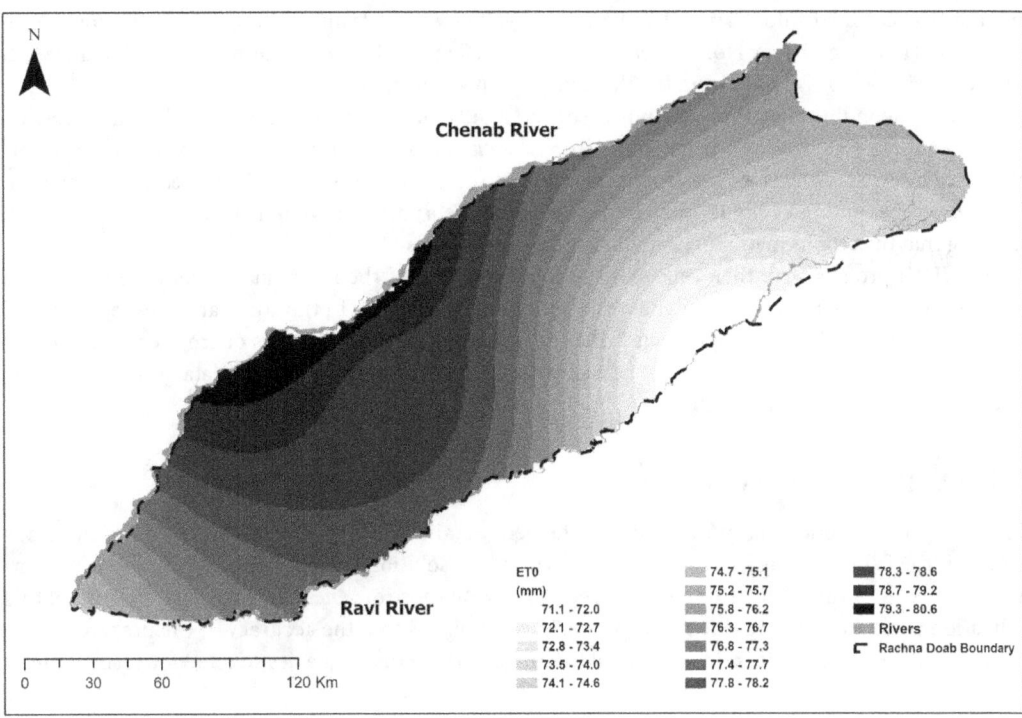

FIGURE 6.2 Calculated evapotranspiration for summer.

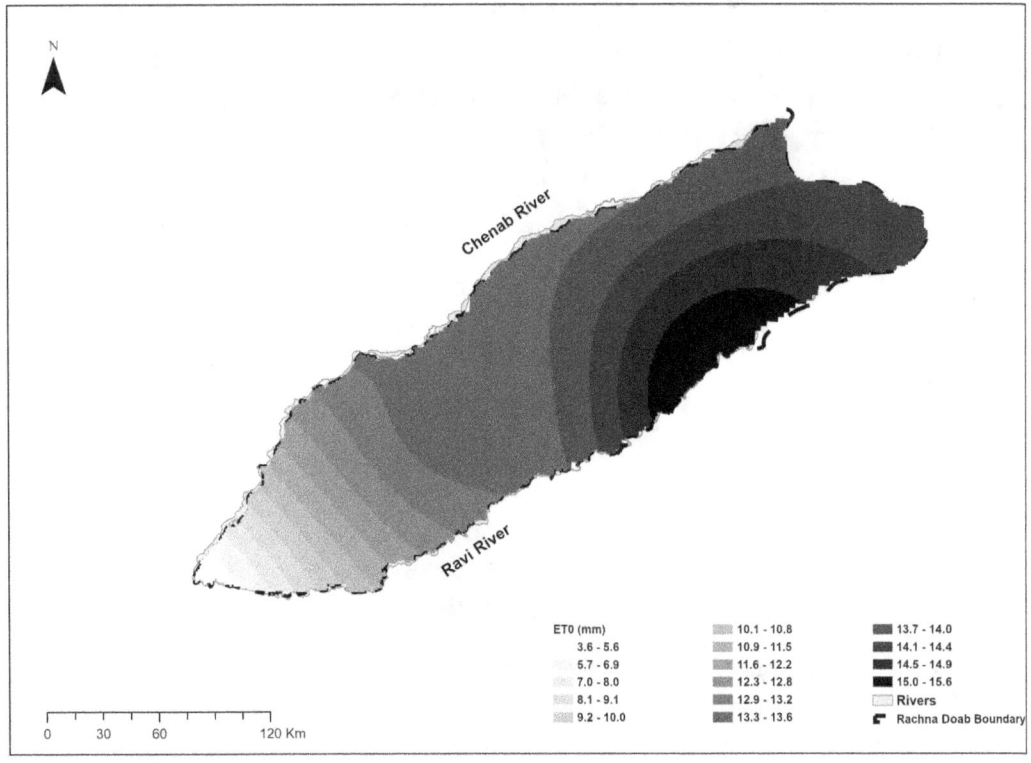

FIGURE 6.3 Calculated evapotranspiration for winter.

and the crop calendar information. Further, the K_c values were obtained for each month by consulting FAO table (Drainage Paper No. 56). The results of ISODATA clustering are presented in Figure 6.4, which shows that 40 classes exist in the Rachna Doab at 250 m resolution.

This corresponding signature was thus used, and all the 40 classes were then related to the practiced crop calendar. An example for behavior of an NDVI class is shown in Figure 6.5. The behavior of this class matches well with the wheat crop grown in that region which is shown by increase in vegetation from November and reaching its peak in the months of February and March. Then it starts to decline, showing that the crop is maturing, and its greenness is decreased.

The NDVI profiles were then linked with crop calendar and the individual crops were mapped as a result. Figures 6.6 and 6.7 show winter and summer crops grown in the study area, respectively. An advantage of using clustering approach is that we can know how many types of crops or mix of crops occur in an area. These groups then can be easily partitioned using crop statistics data or the data generated through field work campaigns.

6.3.7 Validation of Crop Mapping

To validate the generated crop maps, field data were obtained as well as percentage of various crops grown in the field over blocks of 250 m (at par with the resolution of the satellite data). These percentages of crops at 250 m in the field were then compared with those of generated crop maps. Results of this validation procedure are provided in Figure 6.8, which shows that the accuracy of the prepared maps is more than 90% at middle, middle and tail locations of three distributaries selected for verification of results.

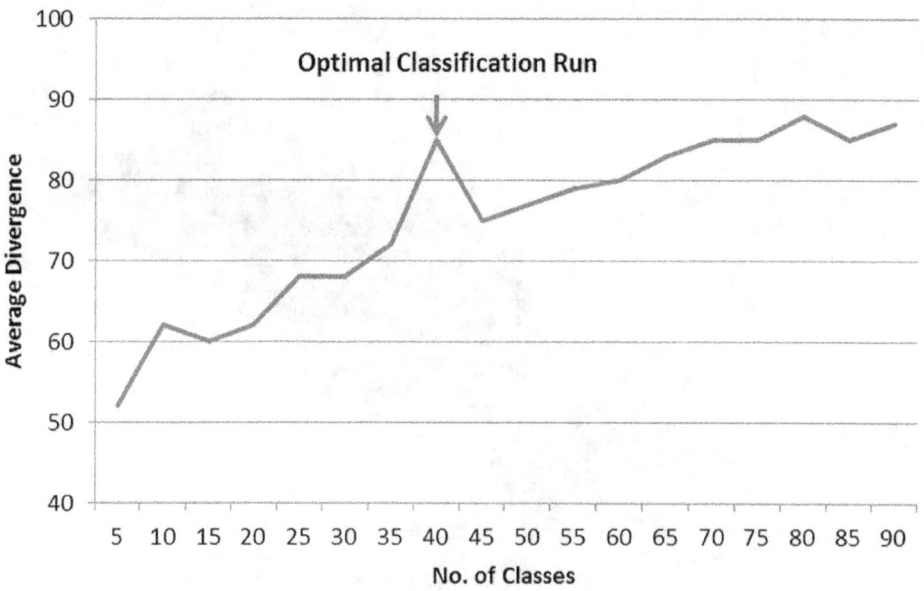

FIGURE 6.4 Divergence statistics of the unsupervised classification runs.

CROP PATTERN

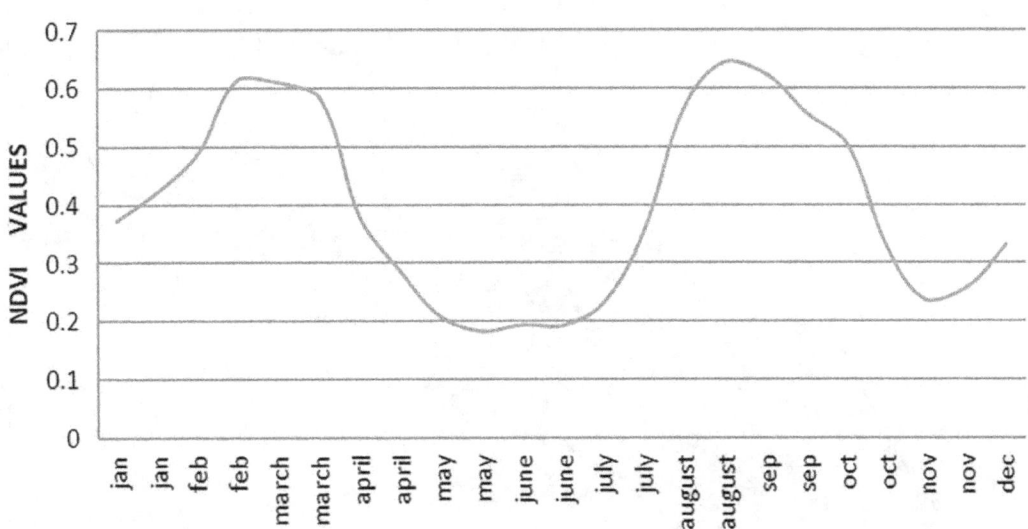

FIGURE 6.5 NDVI profiles over time in the study area.

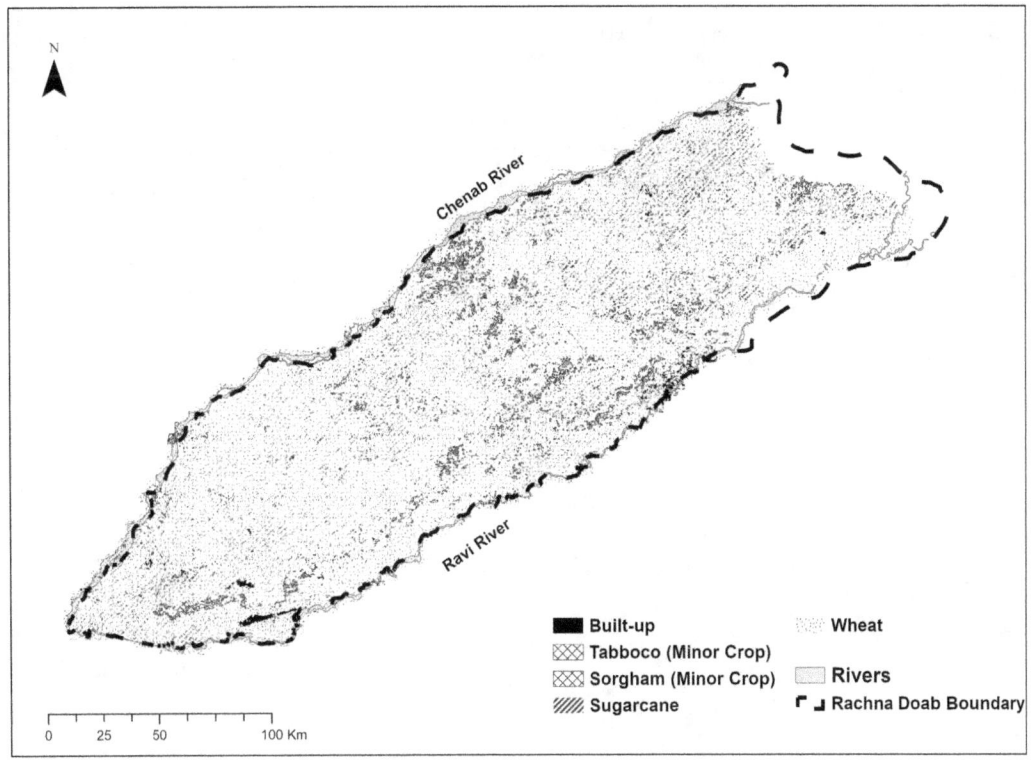

FIGURE 6.6 Map of winter cropping pattern.

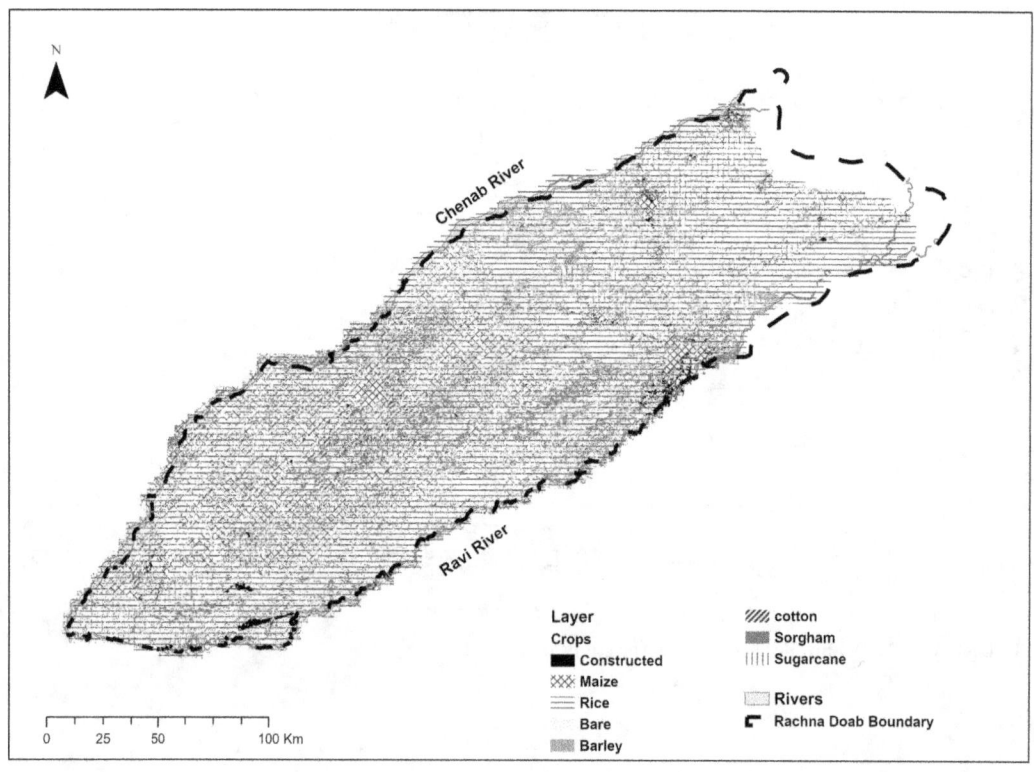

FIGURE 6.7   Map of summer cropping pattern.

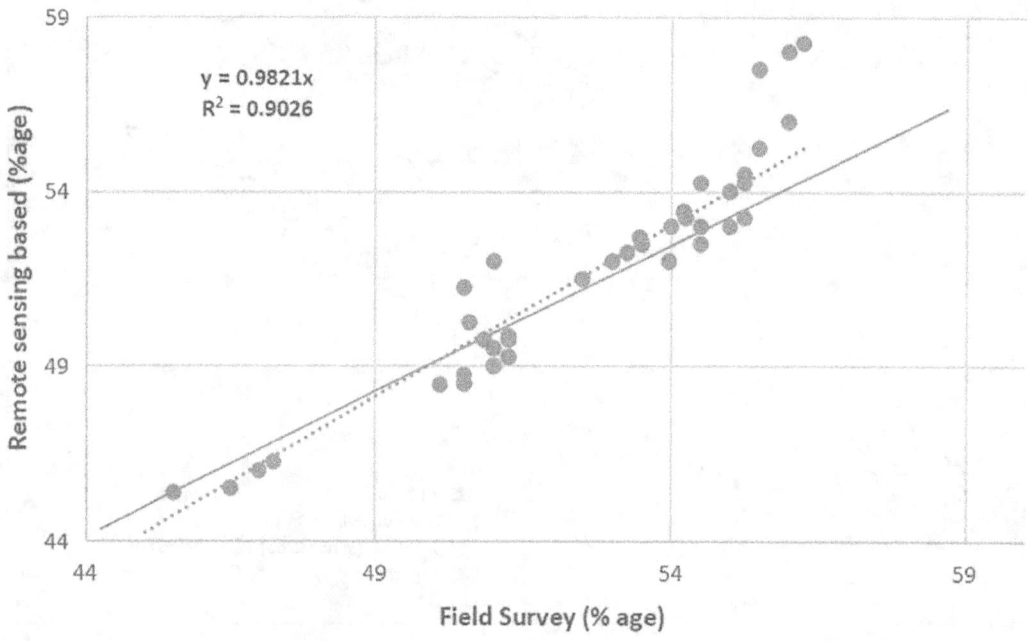

FIGURE 6.8   Validation of prepared crop maps using field data.

6.3.8 Mapping of Actual Evapotranspiration

The outputs of ET_0 and crop area maps were used to calculate the ETA using Eq. 6.3. The results were ETA maps or the crop ET (Figures 6.9 and 6.10).

Over the past few decades, remote sensing has become one of the most valuable data sources and analysis techniques for implementing ET studies. Nowadays, for local to global scales, ET estimation using satellites imagery is the most resourceful and cost-effective technology (Nouri et al., 2013). The major advantage of using satellite imagery for ET estimation is that it can be directly derived without computing intricate hydrological processes. There are two approaches to estimate ET through remote sensing: (1) measuring ET based on VI and (2) computing ET based on surface energy balance (SEB) methods (Trezza et al., 2013). The SEB models are theoretically sound, but their performances are constrained by the estimation of sensible heat flux and soil heat flux (Cleugh et al., 2007). While using first option, it is pertinent to mention that knowledge of vegetations' phenology and cropping cycle at the time of image acquisition is crucial for accurate ET estimates. A time series of ET can be helpful in determining crop water requirement at different crop growth stages.

To estimate ET over a large area, numerous remote sensing-based SEB models have been developed. Five widely used single-source SEB models for estimating ET are Surface Energy Balance Algorithm for Land (SEBAL), Mapping ET with Internalized Calibration (METRIC), operational Simplified Surface Energy Balance (SSEBop), Simplified Surface Energy Balance Index (S-SEBI) and Surface Energy Balance System (SEBS) (Wagle et al., 2017). To estimate ET over a vast area, numerous remote sensing-based SEB models have been developed. Five widely used single-source SEB models for estimating ET are Surface Energy Balance Algorithm for Land (SEBAL), Mapping ET with Internalized Calibration (METRIC), operational Simplified Surface Energy Balance (SSEBop), Simplified Surface Energy Balance Index (S-SEBI) and Surface Energy Balance System (SEBS) (Wagle et al., 2017).

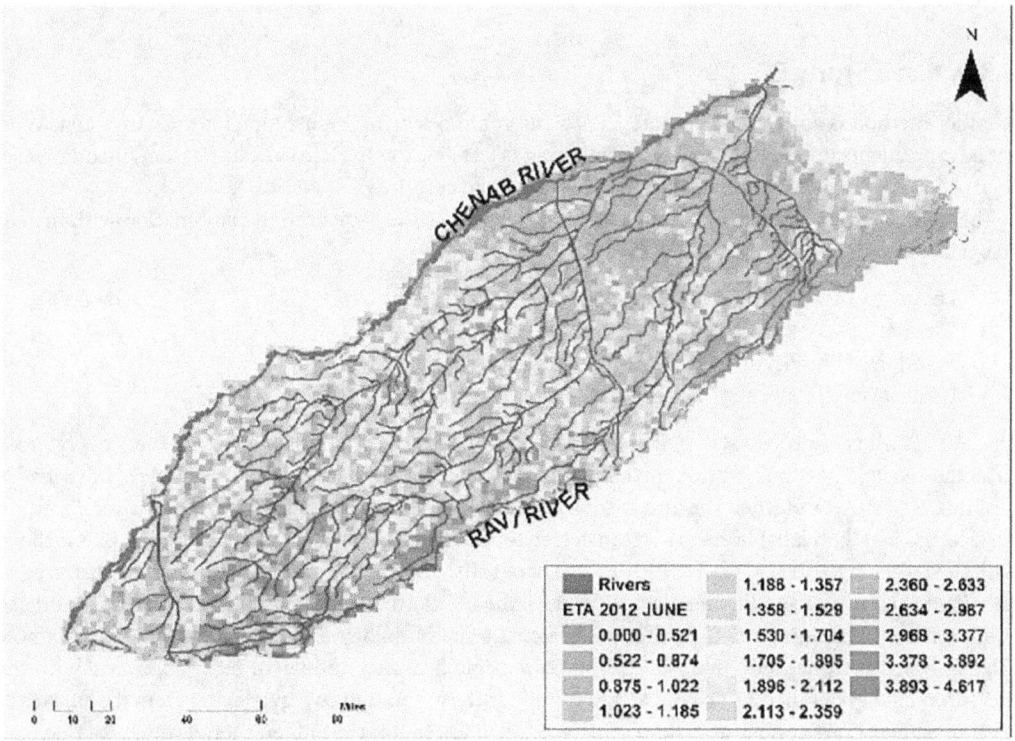

FIGURE 6.9 Map of summer ETA.

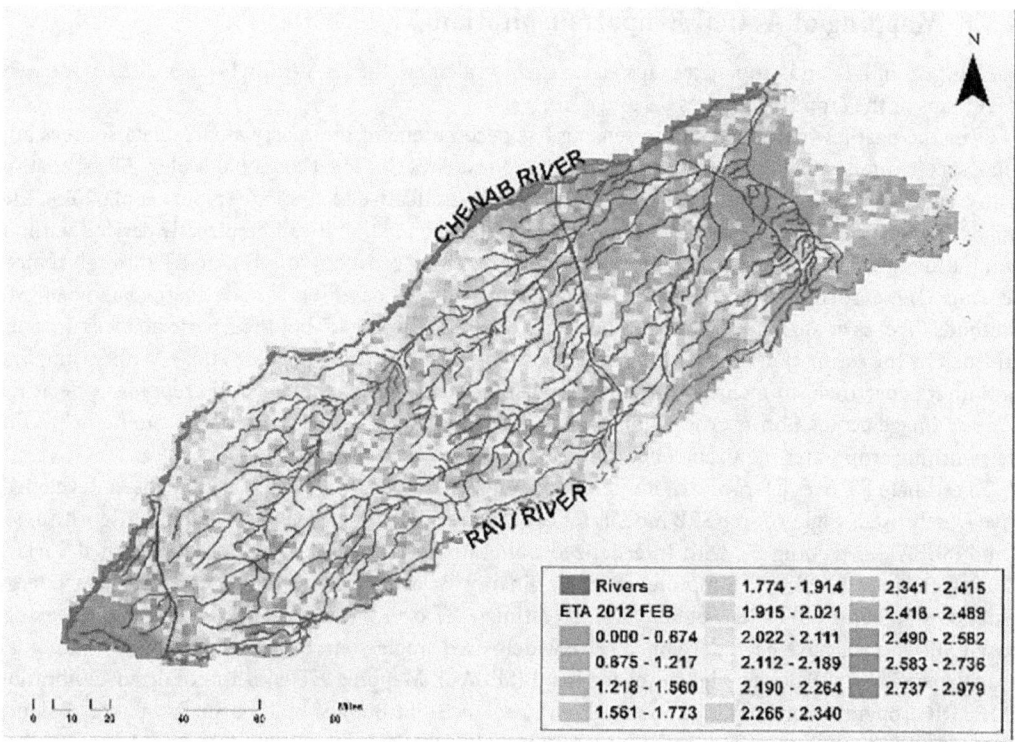

FIGURE 6.10   Map of winter ETA.

6.3.9 Case Study 2

Another method is applied to estimate ETA using remote sensing inputs for Victoria, Australia. With the advancements in remote sensing technology, it is possible to estimate various weather phenomena at high spatial resolution with high level of confidence and certainty.

The following remote sensing products from MODIS satellite were used in combination with meteorological data.

- Leaf Area Index (LAI)
- Albedo
- Fraction of Photosynthetically Active Radiation
- Land cover (classification of NDVI data as explained in case study 1).

Due to heat transfer process, the energy from surfaces is lost to the atmosphere. This energy is referred to as the sensible heat flux. When the energy moves from the surface to the atmosphere, the sensible heat flux is positive and when the heat is transferred from the air to the surface, it is negative. For water surface, most of the available energy is transferred to the air in the form of latent heat which is associated with the ET , and for arid surfaces (e.g. deserts, bare soils), most of the available energy is transferred to the air in the form of sensible heat, which warms the air close to the surface. This process is modeled using remote sensing inputs. Case Study 2 presents a methodology for estimating ETA using remote sensing-based LAI, net radiation, soil heat flux, sensible heat flux and latent heat flux, along with the measurement of reference ET (ET_0). Latent heat flux is the amount of energy moving from the surface to the atmosphere due to evaporation and transpiration of water. It is the hidden energy which is associated with the phase change of water, i.e. evaporation, condensation and sublimation.

The results of potential ET and the ETA for winter and summer months for the state of Victoria, Australia are presented in Figures 6.11–6.14.

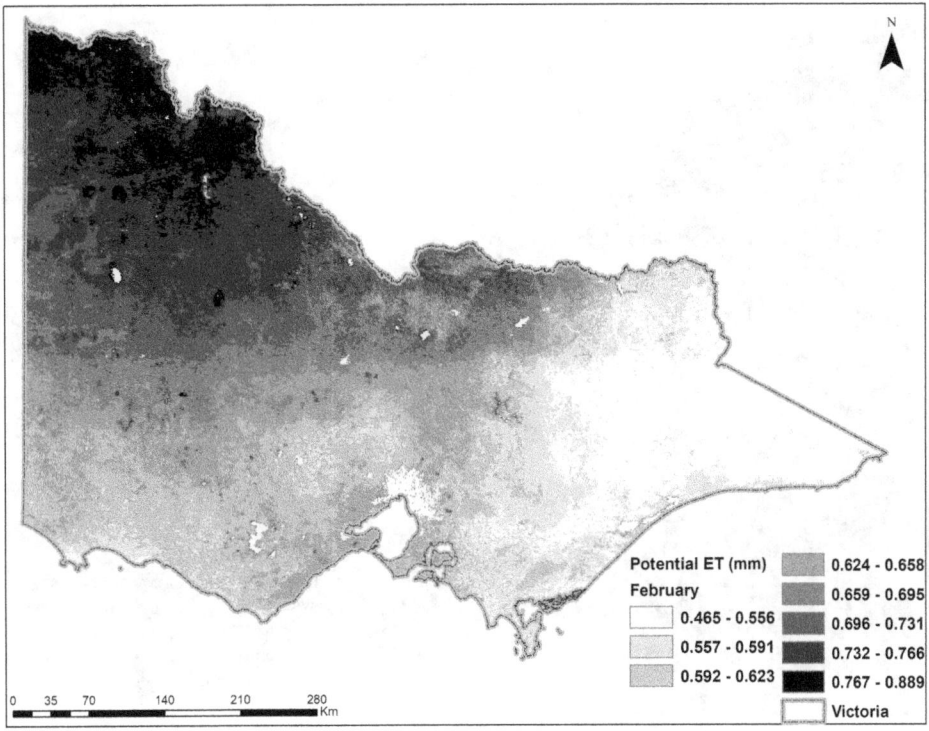

FIGURE 6.11   Calculated evapotranspiration in Victoria during a summer month.

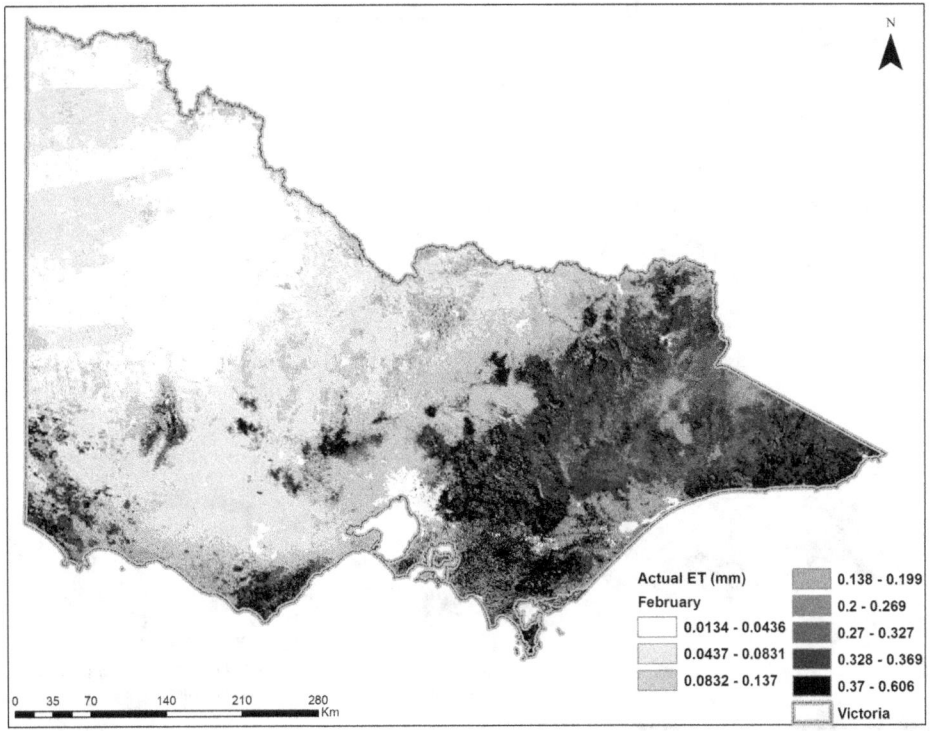

FIGURE 6.12   Actual evapotranspiration in Victoria during a summer month.

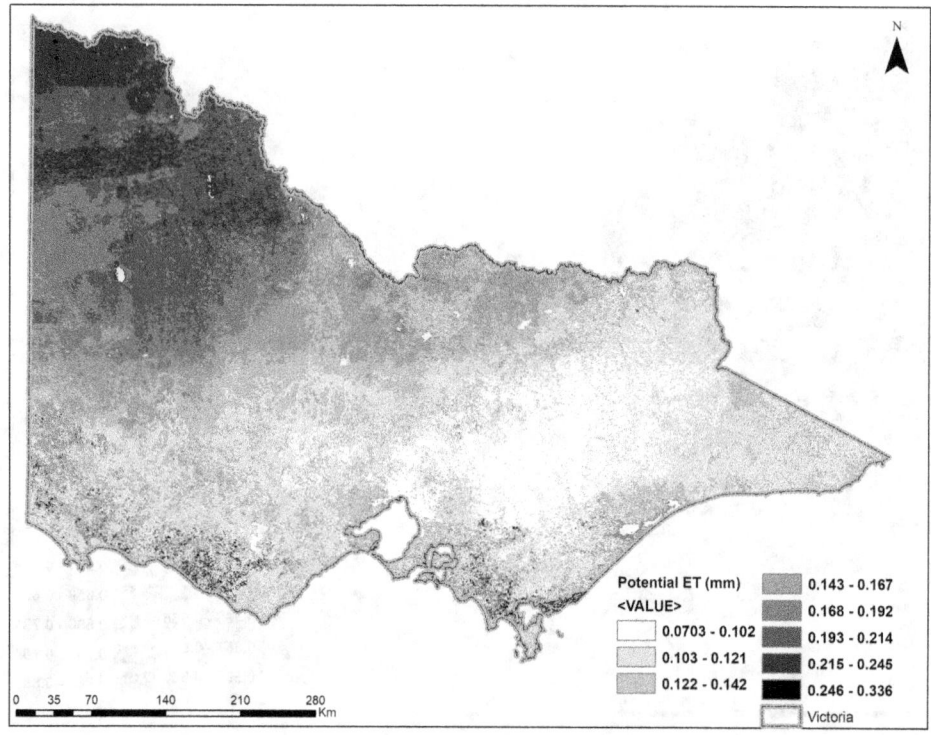

FIGURE 6.13 Calculated evapotranspiration in Victoria during a winter month.

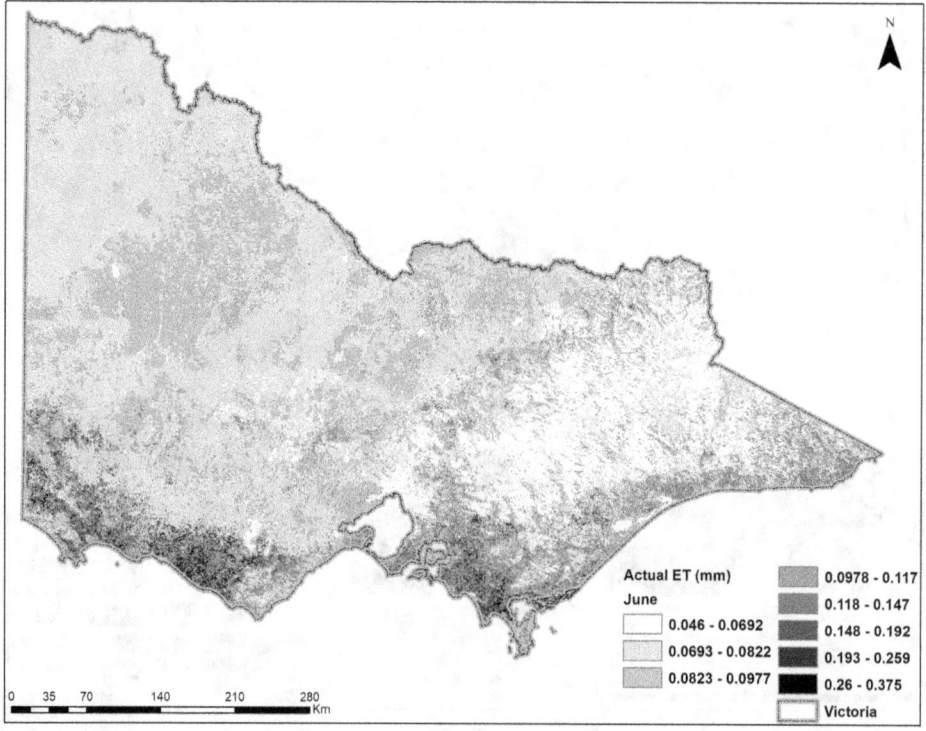

FIGURE 6.14 Actual evapotranspiration in Victoria during a winter month.

6.4 Crop Water Requirement

Crop water requirement is defined here as "the depth of water needed to meet the water loss through ET (ET_{crop}) of a disease-free crop, growing in large fields under nonretracting soil conditions including soil water and fertility and achieving full production potential under the given growing environment". Therefore, ET_{crop} gives the value of crop water requirement.

$$CWR = ET_{crop} \qquad (6.5)$$

To estimate the irrigation water requirement which is sometimes referred to as net irrigation requirement (In), we need to estimate the amount of water being supplied by other resources such as rainfall or precipitation and groundwater subtraction for the purpose of watering the crop. Not all the rainfall will be utilized by the crop as part of rainfall will be percolated beyond the root zone of the crops and part of the rainfall will be drained as runoff. The rainfall which is remained is known as effective rainfall (Pe). The third significant portion of water which is available to the crop is stored in the soil. Thus, net irrigation requirement can be estimated by using the following equation:

$$In = CWR - Pe - Ge - Wb, \qquad (6.6)$$

where,

In = net irrigation requirement
CWR = crop water requirement which is equivalent to ET_{crop}
Pe = effective rainfall
Wb = soil moisture available for crops
Ge = groundwater

At this point, it is worth mentioning to introduce the concepts of irrigation losses and efficiency of the irrigation systems. Less irrigation losses will require less water and less efficient irrigation systems (e.g. center pivots in the field) will use more water to irrigate the fields. Therefore, Eq. 6.6 can be written in terms of net irrigation water requirement as:

Net irrigation water requirement = (crop evapotranspiration + water available in the soil + deep percolation + runoff – effective rainfall – groundwater supply)/efficiency of the irrigation method

6.5 Conclusions

Food security and poverty eradication can be achieved by enhanced water productivity through sustainable management. For sustainable water resource management in the domain of food production, quantification of ET is essential as it is one of the most important constituents of water cycle and plays a vital role in irrigation efficiency by providing precise amounts of water need by the food-producing areas The rate of ET is affected by many conditions including the soil type, climate, plant species and their growing stage. The lower temperature, high humidity and lower winds decrease the amount of water loss allowing the water to be absorbed by the plant while the higher temperature, low humidity and higher winds all increase the amount of water loss from the soil and plant. As a result, the crop's water requirement increases to fulfill the water need for plant growth and production.

Calculating ET is crucial for crops to get enough water during their various growth stages. By analyzing the data regarding the different plant species, along with information regarding soil makeup and current climate conditions, it can be more accurately determined how much water agricultural fields require over a specified period of time. This knowledge can enable the farmers, researchers and policy makers to efficiently use irrigation water which is available in limited amount. The model can be applied to a wide spatial range from global to river basin to region to farm and to field level as it calculates the energy balance on pixel level. As satellite images come in different spatial and temporal resolutions, the ET estimation via energy balance algorithms can provide a wide variety of ET maps with different spatial and temporal scales. Moreover, the time series of ET maps with a good spatial scale can provide a sufficient knowledge to researchers, farmers and policy makers for the efficient use of irrigation water especially in arid and semi-arid regions which are facing shortage of water problems. In addition to remote sensing, dense network of meteorological observatories for measuring ET is also needed, which is lacking in developing countries.

One of the challenges put forth by the climate change is sustaining the agricultural productivity. Variability and scarcity of rainfall in the country produce the uncertainty conditions for agriculture. Therefore, it is required for optimal planning to know what is grown, where it is grown and how much water is needed to efficiently manage the water resources. Further, geographic information system aided by satellite remote sensing provides options to timely availability of accurate information on crop water requirements.

References

Acharya, T. D., and Yang, I. (2015). Exploring LANDSAT 8. International Journal of IT, Engineering and Applied Sciences Research, 4, 4–10. Retrieved from http://www.irjcjournals.org/ijieasr/Apr2015/2.pdf.

Allen, R.G., Pereira, L.S., Raes, D., and Smith, M. (1998). Crop evapotranspiration – Guidelines for computing crop water requirements. *FAO Irrigation and Drainage Paper 56*. Rome: Food and Agriculture Organization of the United Nations. Retrieved from: http://www.fao.org/docrep/X0490E/x0490e00.htm.

Bastiaanssen, W., Pelgrum, H., Wang, J., Ma, Y., Moreno, J., Roerink, G., and Van der Wal, T. (1998). A remote sensing surface energy balance algorithm for land (SEBAL): 2. Validation. *Journal of Hydrology*, 212–213, 213–229. https://doi.org/10.1016/s0022-1694(98)00254-6.

Bates, B. C., Kundzewicz, Z. W., Wu, S., and Palutikof, J. P. (2008). Climate change and water. *Technical Paper of the Intergovernmental Panel on Climate Change*. Geneva: IPCC Secretariat, Switzerland.

Cleugh, H. A., Leuning, R., Mu, Q., and Running, S. W. (2007). Regional evaporation estimates from flux tower and MODIS satellite data. *Remote Sensing of Environment*, 106(3), 285–304. https://doi.org/10.1016/j.rse.2006.07.007.

de Bie, C. A. J. M., Khan, M. R., Smakhtin, V. U., Venus, V., Weir, M. J. C., and Smaling, E. M. A. (2011). Analysis of multi-temporal SPOT NDVI images for small-scale land-use mapping. *International Journal of Remote Sensing*, 32(21), 6673–6693. https://doi.org/10.1080/01431161.2010.512939.

Eslamian, S., Abedi-Koupai, J. and M. J. Zareian. (2012). Measurement and modelling of the water requirement of some greenhouse crops with artificial neural networks and genetic algorithm. *International Journal of Hydrology Science and Technology*, 2(3), 237–251.

FAO (2017). *Water for Sustainable Food and Agriculture, A report produced for the G20 Presidency of Germany*. Rome, Italy: Food and Agriculture Organization of the United Nations. http://www.fao.org/3/a-i7959e.pdf.

Hoogeveen, J.; Faurès, M; Peiser, L.; Burke, J., and van de Giesen, N. (2015). GlobWat – A global water balance model to assess water use in irrigated agriculture. *Hydrology and Earth System Sciences*, 19, 3829–3844. https://doi.org/10.5194/hess-19-3829-2015.

Irmak, A., Ratcliffe, I., Ranade, P., Hubbard, K., Singh, R. K., Kamble, B., and Kjaersgaard, J. (2011). Estimation of land surface evapotranspiration with a satellite remote sensing procedure. *Great Plains Research*, 73–88.

Khan, M. R., de Bie, C. A. J. M., van Keulen, E. M. A., and Smaling, R. (2010). Disaggregating and mapping crop statistics using hypertemporal remote sensing. *International Journal of Applied Earth Observation and Geoinformation*, 12, 36–46, https://doi.org/10.1016/j.jag.2009.09.010.

Kite, G., and Droogers, P. (2000). Comparing estimates of actual evapotranspiration from satellites, hydrological models, and field data: A case study from Western Turkey. *IWMI Research Report 042*. Colombo, Sri Lanka: International Water Management Institute (IWMI). 32p. https://doi.org/10.3910/2009.049.

Kløve, B., Ala-Aho, P., Bertrand, G., Gurdak, J. J., Kupfersberger, H., Kværner, J., Muotka, T., Mykrä, H., Preda, E., Rossi, P., Uvo, C. B., Velasco, E., Pulido-Velazquez, M. (2014). Climate change impacts on groundwater and dependent ecosystems. *Journal of Hydrology*, 518(PB), 250–266.

Kustas, W.P., Norman, J.M., 1996. Use of remote sensing for evapotranspiration monitoring over land surfaces. *Hydrological Sciences Journal*, 41(4), 495–516. https://doi.org/10.1080/02626669609491522

Liu, W., Hong, Y., Khan, S., Huang, M., Vieux, B., Caliskan, S., and Grout, T. (2010). Actual evapotranspiration estimation for different land use and land cover in urban regions using Landsat 5 data. *Journal of Applied Remote Sensing*, 4(1), 041873. https://doi.org/10.1117/1.3525566.

MacAlister C., and Subramanyam, N. (2018). Climate change and adaptive water management: Innovative solutions from the global South, *Water International*, 43(2), 133–144, https://doi.org/10.1080/02508060.2018.1444307.

Mu, Q. Z., Heinsch, F. A., Zhao, M., and Running, S. W. (2007). Development of a global evapotranspiration algorithm based on MODIS and global meteorology data. *Remote Sensing Environment*, 111, 519–536.

Magliulo, V., d'Andria, R. and Rana, G. (2003) Use of the modified atmometer to estimate reference evapotranspiration in Mediterranean environments. *Agricultural Water Management*, 63, 1–14.

McMahon, T. A., Peel, M. C., Lowe, L., Srikanthan, R., and McVicar, T. R. (2013). Estimating actual, potential, reference crop and pan evaporation using standard meteorological data: A pragmatic synthesis, *Hydrological Earth Systems Science*, 17(4), 1331–1363.

Nouri, H., Beecham, S., Kazemi, F., Hassanli, A. M., and Anderson, S. (2013). Remote sensing techniques for predicting evapotranspiration from mixed vegetated surfaces, *Hydrology and Earth System Sciences Discussions*, 10(3), 3897–3925. https://doi.org/10.5194/hessd-10-3897-2013.

Trezza, R., Allen, R., and Tasumi, M. (2013). Estimation of actual evapotranspiration along the middle Rio Grande of New Mexico using MODIS and Landsat imagery with the METRIC model. *Remote Sensing*, 5(10), 5397–5423. https://doi.org/10.3390/rs5105397

Wagle, P., Bhattarai, N., Gowda, P. H., and Kakani, V. G. (2017). Performance of five surface energy balance models for estimating daily evapotranspiration in high biomass sorghum. *ISPRS Journal of Photogrammetry and Remote Sensing*, 128, 192–203. https://doi.org/10.1016/j.isprsjprs.2017.03.022.

Ullah, M. K., Habib, Z., and Muhammad, S. (2001). Spatial distribution of reference and potential evapotranspiration across the Indus basin irrigation systems. *International Water Management Institute (IWMI) Working Paper 24*. Sri Lanka.

United Nations. (2017). *World Population Prospects – Population Division*. United Nations. https://esa.un.org.

Xystrakis, F. and Matzarakis, A. (2011). Evaluation of 13 empirical reference potential evapotranspiration equations on the island of Crete in southern Greece. ASCE Journal of Irrigation and Drainage Engineering, 137, 211–222. http://dx.doi.org/10.1061/(ASCE)IR.1943-4774.0000283.

III

Environmental and Economical Impacts

7

Environmental Impacts of Irrigation

Nathaniel A.
Nwogwu
*Federal University of
Technology Owerri,
Nigeria;
Auburn University US*

Oluwaseyi A. Ajala
*Punjabi University India;
University of
Ibadan, Nigeria*

Ifeoluwa F. Omotade
*Federal University of
Technology Akure, Nigeria*

Toju E. Babalola
*Federal University
Oye-Ekiti, Nigeria*

Temitope F.
Ajibade and
Kayode H. Lasisi
*Federal University of
Technology Akure, Nigeria;
University of Chinese
Academy of Sciences,
PR China;
Chinese Academy of
Sciences, PR China*

Bashir Adelodun
*University of Ilorin, Nigeria;
Kyungpook National
University, South Korea*

Fidelis O. Ajibade
*Federal University of
Technology Akure, Nigeria;
University of Chinese
Academy of Sciences,
PR China;
Chinese Academy of
Sciences, PR China*

7.1 Background

Irrigation is instrumental in increased agricultural production (Kukal and Irmak, 2020; Usman and Gerber, 2019; Burney et al., 2010; Asayehegn et al., 2012). Like every other developmental stride, irrigation projects bring about social, economic, and environmental changes. Development targets achievement of positive and sustainable change, though some conflicting factors may arise due to development. To improve the well-being of both human and the environment, the developmental program should be evaluated before implementation to ensure minimal adverse social and environmental impacts while delivering its purpose (Hassanvand et al., 2017). Gone are the days when ecological degradation was ignored (Perret and Payen, 2020). In this regard, more awareness has been supported to educate people on better irrigation and drainage practices. Although the advantages of irrigation prevail over its disadvantages and costs, it is still necessary to discuss those adverse impacts that come with it. This would help decision/policy makers, irrigators, and farmers adopt ideal irrigation plans, designs, and management practices that could promote ecological sustainability. The need to establish lasting benefits and reduce adverse effects prompted the support of the concept of sustainability. Regarding agricultural development, water is generally insufficient in different regions of the world, yet agriculture is recorded as one of the significant withdrawals of water.

In addition, freshwater is not evenly distributed across the globe. For this reason, with increasing human population and human activities that require water has led to the continuous search for enough quality water to carry out these activities. The problem of water scarcity has prompted people in various parts of the world, especially the middle east, to resort to the use of wastewater (treat and untreated or poorly treated) for irrigation as a means of ameliorating their situation (Guadie et al., 2021; Thiruchelve et al., 2020; Al-Reyami et al., 2020). This reutilizing of wastewater has earned importance recently, particularly in the arid and semi-arid areas (El Moussaoui et al., 2019; Villamar et al., 2018; Ofori et al., 2021). Although many studies have suggested some benefits of wastewater reuse in irrigation, the confirmation of its appropriateness and general acceptance for irrigation is still a debate among researchers. This is owing to the various drawbacks associated with wastewater irrigation. Therefore, the problem with wastewater reuse is that while trying to solve water scarcity in agriculture, it equally brings about harmful effects to both human and the environment (Guadie et al., 2021; Gohil, 2000; Rehman et al., 2019). This is likely to manifest with the continuous use of wastewater (e.g., municipal wastewater) as it would lead to heavy metal accumulation in the irrigated area, plants, and groundwater (Thiruchelve et al., 2020). For instance, when there is a high level of heavy metals in agricultural soils and crops, the lives of the consumers of the farm produce (crops and vegetables) are threatened (Chaoua et al., 2019). Moreover, due to low expertise and funding, many developing countries still irrigate their farms with untreated or unwell-treated wastewater, which could be dangerous to the environment, health, and economy (Adewumi and Ajibade, 2019). A study on the environmental impacts of wastewater for irrigation at Dhamar City, Republic of Yemen, carried out by Rageh et al. (2017) suggested that direct use of wastewater is not suitable for irrigation because of the risks it presents to the environmental resources (Al-Reyami et al., 2020). Therefore, it is imperative to treat wastewaters from chemical contaminants, pathogens, and salinity to avoid the adverse effects it would present on the groundwater and soil conditions (Al-Reyami et al., 2020). Generally, it is necessary to conduct wastewater characterization to determine the possibility of using such wastewater for irrigation without jeopardizing human and environmental health (Ajibade et al., 2014a,b; Braga and Varesche, 2014; Ilori et al., 2019; Adewumi and Ajibade, 2019). Soil characteristics should also be considered while resorting to wastewater irrigation.

To evaluate the imports of any anthropological exploitation practice on the environment and establish effective actions to prevent and lessen negative impacts and enhance encouraging results, a formal process known as environmental impact assessment (EIA) is adopted. The EIA offers such benefits as predicting problems, figuring out ways to avoid the issues, and enhancing the positive impacts of a particular project (Dougherty et al., 1995). The adverse environmental effects of irrigation take some time to surface (Yusuf, 2008); hence, ecological impact assessment of every project should be carried out, bearing in mind the eruption of its adverse impacts in the future. This is imperative as it is inevitable that anything that affects the environment would directly or indirectly affect lives as well.

Generally, evaluating how sustainable an irrigation project can be is dependent upon the availability of local water and other water uses, the historical background of how irrigation systems have developed, and the peculiar attributes associated with particular irrigation practices considered (Baldock et al., 2000). Factors such as topography, rainfall characteristics, type of source available, and soil profile are essential in the selection of an irrigation scheme (Asawa, 2005). Also, combined use of surface and ground waters for irrigation is considered to yield maximal efficiency and benefits (Asawa, 2005). Different irrigation types are in use in the various parts of the world concerning the various determining factors and economic considerations. Therefore, environmental impacts attributed to irrigation would also vary with county and country (Baldock et al., 2000). Agriculture is considered a high consumer of water globally; in some cases, polluted water is used to meet the agricultural demand in irrigation (Ajibade et al., 2021). Baldock et al. (2000) further classified the ways by which irrigation affects the environment into four basic categories, viz.:

i. Direct impacts on water sources: irrigation affects the quality and quantity of surface and ground waters.
ii. Direct impacts on soil: irrigation affects the quality (via pollution) and quantity (via erosion) of the soil.
iii. Direct impacts on biodiversity and landscape: irrigation displaces earlier habitats. It creates new ones, either by debasing or preserving the existing habitats and affecting the diverseness and constitution of landscapes.
iv. Secondary impacts resulting in intensified agrarian production from irrigation (e.g., an increase in fertilizer usage).

These effects may take a gradual process to unfold (for instance, reduction in certain species as a result of contamination) or manifest dramatically (as can be seen when a valley gets flooded and thus creates a reservoir for irrigation, or when there are canal formations in a river which reduce and destabilize the flow of such river) (Baldock et al., 2000). Although irrigation has numerous advantages to both the environment and agricultural production, this chapter focuses more on the adverse impacts of irrigation to the environment and other components therein.

7.2 Soil Degradation, Erosion, and Sedimentation

Soil degradation is one of the significant environmental impacts of irrigation. It is, therefore, important that proper monitoring and management of irrigated soils are regularly done to help proffer solutions to the various soil degradation issues associated with multiple types of irrigation practices. There are varieties of agricultural activities resulting from an increase in intensified production that can lead to depletion of soil fertility. For instance, monocropping without enough time to fallow, rise in soil salinity resulting from continuous irrigation, reduced organic matter (OM) content of the soil, and increase in soil toxicity resulting from increased use of agrochemicals (e.g., fertilizers, pesticides, and herbicides), all contribute to soil degradation (Baldock et al., 2000; Dougherty et al., 1995; Tilman et al., 2002). In addition, irrigating with wastewater and using polymer-based fertilizers and pesticides often contribute to microplastic pollution in agricultural soils, thereby degrading the soil health and quality (Kumar et al., 2020).

FAO (1992) affirms that about 0.5% (i.e., about 5–7 million ha) of arable land becomes unproductive yearly due to soil degradation. Soil degradation as a result of irrigation covers many aspects of environmental degradation, including erosion, sedimentation, soil salinity, alteration of soil properties, and river morphology. Among these environmental impacts, soil salinization remains the primary cause of soil being unproductive. It affects production in many ways, such as in crop selection, crop germination, and yield, and making the soil tough to work on (Dougherty et al., 1995). Therefore, it is necessary to evaluate irrigation water quality and its compatibility with a particular soil type before irrigation is carried out to avoid or reduce associated adverse environmental impacts if not adequately assessed (Ayers and Westcot, 1985). For instance, a low-quality water might be good for irrigation on a sandy or permeable soil while such water could be considered unsafe for irrigation on clayey soil. Such difference in choice or acceptability of a particular water quality for a type is important. These soil degrading impacts of irrigation are discussed one after the other as follows.

7.2.1 Erosion

The exposure of irrigated lands to erosion is highly dependent on the type of irrigation adopted (Dougherty et al., 1995). The regular introduction of water into an area makes it wetter, hence making it get saturated when there is a raindrop. This in turn reduces the absorption or infiltration of rainwater into the soil, thereby building up more runoff which carries both soil particles and essential nutrients alongside. Aspects such as field size, stream size, gradient, and field layout do not change easily and they have a significant influence on erosion rates (Dougherty et al., 1995). Irrigation is a technical project that needs to be implemented and handled with adequate care to avoid extreme environmental hazards. Like other agricultural practices, irrigation affects soil structure and soil erodibility; therefore, proper irrigation practices should be adopted to eliminate soil degradation from irrigation. For instance, the ancient in-field patterns for managing water involving inadequate land development (cut and fill) practices by waterway embarkments could lead to severe local erosion at the upper-end of the watered land and simultaneously causing sedimentation at the middle- or lower-end areas of the farmland. Moreover, the field's microtopographic configuration will be altered, which may result to uneven distribution of water on the irrigated field. The increased erosion rate common in some irrigation projects' surrounding is often due to increased economic activities which the project attracts.

Furthermore, Ofori et al. (2021) affirm that wastewaters have more OM content compared to other sources of irrigation water (e.g., groundwater), and this has made wastewater a good and valued source of irrigation water since it could be rich in OM necessary for plant growth and yield. An increase in the soil OM content improves soil compatibility, soil erodibility, soil buffering capacity against acidification, nutrient cycling, and availability of nutrients such as nitrogen, phosphorus, and sulfur (Murphy, 2015). However, it has been reported that sodium adsorption ratio (SAR), an important factor considered in irrigation, relatively increased in wastewater irrigated soils those irrigated with conventional irrigation water (Ofori et al., 2021). SAR helps to show how sodium ions affect the soil structure (Marchuk and Rengasamy, 2010). Some adverse effects on both crops and soil occur when SAR becomes high in the ground, viz. decreased crop growth and increase in plant toxicity (Maqsood et al., 2015), calcium deficiency in plants (Imran et al., 2010), reduction in dissolved organic carbon (OC) sorption (Mavi et al., 2012), reduced soil fertility and agricultural productivity, damage to soil structural stability (Ofori et al., 2021), and decreased soil permeability and hydraulic conductivity (Arienzo et al., 2012; Oster et al., 2016). Therefore, proper design, good choice of irrigation type/water with regards to the soil type and topography, moral land development, irrigation frequency, and good water management practices associated with surface irrigation methods (e.g., use of gates, siphons, and checks) can be effective in reducing erosion and water distribution hazards on irrigated fields.

7.2.2 Surrounding Area Impact

As earlier said, irrigation is a development project, leading to a rise in anthropogenic activities within the surrounding establishment, especially in developing nations. This increase in human activities could result from the increased economic activities associated with irrigation development in an area. For instance, farmers and their families may engage directly or indirectly in irrigation activities. Irrigation development promotes anthropogenic activities such as grazing due to the increased number of live-stock, and the exploitation of forest resources, mainly for wood in the surrounding areas. These activities, in turn, make the soil more prone to erosion since the vegetation is adversely affected, negatively affecting the area's fertility and ecology.

Moreover, clearing more significant non-irrigated parts of the watershed results in rise in the water table downstream; for areas that have high salinity levels in groundwater, more recharge may lead to more excellent salt points in the streams and increase the degree of pressure in the lower watered parts of the field, thereby hindering leaching; this occurrence has been reported in South-eastern Australia, and it could be controlled by planting deep-rooted crops and trees in the upland areas (Dougherty et al., 1995).

To eliminate these associated problems, necessary plans and actions that can take care of such problems that might come up should be put in place from the outset of the project. This could be the provision of allowance for farm animals, firewood, or veg plots within the surrounding area of an irrigation scheme, and protecting unprotected expanses may be required.

7.2.3 River Morphology and Configuration

River morphology deals with the movement, direction, and shape of rivers and how they change in these aspects over time as controlled by rain, other weather conditions, floods and sediment transport. The capacity and shape of a river are a function of its current, the bed and bank material, and the sediments transported by the current. More energy and sediment loads go with the fast-flowing river than with a slow-moving one. Therefore, sediments settle more in reservoirs and deltas where flow velocity weakens (Khan et al., 2014). On the other hand, a river is said to be in a regime when the flow is moderately constant such that it is not erosive and the same time not slow to cause deposition of sediment (Dougherty et al., 1995). Khan et al. (2014) further stated that when the sediment transport capacity of the flow becomes greater than the sediment input to the river, channel bed erosion develops, while the reverse leads to channel bed aggradation. However, this condition changes with time. When flow reduces, river morphology is affected as the capacity of the flow to carry sediments reduces leading to their deposition in the slower moving reaches, thereby causing shrinkage of the river channel, and the reverse becomes the case for fast flow. Similarly, for a section where sediment balance is altered probably by reservoir or scouring of a dregs controlling structure, river morphology is significantly affected. Petts and Gurnell (2005) suggested that riparian vegetation equally influences river morphology since plants growth along the river floodplains leads to narrower channels. Also, when water is released from a reservoir, scouring may occur which would result in the dropping of the bed level proximately downriver of the dam which can be seen as an opposite of the possible resultant impact of low flow. It can be deduced that the major causes of change in river morphology within the area of irrigation scheme are erosion and sedimentation. Therefore, alteration of river morphology would affect the downstream uses such as navigation, withdrawal for drinking, industry, and irrigation. It may also have a great negative effect on the river ecology.

7.2.4 Channel Structures

Exposure of canal to impairment is greatly associated with alteration of the channel morphology and regime condition (Dougherty et al., 1995). This is because there would likely be some problems of siltation, and others related to pump and filtration operation at the intake structures resulting from increased suspended sediments. Water withdrawal assemblies may get blocked by sediments or water may be left beyond their reach. Furthermore, the debasement of the channel equally endangers the wholeness of the hydraulic structures (e.g., intakes, headworks, flood protection) and bridges; when structures are newly constructed, existing neighboring constructions are affected since the flow condition is changed or affected (Dougherty et al., 1995).

7.2.5 Sedimentation

Sedimentation is unavoidable in irrigated areas as they are very vulnerable to erosion. It is a global issue that affects both the irrigation systems design and the operational performance (de Sousa et al., 2019). Ensuring enhanced water and sedimentation management in irrigation is imperative to ensure sufficient water supply as well as to promote food production (Kisi, 2012; Kuscu et al., 2009). As earlier discussed, irrigated area gets wetter, hence reducing its capacity to absorb rainwater. This often leads to building up of runoff which flows along with sediments. Also, in an event of over-irrigation, runoff may be observed causing scouring on the field. These sediments are thereafter transported downstream and even to the river and reservoir bed, causing reduction of the river and reservoir capacity as well as blockage of hydraulic channels (de Sousa et al., 2019). In addition, irrigated areas are commonly having less wind-protecting vegetation and this sometimes allows for wind erosion blowing soil particles which in turn fill the canals (Dougherty et al., 1995). Failure of irrigation structures can occur when the dregs load of the water source exceeds the volume of the irrigation waterways for transporting the sediments.

Some commonly known irrigation schemes that have been reported to be challenged by high sedimentation problems are: Coromandel region in New Zealand (Ballantine et al., 2014; Marden and Rowan, 2015), Khoshi river system and Sunsari Morang Irrigation Scheme in Nepal (Depeweg and Paudel, 2003; Paudel, 2010), Elkhorn Slough Watershed and Upper North Santiam River Basin, Oregon in USA (Ouellet-Proulx et al., 2016; Spear et al., 2008; Uhrich and Bragg, 2003), Jatiluhur irrigation system, at Bekasi Weir Irrigation Scheme in Indonesia (Sutama, 2010), Magdalena river in Colombia (Higgins et al., 2015), and Iguatu Experimental Watershed in Brazil (Santos et al., 2017). In Africa, most pronounced irrigation schemes in terms of high sedimentation issues include: Southwest Kano Irrigation Scheme in Kenya (Ochiere et al., 2015), Metahara Scheme in Ethiopia (Ali et al., 2014; Bishaw and Kedir, 2015; Munir, 2011), Gezira Irrigation Scheme in Sudan (Osman, 2015), in suburban tropical basin in Congo (Lootens and Lumbu, 1986), and irrigation schemes in South Africa (de Sousa et al., 2019).

In order to arrest this menace as well as avoid the high cost of desilting canals and reservoir, some effective designs, planning and sedimentation monitoring tools should be employed from the onset of the project. For instance, the installation of sediment excluders/extractors at the headworks can assist in mitigating sedimentation and failure of irrigation canals.

7.2.6 Soil Salinity

Increase in soil salinity is one of the major environmental problems caused by irrigation (Manasa et al., 2020; Fathizad et al., 2020; Pulido-Bosch et al., 2018; Singh, 2018, 2021). FAO (2020) (as cited in Singh, 2021) affirms that over 3% of the global soil resources have been affected by salt. Many researchers have identified the common means by which soil salinity is increased on an irrigation scheme (Dougherty et al., 1995), and they include:

i. *Accumulation of salt content in the soil resulting from salts in the irrigation water:* irrigation water often has some salt content; part of the water introduced on the field is absorbed by plants, some infiltrate into the soil, and the other evaporate into the atmosphere. This water redistribution occurs while its salt content remains on the field but some infiltrate with water through the soil profile. Hence, the salt content of the irrigated soil increases with time as irrigation continues.

ii. *Use of agrochemicals:* soil salinity increases from the increased use of solutes or agrochemicals in the form of natural or artificial fertilizers, pesticides, etc. as not all the constituent salts would be used by crops.

iii. *Movement of naturally occurring salt in the soil:* salts occurring naturally in the soil may form solution with soil water or exist naturally as salty groundwater. Such challenge is more pronounced in the desert or dry zones where occurrence of leaching is almost impossible. In areas with both high groundwater level and salinity, water rises by capillarity and evaporates into the atmosphere while salts are left on the soil surface or within the subsoil.

iv. *Change or transfer of irrigation method:* humidity/salinity bridge may occur when there is a change from rainfed to irrigation of a single crop, or change from one to dual irrigation. The bridge is usually created between the deep groundwater and the salt-free surface layers of the soil. Careful soil monitoring is suggested in this regard especially when the irrigated regime is increased.

Moreover, increase in groundwater salinity may exist as result of waterlogging. However, saline groundwater is mainly pronounced in coastal areas. Salinization in soil and water resources is commonly caused by poor irrigation and drainage systems or practices in arid conditions (Perret and Payen, 2020). Fortunately, proper land development (cut and fill) and properly sustained drainage networks could help efficiently in ensuring even water distribution as well as mitigating salinity effects. Furthermore, the position of the freshwater and saline water boundary within the coastline is dependent upon the hydraulic potential of the freshwater (Dougherty et al., 1995). For lower water table, the boundary moves inland as the pressure decreases. This consequently leads to high increase in salinity of freshwater which adversely affects the soil, crop yield, local ecology, and exposes the people using such water (especially

FIGURE 7.1 Soil degradation resulting from waterlogging and salinization in irrigated areas (Singh, 2021).

for drinking) to health risk. Also, the use of wastewater for irrigation increases rate of salinization at 0–30 cm and 30–60 cm depths of the soil, although the salinity of freshwater irrigated soils is relatively low, especially at the deep soil layers (Khanpae et al., 2020). In a similar way, Niemeyer et al. (2020) reported that irrigation with produced water negatively affects the soil habitat function, and aquatic testing with leachates showed likely impacts to adjacent water bodies. Figure 7.1 shows soil degradation because of waterlogging and salinization in irrigated areas.

7.2.7 Soil Properties

Irrigation water affects soil structure and other soil properties, especially when long-term irrigation regimes is practiced; these effects of irrigation water on the soil properties are dependent upon the water quality, method of application, and nature of the irrigated soil, and may be beneficial or negative (Adejumobi et al., 2014), but wastewaters (treated/untreated) have been reported to cause more adverse effects than tap-water (Ofori et al., 2021). The buildup of salts in the soil is capable of causing permanent impairment to the soil structure crucial for irrigation and crop productivity; such negative impact is observed more in clayey soils where the presence of sodium causes the failure of soil structure (Dougherty et al., 1995). This effect brings about some difficulties such as rendering the area less productive for growing crops, making the soil difficult to work, and making reclamation difficult (if not impossible) for reclamation by standard techniques. However, the practice of mixing irrigation water or soil with gypsum prior to irrigation is employed to lessen the sodium content and effect in sodic soils (Dougherty et al., 1995).

In some areas (e.g., coastal swamps), acidification may be a great problem. The transition from rainfed to irrigated crop production needs adequate availability of nutrient as some amount of water and nutrients are lost through the soil cross section. If proper attention is ignored, the irrigation project becomes inefficient. When the water loss through the soil profile gets high, there would likely be an increased loss of beneficial cations. Also, a reduction in soil pH may lead to inability of plants to absorb nutrients and/or a rise in heavy metals presence in the soil; similarly, decrease in soil pH may equally cause a decrease in OM content of the soil and this consequentially leads to degradation of soil structure as well as decrease in soil fertility. For instance, a study carried out on the soils of the Omi Irrigation Scheme, Kogi State, Nigeria after 13 years of its operation showed that there was a change (reduction) from the initial neutral soil pH level (pH = 6.65–7.00) to at the beginning of the scheme to slightly acidic pH level (pH = 6.53–6.60) (Adejumobi et al., 2014). They equally observed a significant increase in some biological, physical and chemical of the soil including cation exchange capacity, OM, and OC. Hence, proper attention and monitoring should be given in this regard as remedying soil deterioration and acidification complications is very expensive.

7.3 Water and Air Quality

The environment and human health are affected by the quality of air. Therefore, efforts are being made globally to develop various ways to improve the quality of air. The largest environmental health risk is air pollution. It is a global crisis that has become increasingly significant in recent times and will continue well into the future. Poor air quality directly affects all living things. Some air pollutants contribute to greenhouse gas (GHG) emission and disrupt the processes in the ozone protective layer. It is thus necessary to protect this resource by ensuring that it is of the highest possible quality.

Water is another vital natural resource. Though the earth surface seemingly has an abundance of water, only some portion of the global water available is usable, making it a limited resource that must be used carefully. Water is used for varying purposes; its quality must, therefore, be determined before use. Furthermore, regular monitoring of the different sources of water is required to assess its quality.

7.3.1 Toxicity

Toxicity usually occurs when certain salts present in irrigation water are taken up by plants and accumulated in the leaves, causing damage to the plants. A high concentration of boron, chloride, and sodium in irrigation water leads to toxicity and harm plants. The concentration of salts in irrigation water that would lead to toxicity differs depending on the crop sensitivity. Testing of water can assist in ascertaining any components that might be toxic to plant.

Wastewater reuse in irrigation agriculture has recently become common in areas where water shortages are more pronounced (low-income, arid and semi-arid countries) (Hajjami et al., 2012). Particularly in the developing countries, reuse of untreated wastewater is becoming widespread (WWAP [United Nations World Water Assessment Programme], 2017). Wastewater irrigation is gaining broader recognition in many parts of the world because of its many advantages, water scarcity, and climate change. However, aside from the multiple benefits of this resource, reuse water also comprises heavy metals that create problems for agricultural production (Alghobar and Suresha, 2017). Therefore, long-term reuse of industrial or municipal irrigation wastewater can contribute to toxic pollutant accumulation in agricultural soils and plants (Singh et al., 2010). A study by Arora et al. (2008) suggested that irrigating farmlands with wastewater led to a high concentration of heavy metals in the growing plants' edible portions (Arora et al., 2008).

Excessive heavy metal deposition on agricultural soils by irrigating with wastewater is dangerous to the lives of inhabitants who consume crops grown in polluted areas (Chaoua et al., 2019). Toxicity effects on human health by heavy metal intake through food is already a concern reported worldwide (Malan et al., 2015).

7.3.2 Pollution

Pollution is the contamination of the environments with harmful substances to the health and well-being of its inhabitants. Pollutants are toxic substances produced in higher-than-average concentrations, which harm our ecosystems. Pollution of water is a global crisis faced by both developing and developed countries; it undermines economic development and the socio-environmental resilience and well-being of billions of people (Mateo-Sagasta et al., 2017). Due to excessive water pollution and global scarcity of water resources, protecting river water quality is very critical (Dwivedi and Shikha, 2016).

Irrigation has often been linked to water pollution caused by runoff and leaching of salt, pesticides, and fertilizers (Mateo-Sagasta et al., 2017). In many regions, the return flow of irrigation is a significant cause of water pollution as it impacts water quality for subsequent uses (e.g., domestic, industrial, and supply for irrigation). More so, institutions like the US Environmental Protection Agency consider irrigation to be the critical cause of water pollution (US EPA, 1992).

Use of polluted water from waste, agricultural, or industrial sources in irrigation adversely affects soil quality, increases the number of trace elements present in plants and soils, and harbors harmful pathogens (Allende and Monaghan, 2015; Hass et al., 2010). Polluted water is a potential cause of both direct and indirect contamination of agricultural produce (Singh et al., 2018), which contributes to increased soil and water degradation (Islam et al., 2018). Contaminants in soils are leached to the unsaturated zone in regions where over-irrigation is practiced, where they can lead to geogenic contaminant mobilization and possibly raise the levels of contaminants in local groundwater (Nolan and Weber, 2015).

7.3.3 Anaerobic Effects

Appropriate irrigation management is needed to maintain adequate soil moisture in the active root zone for optimal development of crops. Soil moisture in the crop root stone is increased above field capacity by over-irrigation. This excessive water application disrupts the soil's oxygen balance, inundates roots, lowers uptake of plant water, and thereby stresses plants (Irmak and Rathje, 2008). The restriction of oxygen in the soil leads to anaerobic conditions, affecting plant development and OM constituents. Shortage of oxygen in the soil profile damages the root, shoots, and germinating seeds they support (Vartapetian and Jackson, 1997), which are not adapted to this type of environment. Providing a medium for plant growth is an important function of soil; therefore, changes in its physical and chemical characteristics have a significant effect on the production of root biomass and, hence, on plant vegetation (Pierret et al., 2007). For the optimal development of plants, air and water have to be available in adequate proportion within the soil. The anaerobic condition affects both the chemical and biological properties of the soil, subsequently impacting biological activities, leading to many microbes not being able to survive without oxygen. The reduction of O_2 and the related mechanisms affect plant survival, growth, and functioning in waterlogged soils (Pezeshki and DeLaune, 2012). An increased population of putrefying microorganisms producing toxins and high levels of soil-borne diseases are also anaerobic effects caused by poor irrigation management.

7.3.4 Gas Emission

Intensified agriculture has proved to play an important role in increasing food production through different irrigation practices (important et al., 2020). Irrigation increases the production of crops; however, its operation also increases machinery and energy demand, which consumes significant fossil energy quantities and potentially contributes to GHG emissions (Pimentel and Pimentel, 2003). According to the Intergovernmental Panel on Climate Change, 2014 (IPCC, 2014), irrigation increases the agricultural impacts of global warming which is a major contribution of GHG. Increase in soil water content, C mineralization, respiration, and microbial activity of soils by irrigation can increase CO_2 and N_2O emission

(Sainju et al., 2012). Soils flooded by over-irrigation can release N_2O and CH_4 emissions through the process of denitrification and anaerobic decomposition of OM in the soil (Sainju et al., 2012).

Furthermore, the common practice of continuous flooding irrigation, used for growing rice, releases a significant quantity of CH_4. Continuous flooding leads to anaerobiosis condition in the soil, causing anaerobic breakdown of complex organic substrates by methanogens and the release of CH_4 (Islam et al., 2020). Aggravating climate change by increase in GHG emissions, through intensified agricultural practices, results in reduced agricultural productivity (Hu et al., 2019). Various other negative effects on the environment include loss of biodiversity, pollution of water, soil erosion, and degradation.

7.3.5 Groundwater Salinity

Roughly 30% of water withdrawn globally for irrigation is estimated to flow back to rivers and groundwater via conveyance losses and return flows (irrigation water above plant need, drained from the soil and returned to ground/surface water) (Scanlon et al., 2007). The quality of groundwater is impacted by return flow as it is usually composed of dissolved salts, concentrated by evapotranspiration. Irrigated agriculture has critical impacts on groundwater quality and can change the quantity of groundwater resources (Bouwer, 1987). Groundwater salinization in irrigated soils can occur by: concentration of salts through the uptake of plant water; irrigation with wastewater, saltwater intrusion due to pumping of groundwater; leaching down of salts into groundwater (Suarez, 1989; Van Weert et al., 2009). Several studies have also shown that intensive abstraction of groundwater for irrigated agriculture has led to the deterioration and depletion of aquifers globally (Faunt et al., 2016; Pulido-Bosch et al., 2018; Wada et al., 2010).

The salinity of groundwater occurs when the level of concentration of dissolved solids is over a pre-defined limit (total dissolved solids, or a thousand milligrams of dissolved solids in a liter of water, mg/L) (Freeze and Cherry, 1979). Saltwater intrusion is a common issue for coastal aquifers worldwide, mainly where groundwater abstraction is substantial (Werner et al., 2013). Over-exploitation and poor management have raised the risk for saline water intrusion, which can harm agricultural productivity of coastal crops by salt accumulation, resulting in harmful effects on soils and plants (El-Fadel et al., 2018). High salts concentration in groundwater increases salts content in the root zone, thereby exerting an osmotic effect on plants, causing them to expend and consume more energy in extracting water rather than using it for plant development and hence stunting plant growth (Blaylock, 1994).

Several past studies also found evidence that high salt concentrations have a detrimental effect on agricultural potential by lowering crop yields and causing additional costs for salinity management (Foster et al., 2000; George et al., 2008; Lee and Howitt, 1996).

7.4 Hydrology

Water is an essential aspect of the ecosystem, about 70% of the earth is covered with water while only about 2.5% is freshwater available for human use. Out of the 2.5% of freshwater available about 30% is found in groundwater and 70% in rivers, lakes, and glaciers (Salas et al., 2014). This is why a comprehensive understanding of hydrology which is the science that encompasses the existence, occurrence, distribution, properties, and storage of water above, on, and below the earth's surface is of utmost important in other to enable conservation of water resources and best management practices for its sustainability in the ecosystem. A process by which water flows continuously from the earth surface and water bodies to the atmosphere is known as hydrologic cycle. Figure 7.2 shows the hydrological cycle illustrating how water is circulated through the ecosystem. The limited availability of freshwater has resulted to its competitive demand, which has widely been increasing, thereby making it difficult to satisfy water for human consumption, agricultural practices, and other uses (Omotade et al., 2019). Globally, agriculture is the largest demand for water (OECD, 2010; Wallace, 2000) because water use in agriculture is for cultivation of crops and rearing of livestock; United States Geological Survey (USGS) reports that

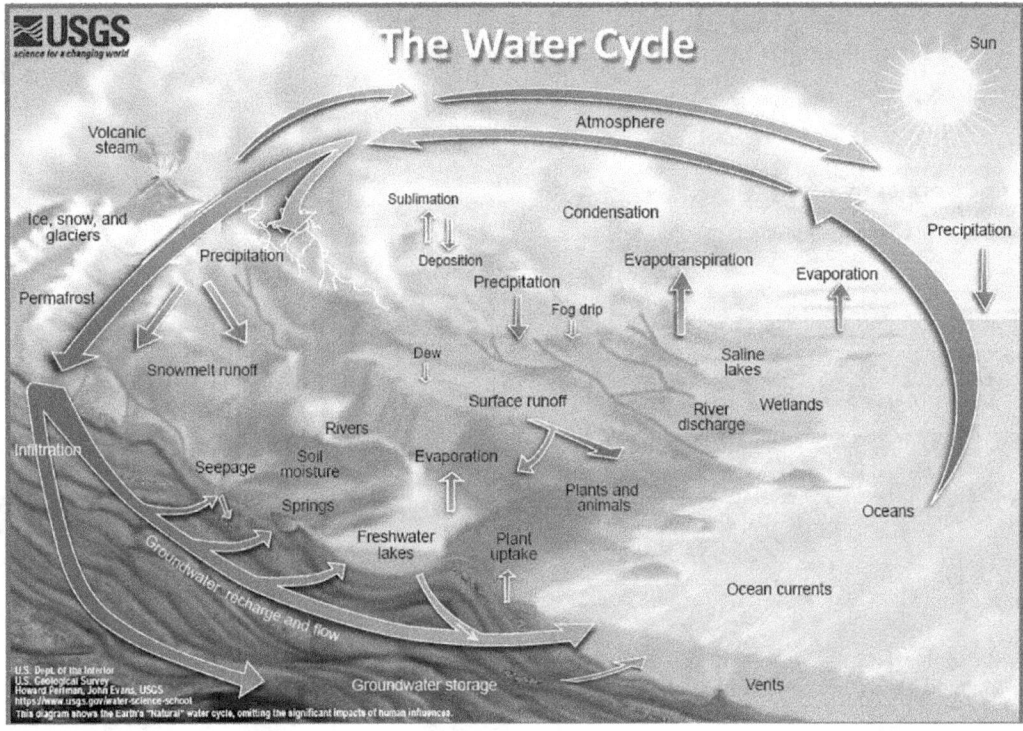

FIGURE 7.2 Hydrological cycle (Water cycle, 2020).

agricultural irrigation alone accounts for about 65% of the global freshwater available (Abunnour et al., 2016). Irrigation is important because it contributes to increase in agricultural production through the supply of water to crops to supplement limited rainfall and in dry regions to enable optimum yield. Recognizing the environmental impacts of irrigation on the hydrological cycle is important in ensuring effective management of groundwater resources and in the enhancement of optimum agricultural productivity (Foster et al., 2018).

7.4.1 Flow Characteristics Alteration

Irrigation affects the flow characteristics of rivers and groundwater discharge, because whenever new irrigation schemes are constructed or old schemes rehabilitated, the hydrological regime will be altered (Dougherty et al., 1995). During the process of fertigation, irrigation results in contamination of surface water through return flow and pollution of groundwater through seepage. This is because irrigation is an anthropogenic factor in the hydrological cycle which makes it highly significant in carrying salt and various contaminants in soil and water (Holy, 1993). Litskas et al. (2010) reported an increase in the concentration of nitrate due to the use of fertilizer during irrigation. This outcome impairs water quality, and decreases the sustainability of agricultural ecosystem and the environment (Holy, 1993) as highlighted in Figure 7.3. Also, poor sustainability of irrigation systems leads to aquifer exhaustion by abstraction of irrigation water, salinization, or contamination of water by minerals. Intrusion of salt into river bodies and infiltration of water with high content of salts endanger neighboring soil water and cause deterioration of watercourses (Foster et al., 2018; Holy, 1993). Impaired water quality in the downstream regions will have negative impacts on human health, recreation amenities, and make the ecosystem vulnerable (Brown and Harris, 2005).

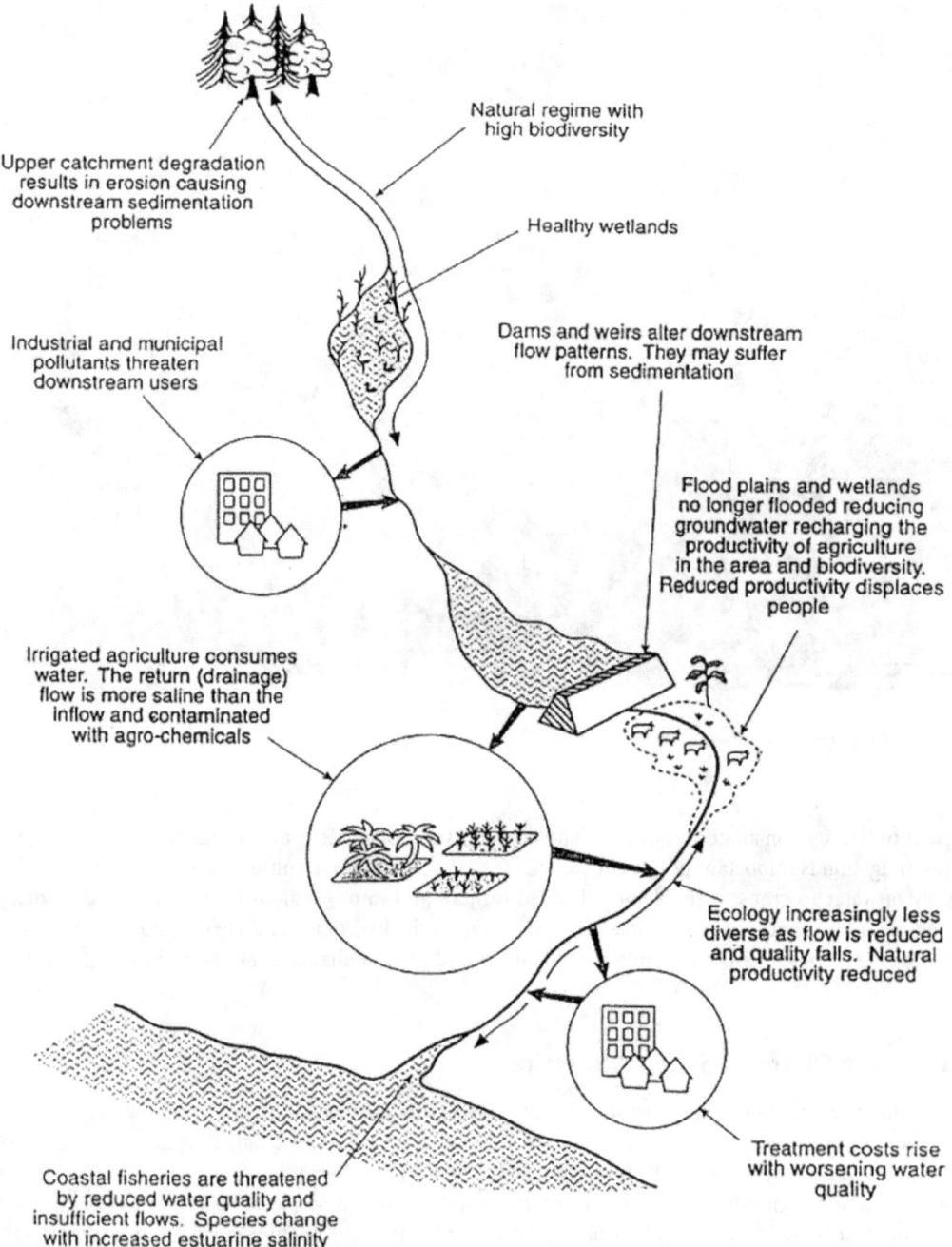

Natural regime with
high biodiversity

Upper catchment degradation
results in erosion causing
downstream sedimentation
problems

Healthy wetlands

Industrial and municipal
pollutants threaten
downstream users

Dams and weirs alter downstream
flow patterns. They may suffer
from sedimentation

Flood plains and wetlands
no longer flooded reducing
groundwater recharging the
productivity of agriculture
in the area and biodiversity.
Reduced productivity displaces
people

Irrigated agriculture consumes
water. The return (drainage)
flow is more saline than the
inflow and contaminated
with agro-chemicals

Ecology increasingly less
diverse as flow is reduced
and quality falls. Natural
productivity reduced

Treatment costs rise
with worsening water
quality

Coastal fisheries are threatened
by reduced water quality and
insufficient flows. Species change
with increased estuarine salinity

FIGURE 7.3 Origins and effects of poor water quality in a river body (Dougherty et al., 1995).

Continuous groundwater-based irrigation also negatively affects the flow between surface and groundwater, which results in drawdown and other issues, including soil compaction, hence, creating a disturbance of natural flow regimes as a result of the altered active and passive groundwater recharge zone (Foster et al., 2000; Leduc et al., 2001). More so, water flow is altered through the reduction of soil water infiltration capacity, which is as a result of solid particles present in irrigation water which thus influence the aeration rate of the soil. In the process of construction of irrigation systems on the land area, vegetative cover and top soil are removed, thus causing adverse effects on the hydrological balance in the watershed. Massive irrigation projects have the possibility of causing environmental disturbances as a result of changes in the hydrology and limnology of river basins. In addition, diverting water for irrigation use reduces river base flow which results in reduction in water supply for downstream users, and thereby increase in the pollution rate from municipal and industrial at the downstream; these pose health hazards.

7.4.2 Flood Characteristics Alteration

Flood can be defined as the presence of excess water on land as a result of high flow or overflow of water in an established watercourse (Magami et al., 2014). During rainfall or irrigation, excess water remains after infiltration and evaporation have taken place when the soil, streams, and artificial reservoirs are fully saturated with water; this excess water moves along the land surface as runoff. Most times, flood occurrences are caused by natural factors, but in some cases, anthropogenic (man-made) factors such as irrigation can also cause flooding. Human activities such as dam construction, irrigation, bridges, and others have impacted on free flow of water in the drainage channels, rivers, and streams (Aderogba, 2012). Altered flood regime can have adverse effect and negative impacts on the environment, as it reduces groundwater recharge through floodplains and causes loss of periodic or permanent wetlands. Irrigation schemes can lead to reduction in the amount of bare ground, thereby making an area to be prone to flooding. Also, it can affect the natural drainage pattern, thereby causing localized flooding. Flow alteration as a result of human activities emanating from dam construction, stream channelization, canal construction, irrigation schemes, etc., will lead to flooding and impair the ecosystems (Hupp et al., 2010).

Also, another positive significant impact of irrigation is the reduction in wind erosion risk (Brown and Harris, 2005). However, the negative impacts of flood are devastating; it can destroy lives and properties. Flood is perceived as the most dangerous in the physical environment that destroys what people have labored for in many years (Memon and Sharjeel, 2016). Also, reduce the water quality of downstream rivers through the introduction of huge quantity of eroded materials and damage of the environment through the leakage of harmful substances which affect the ecosystem (Zeleňáková et al., 2016; Hickey and Salas, 1995). Flood is the most dangerous in the physical environment that destroys what people have labored for in many years (Memon and Sharjeel, 2016).

7.4.3 Water Table Level Change

Water gets below the earth's surface through the infiltration of surface water down the soil profile to become groundwater (Winter et al., 1998). Water is stored within cracks in the rocks and tiny pores between soil aggregates which makes up of sand, gravel, rock, etc., held together by force of cohesion and fragmentation processes (Nimmo, 2013). The amount of water held in a soil is dependent on many factors which include how well the soil particles are loosely arranged, porosity of soil, etc. There are two main zones that describe where water is found in the ground; they are: unsaturated zone and saturated zone. Unsaturated zone is the region which lies immediately below the soil surface and contains air and water in the soil pores and rock fractures. This zone, also known as the zone of aeration, is between the land surface and aquifer which contains different natural constituents; also, flow processes that occur at this zone contribute to broad variation of hydrologic processes (Nimmo, 2005).

The saturated zone is the region underlying the unsaturated zone; it has water completely filled in all its pores and rock fractures. Water table is the upper layer of the zone of saturation; this layer lies between the unsaturated and saturated zones. Depth of water table varies seasonally depending on the attributes of the land area; it can be directly below the land surface if such area is close to permanently existing water bodies such as wetlands, rivers, lakes, or hundreds of meters below the land surface in some areas (Winter et al., 1998).

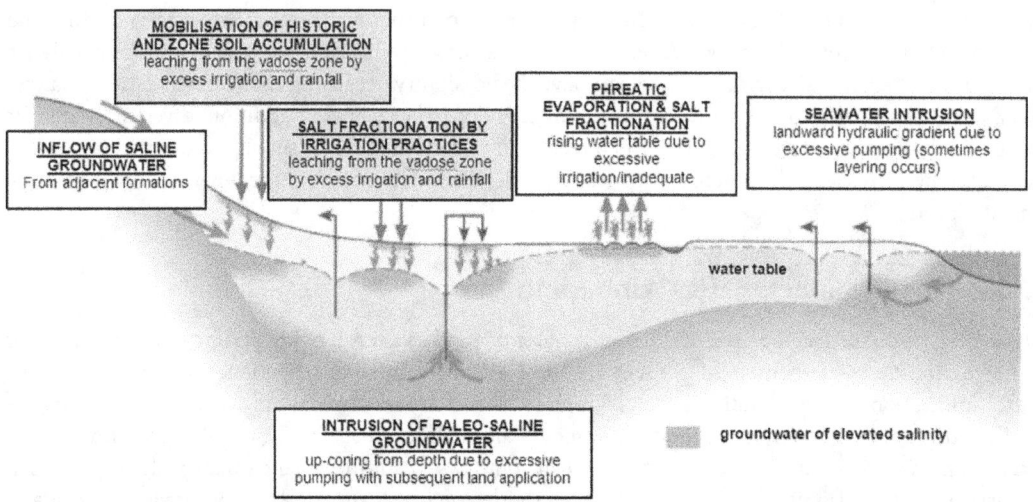

FIGURE 7.4   Processes of groundwater salinization (Foster et al., 2018).

According to Winter et al. (1998), irrigation contributes a larger percent than precipitation to groundwater recharge, especially in regions with low precipitation and low natural recharge where large irrigation systems are used, thus leading to rise in water table. Also, continuous recharge from irrigation return flow can result in permanent rise in the depth of water table, thereby causing increased outflow from irrigated area to surface water bodies from shallow groundwater (Winter et al., 1998). In areas with flat land having low hydraulic gradient, irrigation tends to raise the depth of water table as a result of losses of irrigation water through leakages from canals and percolation from irrigated fields. This results in an increase in groundwater recharge and capillary rise of water through the soil profile, thereby causing the problem of waterlogging of agricultural land and salinity of groundwater. Figure 7.4 shows a schematic representation of rising water table contributing to the processes of groundwater salinization. Groundwater-based irrigation can cause fall of water table; when irrigation water is drawn, there is a fall of water table which depletes the groundwater system, wells, springs, or river flow. Decrease in water table will cause springs that are fed by groundwater to dry up and also reduce low flow in rivers. Likewise, another problem associated with groundwater depletion is saline intrusion along the coast which leads to severe environmental and economic consequences (Dougherty et al., 1995).

7.4.4 Dam Siting and Operation Impacts

Dam is a hydraulic barrier or structure built across a stream, river to confine and check the flow of water, use for human consumption, irrigation, flood control, and electric power generation (Graham, 2002). The importance of dam to the construction of irrigation system is enormous because of the numerous benefits of impounding water for use at any seasonal period. However, there have been reports of alteration of natural flow regimes of watersheds all around the globe as a result of construction of dams (Mailhot et al., 2018; Yan et al., 2010). Dam construction is another direct means of measuring the negative impacts of irrigation on the environment. Dam siting in an area will alter the hydrological, geographical, and topographical factors in the neighboring areas, thereby causing unsuitable environmental conditions. Ajibade et al. (2020) pointed out that safety is the most significant factor over cost and capacity because a dam failure usually results in severe loss of life and property, necessitating the consideration of some crucial factors for an appropriate dam location. Penvenne (1996) reported that the impact of the construction of High Aswan Dam decreased sediment transport which led to erosion of wetlands in the Nile Delta 800 km downstream. Also, during the construction of dam, there will be alterations in the flow of water and changes in water quality downstream. The construction of dam and servicing of irrigation systems cause disruption of natural flow regime, which can significantly cause changes in siltation and erosion pattern. Globally, different studies have shown that dam construction and operations alter frequency, duration, and magnitude of flow regime (Sakaris, 2013; Magilligan and Nislow, 2005; Magilligan et al., 2003) and reduce peak flows and alter year-to-year flow variability (Magilligan and Nislow, 2005). During dam construction, likely risk which occurs is overtopping which results in erosion; this erosion causes increased inundation and sedimentation of downstream areas.

7.5 Biotic and Ecological Effects

Intensive agriculture via irrigation has been a key factor for the loss of biodiversity and environmental deterioration globally (Benton et al., 2003; Tilman et al., 2002; Green et al., 2005). And together with the increased in farm mechanization, agrochemical application, monocropping, and change in farming landscapes (José-María et al., 2010; Benton et al., 2003), irrigation scheme agriculture reduces ecological heterogeneity, causes valley flooding and aquifer drainage, and encourages increased chemical application; increased salt content in the soil and fallow lost due to crop rotation (Dover and Sparks, 2000; Ruiz, 1990; Stoate et al., 2001; Herrero and Snyder, 1997). Since intensive agriculture via irrigation is encouraged globally to increase food production particularly in dry landscapes, agricultural production has increased in two-to-three folds over the last 50 years with 12% increase in arable land (FAO, 2011). Therefore, this enhanced agricultural production due to irrigation which is still growing 117% globally and may implies several biotic and ecological effects.

7.5.1 Biodiversity and Landscapes

Information on the understanding of the impacts of irrigation on biodiversity is still very limited. Studies, especially those focusing on the creation of irrigated land from dryland, recorded the adverse impact of such transformation on birds (de Frutos and Olea, 2008; Ursúa et al., 2005; Tella and Forero, 2000; Laiolo, 2005; Brotons et al., 2004). Records on the adverse effect of irrigation on arthropods have also been highlighted (Holzschuh et al., 2007; Ponce et al., 2011; Hyvönen and Salonen, 2002; Krauss et al., 2011; Roth et al., 2008; Prieto-Benítez and Méndez, 2011; Herzog et al., 2005; Tybirk et al., 2004; Knop et al., 2006), while positive effects such as abundance, richness, and diversity have also been published (González-Estébanez et al., 2011; Kleijn et al., 2001; Weibull and Östman, 2003; Melnychuk et al., 2003; Weibull et al., 2000). However, such dissimilarities were attributed to the landscape complexity. González-Estébanez et al. (2011) explored the complexity of landscape on butterfly diversity due to irrigation. Butterflies are known as a very important environmental indicator as the result of their sensitivity even to subtle ecological changes (Kremen, 1992; Erhardt, 1985; Stefanescu et al., 2004; Thomas, 2005). In a dry cereal farming system compared to conventional irrigation system, increased numbers of butterflies were recorded (González-Estébanez et al., 2011). This was attributed to the landscape complexity due to the irrigation scheme.

7.5.2 Impacts on Biotas and Habitats

Irrigation activities expose surface water to diverse effect ranging in different intensity, scales, duration, and strength, thus depleting biotas and causing environmental disruption (Lake, 2000; Giller and Myers, 1996). Generally, freshwaters including lakes and rivers are exposed to such disturbances especially in irrigation-fed agriculture, grazing by livestock and land clearance, which generate significant sediments into water bodies (Hornung and Reynolds, 1995; Waters, 1995; Harding et al., 1998), thus, reducing the water biodiversity, interstitial and habitat spaces (Waters, 1995; Boulton, 1999). An example is the case of the decreased diversity and abundance of benthic invertebrates such as Ephemeroptera, Plecoptera, and Trichoptera, and burrow animals such as oligochaetes and chironomids (Lake et al., 2000).

7.5.3 Water Bodies

Irrigation-fed agriculture depends on the water supply from water bodies. As a result, water characteristics and method with which the irrigation is done impact the environment (Stockle, 2001). The removal of groundwater for irrigation can cause the land to subside, aquifers to become saline, and can accelerate other adverse ecological impacts (Zektser et al., 2005; Wang et al., 2018; Greene et al., 2016). An example is the Colorado river which has been reported to hold nearly no water due to irrigation schemes (Fradkin, 1996; Nagler et al., 2008; Gertner, 2007). The same was reported for the San Joaquin which beds have been recommended by designers for construction of houses. Cotton flooding has over the past 33 years, reduced the Aral Sea to 75% of its volume causing elevated saltiness.

7.5.4 Pests, Weeds and Animal Diseases

Irrigation commonly provides favorable environments that enable the development of crop diseases, specifically, fungal and bacterial foliage diseases (Dougherty et al., 1995). Illnesses and unwanted plants often outspread easily via wastewater reuse and drainage water. Yusuf (2008) has reported some diseases such as malaria, schistosomiasis, and dysentery/diarrhea within the area of the Bakolori Irrigation Project. Irrigation brings about some favorable environmental changes to some species of pests such as rodents and insects. It may equally provide a good environmental condition for proper growth and yield of local and imported varieties of weeds. However, through the common switching of land uses and intensified application of pest-killing chemicals associated with irrigation, habitats of some natural predators like snakes, birds, and spiders may be lost or reduced (Dougherty et al., 1995). Furthermore, there is usually a problem of aquatic weeds which cause some issues such as reduction in storage and conveyance volume of reservoirs, channels and drainage systems, increased water loss via evapotranspiration, and provision of prosperous and protective environment for disease vectors (e.g., snails and mosquitoes). All these problems arise due to the presence of large water bodies that serve as breeding sites for diseases vectors and pests (Yusuf, 2008).

7.6 Ensuring Environmental Quality and Sustainability: Sustainable Ways and Techniques to Mitigate Negative Impacts of Irrigation on the Environment

The continuous growth in the world's population will lead to the construction of more dams, aqueducts, and other kinds of infrastructure for irrigation, especially in developing nations where an increase in agricultural productivity is needed. More consideration should be given to the local people and their environment by building higher standard infrastructures with minimal environmental impact and suitability than in the past. Even in regions where these infrastructures are most needed, minimal ecological disruption and cheaper resources should be considered. Based on physiography and soil conditions, the FAO has estimated the potential for an additional 400 million hectares of land, of which three-quarters of developing countries will be added to irrigation land. The belief that irrigation land is more productive than rainfed soil has created the expectation for an increase. Although, inadequate suitable lands, shortage of water supplies, and the high installation cost limit the growth of the current irrigation schemes in developing countries. Hence, improving the existing irrigation structure is more advantageous for such a region than giving room to new constructions.

Agriculture is one of the largest consumers of water. Water via irrigation is largely lost and inefficient as it moves from farmers to crops. As a result, more than half of the water diverted into agriculture never yielded to any food. Thus, a modest improvement in agricultural efficiency can save lot of waters for other purposes. Also, tremendous gains in water efficiency can be gained by shifting where people use water. For instance, supporting 100,000 innovative California occupations requires approximately 250 million gallons of water a year; a similar measure of water utilized in the horticultural area supports less than ten positions, a shocking distinction. Comparable figures apply in numerous different nations. Eventually, these variations will prompt more strain to move water from agrarian utilizations to other financial areas.

Advanced ways to deal with addressing water needs will not be anything but difficult to execute because financial institutions are often empowering the wastage of water and damaging environmental systems. Effective water management and utilization are hampered by several factors, including inappropriately low water costs, a lack of knowledge about new, better technologies, inadequate water budgets, and government subsidies for the development of dams or water-intensive crops in desert regions. Different interventions are practicable for forestalling, alleviating, or reversing soil and water degradation at various levels with irrigation-fed farms. A summary of approaches to solving water issues are listed in different categories below:

Policy Interventions

- Introduction of better market value for water and power pricing.
- Transferable water benefits could be encouraged.
- Introduction of standards to foster groundwater recharge and penalties for defaulters.
- Incentives for land reclamation should be provided.
- Comprehensive impact assessment should be carried out for new irrigation schemes.
- Improvement and government incentives for maintaining existing irrigation schemes should be provided.

Engineering Interventions

- EIA should be incorporated in designing, construction, and operations of new irrigation schemes.
- Provision of maintenance scheme for irrigation structures.
- Construction of new and improvement of maintenance for existing drainage facilities.
- Effective alternatives to waste effluents disposal.
- Reduction and prevention of canal seepage via lining.

Irrigation/Agronomic Practices Interventions

- Water loss reduction from on-farm discharge.
- Minimizing deep percolation and surface runoff via improved irrigation systems performance.
- Improvement and precision land leveling of on-farm watercourse.
- Application of effective irrigation methods such as drip irrigation method.
- Sediment concentration in runoff water should be minimized.
- Cultivation of less water absorbing crops.
- Effective of chemical fertilizers.
- Soil amendments and reclamation practices application.

References

7.7 Conclusion

Irrigation has played significant impacts to solve agricultural water shortage for increasing food and environmental security. Due to the rapid growth of water demand from other non-agricultural sectors, the utilization of treated wastewater has been adopted as a supplement to rainfall. However, the poor wastewater treatment quality has led to adverse effects on hydrology, soil fauna communities and surrounding ecological resources. To this end, this chapter sums up the various impacts of irrigation development and practices on the environment including soil degradation, channel configuration, soil physicochemical properties, air and water quality, general hydrology of the region of operation, biotic and general ecological community, as well as their resultant direct/indirect negative impacts to the human society. Furthermore, some promising methods and practices spanning through policy intervention, engineering technology, and improved irrigation practices/techniques were suggested to help educate farmers, irrigators, policy makers, and the general public on the better intervention developments that could be employed. This will not only increase agricultural output at less cost but also help in ensuring human health safety and the sustainability of the environment.

References

Abunnour, M.A., Hashim, N.B. and Jaafar, M.B. (2016). Agricultural water demand, water quality and crop suitability in Souk-Alkhamis Al-Khums, Libya. *IOP Conference Series: Earth and Environmental Science*, 37(012045). https://doi.org/10.1088/1755-1315/37/1/012045.

Adejumobi, M.A., Ojediran, J.O., & Olabiyi, O.O. (2014). Effects of irrigation practices on some soil chemical properties on OMI irrigation scheme. *International Journal of Engineering Research and Applications*, 4(10), 29–35.

Aderogba, K.F. (2012). Qualitative studies of recent floods and sustainable growth and development of cities and towns in Nigeria. *International Journal of Academic Research in Economics and Management Sciences*, 1(3), 1–25, ISSN: 2226-3624

Adewumi, J.R., & Ajibade, F.O. (2019). Periodic determination of physicochemical and bacteriological characteristics of wastewater effluents for possible reuse as irrigation water. *International Journal of Energy and Water Resources*, 3, 269–276. https://doi.org/10.1007/s42108-019-00036-6.

Ajibade, F.O., Adelodun, B., Lasisi, K.H., Fadare, O.O., Ajibade, T.F., Nwogwu, N.A., Sulaymon, I. D., Ugya, A. Y., Wang, H. C., & Wang, A. (2021). Environmental pollution and their socio-economic impacts. In A. Kumar, V. K. Singh, P. K. Singh, & V. Mishra, *Microbe Mediated Remediation of Environmental Contaminants* (pp. 321–354). Cambridge: Woodhead Publishing, Elsevier. https://doi.org/10.1016/B978-0-12-821199-1.00025-0.

Ajibade, F.O., Adewumi, J.R., & Oguntuase, A.M. (2014a). Sustainable approach to wastewater management in the Federal University of Technology, Akure, Nigeria. *Nigerian Journal of Technological Research*, 9(2), 27–36.

Ajibade, F.O., Adewumi, J.R., & Oguntuase, A.M. (2014b). Design of improved stormwater management system for the Federal University of Technology Akure. *Nigerian Journal of Technology (NIJOTECH)*, 33(4), 470–481.

Ajibade, T.F., Nwogwu, N.A., Ajibade, F.O., Adelodun, B., Temitope, E.I., Ojo, A.O., … Akinmusere, O.K. (2020). Potential dam sites selection using integrated techniques of remote sensing and GIS in Imo State, Nigeria. *Sustainable Water Resources Management*, 6(57). https://doi.org/10.1007/s40899-020-00416-5.

Alghobar, M.A., & Suresha, S. (2017). Evaluation of metal accumulation in soil and tomatoes irrigated with sewage water from Mysore city, Karnataka, India. *Journal of the Saudi Society of Agricultural Sciences*. https://doi.org/10.1016/j.jssas.2015.02.002.

Ali, Y.A., Crosato, A., Mohamed, Y.A., Abdalla, S.H., & Wright, N.G. (2014). Sediment balances in the Blue Nile River Basin. *International Journal of Sediment Research*, 29, 316–328.

Allende, A., & Monaghan, J. (2015). Irrigation water quality for leafy crops: A perspective of risks and potential solutions. *International Journal of Environmental Research and Public Health*. https://doi.org/10.3390/ijerph120707457.

Al-Reyami, N.S., Shaik, F., & Lakkimsetty, N.R. (2020). Environmental impact on usage of treated effluents from ISTP for irrigation. *Journal of Water Process Engineering*, 36, 101363.

Arienzo, M., Christen, E.W., Jayawardane, N.S., & Quayle, W.C. (2012). The relative effects of sodium and potassium on soil hydraulic conductivity and implications for winery wastewater management. *Geoderma*, 173–174, 303–310. https://doi.org/10.1016/j.geoderma.2011.12.012.

Arora, M., Kiran, B., Rani, S., Rani, A., Kaur, B., & Mittal, N. (2008). Heavy metal accumulation in vegetables irrigated with water from different sources. *Food Chemistry*. https://doi.org/10.1016/j.foodchem.2008.04.049.

Asawa, G.L. (2005). *Irrigation and Water Resources Engineering*. New Delhi: New Age International (P) Limited, Publishers.

Asayehegn, K., Yirga, C., & Rajan, S. (2012). Effect of small-scale irrigation on the income of rural farm households: The case of Laelay Maichew District, Central Tigray, Ethiopia. *Journal of Agricultural Sciences*, 7(1), 43–57. https://doi.org/10.4038/jas.v7i1.4066.

Ayers, R.S., & Westcot, D.W. (1985). *Water quality for agriculture. Irrigation and Drainage Paper 29 (Revised)*. Rome, Italy: FAO.

Baldock, D., Caraveli, H., Dwyer, J., Einschütz, S., Petersen, J.K., Sumpsi-Vinas, J., & Varela-Ortega, C. (2000). *The Environmental Impacts of Irrigation in the European Union*. London: Institute for European Environmental Policy.

Ballantine, D.J., Hughes, A.O., & Davies-Colley, R.J. (2014). Mutual relationships of suspended sediment, turbidity and visual clarity in New Zealand Rivers. *Sediment Dynamics from the Summit to the Sea Symposium: New Orleans, LA, USA*.

Benton, T.G., Vickery, J.A., & Wilson, J.A. (2003). Farmland biodiversity: Is habitat heterogeneity the key? *Trends in Ecology & Evolution*, 18, 182–188.

Bishaw, D., & Kedir, Y. (2015). Determining sediment load of Awash River entering into Metehara Sugarcane Irrigation Scheme in Ethiopia. *Journal of Environmental & Earth Sciences*, 5(13), 110–117.

Blaylock, A.D. (1994). *Soil Salinity, Salt Tolerance, and Growth Potential of Horticultural and Landscape Plants*. Laramie, WY: University of Wyoming, Cooperative Extension Service, Department of Plant, Soil, and Insect Sciences, College of Agriculture.

Boulton, A.J. (1999). The role of subsurface biological filters in gravel-bed river rehabilitation strategies. In: I. Rutherfurd and R. Bartley (eds) *Second Australian Stream Management Conference: The Challenge of Rehabilitating Australia's Streams* (pp. 81–86). Melbourne: CRC for Catchment Hydrology.

Bouwer, H. (1987). Effect of irrigated agriculture on groundwater. *Journal of Irrigation and Drainage Engineering*. https://doi.org/10.1061/(asce)0733-9437(1987)113:1(4)

Braga, J.K., & Varesche, M.A. (2014). Commercial laundry water characterization. *American Journal of Analytical Chemistry*, 5(01), 8–16.

Brotons L., Mañosa S., Estrada J. (2004) Modelling the effects of irrigation schemes on the distribution of steppe birds in Mediterranean farmland. *Biodiversity and Conservation*, 13, 1039–1058

Brown, I. and Harris, S. (2005). *Environmental, Economic and Social Impacts of Irrigation in the Mackenzie Basin*. Wellington, New Zealand. www.waitaki.mfe.govt.nz

Burney, J., Woltering, L., Burke, M., Naylor, R., & Pasternak, D. (2010). Solar-powered drip irrigation enhances food security in the Sudano–sahel. *Proceedings of the National Academy of Sciences*, 107(5), 1848–1853. https://doi.org/10.1073/pnas.0909678107.

Chaoua, S., Boussaa, S., El Gharmali, A., & Boumezzough, A. (2019). Impact of irrigation with wastewater on accumulation of heavy metals in soil and crops in the region of Marrakech in Morocco. *Journal of the Saudi Society of Agricultural Sciences*, 18(4), 429–436. https://doi.org/10.1016/j.jssas.2018.02.003.

de Frutos A., Olea P.P. (2008) Importance of the premigratory areas for the conservation of lesser kestrel: space use and habitat selection during the post-fledging period. *Animal Conservation*, 11, 224–233.

de Sousa, L.S., Wambua, R.M., Raude, J.M., & Mutua, B.M. (2019). Assessment of Water Flow and Sedimentation Processes in Irrigation Schemes for Decision-Support Tool Development: A Case Review for the Chókwè Irrigation Scheme, Mozambique. *AgriEngineering*, 1(8), 101–118. https://doi.org/10.3390/agriengineering1010008.

Depeweg, H.T., & Paudel, K.P. (2003). Sediment transport problems in Nepal evaluated by SETRIC model. *Wiley InterScience*, 52, 260–274.

Dougherty, T.C., Hall, A.W., & Wallingford, H.R. (1995). Environmental impact assessment of irrigation and drainage projects. *FAO Irrigation and Drainage Paper 53*.

Dover, J., Sparks, T., 2000. A review of the ecology of butterflies in British hedgerows. *Journal of Environmental Management*, 60, 51–63.

Dwivedi, S., & Shikha, D. (2016). Water pollution: Causes, effects and control. *Biochemical and Cellular Archives*, 16(1), 96–102.

El Moussaoui, T., Wahbi, S., Mandi, L., Masi, S., & Ouazzani, N. (2019). Reuse study of sustainable waste-water in agroforestry domain of Marrakesh city. *Journal of the Saudi Society of Agricultural Sciences*, 18(3), 288–293.

El-Fadel, M., Deeb, T., Alameddine, I., Zurayk, R., & Chaaban, J. (2018). Impact of groundwater salinity on agricultural productivity with climate change implications. *International Journal of Sustainable Development and Planning*. https://doi.org/10.2495/SDP-V13-N3-445-456.

Erhardt, A., 1985. Diurnal Lepidoptera: Sensitive indicators of cultivated and abandoned grassland. *Journal of Applied Ecology*, 22, 849–861.

FAO. (1992). *Les périmètres irrigués en droit comparé afraicain (Madagascar Maroc, Niger, Sénégel, Tunisie)*. Rome, Italy: FAO, (French only).

FAO. (2011) *The State of the World's Land and Water Resources for Food and Agriculture (SOLAW) – Managing systems at risk*. London, UK: Food and Agriculture Organization of the United Nations, Rome and Earthscan.

FAO. (2020). *Management of salt affected soils: 'soil management' under 'FAO SOILS PORTAL'*. In: *Food and Agriculture Organization' of the 'United Nations'*. Rome. Retrieved from http://www.fao.org/soils-portal/soil-management/management-of-some-problemsoils/salt-affected-soils/more-information-on-salt-affected-soils/en/.

Fathizad, H., et al. (2020). Investigation of the spatial and temporal variation of soil salinity using random forests in the central desert of Iran. *Geoderma* 365, 114233. https://doi.org/10.1016/j.geoderma.2020.114233.

Faunt, C.C., Sneed, M., Traum, J., & Brandt, J.T. (2016). Water availability and land subsidence in the Central Valley, California, USA. *Hydrogeology Journal*. https://doi.org/10.1007/s10040-015-1339-x.

Foster, S., Chilton, J., Moench, M., Cardy, F., & Schiffler, M. (2000). Groundwater in rural development: Facing the challenges of supply and resource sustainability. *World Bank Technical Paper 463*. Washington, DC: World Bank.

Foster, S., Pulido-Bosch, A., Vallejos, Á., Llop A., & MacDonald, A.M. (2018). Impact of irrigated agriculture on groundwater-recharge salinity: A major sustainability concern in semi-arid regions. *Hydrogeology Journal*, 26, 2781–2791. https://doi.org/10.1007/s10040-018-1830-2.

Fradkin, P.L. (1996). *A River No More: The Colorado River and the West*. University of California Press.

Freeze, R.A., & Cherry, J.A. (1979). *Groundwater*. Englewood Cliffs, NJ: Prentice-Hall.

George, R., Clarke, J., & English, P. (2008). Modern and palaeogeographic trends in the salinisation of the Western Australian wheatbelt: A review. *Australian Journal of Soil Research*, 46(8), 751–767. https://doi.org/10.1071/SR08066.

Gertner, J. (2007). The future is drying up. *New York Times Magazine*, 21.

Giller, P.S., & Myers, A.A. (1996). *Disturbance and Recovery of Ecological Systems*. Royal Irish Academy.

Gohil, M.B. (2000). *Land Treatment of Wastewater* (1st ed.). New Delhi, India: New Age International.

González-Estébanez, F.J., García-Tejero, S., Mateo-Tomás, P., & Olea, P.P. (2011). Effects of irrigation and landscape heterogeneity on butterfly diversity in Mediterranean farmlands. *Agriculture, Ecosystems & Environment*, 144(1), 262–270.

Graham, K.C. (2002). *Construction of Hydraulic Structures for Water Supply Sustenance*. pp. 56–76. ISBN: 0-916122-51-4.

Green, R.E., Cornell, S.J., Scharlemann, J.P.W., & Balmford, A. (2005). Farming and the fate of wild nature. *Science*, 307, 550–555.

Greene, R., Timms, W., Rengasamy, P., Arshad, M., & Cresswell, R. (2016). Soil and aquifer salinization: Toward an integrated approach for salinity management of groundwater. In *Integrated Groundwater Management* (pp. 377–412). Cham: Springer.

Guadie, A., Yesigat, A., Gatew, S., Worku, A., Liu, W., Ajibade, F.O., & Wang, A. (2021). Evaluating the health risks of heavy metals from vegetables grown on soil irrigated with untreated and treated wastewater in Arba Minch, Ethiopia. *Science of the Total Environment*, 761, 143302. https://doi.org/10.1016/j.scitotenv.2020.143302.

Hajjami, K., Ennaji, M.M., Fouad, S., Oubrim, N., & Cohen, N. (2012). Wastewater reuse for irrigation in Morocco: Helminth eggs contamination's level of irrigated crops and sanitary risk (a case study of Settat and Soualem regions). *Journal of Bacteriology & Parasitology*, 4, 1–5. https://doi.org/10.4172/2155-9597.1000163

Harding, J.S., Benfield, E.F., Bolstad, P.V., Helfman, G.S., & Jones, E.B.D. (1998). Stream biodiversity: The ghost of land use past. *Proceedings of the National Academy of Sciences*, 95(25), 14843–14847.

Hass, A., Mingelgrin, U., & Fine, P. (2010). Heavy metals in soils irrigated with wastewater. In *Treated Wastewater in Agriculture: Use and Impacts on the Soil Environment and Crops*. https://doi.org/10.1002/9781444328561.ch7

Hassanvand, M., Yazdi, M., & Karami, R. (2017). Environmental effects of irrigation and drainage network of Kheirabad area, SW of Iran. *Environmental Earth Sciences*, 76(15), https://doi.org/10.1007/s12665-016-6291-0.

Herrero, J., & Snyder, R.L., 1997. Aridity and irrigation in Aragon, Spain. *Journal of Arid Environments*, 35, 535–547.

Herzog, F., Dreier, S., Hofer, G., Marfurt, C., Schüpbach, B., Spiess, M., & Walter, T. (2005). Effect of ecological compensation areas on floristic and breeding bird diversity in Swiss agricultural landscapes. *Agriculture, Ecosystems & Environment*, 108, 189–204.

Hickey, J.T, and Salas, J.D. (1995). *Environmental Effects of Extreme Floods. U.S. – Italy Research Workshop on the Hydrometeorology, Impacts, and Management of Extreme Floods Perugia (Italy), November 1995.*

Higgins, A., Restrepo, J.C., Ortiz, J.C., Pierini, J., & Otero, L. (2015). Suspended sediment transport in the Magdalena river (Colombia, South America): Hydrologic regime, rating parameters and effective discharge variability. *International Journal of Sediment Research*, 31(1), 25–35. https://doi.org/10.1016/j.ijsrc.2015.04.003.

Holy, M. (1993). Is irrigation sustainable? *Canadian Water Resources Journal*, 18(4), 443–449, https://doi.org/10.4296/cwrj1804443.

Holzschuh, A., Steffan-Dewenter, I., Kleijn, D., & Tscharntke, T. (2007). Diversity of flower-visiting bees in cereal fields: Effects of farming system, landscape composition and regional context. *Journal of Applied Ecology*, 44, 42–49.

Hornung, M., & Reynolds, B. (1995). The effects of natural and anthropogenic environmental changes on ecosystem processes at the catchment scale. *Trends in Ecology & Evolution*, 10(11), 443–449.

Hu, A.H., Chen, C.H., Huang, L.H., Chung, M.H., Lan, Y.C., & Chen, Z. (2019). Environmental impact and carbon footprint assessment of Taiwanese agricultural products: A case study on Taiwanese Dongshan tea. *Energies*, 12(1), 138. https://doi.org/10.3390/en12010138.

Hupp, C.R., Noe, G.B., & Schenk, E.R. (2010). Floodplains, equilibrium, and fluvial geomorphic impacts of human alterations. *2nd Joint Federal Interagency Conference*, Las Vegas, NV, June 27–July 1, 2010.

Hyvönen, T., Salonen, J. (2002). Weed species diversity and community composition in cropping practices at two intensity levels – A six-year experiment. *Plant Ecology*, 154, 73–81.

Ilori, B.A., Adewumi, J.R., Lasisi, K.H., & Ajibade, F.O. (2019). Qualitative assessment of some available water resources in Efon Alaaye, Nigeria. *Journal of Applied Science and Environmental Management*, 23(1), 29–34. https://doi.org/10.4314/jasem.v23i1.5.

Imran, M., Maqsood, M.A., Rahmatullah, Kanwal, S. (2010). Increasing SAR of irrigation water aggravates boron toxicity in maize (Zea mays L.). *Journal of Plant Nutrition* 33 (9), 1301–1306. https://doi.org/10.1080/01904167.2010.484091.

IPCC (2014). Synthesis Report. Contribution of Working Groups I, II and III to the Fifth Assessment Report of the Intergovernmental Panel on Climate Change [Core Writing Team, R.K. Pachauri and L.A. Meyer (eds.)]. IPCC, Geneva, Switzerland, 151 pp.

Irmak, S., & Rathje, W.R. (2008). *Plant Growth and Yield as Affected by Wet Soil Conditions Due to Flooding or Over-Irrigation*. Lincoln: Publications of University of Nebraska-Lincoln Extension.

Islam, M.A., Romić, D., Akber, M.A., & Romić, M. (2018). Trace metals accumulation in soil irrigated with polluted water and assessment of human health risk from vegetable consumption in Bangladesh. *Environmental Geochemistry and Health*, 40(1), 59–85. https://doi.org/10.1007/s10653-017-9907-8.

Islam, S.M.M., Gaihre, Y.K., Islam, M.R., Akter, M., Al Mahmud, A., Singh, U., & Sander, B.O. (2020). Effects of water management on greenhouse gas emissions from farmers' rice fields in Bangladesh. *Science of the Total Environment*, 734, 139382. https://doi.org/10.1016/j.scitotenv.2020.139382.

José-María, L., Armengot, L., Blanco-Moreno, J.M., Bassa, M., & Sans, F.X. (2010). Effects of agricultural intensification on plant diversity in Mediterranean dryland cereal fields. *Journal of Applied Ecology*, 47, 832–840.

Khan, O., Mwelwa-Mutekenya, E., Crosato, A., & Zhou, Y. (2014). Effects of dam operation on downstream river morphology: The case of the middle Zambezi River. *Water Management. Proceedings of the Institution of Civil Engineers (ICE)*, 1300122.

Khanpae, M., Karami, E., Maleksaeidi, H., & Keshavarz, M. (2020). Farmers' attitude towards using treated wastewater for irrigation: The question of sustainability. *Journal of Cleaner Production*, 243, 118541.

Kisi, O. (2012). Modeling discharge-suspended sediment relationship using least square support vector machine. *Journal of Hydrology*, 456, 110–120.

Kleijn, D., Berendse, F., Smit, R., & Gilissen, N. (2001). Agrienvironment schemes do not effectively protect biodiversity in Dutch agricultural landscapes. *Nature*, 413, 723–725.

Knop, E., Kleijn, D., Herzog, F., & Schmid, B. (2006). Effectiveness of the Swiss agri-environment scheme in promoting biodiversity. *Journal of Applied Ecology*, 43, 120–127.

Krauss, J., Gallenberger, I., & Steffan-Dewenter, I. (2011). Decreased functional diversity and biological pest control in conventional compared to organic crop fields. *PLoS One*, 6, e19502.

Kremen, C. (1992). Assessing the indicator properties of species assemblages for natural areas monitoring. *Ecological Applications*, 2, 203–217.

Kukal, M.S., & Irmak, S. (2020). Impact of irrigation on interannual variability in United States agricultural productivity. *Agricultural Water Management*, 234, 106141.

Kumar, M., Xiong, X., He, M., Tsang, D.C., Gupta, J., Khan, E., … Bolan, N.S. (2020). Microplastics as pollutants in agricultural soils. *Environmental Pollution*, 265, 114980.

Kuscu, H., Bölüktepe, F.E., & Demir, A.O. (2009). Performance assessment for irrigation water management: A case study in the Karacabey irrigation scheme in Turkey. *African Journal of Agricultural Research*, 4, 124–132.

Laiolo, P. (2005). Spatial and seasonal patterns of bird communities in Italian agroecosystems. *Conservation Biology*, 19, 1547–1556.

Lake, P.S. (2000). Disturbance, patchiness, and diversity in streams. *Journal of the North American Benthological Society*, 19(4), 573–592.

Lake, P.S., Palmer, M.A., Biro, P., Cole, J., Covich, A.P., Dahm, C., … Verhoeven, J.O.S. (2000). Global change and the biodiversity of freshwater ecosystems: Impacts on linkages between above-sediment and sediment biota: All forms of anthropogenic disturbance—changes in land use, biogeochemical processes, or biotic addition or loss—not only damage the biota of freshwater sediments but also disrupt the linkages between above-sediment and sediment-dwelling biota. *BioScience*, 50(12), 1099–1107.

Leduc, C., Favreau, G., & Schroeter, P. (2001). Long-term rise in the Sahelian water-table: The continental terminal in South-West Niger. *Journal of Hydrology*, 243, 43–54.

Lee, D.J., & Howitt, R.E. (1996). Modeling regional agricultural production and salinity control alternatives for water quality policy analysis. *American Journal of Agricultural Economics*, 78(1), 41–53. https://doi.org/10.2307/1243777.

Litskas, V.D., Aschonitis, V.G., & Antonopoulos, V.Z. (2010). Water quality in irrigation and Drainage networks of Thessaloniki plain in Greece related to land use, water management, and agroecosystem protection. *Environmental Monitoring and Assessment*, 163, 347–359.

Lootens, M., & Lumbu, S. (1986). Suspended sediment production in a suburban tropical basin (Lubumbashi, Zaire). *Hydrological Sciences Journal*, 31(3), 39–49. https://doi.org/10.1080/02626668609491026.

Magami, I.M., Yahaya, S., & Mohammed, K. (2014). Causes and consequences of flooding in Nigeria: A review. *Biological and Environmental Sciences Journal for the Tropics*, 11(2), 154–162.

Magilligan, F.J., & Nislow, K.H. (2005). Changes in hydrologic regime by dams. *Geomorphology*, 71(1–2), 61–78, https://doi.org/10.1016/j.geomorph.2004.08.017.

Magilligan, F.J., Nislow, K.H., & Graber, B.E. (2003). Scale-independent assessment of discharge reduction and riparian disconnectivity following flow regulation by dams. *Geology*, 31(7), 569–572, https://doi.org/10.1130/0091-7613(2003)031<0569:saodra>2.0.co;2.

Mailhot, A., Talbot, G., Ricard, S., Turcotte, R., & Guinard, K. (2018). Assessing the potential impacts of dam operation on daily flow at ungauged river reaches. *Journal of Hydrology: Regional Studies*, 18, 156–167. https://doi.org/10.1016/j.ejrh.2018.06.006.

Malan, M., Müller, F., Cyster, L., Raitt, L., & Aalbers, J. (2015). Heavy metals in the irrigation water, soils and vegetables in the Philippi horticultural area in the Western Cape Province of South Africa. *Environmental Monitoring and Assessment*. https://doi.org/10.1007/s10661-014-4085-y.

Manasa, M.R., Katukuri, N.R., Nair, S.S., Haojie, Y., Yang, Z., & Guo, R.B. (2020). Role of biochar and organic substrates in enhancing the functional characteristics and microbial community in a saline soil. *Journal of Environmental Management*, 110737. https://doi.org/10.1016/j.jenvman.2020.110737.

Maqsood, M.A., Khan, M.K., Naeem, M.A., Hussain, S., Aziz, T., & Schoenau, J. (2015). High sodium in irrigation water caused B toxicity at low soil solution and shoot B concentration in maize (*Zea mays* L.). *Journal of Plant Nutrition*, 38(5), 728–741. https://doi.org/10.1080/01904167.2014.939286.

Marchuk, A.G., & Rengasamy, P. (n.d.). Cation ratio of soil structural stability (CROSS). *2010 19th World Congress of Soil Science, Soil Solutions for a Changing World. 1–6 August 2010, Brisbane, Australia.*

Marden, M., & Rowan, D. (2015). The effect of land use on slope failure and sediment generation in the Coromandel region of New Zealand following a major storm in 1995. *New Zealand Journal of Forestry Science*, 45, 1–18.

Mateo-Sagasta, J., Zadeh, S.M., & Turral, H. (2017). Water pollution from agriculture: A global review. Executive Summary.

Mavi, M.S., Sanderman, J., Chittleborough, D.J., Cox, J.W., & Marschner, P. (2012). Sorption of dissolved organic matter in salt-affected soils: Effect of salinity, sodicity and texture. *Science of the Total Environment*, 435–436, 337–344. https://doi.org/10.1016/j.scitotenv.2012.07.009.

Melnychuk, N.A., Olfert, O., Youngs, B., & Gillott, C. (2003). Abundance and diversity of Carabidae (Coleoptera) in different farming systems. *Agriculture, Ecosystems and Environment*, 95, 69–72.

Memon, F. S., & Sharjeel, M. Y. (2016). Catastrophic effects of floods on environment and health: Evidence from Pakistan. Pakistan Journal of Engineering, Technology & Science, 5(2). https://doi.org/10.22555/pjets.v5i2.903.

Munir, S. (2011). *Role of Sediment Transport in Operation and Maintenance of Supply and Demand Based Irrigation Canals—Application to Machai Maira Branch Canals*. Delft, The Netherlands: Wageningen University and UNESCO-IHE Institute for Water Education.

Murphy, B. (2015). Key soil functional properties affected by soil organic matter – Evidence from published literature. *IOP Conference Series: Earth and Environmental Science*, 25, 012008.

Nagler, P.L., Glenn, E.P., Hinojosa-Huerta, O., Zamora, F., & Howard, K. (2008). Riparian vegetation dynamics and evapotranspiration in the riparian corridor in the delta of the Colorado River, Mexico. *Journal of Environmental Management*, 88(4), 864–874.

Niemeyer, J.C., Medici, L.O., Correa, B., Godoy, D., Ribeiro, G., Lima, S. d.-O., … de Carvalho, D.F. (2020). Treated produced water in irrigation: Effects on soil fauna and aquatic organisms. *Chemosphere*, 240, 124791.

Nimmo, J.R. (2013). Aggregation: Physical aspects. *Reference Module in Earth Systems and Environmental Sciences*. Elsevier. ISBN 9780124095489, https://doi.org/10.1016/B978-0-12-409548-9.05087-9.

Nimmo, J.R., (2005). Unsaturated zone flow processes, In Anderson, M.G., & Bear, J., (eds.), *Encyclopedia of Hydrological Sciences: Part 13 – Groundwater* (Vol. 4, pp. 2299–2322). Chichester, UK: Wiley, https://doi.org/10.1002/0470848944.hsa161, http://www.mrw.interscience.wiley.com/ehs/articles/hsa161/frame.html.

Nolan, J., & Weber, K.A. (2015). Natural uranium contamination in major U.S. aquifers linked to nitrate. *Environmental Science and Technology Letters*. https://doi.org/10.1021/acs.estlett.5b00174.

Ochiere, H.O., Onyando, J.O., & Kamau, D.N. (2015). Simulation of sediment transport in the canal using the HEC-RAS (Hydrologic Engineering Centre—River Analysis System) in an underground canal in southwest Kano irrigation scheme—Kenya. *International Journal of Science and Engineering Invention*, 4, 15–31.

OECD (2010), "Water and agriculture: Managing water sustainably is key to the future of food and agriculture" [Online]. https://www.oecd.org/agriculture/topics/water-and-agriculture/

Ofori, S., Puskacova, A., Ruzickova, I., & Wanner, J. (2021). Treated wastewater reuse for irrigation: Pros and cons. *Science of the Total Environment*, 760, 144026.

Omotade, I.F., Alatise, M.O., & Olanrewaju, O.O. (2019). Recycling of aquaculture wastewater using charcoal based constructed wetlands, *International Journal of Phytoremediation*, 21(5), 399–404, https://doi.org/10.1080/15226514.2018.1537247.

Osman, I.E. (2015). *Impact of improved operation and maintenance on cohesive sediment transport in Gezira scheme, Sudan*. Delft, The Netherlands: Wageningen University. UNESCO-IHE Institute for Water Education.

Oster, J.D., Sposito, G., & Smith, C.J. (2016). Accounting for potassium and magnesium in irrigation water quality assessment. *California Agriculture*, 70(2), 71–76. https://doi.org/10.3733/ca.v070n02p71.

Ouellet-Proulx, S., St-Hilaire, A., Courtenay, S.C., & Haralampides, K.A. (2016). Estimation of suspended sediment concentration in the Saint John River using rating curves and machine learning approach. *Hydrological Sciences Journal*, 61(10), 1847–1860. https://doi.org/10.1080/02626667.2015.1051982.

Paudel, K.P. (2010). *Role of Sediment in the Design and Management of Irrigation Canals Sunsari Morang Irrigation Scheme, Nepal*. Delft, The Netherlands: Wageningen University and UNESCO-IHE Institute for Water Education.

Penvenne, L.J. (1996). Disappearing delta. *American Scientist*, 84(5), 438.

Perret, S.R., & Payen, S. (2020). Irrigation and the environmental tragedy: Pathways towards sustainability in agricultural water use. *Irrigation and Drainage*, 69, 263–271.

Petts, E., & Gurnell, A.M. (2005). Dams and geomorphology: Research progress and future directions. *Geomorphology*, 71(1–2), 27–47.

Pezeshki, S.R., & DeLaune, R.D. (2012). Soil oxidation-reduction in wetlands and its impact on plant functioning. *Biology*. https://doi.org/10.3390/biology1020196.

Pierret, A., Doussan, C., Capowiez, Y., Bastardie, F., & Pagès, L. (2007). Root functional architecture: A framework for modeling the interplay between roots and soil. *Vadose Zone Journal*. https://doi.org/10.2136/vzj2006.0067.

Pimentel, D., & Pimentel, M. (2003). World population, food, natural resources, and survival. *World Futures*. https://doi.org/10.1080/02604020310124.

Ponce, C., Bravo, C., García de León, D., Magaña, M., & Alonso, J.C. (2011). Effects of organic farming on plant and arthropod communities: A case study in Mediterranean dryland cereal. *Agriculture, Ecosystems & Environment*, 141, 193–201.

Prieto-Benítez, S., & Méndez, M. (2011). Effects of land management on the abundance and richness of spiders (Araneae): A meta-analysis. *Biological Conservation*, 144, 683–691.

Pulido-Bosch, A., Rigol-Sanchez, J.P., Vallejos, A., Andreu, J.M., Ceron, J.C., Molina-Sanchez, L., & Sola, F. (2018). Impacts of agricultural irrigation on groundwater salinity. *Environmental Earth Sciences*, 77, 197. https://doi.org/10.1007/s12665-018-7386-6.

Rageh, A.Y., Al-Garadi, M.A., & Al-Mashreki, M.H. (2017). Environmental effects of wastewater use in agricultural irrigation at Dhamar city, Republic of Yemen. *SCIREA Journal of Environment*, 2(2), 38–52.

Rehman, K., Bukhari, S.M., Andleeb, S., Mahmood, A., Erinle, K.O., Naeem, M.M., & Imran, Q. (2019). Ecological risk assessment of heavy metals in vegetables irrigated with groundwater and wastewater: The particular case of Sahiwal district in Pakistan. *Agricultural Water Management*, 226 (September).

Roth, T., Amrhein, V., Peter, B., & Weber, D. (2008). A Swiss agri-environment scheme effectively enhances species richness for some taxa over time. *Agriculture, Ecosystems & Environment*, 125, 167–172.

Ruiz, M., (1990). Development of Mediterranean agriculture: An ecological approach. *Landscape Urban Plan*, 18, 211–220.

Sainju, U.M., Stevens, W.B., Caesar-TonThat, T., & Liebig, M.A. (2012). Soil greenhouse gas emissions affected by irrigation, tillage, crop rotation, and nitrogen fertilization. *Journal of Environmental Quality*. https://doi.org/10.2134/jeq2012.0176

Sakaris, P.C. (2013). A review of the effects of hydrologic alteration on fisheries and biodiversity and the management and conservation of natural resources in regulated river systems. In: P. M. Bradley (ed). *Current Perspectives in Contaminant Hydrology and Water Resources Sustainability*. IntechOpen, https://doi.org/10.5772/55963.

Salas, J.D., Govindaraju, R.S., Anderson, M., Arabi, M., Frances, F., Suarez, W., Lavado-Casimiro, W.S., & Green, T.R. (2014). Volume 15: Modern Water Resources Engineering. L.K. Wang and C.T. Yang (eds.). *Handbook of Environmental Engineering*. https://doi.org/10.1007/978-1-62703-595-8_1.

Santos, J.N., Andrade, E.M., Medeiros, P.A., Palácio, H.d, & Neto, J.R.d.A. (2017). Sediment delivery ratio in a small semi-arid watershed under conditions of low connectivity. *Revista Ciência Agronômica*, 48, 49–58.

Sapkota, A., Haghverdi, A., Avila, C.C.E., & Ying, S.C. (2020). Irrigation and greenhouse gas emissions: A review of field-based studies. *Soil Systems*, 4(2), 1–21. https://doi.org/10.3390/soilsystems4020020

Scanlon, B.R., Jolly, I., Sophocleous, M., & Zhang, L. (2007). Global impacts of conversions from natural to agricultural ecosystems on water resources: Quantity versus quality. *Water Resources Research*. https://doi.org/10.1029/2006WR005486.

Singh, A. (2018). Managing the environmental problems of irrigated agriculture through the appraisal of groundwater recharge. *Ecological Indicators*, 92, 388–393.

Singh, A. (2021). Soil salinization management for sustainable development: A review. *Journal of Environmental Management*, 277, 111383.

Singh, A., Sharma, R.K., Agrawal, M., & Marshall, F.M. (2010). Risk assessment of heavy metal toxicity through contaminated vegetables from waste water irrigated area of Varanasi, India. *Tropical Ecology*.

Singh, S., Ghosh, N.C., Gurjar, S., Krishan, G., Kumar, S., & Berwal, P. (2018). Index-based assessment of suitability of water quality for irrigation purpose under Indian conditions. *Environmental Monitoring and Assessment*. https://doi.org/10.1007/s10661-017-6407-3

Spear, B., Smith, D., Largay, B., & Haskins, J. (2008). *Turbidity as a Surrogate Measure for Suspended Sediment Concentration in Elkhorn Slough, CA*. Monterey Bay, CA: The Watershed Institute, Division of Science and Environmental Policy—California State University.

Stefanescu, C., Herrando, S., & Páramo, F. (2004). Butterfly species richness in the northwest Mediterranean Basin: The role of natural and human-induced factors. *Journal of Biogeography*, 31, 905–915.

Stoate, C., Boatman, N.D., Borralho, R.J., Rio Carvalho, C., de Snoo, G.R., & Eden, P., 2001. Ecological impacts of arable intensification in Europe. *Journal of Environmental Management*, 63, 337–365.

Stockle, C.O. (2001). *Environmental Impact of Irrigation: A Review*. Washington State University.

Suarez, D.L. (1989). Impact of agricultural practices on groundwater salinity. *Agriculture, Ecosystems and Environment*. https://doi.org/10.1016/0167-8809(89)90014-5

Sutama, N.H. (2010). *Mathematical Modelling of Sediment Transport and Its Improvement in Bekasi Irrigation System, West Java, Indonesia*. Delft, The Netherlands: UNESCO-IHE, Institute for Water Education.

Tella J.L., & Forero M.G. (2000) Farmland habitat selection of wintering lesser kestrels in a Spanish pseudosteppe: Implications for conservation strategies. *Biodiversity and Conservation*, 9, 433–441.

Thiruchelve, S.R., Chandran, S., Kumar, P.L., & Veluswamy, K. (2020). Assessment of environmental impacts in the untreated urban wastewater irrigated sites – A case study. *ASABE 2020 Annual International Meeting*, 2001193. https://doi.org/10.13031/aim.202001193.

Thomas, J.A. (2005). Monitoring change in the abundance and distribution of insects using butterflies and other indicator groups. *Philosophical Transactions of the Royal Society B*, 360, 339–357.

Tilman, D., Cassman, K.G., Matson, P.A., Naylor, R., & Polasky, S. (2002). Agricultural sustainability and intensive production practices. *Nature*, 418, 671–677.

Tybirk, K., Alroe, H.F., Frederiksen, P. (2004). Nature quality in organic farming: A conceptual analysis of considerations and criteria in European context. *Journal of Agricultural and Environmental Ethics*, 17, 249–274.

Uhrich, M.A., & Bragg, H. m. (2003). *Monitoring Instream Turbidity to Estimate Continuous Suspended-Sediment Loads and Yields and Clay-Water Volumes in the Upper North Santiam River Basin, Oregon, 1998–2000*. Portland, ON: U.S. Department of the Interior and U.S. Geological Survey.

Ursúa, E., Serrano, D., & Tella, J.L. (2005). Does land irrigation actually reduce foraging habitat for breeding lesser kestrels? The role of crop types. *Biological Conservation*, 122, 643–648.

US EPA (US Environment Protection Agency), 1992. *Managing Nonpoint Source Pollution*. Washington, DC: US EPA, Office of Water.

Usman, M.A., & Gerber, N. (2019). Assessing the effect of irrigation on household water quality and health: A case study in rural Ethiopia. *International Journal of Environmental Health Research*, https://doi.org/10.1080/09603123.2019.1668544.

Van Weert, F., Van der Gun, J., & Reckman, J.W.T.M. (2009). Global overview of saline groundwater occurrence and genesis. *IGRAC Report GP 2009-1*.

Vartapetian, B.B., & Jackson, M.B. (1997). Plant adaptations to anaerobic stress. *Annals of Botany*. https://doi.org/10.1006/anbo.1996.0295

Villamar, C.A., Vera-Puerto, I., Rivera, D., & De la Hoz, F. (2018). Reuse and recycling of livestock and municipal wastewater in Chilean agriculture: A preliminary assessment. *Water*, 10(6), 817. https://doi.org/10.3390/w10060817.

Wada, Y., Van Beek, L.P.H., Van Kempen, C.M., Reckman, J.W.T.M., Vasak, S., & Bierkens, M.F.P. (2010). Global depletion of groundwater resources. *Geophysical Research Letters*. https://doi.org/10.1029/2010GL044571.

Wallace, J.S. (2000). Increasing agricultural water use efficiency to meet future food production. *Agriculture, Ecosystems & Environment*, 82, 105–119.

Wang, Y., Zheng, C., & Ma, R. (2018). Safe and sustainable groundwater supply in China. *Hydrogeology Journal*, 26(5), 1301–1324.

Water Cycle. (2020, June 8). *New World Encyclopedia*. Retrieved 09:32, June 25, 2022 from https://www.newworldencyclopedia.org/p/index.php?title=Water_cycle&oldid=1037916.

Waters, T.F. (1995). *Sediment in Streams: Sources, Biological Effects, and Control*. American Fisheries Society.

Weibull A-C., Bengtsson J., Nohlgren E., (2000) Diversity of butterflies in the agricultural landscape: The role of farming system and landscape heterogeneity. *Ecography* 23: 743–750.

Weibull, A.C., & Östman, Ö. (2003) Species composition in agroecosystems: The effect of landscape, habitat and farm management. *Basic and Applied Ecology*, 4, 349–361.

Werner, A.D., Bakker, M., Post, V.E.A., Vandenbohede, A., Lu, C., Ataie-Ashtiani, B., Simmons, C.T., & Barry, D.A. (2013). Seawater intrusion processes, investigation and management: Recent advances and future challenges. *Advances in Water Resources*. https://doi.org/10.1016/j.advwatres.2012.03.004

Winter, T.C., Harvey, J.W., Franke, O.L., & Alley, W.M. (1998). *Ground Water and Surface Water a Single Resource*. U.S. Geological Survey Circular 1139. U.S. Government Printing Office. ISBN 0-607-89339-7.

WWAP (United Nations World Water Assessment Programme). (2017). *The United Nations World Water Development Report 2017*. Paris: UNESCO.

Yan, Y., Yang, Z., Liu, Q., Sun, T. (2010). Assessing effects of dam operation on flow regimes in the lower Yellow River. *Procedia Environmental Sciences*, 2(6), 507–516, https://doi.org/10.1016/j.proenv.2010.10.055.

Yusuf, R.O. (2008). Environmental impacts of an irrigation scheme: The Bakolori irrigation project. *Journal of Environmental Research and Policies*, 3(4), 75–80.

Zektser, S., Loáiciga, H.A., & Wolf, J.T. (2005). Environmental impacts of groundwater overdraft: Selected case studies in the southwestern United States. *Environmental Geology*, 47(3), 396–404.

Zeleňáková, M., Blišťan, P., Alkhalaf, I., Gaňová, L., & Zvijáková, L. (2016). Assessment of environmental damages in case of flood in Bodva River Basin, Slovakia. *International Journal of Safety and Security Engineering*, 6(3), 498–507. https://www.usgs.gov/water-science-school.

8

Environmental Impact Assessment of Irrigation

Never Mujere
University of Zimbabwe

Nelson Chanza
University of Johannesburg

8.1 Introduction

The global population is projected to increase by 33% to 8 billion between 2000 and 2030 while food production is estimated to increase at a slower rate of 1.3% per year (FAO, 2004). With the rate of urbanisation expected to increase by 61% during the same period, food demand in regions like sub-Saharan Africa will shift from maize to rice and wheat. At the same time, there would be rising demand for maize as animal feed. As food production fails to increase at same rate as population growth, most of the world's population will experience food shortages if unsustainable food security measures are put in place. Nevertheless, food security can be improved by irrigation of crop, fibre, fruit and animal pastures throughout the whole year. Also, adoption of irrigation will increase the global food production through crop diversification, extensification and intensification. FAO (2004) estimated that irrigated land area in developing countries is expected to increase by 27% between 1996 and 2030. In sub-Saharan Africa, a 75% growth in crop production required by 2030 would come from an increase of 62% in crop yield, 13% in cropping intensities and 25% expansion of arable land.

In addition to rainfall, irrigation projects obtain most of their water requirements from other sources such as rainwater harvesting sites, river flows and groundwater. Unlike water harvested dams, water from underground and river flows is stored in temporary night storage dams before being delivered to fields via canals or pipes. Schemes like Hippo valley in Zimbabwe and Gezira in Sudan receive water supply from reservoirs, whereas runoff-river schemes such as Nyanyadzi in Zimbabwe get their water needs directly from rivers.

Irrigation is actively promoted in areas such as the Euphrates and Tigris River basins as a springboard for socio-economic development. It helps to avert the food crises in arid and semi-arid areas, a situation which most developing states have become endemic. About 20% of the world's 1,500 million hectares irrigated provide 40% and 70% of food and cereal production, respectively. As such, most developing countries rely on the food produced from irrigated lands. For example, 50% of Indonesia, 70% of China and 80% of Pakistan's food requirements come from irrigated lands. Sudan's 85,000 ha Gezira irrigation scheme produces almost 70% of the country's cotton crop, 50% of wheat, 30% of beans and 12% of sorghum. On average, 50% of Sudan's total revenue comes from the scheme (FAO, 2004).

The other benefits derived from the irrigation projects include modernising peasant agriculture, reducing drought relief, providing economic growth and employment opportunities in rural areas, export earnings, more varied diets and better health standards. For example, the 28 ha Mutema Irrigation Scheme in Zimbabwe, besides alleviating famine in the area, reduced government food relief by approximately 90–180 tonnes per annum (Manzungu, 1999). Accordingly, there is rising investment in irrigation projects government agencies, the private sector, local and international donors. Examples of state institutions created primarily to the spearhead irrigation development include the National Irrigation Administration in Philippines, International Water Management Institute in Sri Lanka and the Department of Agricultural Engineering in Zimbabwe.

Besides the direct benefits of irrigation to a quarter of the world's population or a third of those living in developing countries, the activity causes the severe environmental impacts if not managed efficiently. These effects manifest themselves in health, economic, social, political and cultural changes in the irrigated landscapes. Thus, environmental impact assessments (EIAs) are required before irrigation projects are developed. This helps to minimise the adverse effects.

This chapter consists of five of sections. Section 8.1 gives an introduction and some overview of appropriate literature on the subject. Section 8.2 focuses on irrigation development and performance in Zimbabwe. Section 8.3 highlights the purpose and relevance of carrying out EIAs. Section 8.4 describes the EIA process and significance of conducting an EIA in irrigation projects. Section 8.5 discusses the socio-economic and ecological impacts of irrigation projects. Lastly, Section 8.6 concludes the chapter.

8.2 Irrigation Agriculture in Zimbabwe

Zimbabwe is predominantly an agricultural country in a semi-arid region of southern Africa. Both rain-fed farming and irrigation account for nearly 45%–60% of the total export earnings, provides employment and incomes to more than 70% of the population, supplies 60% of the raw materials required by the industrial sector and supply food to feed the entire nation of about 13 million people (FAO, 2004). The country has a long history of irrigation development dating back to pre-colonial times. Almost 40% of maize, tobacco and soya beans, 45% of cotton, 70% of coffee, 55% of tea, 45% of cotton, and 100% of wheat, horticulture and sugarcane are produced under irrigation or supplementary irrigation during mid-season dry spells in summer (FAO, 2004). Although more than 550,000 ha are the lands that are suitable for irrigation in the country, the availability of water has reduced this potential to 331,000 ha. Also due to water shortage the winter to summer irrigated land ratio is 1:16 on average (FAO, 2004).

Irrigation projects in the country are divided into two groups, namely, the informal and formal schemes. Informal or traditional irrigation schemes are those initiated and constructed by farmers using local resources available to them. Informal irrigation is characterised by temporary structures and channels not built following formal engineering designs. As a consequence, diversion structures may not contain the optimum grades and cross sections. A notable example is found in hill slopes of eastern highlands where stone terraces watered by paved furrows were constructed to divert water from streams during the 15th century (Manzungu, 1999). By the year 2000, between 5,000 ha and 20,000 ha were irrigated from shallow wells and simple furrows (Department of Water Development, 2000).

Formal schemes comprise the large-scale and small-scale irrigation projects (Table 8.1). They are the key strategies for overall agricultural development in the country. The schemes are constructed and/or initiated by the government and are managed either by individual farmers, communities, private companies or the government. Formal schemes were started in the early 1900s and the irrigated area grew

TABLE 8.1 Status of Formal Irrigation Development in Zimbabwe

Irrigation Sector	Irrigated Area (ha)	%	Number of Farms	Number of Beneficiaries
Large scale	126,000	81	1,500	500
Parastatal	13,500	9	18	>500
Settler outgrowers	3,600	2	8	No data
Communal and resettlement	12,900	8	306	20,600
Total	156,000	100	1,832	22,600

Source: Department of Water Development (2000).

20 times between 1920 and 1991. Between 1968 and 2003, the irrigated areas increased from 60,000 ha to about 200,000 ha. However, from 1982 to 1985, the irrigated area dropped from 165,000 to 136,000 ha as a result of droughts (Manzungu, 1999; FAO, 2004). By 1999, there were about 1,872 formal schemes covering more than 156,000 ha in the country.

The dual nature of the irrigation sector comprising the large-scale and smallholder irrigation is a result of the country's colonial history. From 1890 to 1980, the white colonial settlers gave themselves large pieces of fertile agricultural land in natural region I receiving high rainfall exceeding 1,000 mm/ year spreading across all months of the year, region II with rainfall concentrated in summer months and moderate to high (750–1,000 mm/year), and region III which is characterised by moderate rainfall (650–750 mm/year). However, the smallholder native farmers were confined to poor semi-arid agricultural regions IV (450–650 mm/year) and V (below 450 mm/year). These regions are subject to periodic seasonal droughts and severe dry spells during the rainy season from November to April (Manzungu, 1999; Bolding, 2004). Importantly, most of country's irrigation projects were developed in these dry regions. The projects negatively affect the socio-economic and biophysical environmental conditions in a number of ways which differ from place to place and over time. The adverse effects can be minimised or avoided by undertaking EIAs in areas where these schemes are introduced.

8.3 The EIA Concept

There has been diverse interest and attention given to the concept of EIA. This heterogeneity stems largely from numerous but conflicting players influencing the EIA process and affected by it. The role of EIA is formally recognised in Principle 17 of the Rio Declaration on Environment and Development which states that:

> Environmental impact assessment, as a national instrument, shall be undertaken for proposed activities that are likely to have a significant adverse impact on the environment and are subject to a decision of a competent national authority.

One way to understand its meaning would be to trace its origin, an aspect treated later in this section. Drawing from the working definition of EIA, an EIA can be understood as a decision support tool to make informed choices in the development process (Gilpin, 1995). It is a necessary process which seeks to enhance the ability to make informed decisions on the development projects for sustainable development. This means EIA should be understood within the framework of sustainable development, a term which also has not been immune to definitional challenges. When framed with sustainable development, EIA should be used to guide development that is socially acceptable, economically viable, technically appropriate and environmentally friendly, while considering justice and equity within and between generations.

Most definitions of EIA treat it as a process of identifying, measuring, predicting, evaluating and mitigating or enhancing the likely biological, physical, social, cultural and economic effects of a proposed project (Verma, 1986; Gilpin, 1995; Wood, 1995). The International Association of Impact Assessment

(IAIA, 1999) defines it as the process of identifying, predicting and evaluating the biophysical, social and other relevant effects of proposed development proposals prior to major decisions being taken and commitments being taken. Two definitions of EIA, indicating the nature of the process, are that it is "… an assessment of impacts of a planned activity on the environment" (United Nations), and that it "… is the systematic process of identifying the future consequences of a current or proposed action" (IAIA). When applied to the irrigation projects, the process is intended to minimise negative externalities, while enhancing positive benefits, associated with setting irrigation infrastructure, clearing land for farming, maintenance of irrigation schemes, running irrigation projects and the other related activities.

During the early years, EIAs mainly focused on the biophysical impacts of proposed projects (i.e. water and air quality, flora and fauna, climate and hydrology, etc.). As the process matured, the range of aspects increased and today social, health and economic issues are also examined. However, the integration and linking of biophysical and socio-economic impacts do not occur everywhere and to the same extent. In some countries, the social impacts are only given limited consideration, while in others the EIA process is supplemented by the social and health impact assessments, thereby limiting the extent of integration.

The main focus of EIA is how the development should mitigate the negative environmental impacts. In general, an EIA is done for specific irrigation projects, such as large dam and irrigation scheme command and developments. The purpose of an EIA is:

- To ensure that the environment and socio-economic development projects are properly accounted for, that unwarranted negative impacts are avoided or mitigated, and that potential benefits are realised.
- To determine the potential and known effects of proposed projects on the cultural, social, economic and ecological health of an area affected and put in place measures to avoid negative impacts while enhancing the positive ones.
- To ensure that adequate environmental information is available to the decision makers.
- To identify cultural, social, economic and ecological monitoring and management requirements during construction, operation and decommissioning.
- To improve public participation in government decisions by involving the public at all stages of the EIA.

The benefits of an EIA are:

- Increased accountability and transparency during the development process.
- Reduced environmental damage as measures are already put in place before a project starts.
- Increased project acceptance by the public communities as they are involved in the process.
- Better environmental planning and design of a proposal. Carrying out an EIA entails an analysis of alternatives in the design and location of projects.
- A well-designed project can minimise risks and impacts on the environment and people, and thereby avoid associated costs of remedial treatment or compensation for damage.
- It fosters mutual understanding between organisations, proponents and the people affected by the development. It provides the proponent with more realistic and objective information about the constraints placed on it by authorities.
- Savings in capital and operating costs. EIA can avoid the undue costs of unanticipated impacts. These can escalate if environmental problems have not been considered from the start of the proposal design and require rectification later.

8.4 The EIA Process

The generic EIA process needs to assess the environmental impacts of development projects in the following stages, namely baseline survey, screening, scoping, impact identification, prediction and significance assessment. A baseline study involves the description of the environmental and development project setting. A baseline environmental study of area is also presented. It highlights the existing conditions of the biophysical and socio-economic environments of the area prior to the project implementation. The major thrust is to establish the trends and anticipated future environmental conditions that will result from project implementation and to note the environmental sensitive areas of unique scientific, social, economic and cultural values of the area. Screening is a process of selecting which projects require an EIA. Thus, the proponent needs to consult the national EIA guidelines during screening stage. Scoping is a process of selecting relevant environmental effects. Such information can be obtained from literature and experts.

Impact identification is the listing of all of the impacts whether potential or actual resulting from the irrigation projects without any descriptions or attempt to quantify. Methods commonly used for identification of impacts can be literature reviews, matrices, networks, modelling, checklists and professional judgement. Interaction matrices are widely used for large-scale irrigation projects because they are easy to use, comprehensive, summarise and communicate the impact. The shortcomings of an interaction matrix are that it only considers binary relations between impacts, tends to be biased and is not explicit on significance. More importantly matrices inhibit scrutiny in that the way interpretations and conclusions are reached cannot be reproduced.

Impact prediction involves quantifying of the likely nature of the impacts based on a set of criteria or descriptors. With regard to the irrigation projects, impact prediction is the quantification of changes likely to occur as a consequence of the irrigation project. These include nature and spatial extent of impact, impact duration, intensity, reversibility, degree of certainty and mitigatory potential. These criteria can be classified into the basic, supplementary and quality. Basic criteria concern the magnitude or intensity, spatial extent and duration, supplementary criteria which entail synergism between variables, cumulative effects and controversy surrounding the impacts and quality criteria being the information that supports the prediction of an impact, its probability of occurrence, confidence of predication and the existence of environmental standards. This stage of impact prediction serves as the basis for determining the impact significance. Prediction methods consist of predicting the magnitude of the identified impacts and indications of expected (future) situation. This can be achieved by carrying out surveys, matrices, checklists and overlays in Geographic Information Systems. Lawrence (2007) identified the impact prediction and assessment as areas of weakness in the EIA process, in addition to the lack of consensus on methodologies and flawed interpretation of the terms. Technical difficulties with impact prediction have also been highlighted as one of the difficulties faced by practitioners. Other problematic aspects in the identification and assessment of impacts include vague descriptions, lack of systematic methods, lack of detail on criteria used, failure to evaluate the impacts according to laid down criteria and failure to consider all phases of the project.

The focus of EIA will always narrow down to a judgement on whether the predicted impacts are significant. Impact significance includes deciding on the importance of potentials based on the parameters referred to under impact predication. It involves the assessment or interpretation of impacts to make an estimate of the value of the present or future situation (significance). A significant impact is one where anticipated future environmental conditions, resulting from the proposed action (irrigation), differ from those otherwise expected from normal change, and where this anticipation raises the serious concerns among a professional or lay section of the society. Lawrence (2005) summarised the significance determination as making judgements on what is important, desirable or acceptable. Environmental significance should therefore be viewed as an anthropocentric concept, which uses judgement and values to the same or greater extent than science-based criteria and standards. The degree of significance depends upon the nature (i.e. type, magnitude, intensity, scale, probability and duration) of impacts and the importance communities place on them.

8.4.1 Managing Environmental Impacts

Environmental impacts of irrigation projects can be managed by avoidance, reduction and compensation. Avoiding impacts entails the prevention of possible impacts which are likely to occur during the activity. For example, graves can be left untouched in the fields besides exhuming them. Reduction incorporates activities like selective logging or subtle deforestation as opposed to clear cutting of tress in the fields. Impact compensation is done to pay for the destruction which would have been done such as resettlement of people to pave way for the irrigation development.

8.4.2 EIA and Its Significance in Irrigation Projects

This section shows how the concept of EIA has been embraced in the Zimbabwean context. This outline is important in order to show the significance and adoption of EIA in the irrigation projects.

In Zimbabwe, EIA is now widely recognised as a development tool for guiding project development and management. This recognition stems from the realisation that past development projects that were carried out prior to the EIA legislation have not only failed to yield maximum benefits to society and the economy at large, but have also caused serious and irreversible damage to the environment. The formal EIA process was established through the EIA policy of 1994 and is a requirement for certain projects like irrigation schemes in terms of the Environmental Management Act (Chapter 20:27). It is intended to ensure that development proposals, activities and programmes are environmentally sound and sustainable.

Notwithstanding the significance of the EIA, there appears to be tension in recognising the value of EIA in the development process. In theory, it is assumed that EIA results should lead to abandonment of projects that are not environmentally sound. In reality, however, projects are rarely abandoned. Wood (1995) stressed the need to manage the negative impacts to ensure that projects can still be implemented in an environmentally sound manner. In other words, the EIA as a decision-making tool provides a methodology and system of performing analysis, projection and evaluation of potential impacts of developmental projects with the aim of determining the best course of action. The process prevents and ameliorates the adverse impacts and comes up with monitoring mechanisms, which should be used to track the life of the project on the environment. The world over, EIAs are the successful policy innovations for environmental conservation, preservation, optimum utilisation of resources, saving time and cost of the project and management. The results of EIA should be communicated to decision makers who are in most cases the government agencies, engineers designing the project, and public. In order to safeguard the environment and interests of future generations, the EIA plays an important role in human well-being and environmental protection. The EIA process remains a critical planning and decision-making tool which serves to enhance development and not to inhibit it. Table 8.2 summarises various themes justifying the significance of an EIA.

TABLE 8.2   Reasons Explaining the Significance of EIA in the Irrigation Sector

Theme	Explanation
Impact identification	Identification and projection of potential impacts on the environment, which can be positive, negative or both, allow for appreciating environmental impacts. The EIA process enhances positive impacts while providing mitigatory or avoidance measures of negative impacts during the lifespan of the irrigation project.
Project information for decision making	Assist in providing information for decision making on projects that might have significant impacts on the environment and communities, such as genetically modified related organisms.
Addressing perceptions	The EIA is an essential tool for managing mixed perceptions towards a project by the communities and interested stakeholders since projects create change irrespective of whether they are good or bad.
Conflict management	Lessen and manage conflicts by promoting information sharing, communication and effective community participation in decision making.
Retrospective review	EIA provides a framework for management of projects implemented before the introduction of EIA legislation to ensure rehabilitation legacy of such projects.

8.5 Impacts of Irrigation Projects

In agriculture, the world over, irrigation has a long history of being applied as it significantly contributes towards poverty alleviation, food security, employment creation, general economic growth and improving the quality of life for farming communities (Khan et al., 2006). Many countries are turning towards irrigation development as an adaptation measure against aridity and higher temperatures associated with climate change (Finger et al., 2011; Eshel et al., 2014; IPCC, 2014; Sean et al., 2015). However, the sustainability of irrigated agriculture is being questioned owing to the negative environmental effects. This section presents the various impacts of irrigation projects and attempts to show how an EIA can be used to manage such impacts.

Irrigation projects largely involve clearing up large pieces of land, construction of water reservoirs, application of chemical fertilisers, herbicides and pesticides, among other major environmental aspects. One way to understand the significance of EIA in irrigation would be to examine the broad impacts of projects as illustrated in Table 8.2. It shows that impacts are very broad and falls into biophysical and socio-economic categories. The direction of the impacts can also be positive, which the project should enhance, or negative, which the project should avoid or minimise. These impacts also range from being short-term to medium-term and long-term. Some of the impacts can also be anticipated, while others are unforeseen. The latter complicates impact management as there is little room for containing negative unanticipated impacts. Dougherty and Hall (1995) identified the broad impacts associated with irrigation projects in the form of hydrology, water and air quality, soil properties and safety erects, erosion and sedimentation, biological and ecological change, socio-economic impacts, ecological imbalances and human health. These are indicated in Table 8.3.

8.5.1 Hydrological Impacts

With reference to hydrological impacts, the main issues are related to water flow and flood regimes, operation of dams, and fall or rise in the water table. Changes to the flow characteristics of rivers may negatively affect downstream communities who may be using the water for farming, drinking, transportation, fishing or hydropower (Ruffeis et al., 2008). This means the EIA would need to identify and assess the quantitative demands from both existing and potential future users. In situations where the water is abstracted for municipal or domestic uses, low water flow regimes may affect the dilution of pollutants that are drained from the watershed, particularly if the river is draining from industrial, urban or agricultural catchments. Dougherty and Hall (1995) suggested that as part of the irrigation

TABLE 8.3   Impacts of Irrigation Projects

Impact Type	Impact Description
Hydrology	Changes in flow regime of rivers, flood regime, operation of dams and fall or rise in the water table
Water and air quality	Solute dispersion, toxic substances, agrochemical pollution, anaerobic effects and gas emissions
Soil properties	Soil salinity, soil properties, saline groundwater, saline drainage and saline intrusion
Erosion and sedimentation	Local erosion, hinterland effect, river morphology, channel structures, sedimentation and estuary erosion
Biological and ecological change	Project lands, water bodies, surrounding area, valleys and shores, wetlands and plains
Socio-economic impacts	Population change, income and amenity, human migration, resettlement, gender, minority groups, sites of value, regional effects, user involvement and recreation
Ecological imbalances	Pests and weeds, animal diseases and aquatic weeds
Human health	Disease ecology, specific risks and counter measures, and health opportunities

Source: Dougherty and Hall (1995).

infrastructure, dams can offer opportunities to mitigate the potential negative impacts of changes to the flood flows. Flood plains may be used to allow the groundwater recharge and reduce the peak discharges downstream (Khan et al., 2006). In addition to monitoring groundwater levels, an EIA should define and enforce the abstraction regulations and recommend measures for controlling floods.

8.5.2 Impacts on Water and Air Quality

Irrigation can lead to the reduced water quality which can be costly to drinking water. The common rise in salinity levels may affect the crops and natural food chains, leading to loss in agricultural production. Pesticides and herbicides are a more common source of pollutants associated with the irrigation schemes. Such chemicals pose a threat to ecosystems and humans. Chemical sprays can also affect air quality and contribute to release of greenhouse gas emissions (Eshel et al., 2014). In addition, use of organic and chemical fertilisers may result in an excess of nutrients like nitrates and phosphates, which can affect aquatic ecosystems and public health (Jobin, 1999).

8.5.3 Effects on Soils

Salinisation is reported as the major cause of land degradation with adverse environmental impacts in irrigation schemes. Dougherty and Hall (1995) reported that saline conditions can limit crop choices, affect crop germination and yields and make soils difficult to work. Researchers have also established that accumulation of salts in soils can lead to irreversible damage to soil structure, which is essential for irrigation and crop production. For example, the effects are prevalent in clay soils where the presence of sodium can bring about soil structural collapse (Khan et al., 2006; Wichelns and Qadir, 2015). Soil salinity and sodicity are some of the major factors limiting agricultural productivity in irrigation schemes. A study was conducted by Chemura et al. (2014) to assess the quality of irrigation water used in Mutema Irrigation Scheme located in south-east Zimbabwe showed significant differences in the chemical quality of soils from irrigated and non-irrigated parts of the scheme. The soils in the irrigated blocks had higher pH, Sodium Adsorption Ratio, Exchangeable Sodium Percentage and Electrical Conductivity compared to non-irrigated areas of the scheme, indicating an influence of irrigation water on soils characteristics in the irrigated plots. An EIA should recommend careful management to reduce the rate of salinity build-up (Mirzaie et al., 2021). Some of the management options include altering irrigation methods and schedules, installing sub-surface drainage, changing tillage techniques, adjusting crop patterns and incorporating soil ameliorates. In developing countries, however, the choices of these salinity control measures are constrained by costs (Wichelns and Oster, 2006).

8.5.4 Erosion and Sedimentation Effects

Since irrigation is associated with land cover changes in vegetation, erosion may be a common problem in poorly managed projects. Although erosion may bring in alluvial deposits that improve soil fertility, the gain is indicative of the loss of fertility of upstream eroded lands. Erosion is also associated with sedimentation of silt in reservoirs, affecting irrigation intakes and pumping stations. Sediment load can also change the river morphology and increase the water turbidity, with subsequent effect on aquatic ecology. In order to address these problems and in the interest of prolonging the irrigation projects, there is need to recommend erosion control and desiltation measures (Jobin, 1999; Khan et al., 2006). For example, contour drainage and vegetation cover in steep slopes can slow down surface runoff.

8.5.5 Biological and Ecological Changes

Changes brought about by the irrigation project in the form of land and water use can significantly affect both terrestrial and aquatic ecosystems in the catchment area. The EIA studies should focus on the biological diversity and species displacement and migration. For instance, habitat disturbances can affect the entire ecosystem, animal populations or individual species of mammals, birds, fish, reptiles and insects (Jobin, 1999; Wichelns and Qadir, 2015). From the perspective of Dougherty and Hall (1995), it is also recommended to incorporate the local knowledge in the impact studies since communities usually have enriching knowledge about rare and endangered species and their feeding, breeding and migration patterns.

To deal with the ecological imbalances, an EIA should also recommend the appropriate management measures to address the problems of ecological disturbances within project sites and other externalities. It has been established that clearance of natural vegetation can affect the microclimate of the area and can expose the soil to denudation forces, leading to soil degradation. The removal of roots and vegetation, in particular, disrupts the water cycle as there will be more of surface runoff. This development can affect the flow characteristics of rivers and streams and can lead to downstream siltation, with possible threats on viability of fisheries and aquaculture projects. In situations where there are poor farming practices, the creation of agricultural monocultures can affect the local flora and fauna, reducing biodiversity. In addition, the introduction of foreign plant and animal species may lead to the extinction of native species. Although the application of fertilisers and pesticides can be done to correct the imbalances, these chemicals have been found to contaminate the groundwater and surface waters. The nutrients in fertilisers have also been associated with eutrophication of water bodies and the growth of noxious aquatic weeds (Jobin, 1999; Wichelns and Oster, 2006).

8.5.6 Socio-Economic Impacts

Although irrigation projects are intended to increase the agricultural production with consequent improvement in the economic and social well-being of the people in the targeted communities, there is a need to pay attention to both positive and negative impacts in order to promote successful irrigation schemes in developing countries. Thus, an EIA should concentrate on measures to enhance the positive impacts and mitigate negative impacts. Irrigation projects are blamed for affecting sovereignty of indigenous farming regimes particularly where land is converted to irrigated agriculture (Dougherty and Hall, 1995; Wichelns and Oster, 2006). Cropping practices and land tenure systems and rights are also affected by this conversion. Irrigation has also been blamed for affecting women roles and the interests of minority groups. For example, the introduction of formalised irrigation schemes causes gender, ethnic and social class imbalances. Some groups may be disadvantaged from disturbances in traditional livelihoods like hunting, fishing, grazing and dryland farming if land is converted to irrigation. Research has shown that it is often men and powerful groups that have had a greatest access to the benefits and increased income from the irrigated agriculture, while women, migrant groups and poorer social classes rarely enjoy benefits of irrigation farming (Dougherty and Hall, 1995; Ruffeis et al., 2008). However, there are also cases where increased income and improved nutrition from irrigated agriculture have largely benefited women and children (Dougherty and Hall, 1995). Accordingly, broad consultations and participation of all groups during the design of irrigation schemes, including establishing support mechanisms through extension services, credit schemes and markets need to be in place.

8.5.7 Public Health Impacts

The public health impacts associated with irrigation projects are complex and may require a separate health impact assessment. Although agriculture significantly contributes towards good health, food security and improved infrastructure, which allows rural households a greater purchasing power for drugs and health services, there are significant negative impacts associate with irrigation projects on human health. The human health dimension of the irrigation impacts is examined by Verma (1986) and Yusuf and Yusuf (2008). The expansion of irrigation schemes can also lead to the increased threat of health hazards particularly due to the diseases of malaria and Schistosomiasis. Pesticide residues are hazardous to both public health and ecosystems (Dougherty and Hall, 1995; Jobin, 1999). An EIA is expected to broadly cover these concerns.

The prevalence of Schistosomiasis is naturally associated with existence of water bodies such as rivers and reservoirs, particularly where water movement is sluggish or non-existent. Drop structures or water boxes along field canals create conducive environments for the breeding and survival of host snails which promote the spread of Schistosomiasis. They are ideal for breeding of snails due to the water ponding. The water boxes are also the attractive sources of water particularly for bathing and laundering especially when the flow of water in the canal is low or non-existent. The other potential transmission sites are the night storage dam and badly managed or maintained canal sections which are often laden with the organic debris, earth or infested with weeds, is a potential transmission site as well. In irrigation canals, the host snails feed on algal films on canal lining and live plants whose growth is encouraged by the accumulation of fertilisers, decaying vegetation and human excreta. This results in large populations of snails. Also, the proliferation of snail populations is most successful at 26°C. Under such favourable conditions, snails can double their populations in 2 weeks and multiply over 100-fold in a year.

The infection and transmission of Schistosomiasis are always through the contact with infected water. One of the most important processes by which people are brought into contact with water is irrigation. Thus, the introduction of irrigation schemes has increased the prevalence of Schistosomiasis in many areas throughout the world. Any large-scale extension of irrigated territories provides the bigger and more permanent breeding conditions for the snail intermediate hosts to proliferate and propagate the survival of schistosome parasitic worms.

In Zimbabwe, the importance of irrigation in promoting the spread and the transmission of Schistosomiasis has already been established in both large and smallholder irrigation schemes. However, whereas the extent and magnitude of the problem of Schistosomiasis and the impacts of the disease have been studied in detail in large-scale irrigation schemes, little, if any, effort has been made to address the problem of Schistosomiasis in more than 200 small-scale schemes that are operational. The only familiar study of the prevalence and incidence of Schistosomiasis in smallholder irrigation schemes was for Mushandike smallholder irrigation scheme. The Official Development Assistance (ODA) funded pilot project involved attempts to combat the disease using engineering and environmental techniques. As the number of smallholder irrigation schemes increases, the prevalence of Schistosomiasis follows suit, unless some mitigatory measures are implemented to minimise the spread of the disease. If the problem of Schistosomiasis is left unaddressed, the irrigation agricultural productivity will be seriously impaired.

Some attempts have been made to control the transmission of Schistosomiasis and also to use chemotherapy on people infected by the disease. The reasons why the spread of Schistosomiasis has not been successfully controlled in smallholder irrigation schemes in Zimbabwe include the prohibitive costs of molluscicides, high level of skill and operational costs of maintaining an effective control programme without compromising agricultural productivity.

The 1993 and 1994 statistics on the prevalence of Schistosomiasis show that the magnitude of Schistosomiasis is higher in smallholder irrigation schemes than in areas of dryland farming. Prevalence of Schistosomiasis in irrigation schemes is influenced by physical factors such as high temperature and the accumulation of organic matter in canals and is also controlled by the nature of economic and socio-cultural environment.

8.6 Conclusion and Limitations

In this chapter, it has been shown that although irrigation projects improve the economies of societies, they cause severe adverse socio-economic and biophysical environmental impacts. Accordingly, EIAs are required to evaluate the irrigation projects to determine their impact on the environment and human health and to set out the required environmental monitoring and management procedures and plans. The EIA is thus an important tool based on the precautionary principle that enhances the sustainable development where environmental, economic and social pillars are mainstreamed in the project in a balanced manner.

While this chapter has attempted to examine the significance of EIA of irrigation projects, it has largely dealt with the Zimbabwean experience using the information that is readily available from existing sources. It has concentrated on large and small-scale irrigation projects that have been widely documented in the country. As such, the information may not be up to date and may fail to capture changes that occurred in recent years following the intensification of other small irrigation projects owing to climate change responses in the agriculture sector. Some of the irrigation projects cited in this chapter were developed before EIA became a regulatory requirement. This means that the EIA process discussed here cannot give a true reflection of such projects. However, given that EIA is an evolving process, the Zimbabwean situation used here has helped to show the opportunities and challenges associated with embracing EIA in a developing country perspective, which most countries tend to have similar experiences.

References

Bolding, A. 2004. In hot water. *A Study of the Sociotechnical Intervention Models and Practises of Water Use in Smallholder Agriculture, Nyanyadzi Catchment,* Zimbabwe. Wageningen University, The Netherlands.

Chemura, A., Kutywayo, D., Chagwesha, M.T. and Chidoko, P. 2014. An assessment of irrigation water quality and selected soil parameters at Mutema Irrigation Scheme, Zimbabwe. *Journal of Water Resource and Protection*, 6, 132–140. http://dx.doi.org/10.4236/jwarp.2014.62018

Department of Water Development. 2000. *Irrigation Development in Zimbabwe*. Harare, Zimbabwe.

Dougherty, T.C and Hall, A.W. 1995. Environmental impact assessment of irrigation and drainage projects. *FAO Irrigation and Drainage Paper, Number 53*. FAO, Rome, Italy.

Eshel, G., Shepon, M.T. and Milo, R. 2014. Land, irrigation water, greenhouse gas, and reactive nitrogen burdens of meat, eggs, and dairy production in the United States. *PNAS*, 33, 11996–12001. https://doi.org/10.1073/pnas.1402183111

FAO. 2004. *Socio-Economic Impact of Smallholder Irrigation in Zimbabwe: Case Studies of Ten Irrigation Schemes*. FAO SAFR, Harare, Zimbabwe.

Finger, R., Hediger, W. and Schmid, S. 2011. Irrigation as adaptation strategy to climate change—A biophysical and economic appraisal for Swiss maize production. *Climatic Change*, 105, 509–528. https://doi.org/10.1007/s10584-010-9931-5

Gilpin, A. 1995. *Environment Impact Assessment (EIA). Cutting the Edge for the Twenty-First Century*. Cambridge University Press, Cambridge, UK, pp. 182.

IAIA. 1999. *Principles of Environmental Impact Assessment*, IAIA Special Publication http://www.iaia.org/publicdocuments/specialpublications/Principles%20of%20IA_web.pdf, accessed on 10 May 2020.

IPCC. 2014. *Climate Change 2014: Impacts, Adaptation, and Vulnerability. Part A: Global and Sectoral Aspects. Contribution of Working Group II to the Fifth Assessment Report of the Intergovernmental Panel on Climate Change*. In, Field CB, Barros VR, Dokken DJ, Mach KJ, Mastrandrea MD, Bilir TE, et al. (eds.). Cambridge University Press, Cambridge, UK and New York, NY, 1132 p.

Jobin, W. 1999. *Dams and Disease: Ecological Design and Health Impacts of Large Dams, Canals and Irrigation Systems*. London, UK: Taylor and Francis.

Khan, S., Tariq, R. Yuanlai, C. and Blackwell, J. 2006. Can irrigation be sustainable? *Agricultural Water Management*, 80(1–3), 87–99. https://doi.org/10.1016/j.agwat.2005.07.006.

Lawrence, D.P. 2007. Impact significance determination—back to basics. Environmental Impact Assessment Review, 27(8), 755–769.

Manzungu, E. 1999. *Strategies of Smallholder Irrigation Management in Zimbabwe*. Wageningen University, The Netherlands.

Mirzaie, Z., Fatahi, R., Eslamian, S, Azizi, A. 2021. Comparising and evaluating reaction factor and drainage water quality respect to direction of surface irrigation. *International Journal of Hydrology Science and Technology*, 12(2), 214–222.

Ruffeis, D., Loiskandl, W., Spendlingwimmer, R., Schonerklee, M., Awulachew, S.B., Boelee, E. and Wallner, K. 2008. Environmental impact analysis of two large scale irrigation schemes in Ethiopia. *Proceedings of the Symposium and Exhibition, Addis Ababa, Ethiopia, 27–29 November 2007*, 370–388. https://doi.org/10.22004/ag.econ.246407.

Sean, A., Woznicki, A., Nejadhashemi, P. and Parsinejad, M. 2015. Climate change and irrigation demand: Uncertainty and adaptation. *Journal of Hydrology: Regional Studies*, 3, 247–264. https://doi.org/10.1016/j.ejrh.2014.12.003.

Verma, R.D. 1986. Environmental impacts of irrigation projects. *Journal of Irrigation and Drainage Engineering*, 112(4). https://doi.org/10.1061/(ASCE)0733-9437(1986)112:4(322).

Wichelns, D. and Oster, J.D. 2006. Sustainable irrigation is necessary and achievable, but direct costs and environmental impacts can be substantial. *Agricultural Water Management*, 86(1–2), 114–127. https://doi.org/10.1016/j.agwat.2006.07.014.

Wichelns, D. and Qadir, M. 2015. Achieving sustainable irrigation requires effective management of salts, soil salinity, and shallow groundwater. *Agricultural Water Management*, 157, 31–38. https://doi.org/10.1016/j.agwat.2014.08.016.

Wood, C.M. 1995. *Environmental Impact Assessment: A Comparative Review*. Longman Group Ltd. London, UK.

Yusuf, M.F. and Yusuf, R.O. 2008. Environmental impacts of an irrigation scheme: The Bakolori irrigation project. *Journal of Environmental Research and Policies*, 3(4), 74–80.

9

Economic Viability of Irrigation Techniques

Patricia Angélica
A. Marques
University of São Paulo

Catariny Cabral
Aleman
Federal University of Viçosa

Saeid Eslamian
*Isfahan University
of Technology*

9.1 Introduction

One of humanity's main challenges is to ensure the food security. Although there has been a considerable increase in food production in the recent decades, Food and Agriculture Organization of the United Nations (FAO) estimates indicate that some 821 million people do not have an access to food. Some fundamental factors should be considered in order to meet the demand for food production [1], among which the following should be highlighted: reduction of availability of arable land, demand for irrigation incorporation in agricultural areas, exponential population growth in disagreement with food production and climate change.

Allied to the demands, the need for economic growth linked to sustainable development is one of the major challenges facing us today. Thus, the good performance of the agricultural activity depends, among several factors of production, on the adequate management and availability of water. The objective is to provide the exact amount of water required by the plants to maintain soil moisture at the desired levels, to maximize the benefits of economic development and to maintain the availability and quality of natural resources.

Management is defined, in a general way, as the judicious use of resources available to achieve a particular goal. In the case of irrigation, management aims to maximize plant production with the lowest water consumption, seeking not to compromise the quality of natural resources at the lowest possible cost.

Economic viability consists of defining investment alternatives and predicting their consequences, with a time reference, and considering the value of money over time. The high investment in irrigated agriculture must be paid for by an increase in the irrigated productivity.

Sustainable rural development is an indicator of a country's food security [2]. However, in general, farmers do not have the information on the proper irrigation management, leading to waste of water and energy, and, in some cases, loss of productivity. Thus, research in this field of knowledge is necessary to

DOI: 10.1201/9780429290114-12

generate or adapt technologies that meet the demands of productivity with a rational use. Risk-based economic analysis allows you to define which energy source and mode allow for the natural resource savings and profit.

This chapter aims to provide tools to assist farmers in making the decision to irrigate or not, looking for economic viability considering risks.

9.2 General Information about Agricultural Economics

Irrigated agriculture is responsible for the use of 69% of the water consumed in the world, being a limited resource that is becoming scarce in several agricultural regions. Irrigation is justified as a technological resource indispensable to increase the crop productivity in regions where insufficient or poor distribution of rainfall makes the agricultural exploitation unfeasible. Its adoption must be based on the technical and economic feasibility of the project, obtained by means of a detailed and careful analysis of the climatic, agronomic and economic factors involved.

As agriculture becomes a business, the emphasis on technology is to enable the reliable and secure production management and more effective control over the workforce. For this objective to be achieved, it is necessary that decisions, such as the implementation of an irrigation system or even the choice of the crop to be cultivated, are taken in a safe way and based on previously analyzed data. However, this concept is not applied in the current reality of the most agricultural enterprises.

Agricultural economics addresses the allocation, distribution and utilization of resources for agricultural production. In this context, information such as productivity, selling price of the product, generation of labor impacts on the local economy and the trade balance can be highlighted.

The economic returns of production can subsidize the technological and commercial growth of a country. The biggest limitation of agricultural prices is the market fluctuations that can impact the unstable agricultural incomes. One way to avoid market fluctuations is to establish an analysis of suitable investment alternatives for agricultural areas.

Relations between land, population and agricultural production are complex. With the evolution of technologies and the need to subsidize the demand for food, it became necessary to analyze the economic parameters. In this way, the study of risk and market uncertainty should be incorporated into the decisions related to the choice of agricultural area, agricultural culture, cultural treatments and implicit technologies in the productive process.

The production functions represent the inputs and outputs of a productive process, allowing the application of micro- and macroeconomic tools to evaluate the productivity and cash flows [3]. The production function composes a mechanism to quantify the inputs and outputs of the production system, guaranteeing adequate investment and economic return. A typical set of data that is used in the production function encompasses the revenue from production and all of the expenses included in the process [4,5].

9.3 Modeling and Simulation

In irrigated agriculture, the efficient and profitable production must be the main economic objective, always seeking revenues greater than the costs. In this way, it is important to know the degree of risk involved in the decision to irrigate. These risks are due to the economic uncertainties provided by the variation in the sale price of the product, interest rate, water and energy costs, useful life of the irrigation system and the maintenance rate of the irrigation system, as well as productivity. This implies that the randomness of certain coefficients should not be neglected, and the risk should be introduced in the project analysis.

Agricultural economy has maintained a certain conservatism. Although agriculture is a sector in the continuous transformations, the specialized literature maintains a close relation with the neoclassical economy. However, one of the most important characteristics of agricultural activity is the high risk to which it is subject. Therefore, any mathematical model for the study of production plans that aim to increase the profitability of agricultural production units must, strictly speaking, include risk in its formulation, that is, include the estimation of the degree of uncertainty that has to be respected to obtain the desired future results [6].

Decisions taken at risk are those in which the analyst models the decision problem in terms of known possible future outcomes. In this way, the risk exists when the decision maker can rely on the objective probabilities to estimate the different results, so that his expectation is based on historical data; therefore, the decision is taken from estimates considered acceptable by the borrower's decisions.

Risk is a crucial variable for decision-making about the introduction of a technology, even when its potential results are already widely known and publicized. In addition to the climatic risks associated with agricultural products, the high price variation of various agricultural commodities creates uncertainties for farmers. Small- and medium-sized farmers are particularly susceptible and risk-averse, especially those whose immediate survival depends directly on agricultural production. Producers with access to financial resources or credit have the greater ability to cope with price and production risks, showing a greater tendency and speed in accepting new technologies.

Among the different forms that risks can occur, the main ones applied to agronomy are the price risk and yield risk. Risks tend to act as a deterrent in the adoption of advanced practices, as well as in the efficient use of resources. Price risks arise from the uncertainty between the date on which the farmer makes a production decision and the date on which the product is to be marketed, with agricultural products presenting an inelastic demand curve. Already, the risks of yield occur in advent of the climatic variations and incidence of diseases and plagues. If the farmer seeks to reduce the variability of his income in his production plans, he will select the products and production processes that have a low variability.

One way to measure the risk of irrigation adoption is to place the decision in terms of probability distribution of the future annual net benefit. The smaller the dispersion of probability distribution, the lower the risk. When risk is introduced in the investment analysis, it can no longer be said that two alternatives with the same positive annual net benefit are economically equivalent. The best alternative is the one with the least possibility of variation.

Analysis of potential crop yield data, along with production cost and price data, is crucial in irrigation decision-making. In this sense, the use of the computer models that integrate the simulation of growth and productivity of the crop with economic aspects constitutes a powerful tool to aid decision-making. The simulation models allow researchers to consider a large number of factors, which would be impossible in conventional experiments, thus saving time, financial resources and human resources. It also allows the simulating various scenarios and estimating the repercussions of different courses of action on production systems, resulting in not only a deterministic value but also a frequency distribution with the risk being translated into the numbers by the variance.

A widely used simulation technique is the Monte Carlo method, which is based on the comparison of random numbers with a certain cumulative probability density function, that is, the method allows the generation of the other possible values for the event from a random number. This method results in a frequency distribution of the simulated values. It is based on the fact that the relative frequency of occurrence of the event of a certain phenomenon approaches the mathematical probability of occurrence of the same phenomenon, when the experiment is repeated many times. This method can be applied in cases where the exact mathematical analysis would be intractable or excessively laborious.

Assuming that the distribution of a technical coefficient is triangular, it is enough to know the minimum value, the modal value and maximum value in order to establish the frequency distribution of this technical coefficient. This distribution is widely used when not enough information regarding the variables is available. Once the probability distribution is individualized, the next step of the simulation process is the selection, at random, of a value for each of the variables within the pre-established distributions. To associate a random number with the probability distribution of each variable, the numbers generated from a uniform distribution in the closed range of 0 to 1 are used, which will be compared to a reference value r^* (Eq. 9.1). After the simulations, the calculations of the relative frequencies (Eq. 9.2) and accumulated frequencies (Eq. 9.3) are arranged in classes or categories where the number of elements belonging to each of the classes is determined.

$$r^* = \frac{(m-a)}{(b-a)} \tag{9.1}$$

$$f_i = \frac{N(X_i)}{N} \quad 0 \le f_i \le 1 \tag{9.2}$$

$$F_i = f_i + F_{i-1} \quad 0 \le F_i \le 1 \tag{9.3}$$

where r^* is the calculated reference value, a is the minimum value of the studied economic factor, b is the maximum value of the studied economic factor, m is the modal value of the studied economic factor, f_i is the the relative frequency of X_i (decimal), $N(X_i)$ is the number of times X_i occurs, N is the total number of observations and F_i is the accumulated frequency of X_i (decimal).

9.4 Effects of Water Deficit on Crop Productivity

Irrigation is the agricultural sector with the highest water consumption. In many parts of the world, especially the arid and semi-arid regions, the use of water for irrigation is irregular and impairs the availability of water resources. In this way, management strategies and technologies are necessary to enable the water use efficiency and economy. Irrigation management consists of the rational use of water in order to meet the water needs of plants, increasing water production and productivity. It can be based on evapotranspiration, relating the water consumption of the plant, and on the soil moisture, determining the intervals of humidity in field capacity and permanent wilting point to define the soil storage capacity.

Proper water management is of fundamental importance for crop productivity. Water is present in the absorption, sap flow, water loss through transpiration and in the metabolic processes of cultures. In addition, it is a fundamental component of the water balance for the purpose of determining the irrigation depth. The irrigation depth is obtained considering the balance between the available soil water and crop evapotranspiration.

Two main components can be highlighted in the process of water use: soil evaporation and crop (evapotranspiration) and the water losses resulting from the distribution of water in the soil profile. Restriction of the water supply may decrease transpiration that negatively affects the evaporative environmental demand and reduces the carbon assimilation, causing the water use efficiency to be reduced [7].

Water stress can be defined as inadequate soil moisture that can affect the growth, development and productivity of agricultural crops. Situations of water deficit may have a negative impact on plant growth and development. The water deficiency causes the morphophysiological changes that reflect in the primary metabolism, causing damage to the metabolic apparatus or promoting the synthesis of proteins that attenuate the effects of stress and not producing damage to production [8].

The frequency and intensity of the water deficit are the main factors limiting global agricultural production. Therefore, to avoid the negative impact of water deficit and to use it in a positive way for productivity, it is important to define what the water stress strategy will be for the specific conditions of the region.

Regulated water deficit has been identified as an effective alternative for water saving in an irrigated agriculture. It is defined as the reduction of the maximum water demand of the crop, that is, it is irrigated less than the crop would demand for the conditions of cultivation. Irrigation deficits at low development stages are an effective alternative for water savings with little reduction in crop productivity. This can improve crop development and increase water use efficiency (WUE) (Eq. 9.4) [9].

$$WUE = \frac{CP}{I}$$

(9.4)

where WUE is the water use efficiency (kg/mm), CP is the crop productivity (kg), and I is the depth of irrigation (mm).

9.5 Climatic Changes and Impacts in Irrigated Agriculture

Understanding how climate change interacts and impacts on the food production is critical to defining the proper management in an irrigated agriculture. The water crisis in various parts of the world is the result of climate change, which is due to the variability of precipitation, changes in temperature and an increase in the CO_2 concentration. This may threaten the agricultural productivity and food security, especially in the regions that require full irrigation.

The increase in atmospheric CO_2 is the factor that contributes the most to climate change; however, it has the potential to increase the productivity of water use by increasing photosynthetic activity and reducing the plant transpiration [10]. Temperature changes alone can cause the negative impacts on crop productivity [11]. Crops such as wheat, rice, corn and soybeans undergo reductions in productivity, which can affect the world's food supply. This is due to the increase in atmospheric demand, triggering water stress, reduction of the soil water availability and consequent reduction of production [12,13].

The relationship between the climate change and crop production is complex, and it is necessary to represent the numerous crops to obtain a broad approach to production in different scenarios (Figure 9.1). In this way, it is possible to adjust the perspectives that allow for the evaluation of economic variables in more representative time intervals.

The water deficit can be a relevant environmental factor to limit the crop productivity, especially if highlighted as the result of climate changes that increases drought conditions. The great diversity of the plant species cultivated in climatic regions that include the extreme drought conditions suggests that some crops can tolerate water stress with morphological, physiological and biochemical adaptations.

Drought tolerance mechanism is a term that can be adopted for the plant species with the adaptive characteristics that allow them to escape, avoid or tolerate water stress. Some crops are able to modulate their vegetative and reproductive growth according to the water availability, essentially through two distinct mechanisms: rapid phenological development and developmental plasticity. This allows the plants to produce in different abiotic conditions [15].

9.6 Irrigation Viability

Determining the economic viability of a project is critical to its success. For the study of economic viability, its potential irrigated productivity and productivity without irrigation are considered. Productivity without irrigation can be estimated by the data from government, farmer, or crop response factors (K_y), which relates the relative yield decline and the relative evapotranspiration deficit [16] to the local conditions of the study for a total cycle per crop year (Eq. 9.5).

$$Y_s = Y_i \cdot \left[1 - K_y \cdot \left(1 - \frac{ETra}{ETma} \right) \right]$$

(9.5)

where Y_s is the annual yield of the non-irrigated culture (kg/(ha year)), Y_i is the annual yield of the irrigated culture (kg/(ha year)), K_y is the yield response factor representing the effect of a reduction in evapotranspiration on yield losses, ETra is the annual actual evapotranspiration (mm/year) and ETma is the annual maximum evapotranspiration (mm/year).

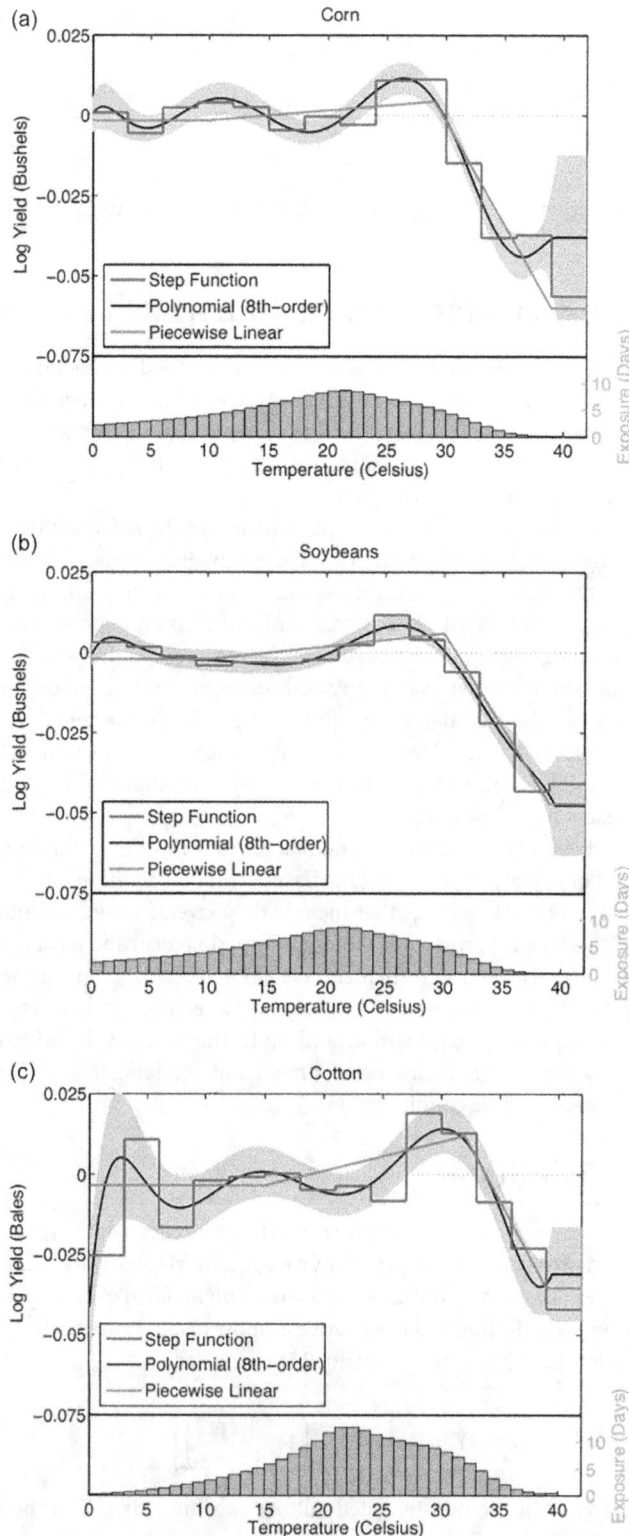

FIGURE 9.1   (a–c) Nonlinear response in relationship of productivity and temperature [14].

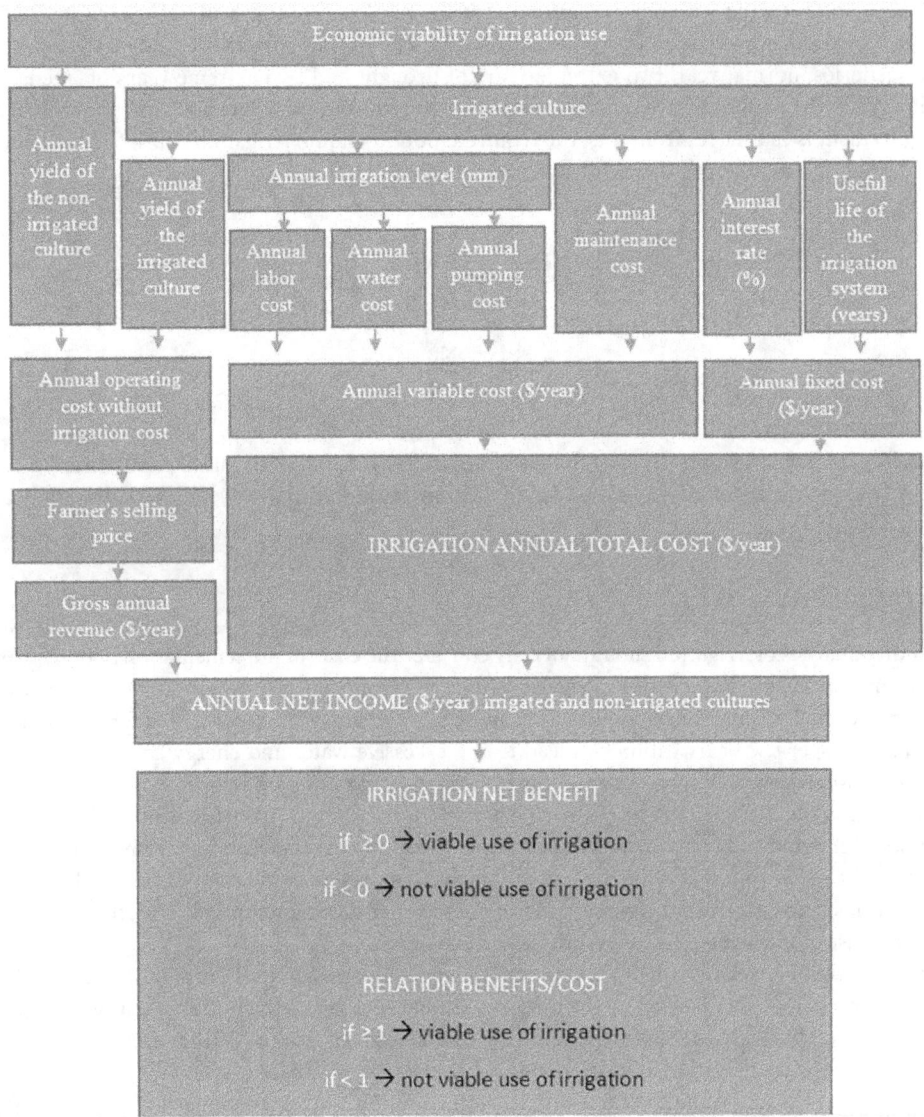

FIGURE 9.2 Flow chart about calculation of economic viability of irrigation use.

All costs involved with the decision to irrigate and their variations should be analyzed and compared at the end to determine whether the decision is feasible or not, that is, if the income gain from the irrigation use paid the fixed and variable costs of irrigation (Figure 9.2). Decision-making should contemplate the consistent choices in order to prioritize the profit maximization, given the constraints imposed. The decision-making process must be logical and objective, defining the problem to be careful, defining the clear and specific goal, and in view of this scenario, select the best alternative to maximize the results of the goal. For the final decision on a particular investment, it is necessary to analyze the data and trends related to making a decision, defining a given project as feasible.

In order to evaluate the variability resulting from the responses of the crop to the climate, it is recommended to perform the daily water balance using climatological data for the studied regions in five climate scenarios: normal year, two extreme years of drought and two extreme years of intense precipitation. With the daily values of irrigation depths required from these scenarios, the monthly irrigation requirement is estimated; from this, the required monthly liquid irrigation depth (Eq. 9.6) and the monthly gross irrigation depth (Eq. 9.7) can be calculated:

$$NI = \sum_{i=1}^{n} (NIdi), \tag{9.6}$$

$$LBm = \frac{NI}{AE}, \tag{9.7}$$

where NI is the net irrigation depth (mm/month), NIdi is the diary irrigation needs (mm), i is the number of days in the month, LBm is the gross irrigation depth (mm/month) and AE is the application efficiency (decimal).

9.7 Irrigation Costs

The viability analysis of irrigation should not only consider the costs of implementation and investment but also consider the other fixed and variable costs arising from expenditures used in the operation of the system (i.e., fuel, maintenance, inputs, labor and electric energy) being calculated for a year and per unit area. Improper use of irrigation can lead to the excessive water and energy consumption. Thus, using simulation technologies, the use of irrigation considering different irrigation depths in agricultural production can be evaluated by means of a probability analysis of irrigation viability.

With local climatic and economic data, it is possible to simulate the economic viability of irrigated agriculture, the risk involved and the influence of the factors studied in the composition of the cost. The components needed to study the economic viability of irrigated crops are the fixed cost, the variable cost and the increase of income obtained with the use of irrigation.

Costs are classified as fixed when they do not change with the quantity produced. The calculation of the fixed cost can be represented by the uniform annual cost, using the initial investment, expected useful life and residual value of the component. The capital recovery factor (CRF) is used, which presupposes the reserve of a sufficient amount of money in each year to enable the asset to be restored in n years plus the interest charges on the capital invested, with common use in economic engineering (Eqs. 9.8 and 9.9). The variability of the expected useful life (Table 9.1) can occur due to the differences in physical conditions of operation, level of repair, operation and maintenance and the total number of hours that the system is used each year [17,18].

$$CRF = \frac{i \cdot (i+1)^n}{(i+1)^n - 1} \tag{9.8}$$

$$FC = CRF \cdot PS - \frac{RV \cdot i}{(i+1)^n - 1} \tag{9.9}$$

where CRF is the capital recovery factor (decimal), i is the annual interest rate (decimal), n is the useful life of the irrigation system (years), FC is the annual irrigation fixed cost (US\$/(ha year)), PS is the price of acquisition and installation of irrigation equipment and/or automation (US\$/ha) and RV is the residual value of the irrigation system (US\$/ha).

TABLE 9.1 Useful Life and Cost of Maintenance of Irrigation Systems

Systems and Compounds	Useful Life (Years)	Annual Maintenance (% Initial Value)
Sprinkler		
Conventional portable	10–15	1.0–4.0
Conventional non-portable	10–18	1.5-3.0
Conventional permanent	15–25	1.0–3.0
Self-propelled mechanized (traveler irrigation)	8–12	5.0–7.0
Center pivot	12–18	4.0–6.0
Linearly movable mechanized	12–18	5.0–7.0
Located		
Drip	10–15	2.0–4.0
Microsprinkler	10–15	1.0–3.0

Source: From Ref. [19].

Advances in irrigation technology have increased the water use efficiency. Despite this fact, farmers are still reluctant to convert the traditional system to an even more efficient irrigation system because of the high initial investment costs. In addition, the decision-making process can be tricky. Producers should evaluate several factors when making their decision, including financing, energy prices, available energy sources, commodity prices and/or product sales, labor availability, water availability, irrigation system efficiency, pressure of operation, and handling and characteristics of the pumping [20,21].

Superficial irrigation systems are those that usually require less initial investment, followed by sprinkler and drip systems. Although efficiency and production gain are better with the use of localized irrigation, their adoption to replace the surface irrigation and sprinkler systems remains low not only because of the high capital costs but also because of higher costs of maintenance and management, among other factors.

In order to decide to carry out the change of irrigation system or energy source, the producer needs an economic study with the local and historical data that facilitates its decision considering the costs and revenues as well as the horizon of possibilities arising from this decision instead of deciding only by the cost of equipment acquisition, disregarding variable costs and the variability of the economic factors involved. Variable costs (Eq. 9.10) are the expenditures used in the operation of the system (i.e., maintenance, inputs, labor and energy).

$$CV = C_{\text{maint}} + C_{\text{pump}} + C_w + C_{\text{labor}} \tag{9.10}$$

where CV is the annual variable cost (US$/(ha year)), C_{maint} is the annual maintenance cost (US$/(ha year)), C_{pump} is the annual pumping cost (US$/(ha year)), C_w is the annual water cost (US$/(ha year)) and C_{labor} is the annual labor cost (US$/(ha year)).

The cost of energy, or cost of pumping, depends on the type of pumping used and is commonly the factor of greater participation in the total costs of irrigation. It will be discussed below. The maintenance cost (Eq. 9.11) corresponds to the expenses to keep the irrigation system in adequate conditions of use and varies according to the number of hours of operation per year. It is common to estimate them as a percentage of the initial investment of the irrigation equipment (Table 9.1).

$$C_{\text{maint}} = \frac{T_m \cdot \text{PS}}{100} \tag{9.11}$$

where T_m is the annual maintenance fee (%) and PS is the price of acquisition and installation of irrigation equipment and/or automation (US$/ha).

TABLE 9.2 Medium Values of Irrigation Efficiency, Energy Use and Labor for Irrigation Systems in Brazil

Irrigation System		Irrigation Efficiency (%)	Energy Use[a] (kWh/(mm ha))	Labor[b] (hours/(ha irrigation))
Surface	Furrow	40–70	0.3–3.0	1.0–3.0
	Corrugation	40–70	0.3–3.0	1.0–3.0
	Flood	50–70	0.3–3.0	0.3–1.2
Sprinkler	Portable conventional	60–75	3.0–6.0	1.5–3.5
	Semi-portable conventional	60–75	3.0–6.0	0.7–2.5
	Permanent conventional	70–80	3.0–6.0	0.2–0.5
	Self-propelled mechanized (traveler irrigation)	60–70	6.0–9.0	0.5–1.0
	Center pivot	75–90	2.0–6.0	0.1–0.7
	Linearly movable mechanized	75–90	2.0–6.0	0.3–1.0
Located	Drip	80–95	1.0–4.0	0.2–1.0
	Microsprinkler	80–90	1.5-4.0	0.1–0.5

Source: From Refs. [19, 22].

Note: Irrigation efficiency values for sprinkler systems are presented for conditions where evaporative and drift losses are less than 1%.

[a] Estimated for a settling height between 0 and 50 m.

[b] Depends on the level of system automation, general efficiency and labor.

Labor cost (Table 9.2) refers to the salaries and social charges for all of the activities involved in the operation of the system. It depends on the irrigation system and varies with the level of automation (Eq. 9.12).

$$C_{labor} = \sum_{\text{irrigated days}=1}^{n} \left\{ R \cdot \left[1 + \left(\text{charges} \right) \right] \cdot \text{HI} \right\} \tag{9.12}$$

where C_{labor} is the is the annual labor cost (US\$/(ha year)), R is the renumeration for worked hours (US\$/hour) and HI is the hours of work (hours/(ha irrigation)).

Among the natural resources, water resources are the ones that have most worried researchers, due to the possibility of a global crisis in water availability, which has required a change in behavior in the use of this resource. The origin of this crisis is associated with the scarcity and degradation of water sources, resulting from the intensification of the processes of population growth, agricultural expansion and industrialization starting in the second half of the 20th century [23]. However, water is a limited good in Brazil, and it is measured within the values of the economy. This should not lead to conduct that allows someone (i.e., for the payment of a price) to use water unscrupulously.

In principle, the charge on water use should be applied to any use, including irrigators, which is another item that should be computed in the costs (Eq. 9.13). This charge has the legal support in Brazil in Federal Law No. 9433/97, which in article 19 considers water as an economic good subject to collection, and the financial resources collected should be used to finance the programs and interventions for the environmental recovery of the river basin where they were generated. The economic valuation of water must consider the price of conservation, recovery and better distribution of this good.

$$C_w = \sum_{\text{month}=1}^{12} \left(y_r \cdot P_w \cdot 10 \right) \tag{9.13}$$

where C_w is the annual water cost (US\$/(ha year)), y_r is the gross irrigation depth (mm/month) and P_w is the water price (US\$/m³).

Among the variable costs of an irrigation system, energy consumption stands out as a major component. In the 1970s and 1980s, the serious problems began to occur from the first oil shock, and fossil fuel prices soared. The energy matrix was predominantly fossil fuel, and electricity prices, which were regulated, began to rise in real terms for the first time since the beginning of its commercialization. As a result, several states of the United States of America and European countries have chosen to diversify their energy matrix [24].

The increase in the energy costs and the growing worldwide concern with water resources led farmers to adopt management strategies that allowed for water savings without sacrificing productivity. In Pakistan, for example, the agricultural development has suffered from the scarcity of irrigation water with about 50% of arable land without irrigation [25].

Thus, the advancement and improvement of new energy sources are necessary because of the worldwide need for energy readjustment. The main drivers of this change are the global energy crisis and the current scenario of climate change in which rainy periods have been increasingly dry, consequently lowering the level of the reservoirs of hydroelectric plants responsible for the energy supply. In this context, several farmers have been using wind energy and photovoltaic solar energy as forms of viability in the generation of alternative electric energy, providing diversification of the energy matrix.

In terms of the composition of installed capacity by source (Figure 9.3), Brazil is significantly different from the countries used in the previous comparison. While in Brazil hydroelectric, wind, solar and biomass sources account for 80% of installed capacity, they are only 45% in other countries (even in Spain and Germany, which strongly subsidized wind and solar generation). In the other countries (Figures 9.4 and 9.5), the predominant sources are natural gas, coal and oil, with emphasis on the United States where these sources correspond to 74% of installed capacity. France would be the exception to the rule, because it has a predominance of nuclear power plants, with approximately half of the country's installed capacity [24].

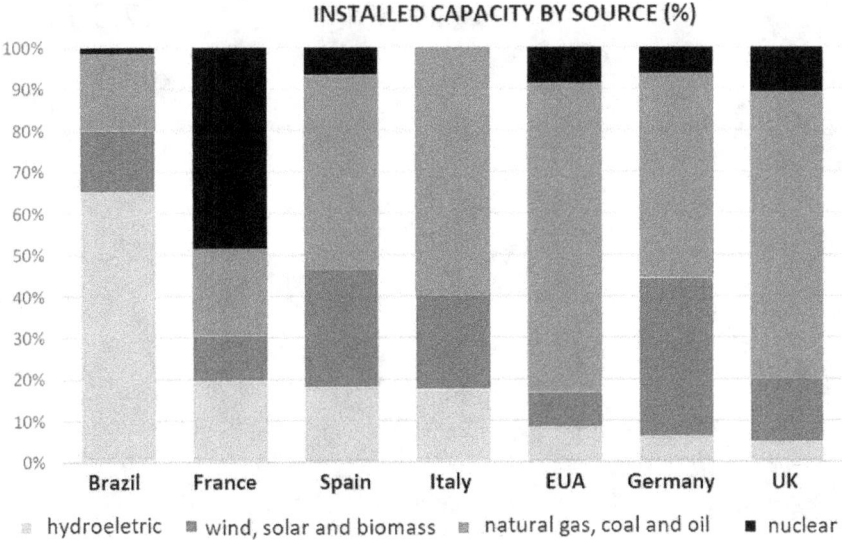

FIGURE 9.3 Composition of installed capacity in several countries [24].

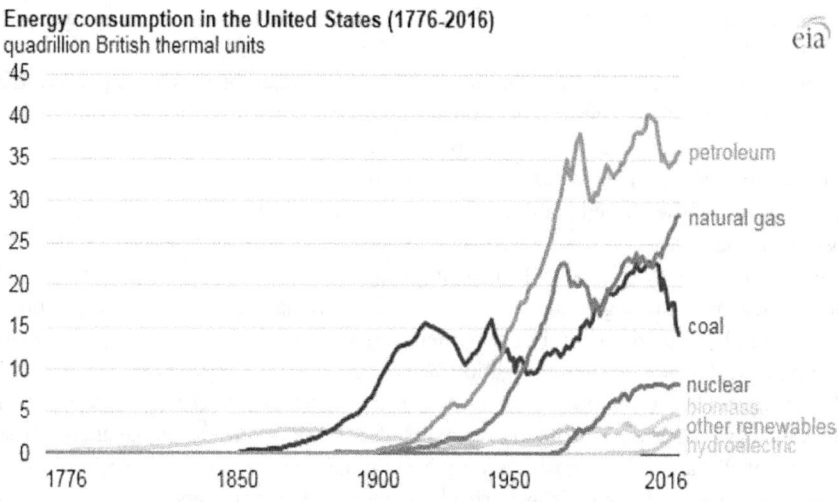

FIGURE 9.4 Energy consumption in the United States from 1776 to 2016 [26].

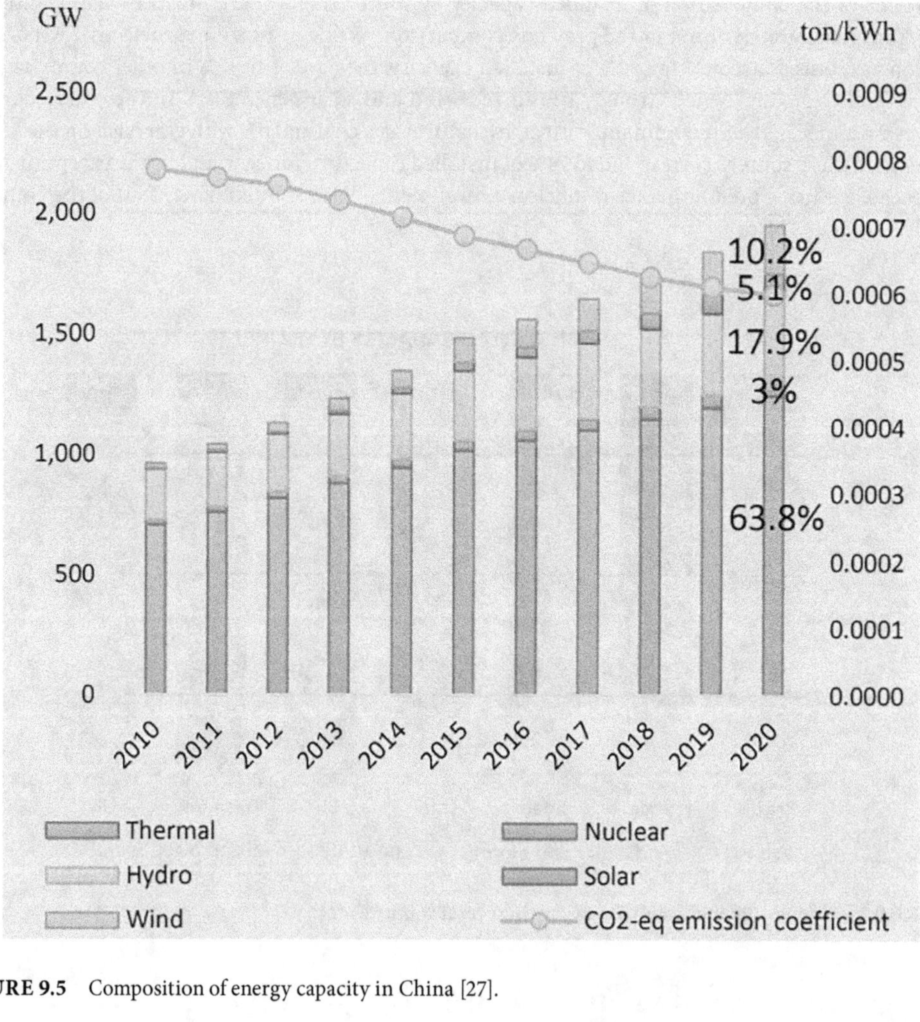

FIGURE 9.5 Composition of energy capacity in China [27].

The application efficiency of the irrigation system can reduce the losses, and the efficiency of the pumping can reduce the use of energy. Producers should also consider changing the type of energy used to irrigate. The authors also point out that electricity and diesel fuel are used in approximately 76% of the irrigated land. Nebraska uses electricity or diesel fuel to pump the irrigation systems on approximately 7.58 million hectares. Natural gas and propane are used in about 20% and 4%, respectively. Few properties use the gasoline engines.

For the calculation of the annual pumping cost (C_{pump}), the type of engine, combustion (Eq. 9.14) or electric (Eq. 9.15), must be considered, since this is a parameter with a great effect on the costs of the engines. In the United States, locational pricing prevails, that is, each point of entry and exit of the system has its price set separately. In Brazil, the irrigators can choose one of three types of electricity tariffs (conventional, blue and green) and two basic components in defining their price, one related to the power demand and the other one related to energy consumption.

$$C_{pump} = C_f \cdot \text{Pot} \cdot C_s \cdot \sum_{month=1}^{12} H \qquad (9.14)$$

$$C_{pump} = \sum_{month=1}^{12} \left(T_d \cdot 0.736 \cdot \text{Pot} + \text{TCc} \cdot H \cdot 0.736 \cdot \text{Pot} \right) \qquad (9.15)$$

where C_f is the cost of fuel in the property (US\$/gal), Pot is the engine power (cv/ha), C_s is the specific consumption of the combustion engine (gal/(cv h)), H is the monthly operating hours (h/(ha month)), T_d is the electric energy demand tariff (US\$/kW) and TCc is the electric energy consumption tariff (US\$/kWh).

9.8 Benefits

Irrigated agriculture requires the other investments that represent significant additional costs, which must be paid for by the increased productivity provided by the water supply to the plants. In this way, the annual net benefit (Eq. 9.16) is the difference between the net revenue obtained for the productivities of each irrigation depth and the net income without irrigation. To be viable, a project must present an annual net benefit greater than zero and a cost–benefit (B/C) ratio greater than the unit, that is, the larger this ratio is, the more attractive the project (Eq. 9.17).

$$\text{BL} = \left(Y_i \cdot P_p - C_p \cdot Y_i - \text{CT} \right) - \left(Y_s \cdot P_p - C_p \cdot Y_s \right) \qquad (9.16)$$

$$B/C = \frac{\text{BL}}{\left(\text{CT} + Y_i \cdot \text{CP} \right)} \qquad (9.17)$$

where BL is the annual net benefit of irrigation (US\$/(ha year)), Y_i is the productivity with irrigation (kg/(ha year)), P_p is the sale price of the product paid to the producer (US\$/kg), C_p is the cost of production independent of irrigation (US\$/(kg year)), CT is the total annual irrigation cost (US\$/(ha year)), Y_s is the yield without irrigation (kg/(ha year)) and B/C is the cost–benefit ratio (dimensionless).

Agricultural production is facing many challenges in relation to production costs, water resource limitations and concerns about the impacts on the environment. Thus, precision tools can be used to define management in the face of varying fertility conditions, water availability and crop characteristics.

The evolution of technologies has allowed an increase in productivity, which contributes to the expansion of economic benefits. Thus, precision irrigation is an alternative for increasing the efficiency of water use and economic viability. This irrigation tool depends on variable rate application (VRI) considering the demand for water from crops. The VRI is an important tool in precision farming favoring increased efficiency of water use in different crops [28].

Precision irrigation allows farmers to reduce the risks and uncertainties of agricultural production, because it presents a great potential for water use and economic viability. The use of the precision tool allows the farmers to reduce the demands of agricultural crops, reducing costs. Depending on the irrigation system in which the precision tools are inserted, the payback period can vary from 5 to 20 years [29].

Many economic benefits can be obtained when adopting precision irrigation, especially if their application in variable rates in considered. The limitation for defining the economic viability of using the precision tools is that it is unclear whether the benefits outweigh the costs by an enough of a margin to justify an adoption. Identifying enough crop varieties, soils and irrigation application systems are critical to determining the maximum benefit that can be achieved, as well as priorities for investment. This will also establish the advantages of the total versus the partial adoption on farms.

9.9 Conclusions

The inclusion of risks provided from economic uncertainties (variation in the sale price of the product, interest rate, water and energy costs, useful life of the irrigation system, maintenance rate of the irrigation system and productivity) allows for an economic viability study of irrigation. Thus, the producer can decide with more assertiveness and predict irrigation costs and benefits.

References

1. Rodrigues, L. N.; Domingues, A. F., 2017. *Agricultura irrigada: desafios e oportunidades para o desenvolvimento sustentável*. Brasília-DF: INOVAGRI, 327p.
2. Usoh, G.; Nwa, E. U.; Okokon, F. B.; Nta, S.; Etim, P. J., 2017. Effects of drip and furrow irrigation systems application on growth characteristics and yield of sweet maize under sandy loam soil. *International Journal of Scientific Engineering and Science*, v.1, n.1, pp. 22–25.
3. Morgan, W., 2016. Agricultural economics. Disponível em: http: www.studyingeconomics.ac.uk/module-options/agricultural-economics/.
4. Ackerberg, D. A.; Caves, K.; Frazer, G., 2015. Identification properties of recent production function estimators. *Econometrica*, v.83, pp. 411–2451.
5. Grieco, P. L. E.; Li, S.; Zhang, H., 2016. Production function estimation with unobserved input price dispersion. *International Economic Review*, v.57, pp. 665–690.
6. Frizzone, J. A.; Andrade Júnior, A. S., 2005. *Planejamento de irrigação. Análise de decisão de investimento*. Brasília: Embrapa Informação Tecnológica, 626p.
7. Fereres, E.; Soriano, M. A., 2007. Deficit irrigation for reducing agricultural water use. *Journal of Experimental Botany*, v.58, pp. 147–159.
8. Taiz, L.; Zeiger, E. 2013. *Fisiologia Vegetal*. Porto Alegre: Artmed, 5.ed, 918p.
9. Geerts, S.; Raes, D., 2009. Deficit irrigation as an on-farm strategy to maximize crop water productivity in dry areas. *Agricultural Water Management*, v.96, pp. 1275–1284.
10. Deryng, D.; Elliott, J.; Folberth, C.; Müller, C.; Pugh, T. A. M.; Boote, K. J., 2016. Regional disparities in the beneficial effects of rising CO_2 concentrations on crop water productivity. *Nature Climate Change*, v.6, pp. 786–790.
11. Asseng, S.; Ewert, F.; Martre, P.; Rötter, R. P.; Lobell, D. B.; Cammarano, D.; Kimball, B. A.; Ottman, M. J.; Wall, G. W.; White, J. W.; Reynolds, M. P.; Alderman, P. D.; Prasad, P. V. V.; Aggarwal, P. K.; Anothai, J.; Basso, B.; Biernath, C.; Challinor, A. J.; De Sanctis, G.; Doltra, J.; Fereres, E.; Garcia-Vila, M.; Gayler, S.; Hoogenboom, G.; Hunt, L. A.; Izaurralde, R. C.; Jabloun, M.; Jones, C. D.; Kersebaum, K. C.; Koehler, A-K.; Müller, C.; Naresh Kumar, S.; Nendel, C.; O'Leary, G.; Olesen, J.

E.; Palosuo, T.; Priesack, E.; Eyshi Rezaei, E.; Ruane, A. C.; Semenov, M. A.; Shcherbak, I.; Stöckle, C.; Stratonovitch, P.; Streck, T.; Supit, I.; Tao, F.; Thorburn, P. J.; Waha, K.; Wang, E.; Wallach, D.; Wolf, J.; Zhao, Z.; Zhu, Y., 2015. Rising temperatures reduce global wheat production. *Nature Climate Change*, v.5, pp. 143–147.

12. Burke, M.; Hsiang, S. M.; Miguel, E., 2015. Global non-linear effect of temperature on economic production. *Nature*, v.527, pp. 235–239.

13. Zhao, C.; Liu, B.; Piao, S.; Wang, X.; Lobell, D. B.; Huang, Y.; Huang, M.; Yao, Y.; Bassu, S.; Ciais, P.; Durand, J.; Elliott, J. Ewert, F.; Janssens, I. A.; Li, T.; Lin, E.; Liu, Q.; Martre, P.; Müller, C.; Peng, S.; Peñuelas, J.; Ruane, A. C.; Wallach, D.; Wang, T.; Wu, D.; Liu, Z.; Zhu, Y.; Zhu, Z.; Asseng, S., 2017. Temperature increase reduces global yields of major crops in four independent estimates. *Proceedings of the National Academy of Sciences*, v.114, pp. 9326–9331.

14. Schlenker, W.; Roberts, M.J. 2009. Nonlinear temperature effects indicate severe damages to U.S. crop yields under climate change. *PNAS*, v.106, n.37, pp. 15594–15598.

15. Basu, S.; Ramegowda, V.; Kumar, A.; Pereira, A., 2016. Plant adaptation to drought stress. F1000Research, 5 (F1000 Faculty Rev), 1554.

16. Steduto, P.; Hsiao, T. C.; Fereres, E.; Raes, D., 2012. *Crop Yield Response to Water*. Roma: FAO, 501p. (FAO Irrigation and Drainage Paper, 66).

17. Marques, P. A. A. Modelo computacional para determinação do risco econômico em culturas irrigadas. Piracicaba, 2005. 142p. Thesis (doctoral dissertation) - Luiz de Queiroz College of Agriculture (ESALQ)", Universidade de São Paulo. Available in: http://www.teses.usp.br/teses/disponiveis/11/11143/tde-09092005-162338/publico/PatriciaMarques.pdf.

18. Paredes, P.; Rodrigues, G. C.; Cameira, M. R.; Torres, M. O.; Pereira, L.S., 2017. Assessing yield, water productivity and farm economic returns of malt barley as influenced by the sowing dates and supplemental irrigation. *Agricultural Water Management*, v.179, n.1, pp. 132–143.

19. Marouelli, W. A.; Silva, W. L. C., 2011. *Seleção de sistemas de irrigação para hortaliças*. Brasília: EMBRAPA, 22p. (Circular Técnica, 98) Available in https://ainfo.cnptia.embrapa.br/digital/bitstream/item/75698/1/ct-98.pdf.

20. Guerrero, B.; Amosson, S.; Almas, L.; Marek, T.; Porter, D., 2016. Economic feasibility of converting center pivot irrigation to subsurface drip irrigation. *Journal of American Society of Farm Managers and Rural Appraisers*, v.12, pp. 77–88.

21. Montazar, A.; Zaccaria, D.; Bali, K.; Putnam, D., 2017. Model to assess the economic viability of alfalfa production under subsurface drip irrigation in California. *Irrigation and Drainage*, v.66, n.1, pp. 90–102.

22. Christofidis, D., 2002. Irrigação, a fronteira hídrica na produção de alimentos. *ITEM*, v.54, pp. 46–55.

23. Zareian, Z.; Eslamian, S., 2018, Using of optimization strategy for reducing water scarcity in the face of climate change. In *Climate Change-Resilient Agriculture and Agroforestry: Ecosystem Services and Sustainability*, by Paula Cristina Castro, Anabela Marisa Azul, Walter Leal Filho, and Ulisses M. Azeiteiro, Climate Change Management Book Series. Springer, Cham pp. 317–331.

24. Moreira, L. C., 2016. Um novo Mercado de Energia Elétrica para o Brasil. Brasília, 160p. *Dissertation (Master Degree)* - Universidade de Brasília. Available in: http://bdtd.ibict.br/vufind/Record/UNB_fe184efb6191958887l6cd28efcc57a0.

25. Ullah, Z.; Ullah, R.; Ali, M.; Jan, J., 2016. Performance evaluation of diesel and electric operated tube wells irrigation system in sub-tropical conditions. *Pure and Applied Biology*, v.5, n.1, pp. 142–148.

26. Marchionna, M., 2018. Fossil energy: From conventional oil and gas to the shale revolution. *EPJ Web of Conferences*, 189, 00004. Available in: https://www.epj-conferences.org/articles/epjconf/abs/2018/24/epjconf_eps-sif2018_00004/epjconf_eps-sif2018_00004.html.

27. LU, K.; Zhu, Y.; Li, Z.; Singh, R. K.; Nozaki, N., 2016. Dynamic simulation assessment of environment friendly vehicles introduction and clean energy promotion policy in China. *Journal of Sustainable Development*, v.9, n.5, pp. 83–99.

28. Almas, L. K.; Amossono, S. H.; Marek, T.; Colette, W. A., 2003. Economic feasibility of precision irrigation in the northern Texas high plains. *Paper Presented at the Southern Agricultural Economics Association Annual Meeting*, Alabama, pp. 1–15.
29. Smith, R.; Baillie, J. N.; McCarthy, A. C.; Raine, S. R.; Baillie, C. P., 2010. Review of precision irrigation technologies and their application. Review of Precision Irrigation Technologies and their Application. NCEA Publication 1003017/1. National Centre for Engineering in Agriculture (NCEA), University of Southern Queensland, Toowoomba.

IV

Earth and Satellite Measurements for Irrigation

10

Irrigation Water Measurement

Mir Bintul Huda
*National Institute of
Technology Srinagar*

Nasir Ahmad
Rather
*Baba Ghulam Shah
Badshah University*

10.1 Introduction

Water being an asset to earth needs to be used judiciously for a sustainable development. Of all the uses of water, irrigation accounts for a substantial portion of this natural reserve. One of the important aspects of irrigation engineering is the amount of irrigation water required. This depends upon the type of crop, its water requirement, stage of the crop, soil conditions, agricultural practices, and the rainfall of the region. Before planning any irrigation scheme, it is a requirement to compute the amount of irrigation water that is sufficed by precipitation so that the deficit can be augmented by the irrigation scheme. Based on different irrigation techniques, water may be conveyed through open channels or pipes. Design of the irrigation system and the water management depends upon the water requirement of the crop and the available water. The measurement of water may be in two states, at rest or in motion. At rest, water is measured in units of volume such as liters, cubic meter, and hectare meter. When in motion, water is measured in terms of flow rate such as cumecs and cusecs. Volumetric measurement is important for ensuring a proper amount of water supply at each irrigation, while flow rate measurement ensures the proper operation of the irrigation system.

This chapter is aimed at presenting the most common methods and techniques that are employed for the measurement of water in the various types of irrigation schemes. An in-depth discussion on the various methods has been presented and suitably illustrated with diagrams and computation steps. This chapter will serve as a reference especially for irrigation engineering students. It gives an overview of the measurement techniques that are frequently employed by the engineers and workers on the field; the chapter is a review of the widely known and in-use techniques and methods. The various reports and works used to compile this document have been referred at the end.

Although most of the techniques and methods have been presented, it needs to be mentioned that with the continuous research in the field of irrigation engineering, many more upcoming and technologically advanced techniques and area-specific research are also available. These have led to improved and efficient means of measurement which cannot be discussed under this review chapter. Those can be separately taken up under different titles and classified as per the area of their development and the specific situations of their requirement.

DOI: 10.1201/9780429290114-14

10.2 Units of Measurement of Water

Water at rest is measured in terms of volume such as liters, cubic meters, and hectare meter. This is useful for design and storage computations of ponds, tanks, reservoirs, or soil water. For water measurement in motion such as rivers, pipelines, canals, and channels, rate of flow is useful, which is commonly expressed in liters per second, cubic meters per second, hectare-meters per day, etc.

10.3 Methods of Water Measurement

Irrigation water measurement methods can be classified as: (1) Volumetric Measurements, (2) Velocity-Area Methods, (3) Structures, (4) Other Devices, and (5) Dilution Method. These are discussed separately.

10.3.1 Volumetric Method

For measurement of small irrigation streams, volumetric method of measurement is generally used. In this method, water is collected in a container of known volume and the time taken to fill the container is noted down with the help of a stopwatch. The rate of flow is given as:

$$\text{Discharge} = \frac{\text{Volume of Container}}{\text{Time required to fill the container}}$$

$$Q = av \tag{10.1}$$

Discharge is normally taken in liters/second, volume in liters and time in seconds. This method is commonly used to determine the discharge rates of pumps and other water lifting devices such as Persian Wheel.

10.3.2 Velocity-Area Method

The rate of flow of water passing a point in open channel is determined by multiplying the cross-sectional area of the flow section at right angles to the direction of flow by the average velocity of water. The cross-sectional area is determined by measuring the depths at various locations. The depth can be measured by different methods like sounding rod, lead line, echo sounder, Haigh's depth meter, or Kelvin tube. The entire cross section is divided into several subsections and the average velocity at each of these subsections is determined by current meters or floats. The accuracy of discharge measurement increases with the increase in the number of segments. Some precautions to be taken in this type of irrigation water measurement are as follows:

- Discharge in the segment should not be more than 10% of the total discharge.
- Difference in velocities between two adjacent sections should not be more than 20%.
- Segment width should be 1/15th to 1/20th of total width.

The total discharge is calculated using the method of mid sections. It is considered that the section is divided into ($N-1$) sections.

$$Q = \sum_{i=1}^{N-1} \Delta Q_i \tag{10.2}$$

where ΔQ_i is the discharge in the ith section.

For $i=2$ to $i=(N-1)$,

$$\Delta Q_i = \left(\text{depth at } i\text{th section}\right) \times \left(\frac{1}{2}\text{width to the left} + \frac{1}{2}\text{width to the right}\right)$$

$$\times \left(\text{average velocity in the } i\text{th section}\right),$$

$$\Delta Q_i = y_i \times \left(\frac{W_i}{2} + \frac{W_{i+1}}{2}\right) \times v_i, \tag{10.3}$$

where W_i is the width of ith section and v_i is the average velocity of ith section.

For the first and last sections,

$$\Delta Q_1 = \Delta A_1 \overline{v_1}, \tag{10.4}$$

$$\Delta A_1 = y_1 \overline{W_1}, \tag{10.5}$$

$$\overline{W_1} = \frac{\left(W_1 + \dfrac{W_2}{2}\right)^2}{2W_1}. \tag{10.6}$$

Similarly,

$$\Delta Q_{N-1} = \Delta A_{N-1} \overline{v_{N-1}}. \tag{10.7}$$

The cross-sectional area is determined as discussed above and the velocity is measured by several methods such as surface floats, velocity rods, sub-surface floats, or a current meter; these are described in the following sections.

10.3.2.1 Float Method

The float method gives a rough estimate; it consists of noting down the rate of movement of a floating body. A long-necked bottle partly filled with water or black wood, an orange or lemon may be used as float. A straight section of the channel about 30 meters long with uniform cross section is selected. Several methods of depth and width are made within the trial section to arrive at average cross-sectional area. A string is stretched across each end of section at right to the direction of flow. The float is placed in the channel, a short distance upstream from the trial section. The float needed to pass from upper end to lower end of the section is recorded. Several trials are made to get average time of travel.

To determine the velocity of water at the surface of the channel, the length of the trial section is divided by the average time taken by the float to cross it. Since the velocity of the float on the surface of the water will be greater than the average velocity of the stream; it is constant factor, which is usually assumed to be 0.85. To obtain the rate of flow, this average velocity (measured velocity × coefficient) is multiplied by the average cross-sectional area of the stream.

$$\text{Discharge} = \text{Area} \times \text{velocity}$$

It is an inexpensive and simple method. This method measures surface velocity. Mean velocity is obtained using a correction factor. The basic idea is to measure the time that it takes for the object to float a specified distance downstream.

$$V_{surface} = \frac{\text{Travel distance}}{\text{Travel time}}$$

Surface velocities are typically higher than mean or average velocities, and therefore,

$$V_{mean} = kV_{surface},$$ (10.8)

where k is a coefficient that generally ranges from 0.8 for rough beds to 0.9 for smooth beds.

The floats are released in such a way to pass in the middle of the river cross section. Certain steps need to be ensured so that the measurements are accurate; these are enumerated as follows:

- A suitable reach of the river is selected which is relatively straight and has a minimum turbulence.
- The start and end points need to be marked clearly.
- Travel time must be ideally restricted to about 20 seconds.
- The float is dropped upstream of the start marker.
- Stopwatch is activated when the float reaches the start marker and stopped exactly at the downstream marker.
- Measurements are repeated at least thrice, and the average velocity is used for discharge calculations.

Despite being an approximate and a primitive method of velocity measurement, floats are suitable for small streams in floods, small streams with rapidly changing water levels, and preliminary surveys of smaller scale.

10.3.2.2 Current Meters

Current meter gives an accurate stream velocity measurement. It is a mechanical device consisting of a rotating element which rotates due to the reaction of the stream current with an angular velocity proportional to the stream velocity. The rotating element may be made of metal, plastic, or rubber, rotating in a vertical or horizontal plane and geared to a totalizer in such a way that a numerical counter can totalize the flow in any desired volumetric units.

The basic requirements for accurate operation of the current meter are as follows:

- Pipe must flow full at all the times.
- Rate of flow must exceed the minimum for the rated range.

Meters are calibrated in the factory and field adjustments are usually not required. When water meters are installed in open channels, the flow must be brought through the pipes of known cross-sectional area. Care must be taken that no debris or other foreign materials obstruct the propeller. There are two main types of current meters:

- Vertical-axis meters
- Horizontal-axis meters

Vertical-axis meters consist of a series of conical cups mounted on a vertical axis. The cups rotate in a horizontal plane and a cam attached to the vertical-axis spindle records generated signals proportional to the revolutions of the cup assembly. The normal range of velocities is 0.15–4.0 m/s; the accuracy of these meters is about 1.50% at threshold value and improves to about 0.30% at speeds more than 1.0 m/s. One of the disadvantages of vertical-axis meters is that they cannot be used in situations where the vertical component of velocities is appreciable.

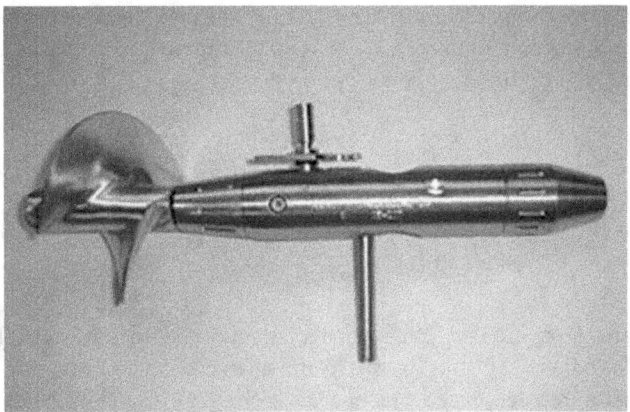

FIGURE 10.1 Horizontal-axis current meter.

Horizontal-axis meters (Figure 10.1) consist of a propeller mounted at the end of the horizontal shaft. These meters are rugged and are not affected by oblique flows of as much as 15°. The accuracy is about 1% at the threshold value and is about 0.25% at a velocity of 0.3 m/s and above.

Some steps to be ensured while measuring the velocity with the help of current meters are enumerated as follows:

- The channel cross section needs to be divided into at least 25 divisions; for deeper channel portions, closer intervals are preferred.
- The measurement needs to proceed at the water's edge and the correct sequence is to call out the distance first, then the depth, and then the velocity.
- Velocity needs to be least affected by the current meter; hence, for measurement, the observer needs to be at the downstream position of the meter.
- The rod needs to be in a vertical position with the meter directly into the water.
- For taking the reading, the meter must be completely under water, facing the current, and free of any kind of interference.
- The meter must be adjusted in a manner to avoid boulders, snags, and any other obstructions.
- At least two velocity measurements need to be taken at each subsection.
- For channel depth less than 60 cm, velocity is measured once for each subsection at 0.6 times the total depth from the water surface.
- For channel depth greater than 60 cm, velocity is measured twice, at 0.2 and 0.8 times the total depth. The average of these two readings is taken as the velocity for the subsection.
- A minimum of 40 seconds for each reading must be allowed.
- Discharge must be calculated in the field. If any section has more than 5% of the total flow, the section is subdivided, and more measurements are taken.

Current meters are designed in a manner such that the rotation speed of the blades varies linearly with the stream velocity. This can be expressed by the following equation:

$$v = aN_s + b, \tag{10.9}$$

where v is a stream velocity at measuring site in m/s, N_s is the revolutions per second of the meter, and a and b are constants of the meter. To determine the constants, which are different for each instrument, the current meter must be calibrated before use. This is done by towing the instrument in a tank at a known velocity and recording the number of revolutions N_s. This procedure is repeated for a range of velocities. It must be kept in mind that for shallow streams the measurement can be taken at a depth of 0.6 times the total depth, whereas for deeper streams two measurements are needed at 0.2 and 0.8 of total depth and then averaged to get the actual velocity.

10.3.3 Structures

The most employed water measuring techniques used in the farms are the structures such as orifices, weirs, and flumes.

10.3.3.1 Orifices

Orifices in open channel are usually circular or rectangular openings in vertical bulkhead through which water flows. The edges of opening are sharp and often constructed of metal. The cross-sectional area of orifice is small in relation to the stream cross section. Orifice may operate under free flow or submerged flow conditions. There are three main types of orifices as follows.

Free flow orifice: Orifices may be used to measure rates of flow when the size and shape of the orifice and head acting upon them are known. Orifices used in measurement of irrigation water are commonly circular or rectangular in shape and are generally placed in the vertical surfaces, perpendicular to the direction of channel flow. The section where contraction of the jet is maximal is known as the vena contracta. The vena contracta of a circular orifice is about half the diameter of the orifice itself.

$$Q = C_d a \sqrt{2gh} \qquad (10.10)$$

where Q is discharge in cumecs; a is cross-sectional area of water, the canal, or orifice in sq. m; C_d is the discharge coefficient which varies from 0.6 to 0.8 or more depending upon the position of orifice relative to the sides and bottom of vessels or the degree of roundness of the edge of orifice; g is an acceleration due to gravity; and h is the height of water level in cistern from the middle of the orifice.

Coefficient of discharge (C_d) is defined as the ratio of the actual discharge from an orifice to the theoretical discharge from the orifice. It can be expressed as:

$$C_d = \frac{\text{Actual discharge}}{\text{Theoretical discharge}} = \frac{Q}{Q_{th}}, \qquad (10.11)$$

$$C_d = \frac{\text{Actual velocity} \times \text{actual area}}{\text{Theoretical velocity} \times \text{theoretical area}}, \qquad (10.12)$$

$$C_d = C_v \times C_c. \qquad (10.13)$$

Coefficient of velocity (C_v) is defined as the ratio of the actual velocity of a jet of liquid at vena contracta and the theoretical velocity of the jet.

$$C_v = \frac{\text{Actual velocity of the jet at vena contracta}}{\text{Theoretical velocity}} \qquad (10.14)$$

$$C_d = \frac{V}{\sqrt{2gh}} \qquad (10.15)$$

where V is the actual velocity and h is the head.

Coefficient of Contraction (C_c) is defined as the ratio of the area of the jet at vena contracta to the orifice.

$$C_d = \frac{\text{Area of jet at vena contracta}}{\text{Area of the orifice}} \qquad (10.16)$$

Submerged orifice: In fully submerged orifice, the outlet side is fully submerged under the liquid, and it discharges a jet of liquid into the liquid of the same kind. It is also called totally drowned orifice. Discharge through fully submerged orifice is calculated as:

$$Q = C_d b \left(H_2 - H_1\right)\sqrt{2gh}, \qquad (10.17)$$

where H_1 is the height of water above the top of orifice on upstream, H_2 is the height of water above the bottom of orifice, h is the water level difference, and b is the width of orifice.

Partially submerged orifice: In the partially submerged orifice (Figure 10.6), the outlet side is partially submerged under liquid. The discharge is calculated as follows:

$$Q = C_d b \left(H_2 - H_1\right)\sqrt{2gh} + \frac{2}{3}C_d b \sqrt{2gh}\left[H_2^{\frac{3}{2}} - H_1^{\frac{3}{2}} \right]. \qquad (10.18)$$

The first term of the equation represents the submerged flow, and the second term represents flow through the free portion.

10.3.3.2 Weirs

A weir means a notch in a well built across a stream. The weir provides the obstruction across the river or channel, which raises the water level and diverts the water. It is aligned at right angles to the direction of flow in the river or channel. The notch may be rectangular, trapezoidal, and 90-degree V (Triangular) notch or weir. A weir consists of a weir wall made of concrete, timber, or sheet metal. Weirs may be built fixed or portable. Weirs may be sharp crested or broad crested; irrigation water is measured on farms mostly by sharp crested weirs. Some terms associated with the weirs are as follows:

Weir pond: Channel portion immediately upstream of the weir
Weir crest: Bottom of the weir notch
Head: Depth of water flowing over the crest
End contraction: Horizontal distance from the ends of the weir crest to the sides of the weir pond
Nappe: Sheet of water which overflows the weir.

Sharp crested weir: Sharp crested weirs (Figure 10.2) are commonly used for water measurement in the farm. They are of three main types as given below. Sharp crested weirs may further be divided into weirs with end contractions and weirs without end contractions.

Rectangular weir: This weir has a notch rectangular in shape (Figure 10.3). It is suitable for measurement of large discharges.

The length of a weir may be equal to width of the upstream channel or less than it. The discharge through a rectangular weir is given by the following equation.

$$Q = CLH^{\frac{3}{2}}, \qquad (10.19)$$

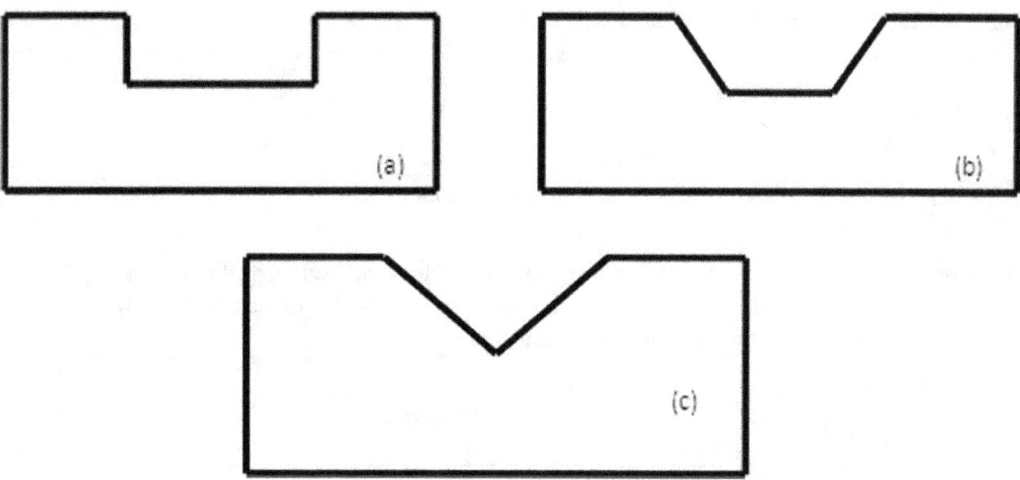

FIGURE 10.2 Sharp crested weirs: (a) rectangular; (b) trapezoidal (Cipoletti); and (c) triangular (V-notch).

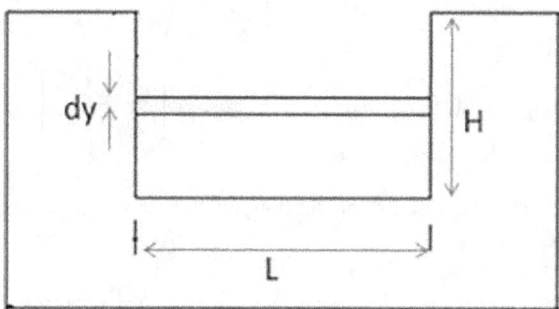

FIGURE 10.3 Rectangular weir.

where Q is the discharge in liters/second, C is the coefficient depending on crest and approach conditions, L is the length of crest in cm, and H is the head over the weir in cm.

$$Q = 0.0184 L H^{\frac{3}{2}} \qquad (10.20)$$

For one side contraction,

$$Q = 0.0184 (L - 0.1H) H^{\frac{3}{2}}. \qquad (10.21)$$

For both side contraction,

$$Q = 0.0184 (L - 0.2H) H^{\frac{3}{2}}. \qquad (10.22)$$

Trapezoidal weir: It is a contracted weir with each side of notch having a slope of 1:4. It is suitable for medium discharges; the side slopes are sufficient to correct the end contractions of nappe and the flow is proportional to the length of weir crest (Figure 10.4).

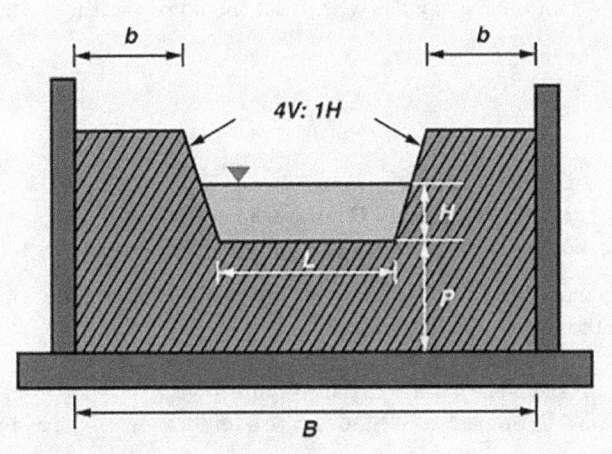

FIGURE 10.4 Trapezoidal or Cipoletti weir.

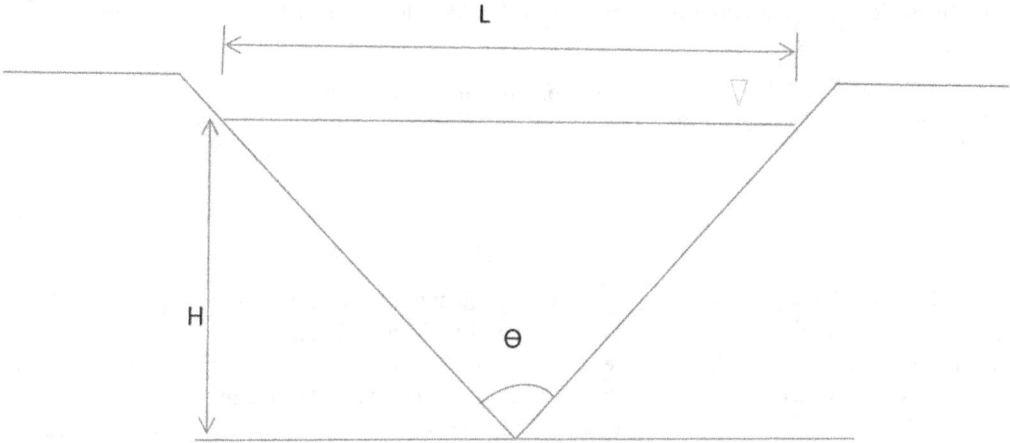

FIGURE 10.5 Triangular weir.

The discharge of water is given by the following equation:

$$Q = 0.0186LH^{\frac{3}{2}},\qquad(10.23)$$

where Q is the discharge in liters/second, L is the length of crest in cm, and H is the head over the crest in cm.

Triangular weir: It has a V-shaped notch (Figure 10.5). It is suitable for small to medium discharges. Its design causes small changes in discharge, hence causing a large change in depth and allowing measurement accurate than the rectangular weir. Head should be measured at a distance at least 4H upstream of the weir.

$$Q = 0.0138 H^{\frac{5}{2}}$$

(10.24)

where Q is the discharge in liters/second and H is the head in cm.

Some general precautions to be taken for installation and operation of weirs are as follows:

- The velocity of approaching flow should not exceed 15 cm/second. Baffles must be used in the weir pond to reduce the velocity and turbulence.
- Weir wall must be vertical.
- Center line of weir should be parallel to the flow direction.
- The weir crest must be properly leveled to ensure a uniform depth of passing water all along the crest.
- Notch should be of a regular shape and must have rigid and straight edges.
- Weir Crest should be above the bottom of the approach channel by at least twice the depth of water flowing over the weir.
- Weir wall should be high enough so that the water falls freely below the weir leaving an air space under the over-falling sheet of water.
- The scale or gauge used for measuring head should be located at a distance about four times the approximate head.

Broad-crested weir: A broad-crested weir is a broad wall set across the channel bed. The discharge is given by the equation as follows:

$$Q = C \cdot C_v C_D b h^{\frac{3}{2}},$$

(10.25)

where Q is the discharge, C is a constant, C_v is the coefficient of velocity depending on approach velocity, C_D is the coefficient of discharge based on friction and turbulence losses, and b is the width of channel. It is assumed that the thickness of the sheet over the crest adjusts itself such that the discharge is a maximum, as seen in the experiments. A broad-crested weir is not constructed in standard dimensions; it is normally calibrated in the field by current meter measurements or by model testing in laboratory. Sediment laden water is a common problem in measuring irrigation water; it causes rapid rounding of sharp weir plate which distorts the discharge measurements, and in such cases broad-crested weir is preferable.

10.3.3.3 Flumes

Flumes are specially shaped and constructed channel sections used to measure the flow of water. The principle of the flumes is based on the concept of specific energy and critical flow in open channels. Some commonly used flumes are Parshall flume and Cut-throat flume. For measuring varying flows of large earthen channels, several alternative shapes of broad-crested weirs and long throated flumes are used, such as broad-crested rectangular weir, truncated flume, and triangular throated flume.

Parshall flume: In 1950, Parshall fabricated a device in which the discharge is obtained by measuring the loss in the head caused by forcing a stream of water through a throat or converged section of a flume with a depressed bottom. The loss in head is very small in this device just about 25% of that for a weir. The accuracy of measurement in the Parshall flume is within allowable limits of 5%. Parshall flume is also known as Venturi flume. It consists of three main sections: (1) converging section at upstream, (2) constricted section known as throat, and (3) diverging section at downstream.

Parshall flume is a self-cleaning device, and presence of sand or silt in the flowing water does not affect its operation. Also, the approaching velocity of the water stream doesn't influence its working. It gives the satisfactorily accurate results even under submerged conditions.

The floor of upstream converging section of the flume is level, and the walls converge toward the throat; walls of the throat section are parallel, and the floor is inclined downward; walls of the downstream section expand toward outlet, and the floor is inclined upward. The size of the flume is determined by the width of throat. Generally, flumes ranging from 3 inches to 10 feet throat width are used, which gives the range of discharge of 1/30 to 200 cusecs. The flumes of 3-, 6-, and 9-inch size are generally used in field measurement. Under turbulent conditions, stilling well is used to measure water surface elevations in the flume. Under smooth flow conditions, plastic scale is fixed to the inside of flume; zero of the scale is set at the floor level of the converging section.

Discharge through the flume can be free flow or submerged. The flow is termed submerged, when the water elevation at the downstream is high enough to retard the discharge rate; in this case, tow scales (H_a and H_b) are provided at upstream and downstream sections of the flume. For free flow, only H_a needs to be measured. The ratio between the reading at H_b and H_a should be carefully studied. This ratio should not exceed 0/3 for 3-, 6-, and 9-inch sized Parshall flumes; otherwise correction needs to be applied. When the ratio is less than 0.6, it is termed as free flow and if it exceeds 0.6, it is called submerged flow.

Cut-throat flumes: Cut-throat flume is an improvement of the Parshall flume. It has a flat bottom, vertical walls, and a zero-length throat section; it was developed by Skogerbee, Hyatt, Anderson, and Eggleston in 1967 (Figure 10.6). Its simple construction is an economical advantage over the Parshall flume.

The flumes have a level floor and are placed in a concrete lined channel or on a channel bed conveniently. These flumes consist of converging inlet section and diverging outlet section. The minimum width (w) of the flume is known as the flume neck or throat. These flumes can operate either as a free flow or a submerged flow structure. In free flow conditions, critical depth occurs near the flume neck. The critical depth helps in determining flow rate with the help of upstream depth h_a only; this is possible because when critical depth occurs in the flume, h_a is not affected by any changes in the downstream depth. The free flow condition is achieved when the ratio of h_a to flume length is less than 0.4. Discharge for free flow conditions is given as:

$$Q = C_1 h_a^{\,n} L, \tag{10.26}$$

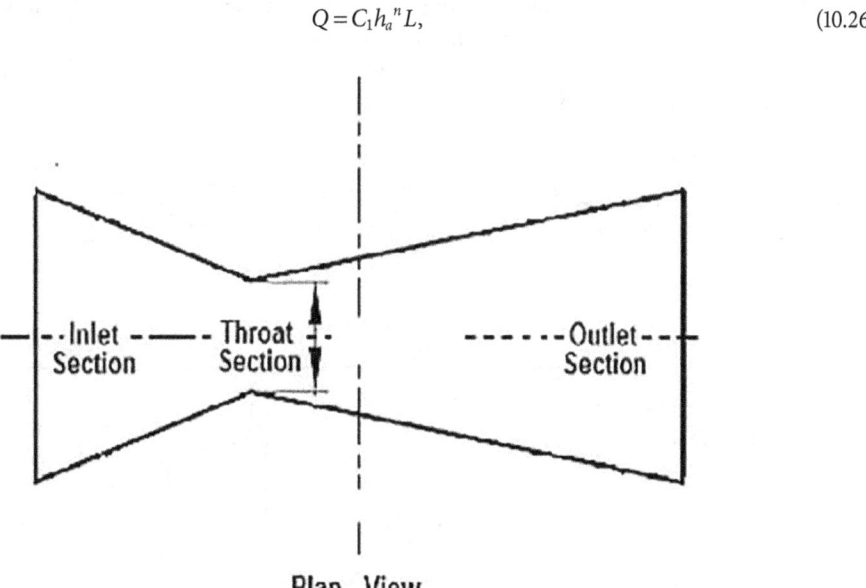

Plan View

FIGURE 10.6 Cut-throat flume.

where Q is the discharge, C_1 is the free flow coefficient whose value is same as that of Q for $h_a = 1$ foot, and n is the exponent whose value depends on the flume length only.

For submerged flow conditions, both upstream (h_a) and downstream depths (h_b) need to be measured. In such condition, submergence (S) is computed, which is defined as the ratio of downstream to upstream depths.

$$S = \frac{h_b}{h_a} \qquad (10.27)$$

Some precautions to be taken for installation of cut-throat flumes are as follows:

- It should be suitably installed in a straight section of the channel.
- Avoid its location near a farm gate which may induce surging effects due to the gate operations.
- It is better to have a flow measuring device operate under free flow conditions.

10.3.4 Other Devices

The accurate water measurement is an important part of the water use efficiency; it helps in an effective water management. The measured flow should be as close as possible to the actual amount of water flowing. Such accuracy is maintained by the various types of meters that have evolved over the time. The irrigation meters must have accuracy around +2.0% of the actual amount. Flow meters are commonly used in micro-irrigation systems, where water management and the system performance are important aspects. They provide sufficient and accurate information about the amount and time of irrigation. Some commonly employed flow meters are discussed as follows.

10.3.4.1 Propeller Meters

They are the most common type of flow meter. They record the cumulative flow of water. They need to be installed in straight sections of pipeline. The accurate measurement is ensured by full flow condition of the pipe. Turbulence must be avoided at the upstream and downstream of the meter installation. The rate of flow must exceed the minimum for the rated range. The flow from the canal outlet is allowed to pass through a pipe into a basin. A propeller which rotates due to the flow of water is installed at the pipe outlet. The number of rotations indicated by a counter gives the cumulative water flow. To ensure accurate readings, calibration and maintenance are necessary.

10.3.4.2 Pitot Tube

A pitot tube measures the velocity of flow in a pipe (Figure 10.7). It is an open 'L'-shaped tube useful for measuring velocity of flows in an open channel as well as in pipes. The velocity is measured at several

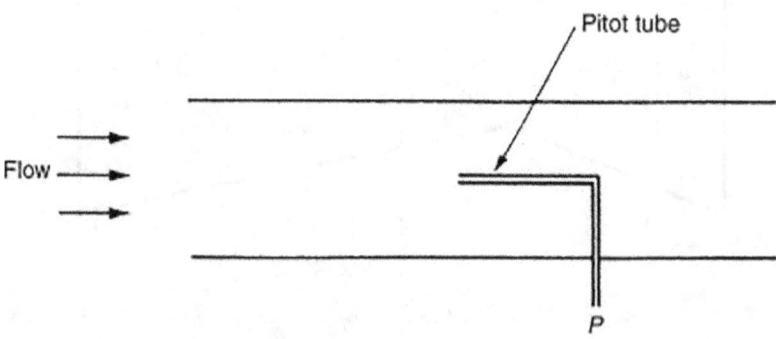

FIGURE 10.7   Pitot tube.

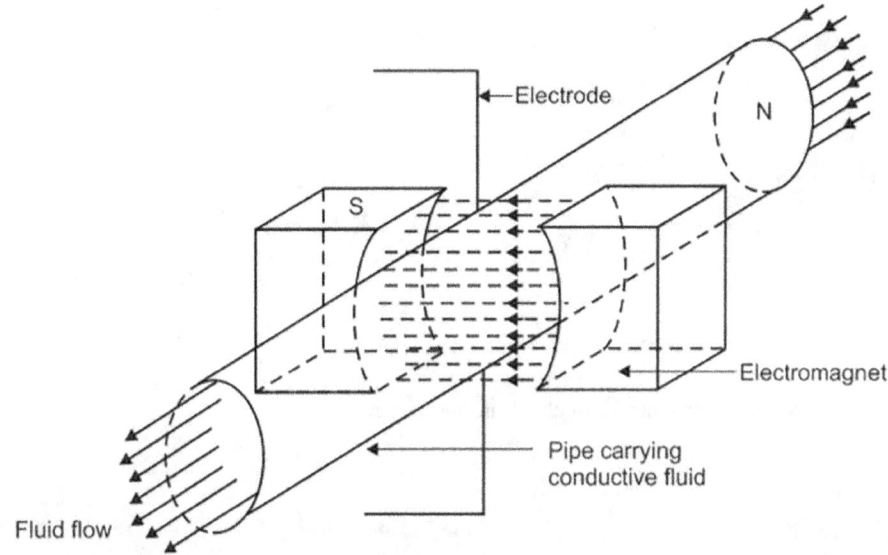

FIGURE 10.8 Schematic representation of magnetic flow meter.

points across the pipe diameter and an average is computed to calculate the flow rate with the help of cross-sectional area. The velocity of flow is given as follows:

$$V = C\sqrt{2gh}, \tag{10.28}$$

where V is the velocity, C is the pitot tube coefficient (usually 0.95 to 1.0), h is the loss of head, and g is an acceleration due to gravity.

10.3.4.3 Magnetic Flow Meters

Magnetic flow meters do not cause obstructions in the pipe. They measure the flow of conductive liquids flowing in filled pipes. The sensor creates a pulsating, alternating magnetic field on the inside of pipe. The liquid in the pipe will move through this field and generate a signal proportional to its velocity (Figure 10.8). This is collected by electrodes and processed by a microprocessor. These meters require the less maintenance and have the higher accuracy, but they have a high initial cost and need external power supply.

10.3.4.4 Ultrasonic Flow Meters

These flow meters are based on the Doppler effect (frequency shift) of the ultrasonic signal reflected by suspended particles or gas bubbles in the water stream. They direct ultrasonic pulses diagonally across the pipe both upstream and downstream (Figure 10.9). The time travel of the signal is converted into flow velocity.

Ultrasonic meters are accurate, require little maintenance, and are portable, but cost more than all other meters.

10.3.4.5 Paddlewheel Meters

They are suitable for measurement in pipes ranging from 1 to 48 inches in diameter. Water in the pipe causes the paddle wheel to rotate; this rotational velocity is proportional to the flow rate in the pipe (Figure 10.10). The rotation of the wheel induces a voltage pulse and is converted by an electronic unit

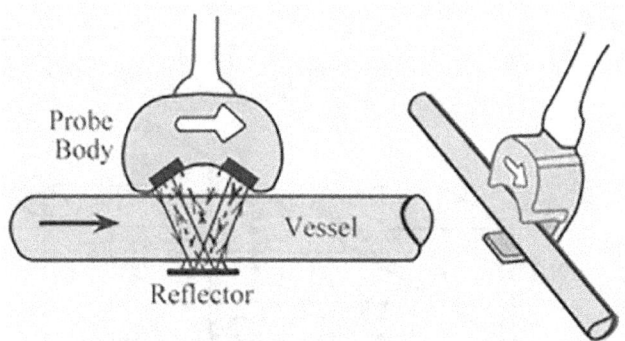

FIGURE 10.9 Schematic representation of ultrasonic flow meter.

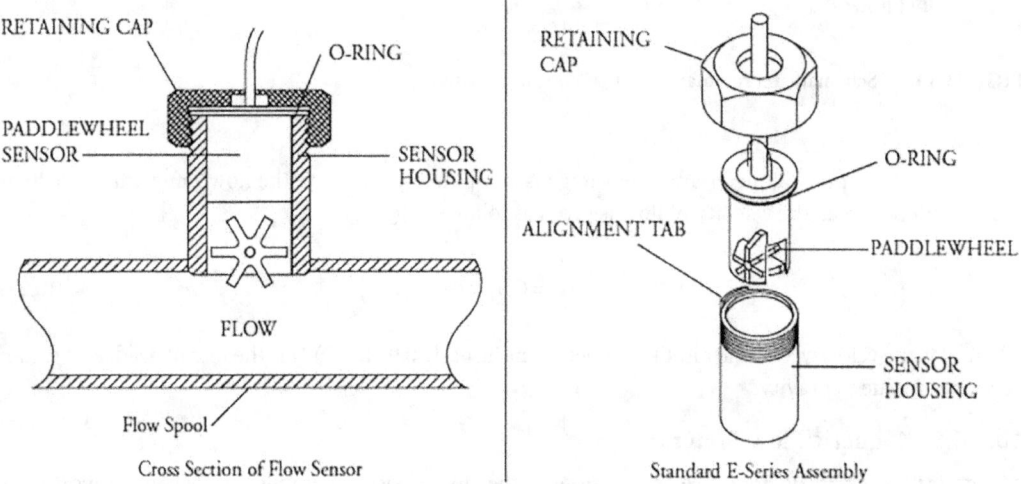

FIGURE 10.10 Paddlewheel meter.

to the flow units depending upon the pipe diameter and flow characteristics of the pipe. The insertion depth of the rotor and the proper flow profile is critical for accuracy of the measurements.

10.3.4.6 Venturi Meters

Venturi meters measure water flow in pipes under pressure. They are based on venturi principle. The flow while passing the converging section accelerates and the pressure head is lowered (Figure 10.11). By knowing the pipe cross-sectional area, the constriction area, and the pressure drop, the water passing through the pipe can be computed. The rate of flow can be computed by following equation using consistent units:

$$Q = CKd\sqrt{h_1 - h_2},\tag{10.29}$$

where Q is the rate of flow, C is the discharge coefficient, K is a factor corresponding to the ratio of the throat diameter to the diameter of entrance section, d_1 is the diameter of the entrance section, d_2 is the

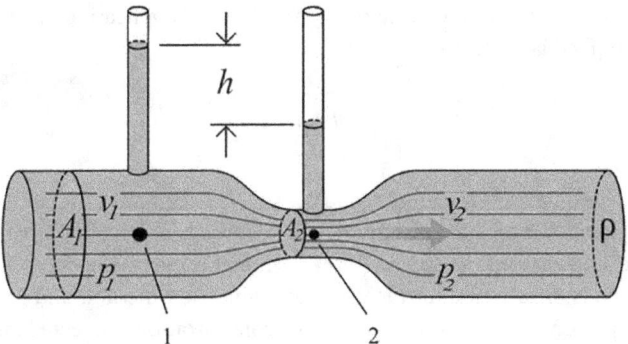

FIGURE 10.11   Venturi flow meter.

diameter of the throat section, h_1 is the pressure head measured above the axis of the meter, and h_2 is the pressure head measured above the axis of the meter at the throat.

K is determined by the following equation:

$$K = \frac{\pi}{4}\sqrt{\frac{2g}{1 - \left(\dfrac{d_2}{d_1}\right)^4}}, \qquad (10.30)$$

where g is acceleration due to gravity.

10.3.4.7 Shunt Meters

In shunt meters, a fraction of the water in the main irrigation line is shunted through a smaller diameter pipe. The smaller pipe is parallel to the main line for a short distance for measurement purpose and then joins back the main pipeline. The smaller pipe is equipped with a totalizing flow meter and the flow in main pipe is proportional to the amount of flow in the shunt pipeline. A head loss is required in the main pipe to create pressure difference that causes water to flow through shunt pipe. The flow rate in the shunt pipe depends upon the head loss in the main pipe and this relationship is determined by calibration.

10.3.4.8 Dethridge Meter

It is widely used in Australia. It consists of an under-shot water wheel turned by water discharging over its implacement, which is a short concrete outlet specially formed to provide only the minimum practicable clearance of the lower half of the wheel at its sides and around its circumference. The meter gives a positive measurement of discharge, recorded directly in volumetric units by a revolution counter driven by the wheel axle. The accuracy of the meter under normal conditions is within 5%. The main advantage of this meter is that it operates with small head loss, simple design, and robust construction, and it passes the ordinary floating debris without any damage or stoppage of the wheel.

10.3.5 Dilution Method

The dilution method is also known as the chemical method. It is based on the continuity principle applied to a tracer which is allowed to mix completely with the flow. Tracer is a substance that does not react with the fluid or the boundary; it is introduced in concentrated form into flowing water and allowed to mix thoroughly. The concentration of the tracer is measured at downstream section. Since only the quantity of water is necessary to accomplish the dilution involved, there is no need to measure velocity, depth, and head, cross-sectional or any other hydraulic factor usually considered in discharge measurement. The relationship between size of stream, time of application, area to be irrigated, and depth of water to be applied is as follows:

$$Qt = ad, \tag{10.31}$$

where Q is the discharge of stream in (liter/second) or (ha. cm per hour), t is the time of application of water in (seconds or hour), a is area in (sq. m or hectare), and d is depth in cm that the volume of water used would cover the land irrigated, if quickly spread uniformly over its surface.

There are two methods of injection of tracers: sudden injection method and constant rate injection method. In sudden injection method, tracer of a known concentration is injected into the stream and its concentration after a certain point at a section sufficiently downstream is observed. Let C_0 be the small initial concentration of the tracer in the stream flow. At a section, say 1, a small quantity ($\forall$) of the high concentration C_1 of the tracer is added. Another section downstream, say 2, is selected such that the tracer mixes thoroughly with the fluid due to turbulent mixing process while passing through the reach. The concentration will have a base value of C_0, increases from time t_1 to a peak value, and gradually reaches the base value of C_0 at time t_2. The main assumption of this method is that the stream flow is taken as steady. Applying the continuity principle and simplifying, the discharge of stream is obtained as follows.

$$Q = \frac{\forall_1 C_1}{\int_{t_1}^{t_2} (C_2 - C_0)\,dt} \tag{10.32}$$

The discharge Q in the stream can be estimated if for a known mass of tracer M_1 the variation of C_2 with time at section 2 and C_0 are determined.

Some commonly employed chemicals for the dilution tests are sodium chloride and sodium dichromate; concentration can be determined by chemical analysis or flame photometer. Dyes have also been useful in the dilution tests. Instead of chemicals and dyes, radioisotopes may also help in determining the degree of dilution based on counting the gamma ray emissions using Geiger counter or scintillation counter. The following relationship is applied for a radioisotope method:

$$Q = \frac{FA}{N}, \tag{10.33}$$

where Q is the volume of water flowing in the channel per unit time, F is the counts per unit of radioactivity per unit volume of water per unit of time, A is the total units of radioactivity to be introduced for each discharge measurement, and N is the total counts.

The most important precaution for this method of measurement is to handle the radioactive material with great care and only after obtaining required permissions and used by licensed personnel only.

10.4 Conclusions

Irrigation water is the major source for fulfilling the water demand of crops throughout the world. Irrigation schemes are designed as per the water requirement of crops in the command area. The supply of irrigation water to the crops depends mainly on the efficiency of the irrigation scheme. The efficiency of irrigation schemes can be determined by irrigation water measurements. The various methods for irrigation water measurements like Volumetric Measurements, Velocity-Area Methods, Structures, Other Devices, and Dilution Method are discussed in detail in this chapter.

Bibliography

1. Boman, B. and Shukla, S. 2006. Water Measurement for Agricultural Irrigation and Drainage Systems. Circular 1495, Agricultural and Biological Engineering Department, UF/IFAS Extension Service, University of Florida, USA.

2. Boss, M. G. 1989. Discharge Measurement Structures. International Institute for Land Reclamation and Improvement, Wageningen, the Netherlands.

3. Christiansen, J. E. 1935. Measuring Irrigation Water. University of California, Agri. Expt. tn. Berkeley, USA.

4. Hansen, V. E., Israelsen, O. W. and Stringham, G. E. 1979. *Irrigation Principles and Practices*, 4th Edition. John Wiley and Sons Inc., New York.

5. Michael, A. M. and Ojha, T. P. 2003. *Principles of Agricultural Engineering, Volume II.* Jain Brothers, Karol Bagh, New Delhi.

6. Michael, A. M. 2009. *Irrigation Theory and Practice*, 2nd Edition. Vikas Publishing House Pvt Ltd., New Delhi.

7. Replogle, J. A. and Bos, M. G. 1982. Flow measurement flumes: Application to irrigation water management. In: *Advances in Irrigation*, Volume I, D. Hillel, Ed., Academic Press, New York.

8. Rogers, D. H. and Black, R. D. 1993. Irrigation Water Measurement. Irrigation Management Series, L-877, Cooperative Extension Service, Kansas State University, Manhattan, Kansas, USA.

9. Schuster, J. C. 1970. Water Measurement Procedures Irrigation Operators' Workshop. Engineering and Research Centre, Bureau of Reclamation, Denver, Colorado, USA.

10. Schwab, D. Irrigation Water Measurement. F-1502, Division of Agricultural Sciences and Natural Resources, Oklahoma Cooperative Extension Service, Oklahoma State University, USA.

11. Sihag, P., Darsun, O. D. et al. 2021. Prediction of aeration efficiency of Parshall and modified venture flumes: application of soft computing versus regression models. *Water Supply.* doi: 10.2166/ws.2021.161.

12. Skogerboe, G. V., Bennet, R. S. and Walker, W. R. 1973. Selection and Installation of Cut-throat flumes for measuring Irrigation and Drainage Water. Colorado State University, Expt. Std, USA.

13. Stock, E. M. 1955. Measurement of Irrigation Water. Bulletin No. 5, Utah Cooperative Extension Service, Utah State Engineering Experiment Station, USA.

14. Subramanya, K. 2008. *Engineering Hydrology*, 3rd Edition. The McGraw Hill Companies, New Delhi.

11

Irrigation and Agrometeorology: Innovative Remote Sensing Applications in Crop Monitoring

Nicolas R. Dalezios
University of Thessaly

Nicholas Dercas
Agricultural University of Athens

Ioannis N. Faraslis, Marios Spiliotopoulos, Pantelis Sidiropoulos, and Stavros Sakellariou
University of Thessaly

Saeid Eslamian
Isfahan University of Technology

11.1 Introduction

Variable and inadequate water supply and the temperature extremes are the two universal environmental risks in agricultural production (Mavi and Tupper, 2004). Rainfall contributes to an estimated 65% of the global food production, while the remaining 35% of global food is produced with irrigation. Not all of the rainfall that falls in a field is effectively used in crop growing, as part of it is lost by runoff, seepage and evaporation. In most parts of the world, rainfall is, for at least part of the year, insufficient to grow crops, and rainfed food production is heavily affected by annual variations in precipitation. For temperate, humid climates, about 50% of precipitation ends up in the groundwater. For Mediterranean-type climates, this figure is 10%–20%, and for dry climates it can be as little as 1% or even less. A major part of the developed global water resources is used for the food production. In most countries, 60%–80% of the total volume of developed water resources is used for agriculture and may reach well over 80% for the countries in arid and semiarid regions (Smith, 2013). Irrigation is an obvious option to increase and stabilize crop production. Major investments were made in irrigation during the latter half of the 20th century by diverting the surface water and extracting groundwater. The irrigated areas in the world, during the last three decades of 20th century, increased by 25% (FAO, 1993). Under irrigated farming, irrigation can be planned using data regarding consecutive periods of rainfall to satisfy the demands for critical phenological stages. With water resources becoming scarce, waters of inferior quality are increasingly used. Excessive use and poor management of such irrigation water have had, in some cases, detrimental effects on soil quality, causing whole areas to be taken out of production or requiring the construction of expensive drainage works. Defining strategies in planning and management of available water resources in the agricultural sector have become a national and global priority.

Agrometeorological information plays a major role in evolving the strategies for improving water use efficiency. The prerequisites for water-supply control and introduction of irrigation scheduling strategies consist of analysis of the climatic conditions of the region and use of weather forecast information. The literature states that weather/climate data should be adopted for crop monitoring in order to maintain farms' profitability and sustainability (Anderson and Kyveryga, 2016; Basso et al., 2016; Duarte et al., 2019). Accurate, localized forecast weather services that can be tailored to specific field demands should be continually improved. Field experiments from different researchers suggest the need of in-field scale agrometeorological data for specific crop types must be coupled within regional agroclimatologies in order to assess the long-term climate effect within field variability (Dalezios, 2015). The concept of agro-climatological classification and zoning can be used as input climate data to simulation models for predicting the climate average risk and forecasts of seasonal weather (Collis and Corbett, 2000; Yobom, 2020). A short-range weather forecast (up to a lead time of 5 days) would be of significant value to farmers, particularly in the surface irrigation. This value comes from advantages, such as better ability to manage waterlogging, particularly in surface irrigation; better ability to manage the soil moisture and plant stress; better time management of sprays for disease and pest control; water use savings and the associated cost savings.

Estimation of spatio-temporal variability of rainfall and evapotranspiration is required for simulation of the expected yield improvements and options for water storage (Smith, 2013). Scientific and technological progress in terms of computer programs has been achieved for the estimation of reference crop evapotranspiration from climatic data and allows the development of standardized information and criteria for planning and management of rainfed and irrigated agriculture. The FAO CROPWAT program is based on previous research (FAO, 1979; Allen et al., 1998) and incorporates procedures for reference crop evapotranspiration and crop water requirements and allows the simulation of crop water use under the various climate, crop and soil conditions. This chapter is organized as follows:

First, background information on agrometeorological aspects and land features is presented. Similarly, irrigation aspects are considered covering irrigation types, efficiencies, simulation methodologies and deficit irrigation (DI). Then, remote sensing applications in crop monitoring are presented, which include the image processing methodology, study area and databases, estimation of irrigation requirements, followed by analysis and discussion of results.

11.2 Agrometeorological Aspects

Agrometeorology is concerned with the meteorological, hydrological, pedological and biological factors that affect agricultural production and with the interaction between agriculture and the environment (Dalezios, 2015; WMO, 2010). Its objectives are to elucidate these effects and to assist farmers in preparing themselves by applying this supportive knowledge and information in agrometeorological practices and through agrometeorological services. In addition to the large and mesoscale climate characteristics and their variations, operational agrometeorology concerns itself with small-scale climate modifications, such as windbreaks, irrigation, mulching, shading, and frost and hail protection. Another important subject is agroclimatic classification. Much attention is paid to the impacts of climate change and climate variability and how to be adapted, including phenological aspects, monitoring, early warning, and estimation of changes in the risks relating to pests, diseases and extreme events, such as drought, desertification and flooding (Mohseni et al., 2021). Moreover, water management is significant to ensure adequate supplies while maintaining the quality of surface sources and groundwater is a key topic. Specifically, in relation to irrigation and crop monitoring there are agrometeorological and land requirements, which are briefly presented.

11.2.1 Agrometeorological Parameters

Key factors in crop monitoring are the meteorological parameters and their interactions. For efficient and effective monitoring of crops, accurate, localizing and tailoring, 3–5-day, forecasts are required (Pelosi et al., 2019; Chirico et al., 2018). Also new protocols based on innovative farming solutions have been developed in order to continuously monitor climatic conditions (Pooja et al., 2017). A brief review of basic meteorological parameters follows.

11.2.1.1 Precipitation

Crop–water interactions are the most important factor concerning yield production (Sadler et al., 2000; Mukherjee and Wang, 2019). Significant temporal rainfall variability means that farm meteorological observations must be collected for crop simulation modeling (O'Neal et al., 2000). Moreover, it is important to know at what degree precipitation is related to soil properties and water availability patterns (Bakhsh et al., 2000). Precipitation has a twofold effect on crop monitoring, namely the available water and the planning season-long management campaign.

11.2.1.2 Temperature

Temperature is a key element in crop growth in conjunction with both available water and radiation. Thermal time (accumulated degree days, °C) is a very important component in crop monitoring, mainly due to its interaction with topography of the field (slope, aspect). Moreover, temperature is related to the emergence of pest populations (Zidon et al., 2015). New technological tools (UAV: unmanned aerial vehicle) and thermal cameras can also be used for plant canopy temperature retrievals (Martínez et al., 2017).

11.2.1.3 Humidity

For crop monitoring, humidity is explicitly related to the development of pest and diseases. It is a key for predicting when infestation problems arise (Roux et al., 2018). One important question is whether it is viable to use humidity information in order to manage a field in a spatially variable manner.

11.2.1.4 Radiation

Solar radiation variability is also a key parameter. The impact of cloud cover, and aspect and slope of solar radiation variation for each field need to be systematically evaluated, especially during crop growth periods (Keane and Collins, 2004).

11.2.1.5 Wind

An important factor, which is influenced by the wind, is the spray management. For effective crop monitoring spray timing in conjunction with variable spay characteristics are critical for chemical efficiency (Giles et al., 1999). Thus, localized wind forecast is valuable. Moreover, field installation sensor systems and machine vision systems are influenced by the wind. Finally, wind is related to pest management. Also, wind direction can be used for forecasting migratory pests in management of crop monitoring systems.

11.2.1.6 Extreme Events

The extreme weather conditions, such as floods, hurricanes, tornadoes, and storms, are the most considerable environmental risks. Remote sensing techniques are used in crop monitoring for two purposes, namely alerts for extreme weather conditions and to assess the impact of extreme weather at the farm level . The latter is already used for farmers' insurance. Lastly, the remote sensing data could be proved very useful in monitoring and quantifying the loss of crop yield due to extreme weather conditions (Powell and Reinhard, 2016).

11.2.2 Land Features

Land features and characteristics have also significant role in crop monitoring and production, along with the agrometeorological parameters. A brief description follows.

11.2.2.1 Terrain and Slope

It is known that the effect of both soil and terrain can explain about 40% of yield variability in a field. The remaining 60% depends on weather-climate effects, biological variability, diseases and similar aspects (Kravchenko et al., 2000). Crop yield is significantly affected by slope and aspect. Indeed, the impact of terrain and slope on crop yield is related to its interaction with weather and consequently with soil water status. Specifically, in dry years the highest yield has been recorded on the concave locations. The basic conclusion is that it is not possible to rely solely on climate and topography for an effective crop monitoring management system. Moreover, a thorough analysis is required whether the data input for crop monitoring provides the appropriate cash benefit to farmers.

11.2.2.2 Soil

There are two approaches, which integrate soil data for field management decisions, namely the soil-based and climate-based approaches (Ansari et al., 2019). Both are complimentary given that climate is the primary soil forming factor. It is recognized that there is a need for interrelated data, such as soil micronutrient maps and climate. It has been reported that significant field differences in soil type and soil water holding capacity exist with respect to crop yield variability (Moore and Tyndale-Bisco, 1999). Soil and precipitation also have a great impact on crop yield (Bakhsh et al., 2000). Moreover, soil water holding capacity leads to high spatial yield variability, especially in relatively dry years. Finally, several researchers suggest that soil classification by using hydraulic properties, such as water content capacity, will be very useful as input data to a decision support system for the farm management (Van Alphen and Stoorvolel, 2000).

11.2.2.3 Water Status

The soil-available water capacity is a significant factor for crop yield. The importance of spatial variability in soil water has been subject for thorough examination (Dalezios et al., 2017, Dalezios et al., 2019). It was found that soil water capacity was related to yield only in drought years. This means that there is a temporal variability in crop yield. Regional climate variability constitutes another important aspect in the soil-precipitation system. Finally, irrigation and water management are closely linked to the soil water capacity and plants' water stress. A variable rate irrigation could be a viable means for optimizing the water consumption (Sadler et al., 2000).

11.2.2.4 Classification of Land Suitability

The classification of land suitability follows the type or pattern of spatial variability. Management planning models tend to be more efficient when few and well distinct soil types are used (Moore and Tyndale-Biscoe, 1999). Another significant factor is that the field applications of different inputs depend on the technical limitations of the available machinery (Giles et al., 1999). This means that in a field with more than two different soil types, errors could appear and it is likely that the small target areas in the field could receive very little of the correct inflows. There is, thus, a need to quantify spatial soil patterns in a field, in management units, which can be addressed with mechanical technology.

11.3 Irrigation Aspects

Irrigation is the supply to plants of controlled amount of water by artificial means, namely furrows, flooding, spraying or dripping. It is known that hydraulic infrastructure has been built for irrigation, which has allowed the development of important civilizations (Egypt, Mesopotamia). The prosperity and decline of these cultures were strongly intertwined with the state of these works. Various irrigation methods have been developed, such as surface (furrows, strips, basins), sprinkler, micro-irrigation and sub-irrigation. From antiquity to the middle of the 20th century, irrigation was carried out by surface methods. Then other methods were developed. In the year 2000, an area of 2,788,000 km² was equipped with irrigation infrastructure worldwide. By 2012, the area of irrigated land had increased to an estimated total of 3,242,917 km², which is nearly the size of India. At the same time, total arable land area is 13,958,000 km² (FAO, 2016). Irrigation schemes in the world use about 3,500 km³ water per year, of which 74% is evaporated by the crops (Sundquist, 2007). This is some 80% of all water used by mankind (4,400 km³ per year).

The characteristic of **surface** methods is the lack of energy costs for pumping, which results in low operating costs, as well as the usually low degree of efficiency of water use (large losses). Surface methods are the world's leading methods with 94% coverage and have been practiced in many areas virtually unchanged for thousands of years. Although fairly high irrigation efficiencies of 70% or more (i.e. losses of 30% or less) can occur by well design and management, in practice the losses are commonly in the order of 40%–60%.

As for the **sprinkler**, it has been developed in countries that have both the technological background and the ability to incur the cost of pumping, which in most cases is necessary. **Micro-irrigation** is a newer method that allows the application of water with small flow rates and small pressures. The most common method is the drip: the water is applied locally (partial moistening of the root layer) with frequent and small doses. This method has the advantage of having small water losses and low pumping costs as the pressures are small. This system allows the use of small flow rates (drills with small capacity). On the contrary, it is characterized by the high installation costs and it requires good filtration. A serious problem as far as this method is concerned is salt accumulation at the root layer. The latter issue can be overcome using significant irrigation doses (e.g. at the end of the growing season) or with systematic leaching in each irrigation. The method is suitable for areas that lack water resources. A typical example is Israel, which uses this as the main irrigation method.

The **sub-irrigation** system has a much more limited use. It can be applied in suitable soil conditions (existence at a shallow depth of impermeable layer, e.g. the case of the Kopaida plain in central Greece, where a clay layer is presented at 2 m depth) or with the use of special underground dripper pipelines in order for the application to take place directly on the roots layer. The method is suitable for the application of processed effluents in agricultural crops and green spaces. This method can lead to greater water savings because the surface does not get wet and thus evaporation is negligible. Table 11.1 shows the indicative application efficiency values for the various irrigation systems and Table 11.2 presents the overall efficiencies of irrigation projects.

TABLE 11.1 Indicative Values of Distribution and Application Efficiencies in Collective Irrigation Networks

Type of Network	Operation and Maintenance	Conveyance and Distribution Efficiency
Surface	Very good up to excellent	0.60–0.75
	Good	0.50–0.60
	Mediocre	0.35–0.50
	Bad	0.20–0.35
Pressure	Good up to excellent	0.80–0.95
Irrigation Method	Efficiency of Application in the Farm	
Basin	0.60–0.80	
Border strip	0.60–0.75	
Furrow	0.50–0.75	
Sprinkler	0.60–0.80	
Travel gun	0.55–0.75	
Micro-irrigation	0.80–0.95	

Source: From Papazafiriou (1999).

TABLE 11.2 Efficiencies According to an Important Number of Projects

	No. of Projects	Efficiency Range	Average Efficiency
Overall or project efficiency	56	7–60	30
Field application efficiency	60	14–88	53
Distribution efficiency	48	50–97	78
Conveyance efficiency	48	26–98	73

Source: From Bos and Nugteren (1990).

About 68% of irrigated areas are in Asia, 17% in the Americas, 9% in Europe, 5% in Africa and 1% in Oceania. The largest contiguous areas of high irrigation density are situated: in Northern India and Pakistan along the Ganges and Indus rivers; in the Hai He, Huang He and Yangtze basins in China; along the Nile River in Egypt and Sudan; and in the Mississippi–Missouri river basin, the Southern Great Plains and in parts of California. Smaller irrigation areas are spread across almost all populated parts of the world (Siebert et al., 2006). As of 2012, China and India are the fastest expanding countries in the field of drip- or the other micro-irrigation systems, while 10 million hectares utilize these technologies worldwide. This amounts to less than 4% of the world's irrigated land (National Geographic, 2012). Today, for the planning of irrigation and its optimization as well as for the general management of the cultivation, Crop Simulation Model (CSM) has been developed that simulates the entire growing season and calculates the evolution of soil moisture, the plant growth and the actual evaporation and water balance at the field level. These tools are very important because if properly calibrated they can assess the impact of various management practices and irrigation planning on crop production.

The **Crop Simulation Model** (**CSM**) is a simulation model that describes the processes of crop growth and development (yield, maturity date, efficiency of fertilizers, etc.) as a function of weather conditions, soil conditions and crop management (USDA, 2014). Typically, such models estimate times that specific growth stages are attained, biomass of crop components (e.g., leaves, stems, roots and harvestable products) as they change over time, and similarly, changes in soil moisture and nutrient status.

The usefulness of these tools becomes even more important if the weather forecast data are used for the next period (5–6 days) that will allow us to predict the actual evapotranspiration in order to make the most appropriate planning of the range and irrigation dose. It should be noted here that estimating the reference evapotranspiration (ET_o) in combination with the crop coefficient (K_c) leads to the calculation of the ETmax. The latter of course will not take place unless the plant is fully irrigated. In practice, this is usually not the case, and it is necessary to assess the actual evapotranspiration with a model of simulation of the soil moisture conditions and the growth stage of the plant. In terms of irrigation amount (doses), these are calculated based on the meteorological conditions, the specific crops' requirements, and the soil conditions of each area. When water resources are available in high quantities, irrigation planning may aim to maximize production. But because this is not the case in many areas and especially under climate change conditions, available water resources will be increasingly limited in many parts of the world; therefore, we need to take a different approach by applying DI.

Deficit irrigation (**DI**) is a watering strategy that aims to maximize crop water productivity instead of maximizing the harvest per unit land, in regions where water resources are restrictive. The correct application of this strategy requires thorough understanding of the yield response to water (crop sensitivity to drought stress) and of the economic impact of reductions in harvest (Fereres and Soriano, 2006). Today, the remote sensing models have been developed that assess the water needs by linking them to plant factors (usually Normalized Difference Vegetation Index or NDVI). In fact, the new satellites with high-resolution (pixel 10 or 2 m) products allow a spatial distribution of water needs with significant accuracy. This new technology has become commercially available and farmers can register on electronic platforms that provide a forecast of the irrigation needs and identify heterogeneities in agricultural parcels (zone establishment) in order to have a specific intervention in every zone. Also, satellite products allow the identification of crop pattern in the wider areas and together with the meteorological data allow the irrigated plots to be located with significant accuracy. This is important for the assessment of water needs but also for the identification of illegal drilling.

Technological improvements and irrigation benefits arising therefrom are significant. New electronic water intakes with cards in pressure networks on demand allow significant water savings (20% in Consortium for the Reclamation and irrigation of Capitanata Bari/Italy, Lamaddalena et al., 2007) and 40% in Northern Greece. Given the large quantities used in agriculture, it is clear these water economies are very important. Precision farming has also improved the application of irrigation, and today the GPS systems allow irrigation to be adapted to the local needs of the field (e.g., Center Pivots can adapt irrigation to local needs). These new precision farming capabilities allow for better assessment of water needs, protection of water resources (reduction of consumption, reduction of deep penetration and loss of fertilizers) and at the same time allow the improved management of collective infrastructure and better adaptation to the demand.

11.4 Applications in Crop Monitoring

It is recognized that a portion of the farmers understands the benefits of recent technological and scientific advances in agriculture, such as the remote sensing and GIS, information, and communication technologies or precision farming. The dissemination of information about precision agriculture is steadily increasing and implementation programs are being developed internationally. However, there is not a clear vision among farmers of what this technology can offer. Contemporary research shows that about a 13% of the farmers in USA and Europe have adopted precision agriculture and new technologies. Globaly in the developed countries, the above percentage drops to 2.5%; however, this percentage is gradually increasing, especially among the young population. There are a lot of reasons of explaining this lagging, such as the perception that precision agriculture is for large farms and fields, the lack of knowledge about the advances of new technologies and the costs of adoption of precision agriculture to small-size farms.

11.4.1 Methodology

11.4.1.1 Image Preprocessing and Rectification

Overall there is a major requirement for calibration, preprocessing, correction and rectification of imagery before it can be used in any qualitative or quantitative analysis. The local different weather conditions occurred during the acquisition of the cloud free imagery, seasonal atmospheric models for the concerned geographic latitude, the type of land-use and 5 m Digital Elevation Model (DEM) are used to run the corrections of the data. Initially a top of the atmosphere reflectance image is created to be followed by haze removal and cloud masking process, converting Digital Numbers (DN) to the reflectance values above the atmosphere. Reflectance values are then transformed to ground reflectance considering the solar azimuth and zenith angles on the acquisition date (Lanzl and Richter, 1991; Dercas et al., 2017). The ATCOR3 Top of the Atmosphere and Ground reflectance workflows, available in Geomatica 2016 system, were used. The subsequent geometric correction follows two different geocoding approaches, by means of image-to-Ground Control Points (GCPs) and image-to-image registration. The first acquired image for the three fields, which are cotton fields, was used as master to slave all the rest. The first WV-2 was acquired on May 17, 2015, and corrected using 10 GCPs, acquired from a differential GPS, while all the other images were registered using as reference for the first image. It should be mentioned that the area of the cotton fields is relatively flat with no terrain fluctuations and the farm sizes were not more than 5 ha. The bilinear algorithm was used to correct the sensor optical aberration satellite instabilities and terrain variations. This algorithm does not alter the DN values of the pixels, a critical decision for

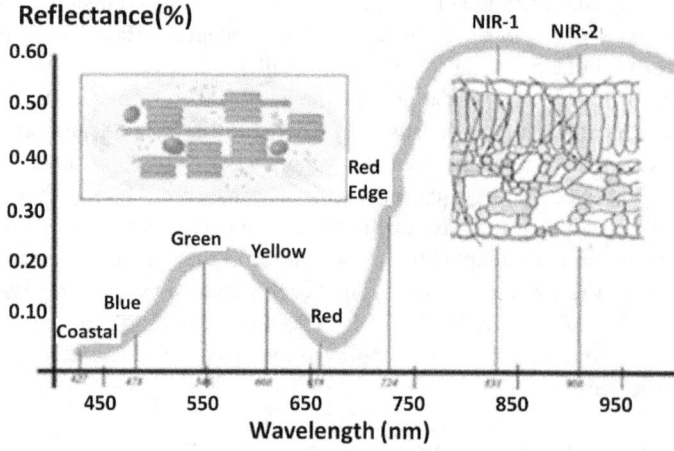

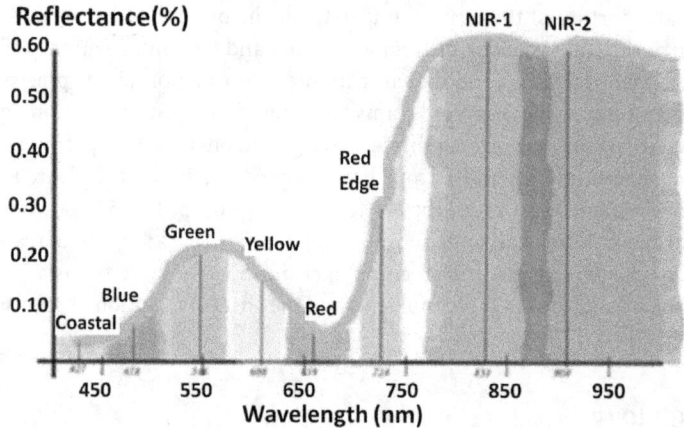

FIGURE 11.1   (a) Typical example of spectral behavior of plants based on pigments and plant structure. (b) WV-2 satellite spectral bands overlaid onto a typical vegetation reflectance curve (Peroni, 2010).

the subsequent qualitative analysis and buildup of vegetation indices. In all cases, a half-pixel root-mean-square error (RMSE) error was achieved in X- and Y-axes. Figure 11.1 presents a typical example of spectral behavior of plants based on pigments and plant structure (upper part) and WV-2 satellite spectral bands overlaid onto a typical vegetation reflectance curve (lower part).

11.4.1.2 Image Processing

The satellite data processing procedures include the computation of the various NDVI indices, Soil Adjusted Vegetation Index (SAVI), Leaf Area Index (LAI), Fraction of Photosynthetically Active Radiation (FPAR) and clustering (Dalezios et al., 2012). The basic principle that followed in the work is the estimation of $ETc_{(satellite)}$, which is based on the $Kc_{(satellite)}$ derived from satellite data acquired during the crop season rather than FAO crop-specific tables. Initially, three different band combinations or False Color Composites (FCCs), of pan-sharpened fusion data were processed and developed to assist the subsequent field interpretation work. Pan-sharpening is the result derived from the fusion of 2 m multispectral channels with the 0.5 m panchromatic band to generate multispectral bands that maintain the spectral information, but in higher spatial resolution of 0.5 m, thus enhancing the information analysis and clarity, which is very important for the fragmented small mixed and multiple land-uses of Mediterranean agricultural fields. The first FCC was the 5,3,2 (RGB: red, green and blue) using the visible part of the electromagnetic spectrum. The second was the 6,5,3 (RGB: redEdge, red and green) and the third one the 8,5,4 (RGB: NIR2, red and yellow). The two latter combinations exhibited and depicted crop and soil variations. The reflectance-calibrated WV-2 data then processed to build the VIs such as Chlorophyll, redNDVI and redEdgeNDVI using red, redEdge and the two near-infrared (NIR) channels NIR1 and NIR2. The use of the second innovative NIR band of WV-2 is the first attempt to reach better-quality VIs, since it is less influenced by the atmosphere.

In Mediterranean agricultural areas, the NDVI derived from very high spatial analysis WV-2 data has shown that it constitutes a significant indicator of green biomass vegetation density and health. NDVI uses leaves chlorophyll properties that demonstrate high energy absorption in the Red part and high reflectance in the NIR part of electromagnetic spectrum, and uses the following formula (Eq. 11.1):

$$NDVI = (NIR - RED) / (NIR + RED).$$ (11.1)

The NDVI takes pixel values from −1 to 1. In general, NDVI lower than 0.2 corresponds to non-vegetated surfaces, while NDVI greater than 0.2 to vegetated ones (Figures 11.1 and 11.2). The redEdge NDVI index (nir − redEdge)/(nir + redEdge) shows the different behavior from NDVI. WV-2 redEdge channel is more sensitive to the change of vegetation reflectance. Thus, it was also generated to monitor cotton phenological cycle. Additionally, Chrorophyll index [(nir/redEdge) − 1] was also produced to extract and monitor the behavior of local cotton farm. In conjunction with meteorological and ground data, as well as models, the imagery was utilized to calculate several biophysical parameters, such as reflectance, NDVI, fractional cover, crop coefficient (Kc) and crop evapotranspiration (ETc) maps over the image coverage.

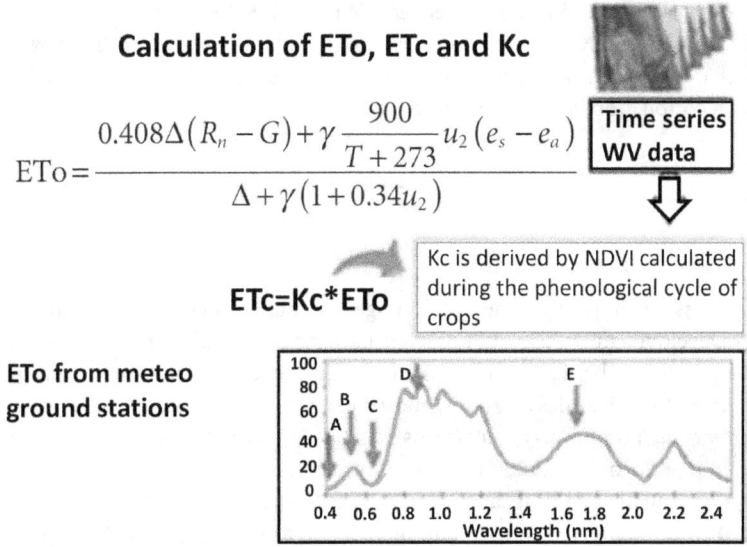

FIGURE 11.2 Methodological architecture on how the extracted Kc by satellite used in ETc satellite estimation (from Spyropoulos et al., 2020).

11.4.1.3 Estimation of the FAO-56 ET₀

First, the following FAO-56 Penman–Monteith equation was used to estimate the reference ET_0 based on meteorological data:

$$ETo = \frac{0.408\Delta\left(R_n - G\right) + \gamma\dfrac{900}{T+273}u_2\left(e_s - e_a\right)}{\Delta + \gamma\left(1 + 0.34u_2\right)},$$ (11.2)

where ET_0 is reference evapotranspiration [mm/day], R_n is the net radiation at the crop surface [MJ/$(m^2 day)$], G is the soil heat flux density [MJ/$(m^2 day)$], T is the mean daily temperature at 2 m height [°C], u_2 is the wind speed at 2 m height [m/s], e_s is the saturation vapor pressure [kP_a], e_a is the actual vapor pressure [kP_a], $e_s - e_a$ is the saturation vapor pressure deficit [kP_a], Δ is the slope vapor pressure curve [kP_a/°C], and γ is the psychrometric constant [kP_a/°C].

11.4.1.4 Estimation of the Irrigation Water Needs from Crop ET from Satellite Data

This methodological architecture is depicted in Figure 11.2. According to this, it is possible to estimate the irrigation water needs of a crop by calculating the ET from satellite data by mainly using the Kc coefficient, which is also derived through the previously described vegetation knowledge, such as the vegetation index.

Calera et al. (2017) estimated the water irrigation needs of a crop by calculating the ET from satellite data by mainly using the K_c coefficient, which is derived by a vegetation index. The full ET estimation equation is based on:

$$ET = \left(Ks * Kcb + Ke\right)ETo.$$ (11.3)

The basal crop coefficient (K_{cb}) is defined as the ratio of the crop evapotranspiration over the reference evapotranspiration (ET_c/ET_0), when the soil surface is dry, but transpiration is occurring at a potential rate, i.e. water is not limiting transpiration. Therefore, the term "$K_{cb}\,ET_0$" represents primarily the maximum transpiration component of ET_c of an unstressed canopy. The $K_{cb}\,ET_0$ does include a residual diffusive evaporation component supplied by soil water below the dry surface (Ke) and by the soil water from the beneath dense vegetation (Ks). Moreover, the term $K_s\,K_{cb}\,ET_0$ represents the actual transpiration of a canopy and the term $K_e\,ET_0$ is the evaporation from bare soil fraction. The term K_c is a spectral crop coefficient that takes the values between 0.15 and 1.20. Dercas et al. (2017) were using the following equation based on cotton crop in central Greece.

$$Kc = 1.33 * redNDVI + 0.21$$ (11.4)

11.4.1.5 Soil Water Balance Modeling

CROPWAT is a decision support tool for crop irrigation management. It has been developed by the Land and Water Development Division of FAO. It is used for the calculation of crop water requirements and irrigation requirements based on soil, climate and crop data. In addition, the program allows the development of irrigation schedules for different management conditions and the calculation of scheme water supply for varying crop patterns. It can also be used to evaluate the farmers' irrigation practices and to estimate crop performance under both the rainfed and irrigated conditions. The program uses a flexible menu system and file handling, and extensive use of graphics. Graphs of the input data (climate, cropping pattern) and results (crop water requirements, soil moisture deficit) can be drawn and printed with ease. Complex cropping patterns can be designed with several crops with staggered planting dates. All calculation procedures used in CROPWAT are based on the two FAO publications of the Irrigation and Drainage Series (FAO, 1979; Allen et al., 1998).

As a starting point, and only to be used when local data are not available, CROPWAT includes the standard crop and soil data, which can be adjusted, or new ones can be created. Likewise, if local climatic data are not available, these can be obtained for over 5,000 stations worldwide from CLIMWAT, the associated climatic database. The development of irrigation schedules in CROPWAT is based on a daily soil water balance using various user-defined options for the water supply and irrigation management conditions. Scheme water supply is calculated according to the cropping pattern defined by the user, which can include up to 20 crops. The estimation of crop water requirements and irrigation requirements could be done at farm and/or basin level.

The main features of CROPWAT are the following: calculation of reference crop evapotranspiration (ET_o) using climatic data at month, 10 days and day-time scale; ability to estimate climate information when no relevant values are available; establishment of irrigation programs (through water balances) at farm and basin area level with extensive user adjustment options; daily Water Balance Scoreboards for the entire growing season; graphic presentation of data and results (water crop needs and irrigation programs); and easily import and export data via Windows clipboard or ASCII files. The basic meteorological parameters needed to calculate the evapotranspiration from CROPWAT are: coordinates (WGS84: World Geodetic System 1984) and altitude (m) of the area or meteorological station; mean minimum temperature (°C); mean maximum temperature (°C); mean relative humidity or saturation pressure deficit (% or kPa); mean wind speed (altitude at 2 m) (km/day or m/s); hours of sunshine, or % of day or ratio of sunshine; and effective rainfall (FAO, 2011).

The necessary data for the cultivation period and agronomic and crop patterns estimation Include crop name, planting date, duration of stages, the Kc values, rooting depth, critical depletion fraction, yield reduction due to crop stress, and crop height. Similarly, the necessary soil data are total available soil moisture, maximum rainfall infiltration rate, maximum rooting depth and initial soil moisture depletion. Using these data, the initial available soil moisture is calculated by CROPWAT (mm/m). Once all of the necessary data (rainfall, evapotranspiration, crop season and agronomic and crop patterns) are available, the CROPWAT produces a table or graph with the water requirements. For this calculation, effective rainfall is deducted from the crop evapotranspiration.

11.4.2 Study Area and Data Sets

Study area: The case study area is the region of Thessaly located in central Greece (Figure 11.3). It covers an area of 14,036 km², and thus the 10.6% of Greece territory. In Thessaly region, three geomorphology types are dominated: the flat areas about 36%, the semi-mountainous about 17% and the mountainous about 44.9%. Generally, Thessaly is divided in two climatic zones. The west part has continental climate with the annual precipitation reaches more than 1,850 mm at the mountain peaks. The east part is characterized as Mediterranean climate with mean annual precipitation of about 400 mm (Dalezios et al., 2014). The Thessaly plain is one of the biggest agricultural areas of Greece. The main crops are cotton, cereals, fruits and olive trees. It is characterized by vulnerable agriculture, due to the rare rainfall from June to August and the growing needs of irrigation reaching 96% of the water consumption at regional level. Over the last 30 years, several studies have revealed that at Thessaly plain the mean annual precipitation has been reduced to about 20% and ranges from 250 to 500 mm (Dalezios, 2011). The experiment fields were three agricultural fields of cotton covering a surface of 1.8, 2.1 and 4.3 ha, respectively.

Data sets: Three WV-2 series of scenes from May, June, July and September of 2015 (4 images); from July and Aug 2016 (2 images); and from May, June, July, August and September 2017 (5 images) over cotton fields were atmospherically corrected, radiometrically enhanced and georeferenced (Lanzl and Richter, 1991; Updike and Comp, 2010). Additionally, the meteorological data records such as rainfall, temperature and wind speed were collected. For the soil water monitoring, suitable sensors were placed to different depths as representative to the root zone. In more detail, data loggers in depths levels spanning from 15 to 90 cm were installed in order to monitor the soil profile moisture at the experimental fields. These records were very important in order to evaluate the stress level of the crops.

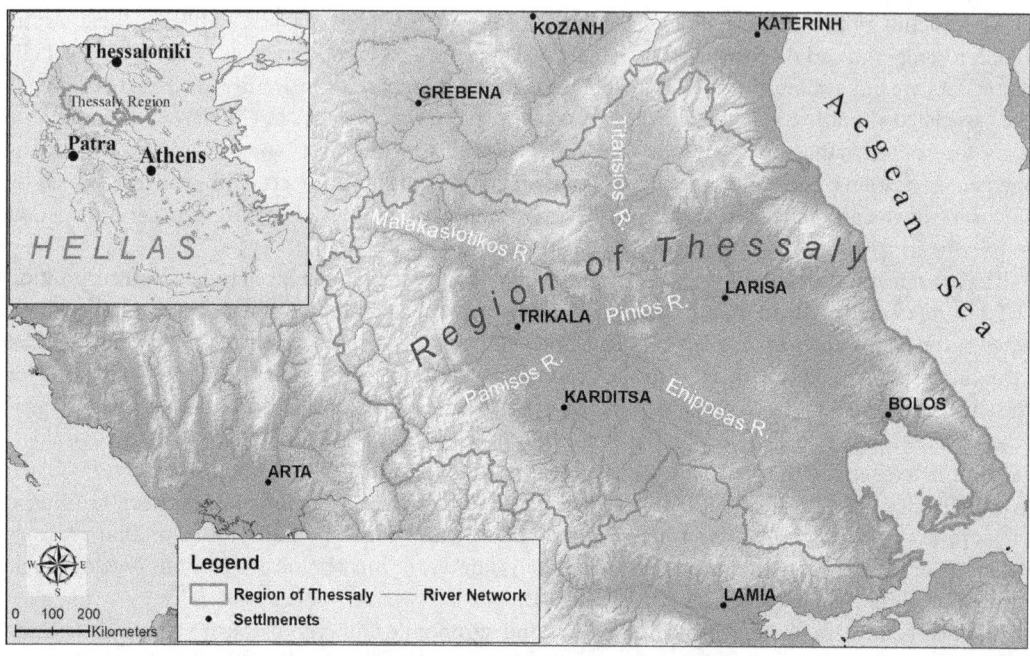

FIGURE 11.3   The region of Thessaly with the primary river network.

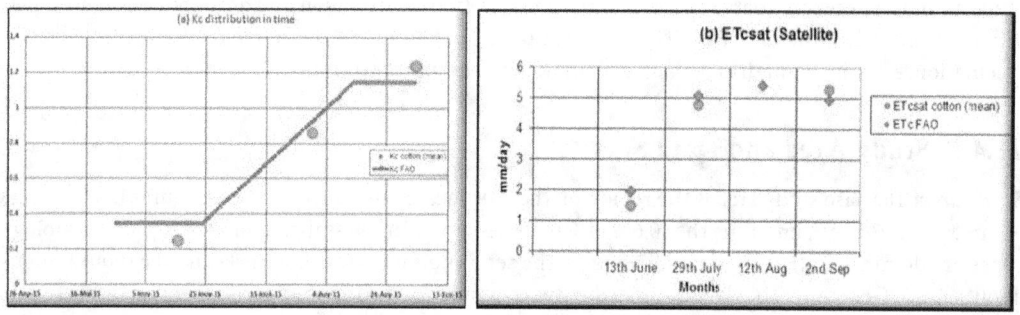

FIGURE 11.4   (a and b) Comparison of Kc$_{sat}$ from WV-2 and Kc produced by FAO for the cotton field in 2015 (from Dercas et al., 2017).

11.5 Results and Discussions

The results are summarized in Figure 11.4. Specifically, Figure 11.4a shows the estimated Kc$_{sat}$ (Kc estimated using satellite data) average values for cotton at the field scale, from Eq. (11.4) and the Kc from FAO for cotton during the growing season of 2015. Figure 11.4b shows the ETc$_{sat}$ average values at the field scale from equation ET$_c$=K$_c$ * ET$_o$ and the ET$_c$ produced by FAO for cotton field in 2015.

The cotton Kc$_{sat}$ values based on Red and NIR spectral bands of WV-2 data, are very close to the Kc values from FAO for 2015 (Figure 11.4a). Indeed, Kc$_{sat}$ is dependent on the type of satellite data and cannot be applied in a generic way. Spatial and spectral resolution are the key components of any sensor, which is used for Kc$_{sat}$ extraction. Sensors with different bandwidths of Red, redEdge and NIR spectral bands create heterogeneous results. Additionally, the type and size of fields, such as small-size

TABLE 11.3 Comparison of Kc_{FAO} and Kc WV-2 Satellite for 2015–2016–2017

Cotton 2015			Cotton 2016			Cotton 2017		
Acquisition date	Kc_{FAO} (mm)	Kc_{sat} (mm)	Acquisition date	Kc_{FAO} (mm)	Kc_{sat} (mm)	Acquisition date	Kc_{FAO} (mm)	Kc_{sat} (mm)
						May 12	0.35	0.57
June 13	0.35	0.35				June 30	1.07	0.80
July 29	0.99	0.91	July 8	1.13	0.99	July 29	1.15	1.21
			August 12	1.15	1.05	August 18	1.15	1.16
September 2	1.15	1.28				September 1	1.11	1.31

Source: From Dercas et al. (2017).

TABLE 11.4 Comparison of ETc_{FAO} and ETc WV-2 satellite for 2015–2016–2017

Cotton 2015			Cotton 2016			Cotton 2017		
Acquisition date	ETc_{FAO} (mm)	ETc_{sat} (mm)	Acquisition date	ETc_{FAO} (mm)	ETc_{sat} (mm)	Acquisition date	ETc_{FAO} (mm)	ETc_{sat} (mm)
						May 12	1.96	3.19
June 13	2.42	2.42				June 30	9.46	7.07
July 29	6.65	6.11	July 8	6.24	5.46	July 29	7.19	7.56
			August 12	5.66	5.17	August 18	6.06	6.11
September 2	6.64	7.39				September 1	5.35	6.31

Source: From Dercas et al. (2017).

farms or land fragmentation, play a role on the satellite data and thus on the Kc equation to be used. Moreover, ETc_{sat} values are very close to ETc_{FAO} (Figure 11.4b), which is a promising sign for the new WV-2 approach.

The results of water requirements are summarized in Figure 11.4 and Tables 11.3 and 11.4. The estimated Kc_{sat} (Kc assessed using satellite data) average values for cotton at the field scale are computed from Eq. (11.4) and the Kc from FAO for cotton during the growing season of 2015. The cotton Kc_{sat} values based on Red and NIR spectral bands of WV-2 data are quite close to the Kc values from FAO for 2015 (Figure 11.4). The Kc obtained by the new proposed Eq. (11.4) (using the NDVI obtained by WV-2 images) for 2015, 2016 and 2017 are presented in Table 11.3. In Table 11.4, the ETc estimated according to Kc_{FAO} and Kc_{sat} are presented. In both cases, the ET_o was estimated using the Hargreaves method due to limited meteorological data. The values of ETc_{sat} and ETc_{FAO} are in good agreement for both methods, which is a promising mark for the new WV-2 approach. The water conditions in the experimental fields were analyzed for all of the cases and were not under any water stress. The actual evapotranspiration was estimated and showed that the crops were well irrigated. This was an essential condition in order to be able to confirm that the Kc evaluated was not a Kc stress.

11.6 Summary and Conclusions

Agriculture is estimated to be highly affected by the adverse impacts of climate variability and change, with a high probability of crop production reductions, due to lack of the irrigation water. Furthermore, trying to decrease the risk of climate change impacts on agriculture, integrated methodologies need to be developed, composing the results of the studies on drought and evapotranspiration calculation and monitoring, for the formulation of the cost-effective mitigation measures and adaptation strategies. The study outcomes the designation that agricultural drought takes place every year at the warm period with increasing severity and areal extent, having its maximum appearance at the end of summer, which is quite typical in the Mediterranean region (Dalezios et al., 2014). The comparison of satellite-based WV-2 earth observation data with in situ measurements has revealed that the utilization of very high-resolution satellite data can definitely be an advantageous tool for precision or operational farming, providing, apart from the extraction of crop area, a number of valuable parameters, such as ETc from Kc, that can be utilized for the assessment of crop water needs. The results of the implemented methodology explain the synergistic use of the WV-2 images with ground-truth data set for monitoring ETc.

The development and implementation of an efficient site-specific management system for reducing the irrigation and chemical inputs are possible by combining field measurements and well-defined satellite data coverage intervals. Satellites proved to be ideal due to their global repetitive coverage over an area of interest (AoI) (every 2–3 days), spatial and spectral resolution. The comparison of satellite data and field measurements has shown that the utilization of the very high spatial and spectral data, such as WV-2, is definitely a useful tool for farm-based stakeholders,), but also the derivation of useful parameters, such as the $Kc_{(satellite)}$ and $ETc_{(satellite)}$, that can be used in the estimation of the crop water and fertilization needs. The $Kc_{(satellite)}$ and $ETc_{(satellite)}$ depict better phenological cycle of the crop and they are more close to what is "happening in the field" and also to weather conditions, and always reflect the agricultural practices, e.g. irrigation and fertilization, as these actions are very dynamic and modified according to crop and weather, farmer behavior and policies. The present work is expected to be the basis for future work, which will provide a broad range of results and is expected to show how the farmers can optimize the water and N fertilization inflows.

Acknowledgments

This research was funded by the INTERREG IIIb PRODIM project, EU FP6 PLEIADES project, (and HORIZON2020 FATIMA project.

References

Allen, R. G., Pereira, L. S., Raes, D., and Smith, M. (1998). Crop evapotranspiration -Guidelines for computing crop water requirements - Irrigation and Drainage Paper 56- Food and Agriculture Organization of the United Nations, Rome, Italy.

Van Alphen, B.J., Stoorvogel, J.J. (2000). A Methodology for Precision Nitrogen Fertilization in High-Input Farming Systems. Precision Agriculture 2, 319–332, https://doi.org/10.1023/A:1012338414284

Anderson, J. Ch. and Kyveryga, M. P. (2016). Combining on-farm and climate data for risk management of nitrogen decisions. *Climate Risk Management*, 13, 10–18.

Ansari, R., Cheema, M., Liaqat, M., Ahmed, S., and Khan, H. (2019). Evaluation of irrigation scheduling techniques: A case study of wheat crop sown over permanent beds under semi-arid conditions. *Journal of Agriculture and Plant Sciences*, 17(1), 9–21.

Bakhsh, A., Jaynes, D. B., Colvin, T. S., and Kanwar, R. S. (2000). Spatio-temporal analysis of yield variability for a corn-soybean field in Iowa. *Transaction of the American Society of Agricultural Engineers*, 43, 31–38.

Basso, B., Liu, L., and Ritchie, T. J. (2016). A comprehensive review of the CERES-wheat, -maize and -rice models' performances. *Advances in Agronomy*, 136. doi: 10.1016/bs.agron.2015.11.004.

Bos, M. G. and Nugteren, J. (1990). On irrigation efficiencies, ILRI publication 19, Wageningen, The Netherlands.

Calera, A., Campos, I., Osann, A., D'Urso, G., & Menenti, M. (2017). Remote sensing for crop water management: From ET modelling to services for the end users. *Sensors*, 17(5), p.1104.

Chirico, G., Pelosi, A., De Michele, C., Falanga Bolognesi, S., and D'Urso, G. (2018). Forecasting potential evapotranspiration by combining numerical weather predictions and visible and near-infrared satellite images: An application in southern Italy. *The Journal of Agricultural Science*, 156 (5), 702–710. doi: 10.1017/S0021859618000084.

Collis, S. N. and J. D. Corbett. (2000). Integration of weather, soil and crop databases for crop simulations in an object-oriented geographic information system. Geospatial information in Agriculture and Forestry. *Proceedings of the Second International Conference*, 10–12 January 2000, Lake Buena Vista, Florida, USA. P II–241.

Dalezios, R. N. (2011). Climatic change and agriculture: Impacts-mitigation-adaptation. *Scientific Journal of GEOTEE*, 27, 13–28.

Dalezios, R. N., Spyropoulos, V. N., and Blanta, A. (2012). Agrometeorological remote sensing of high resolution for decision support in precision agriculture. In Helmis et al. (Eds.) *Advances in Meteorology, Climatology and Atmospheric Physics*. Springer Atmospheric Sciences. doi: 10.1007/978-3-642-29172-2_8, # Springer-Verlag Berlin Heidelberg, pp. 51–56.

Dalezios, R. N, Blanta, A. Spyropoulos, V. N., and Tarquis, M. A., (2014). Risk identification of agricultural drought for sustainable agroecosystems. *Natural Hazards and Earth System Sciences*, Special issue: Advances in meteorological hazards and extreme events, 14(9). doi: 10.5194/nhess-14-2435-2014.

Dalezios, R. N. (2015). *Agrometeorology: Analysis and Simulation (in Greek)*. KALLIPOS: Libraries of Hellenic Universities (also e-book), ISBN: 978-960-603-134-2, 481 pages, Nov 2015.

Dalezios, R. N., Dercas, N., Spyropoulos, V. N., and Psomiadis, M. (2017). Water availability and requirements in precision farming of vulnerable agroecosystems. *European Water*, 59(1), 387–394.

Dalezios, R. N., Dercas, N., Spyropoulos, V. N., and Psomiadis, M. (2019). Remotely sensed methodologies for water availability and requirements in precision farming of vulnerable agriculture. *Water Resources Management*, 33(4), 1499–1519.

Dercas, N., Spyropoulos, N. Dalezios, N., Psomiadis, E., Stefopoulou, A., Mantonanakis, G. and Tserlikakis, N., (2017). Cotton evapotranspiration using very high spatial resolution wv-2 satellite data and ground measurements for precision agriculture. ECOLOGY AND THE ENVIRONMENT (WATER RESOURCES MANAGEMENT IX). 220. 101-107. 10.2495/WRM170101.

Duarte, C. N., Sentelhas, Y., and Paulo, C. (2019). NASA/POWER and daily gridded weather datasets—how good they are for estimating maize yields in Brazil? *International Journal of Biometeorology*, 64, 319–329. https://doi.org/10.1007/s00484-019-01810-1.

FAO (1979). Irrigation and drainage paper 33, Yield response to water: Part A: http://www.fao.org/WAICENT/faoINFO/AGRICULT/agl/aglw/cropwater/parta.stm and part B: http://www.fao.org/WAICENT/faoINFO/AGRICULT/agl/aglw/cropwater/cwinform.stm.

FAO (2011). Effective Rainfall in Irrigated Agriculture. http://www.fao.org/docrep/x5560e/x5560e03.htm#3.4%20usda, %20scs%20.

FAO (2016). FAOSTAT Land Use Module. Food and Agriculture Organization, Italy.

Food and Agriculture Organization (FAO) (1993). *AGROSTAT. PC, Computerized Information Series*. Rome: Food and Agriculture Organization of the United Nations.

Fereres, E. and Soriano, M. A. (2006). Deficit irrigation for reducing agricultural water use. Journal of Experimental Botany, 58(2), 147–159. doi: 10.1093/jxb/erl165.

Giles, D. K., Stone, M. L., and Dieball, K. (1999). Distributed network system for control of spray droplet size and application rate for precision chemical application. In J. V. Stafford (Ed.) *Precision Agriculture '99*. Sheffield Academic Press, Sheffield, England, pp. 867–876.

Keane, T. and Collins, F. J. (2004). *Climate, Weather and Irish Agriculture*, 2nd Edition. Agmet, Dublin, Ireland.

Kravchenko, A. N., Bullock, D. G., and Boast, C. W. (2000). Joint multifractal analysis of crop yield and terrain slop. *Agronomy Journal*, 92, 1279–1290.

Lamaddalena, N., Bogliotti, C., Todorovic, M., Scardigno, A. (Eds.) (2007). Water saving in Mediterranean agriculture and future research needs [Vol. 3]. Bari: CIHEAM, 351 p. (Options Méditerranéennes: Série B. Etudes et Recherches; n. 56 Vol. III). *Proceedings of the International Conference WASAMED Project* (EU contract ICA3-CT-2002-10013), 2007/02/14-17, Valenzano (Italy). http://om.ciheam.org/om/pdf/b56_3/b56_3.pdf.

Lanzl F. and Richter R. (1991). A Fast Atmospheric Correction Algorithm For Small Swath Angle Satellite Sensors. ICO Topical Meeting on Atmospheric Volume and Surface Scattering and Propagation, Florence, Italy, August 27-30, pp.455-458.

Martínez, J. Egea, G., Agüera, J., and Pérez-Ruiz, M. (2017). A cost-effective canopy temperature measurement system for precision agriculture: A case study on sugar beet. *Precision Agriculture*, 18(1), 95.

Mavi, H. S. and Tupper, G. J. (2004). Agrometeorology: Principles and Applications of Climate Studies in Press Agriculture. Food Products, 363p.

Mohseni, F, Sadr, M. K., Eslamian, S., Areffian, A., Khoshfetrat, A. (2021). Spatial and temporal monitoring of drought conditions using the satellite rainfall estimates and remote sensing optical and thermal measurements. *Advances in Space Research*, 67(12), 3942–3959.

Moore, G. A. and Tyndale-Biscoe, J. P. (1999). Estimation of the importance of spatially variable nitrogen application and the soil moisture holding capacity to wheat production. *Precision Agriculture*, 1, 27–38, https://doi.org/10.1023/A:1009973802295

Mukherjee, A. Wang, S.-Y. S., and Promchote, P. (2019). Examination of the climate factors that reduced wheat yield in northwest India during the 2000s. *Water*, 11, 343.

National Geographic (2012). Drip Irrigation Expanding Worldwide, 25 June, https://www.nationalgeographic.com.

O'Neal, M. R., Frankenberger, J. R., Ess, D. R., and Grant. R. H. (2000). Precipitation variability at the farm scale during crop phenological phases. *Transactions of the American Society of Agricultural Engineers*, 46, 1449–1458.

Papazafiriou, Z. (1999). *The Water Needs of Crops*. Publisher Ziti, Thessaloniki, Greece.

Pelosi, A., Chirico, B. G., Bolognesi, F. S., De Michele, C., and D'Urso, G. (2019). Forecasting crop evapotranspiration under standard conditions in precision farming. *2019 IEEE International Workshop on Metrology for Agriculture and Forestry (MetroAgriFor)*, 2019, Portici, Italy, pp. 174–179.

Peroni, G. (2010). New spectral data available for the control in agriculture (CwRS) and for vegetation monitoring. A French experience on the potential benefits in using WorldView-2 new bands. *Paper Presented at the 16th GeoCAP Annual Conference: Geomatics in support of the CAP*, 2010, Bergamo, Italy.

Pooja, S. Uday, V. D., Nagesh, B. U., and Talekar, G. S. (2017). Application of MQTT protocol for real time weather monitoring and precision farming. *International Conference on Electrical, Electronics, Communication, Computer, and Optimization Techniques (ICEECCOT)*, Mysuru, India, pp. 1–6. doi: 10.1109/ICEECCOT.2017.8284616.

Powell, P. J. and Reinhard, S. (2016). Measuring the effects of extreme weather events on yields. *Weather and Climate Extremes*, 12, 69–79.

Roux, J., Fourniols, J., Soto-Romero, G., and Escriba, C. (2018). A connected soil humidity and salinity sensor for precision agriculture. *International Conference on Computational Science and Computational Intelligence (CSCI)*, 2018, Las Vegas, NV, USA, pp. 1028–1031.

Sadler, E. J., Gerwig, B. K., Evans, D. E., Busscher, W. J., and Bauer, P. J. (2000). Site-specific modeling of corn yield in the SE coastal plain. *Agricultural Systems*, 64, 189–207.

Siebert, S., Hoogeveen, J., Döll, P., Faurès, J-M., Feick, S., and Frenken, K. (2006). The digital global map of irrigation areas – Development and validation of map version 4. *Conference on International Agricultural Research for Development*, Bonn, Germany –Tropentag, 2006.

Smith, K. (2013). *Environmental Hazards: Assessing Risk and Reducing Disaster*, 6th Edition. Springer, London, 478p, https://doi.org/10.4324/9780203805305.

Sundquist, B. (2007). Irrigation Overview. in: the Earth's Carrying Capacity, Some Related Reviews and Analysis. On Line: "Archived Copy". https://web.archive.org/web/20120217192619/http://home.windstream.net/bsundquist1/ir1.html

United States Department of Agriculture (2014). *What Are Crop Simulation Models?* Agricultural Research Service, Washington, DC.

Updike, T., and Comp, C., (2010). Radiometric use of WorldView-2 Imagery. Technical Note, rev. 1.0, DigitalGlobe Inc., Longmont CO.

WMO (2010). Guide to Agricultural Meteorological Practices. WMO-No 134, 799p.

Yobom, O. (2020). Climate change and variability: Empirical evidence for countries and agroecological zones of the Sahel. *Climatic Change.* doi: 10.1007/s10584-019-02606-3.

Zidon, R., Hirotsugu, T., Efrat, M., and Shai, M. (2015). Projecting pest population dynamics under global warming: The combined effect of inter- and intra-annual variations. *Ecological Applications*, 26(4), 1198–1210.

V

Irrigation Water Quality Issues

12

Irrigation with Reclaimed Municipal Wastewater: Opportunity or Risks

Tamer A.
Salem, Ahmad
Aboueloyoun
Taha and Nashwa
A. Fetian
Agriculture Research Center

Saeid Eslamian
*Isfahan University
of Technology*

Nabil I. Elsheery
Tanta University

12.1 Introduction

Water is considered as one of the main problems in arid and semi-arid countries, which is becoming an increasingly scarce resource in such areas. To meet the higher demand for water, planners are obliged to consider any source of water that might be used effectively and economically. Whenever high-quality water is rare, water of marginal quality will be considered for use in agriculture and groundwater recharge. Recently, the methodology for managing the reuse of wastewater has shifted from conventional disposal strategies to using value-added products. With the increase of wastewater reuse for different purposes, concerns over the health and environmental effects of this reuse have also increased. For both environmental and economic reasons, the use of treated sewage wastewater has become increasingly important in water resources management. In Egypt, wastewater reuse is an old practice. It has been used since 1930 in sandy soil areas such as Abou Rawash and Al Gabal Al Asfar, near Cairo. Since 1980, the interest in the use of treated wastewater, as a substitute for freshwater in irrigation, has accelerated. Nowadays, 0.7 BCM/year of treated wastewater is being used in irrigation, of which 0.26 BCM is undergoing secondary treatment and 0.44 BCM undergoing primary treatment (Abdel-Wahab and Omar, 2011; MWRI water strategy for 2050). In Tunisia, the available amount of reclaimed municipal wastewater for irrigation purposes will reach 125 Mm^3 in the near future (Bahri, 1988), whereas 82,900 ha are already being irrigated with wastewater in Mexico (Strauss and Blumenthal, 1989). In the USA, the reuse of reclaimed municipal wastewater reached 0.76 Mm^3/day of freshwater (Pescod, 1992). In California, more than 70% of reclaimed municipal wastewater is used for plant irrigation (California State Water

DOI: 10.1201/9780429290114-17

Resources Control Board, 1990). Reclaimed municipal wastewater plays a vital role in Israel's agriculture, where 72% of such water is dedicated to agricultural use (Shelef, 1990; Shelef and Azov, 1995).

There are several reasons for the rising use of reclaimed municipal wastewater, including (1) the lack of freshwater at a competitive price; (2) the potential use of plant nutrients in reclaimed municipal wastewater; (3) the availability of high-quality effluents; (4) a need to establish the comprehensive water resource planning, including water conservation and reuse; and (5) the avoidance of more stringent water pollution control requirements, including advanced wastewater treatment facilities at municipalities (Asano and Pettygrove, 1987). Although irrigation with municipal wastewater is in itself an effective form of wastewater treatment, some additional treatment must be made before such water can be used for the agricultural or landscape irrigation. In the planning, design and management of wastewater irrigation systems, the degree of treatment is an important factor. Pre-application treatment is essential to protect the public health, to avoid the nuisance conditions during application and storage, and to avoid damage to crops, soils and groundwater.

12.2 Municipal Wastewater Treatment Technologies

Wastewater treatment plants (WWTPs) are considered as a part or whole process described below for elimination of different pollutants according to the application of effluents, source of wastewater, economic reasons and local regulation for treatment. This section describes the different stages applied in WWTPs and their mechanisms for pollutant elimination (US EPA, 2004).

12.2.1 Preliminary Treatment

During this stage, the wastewater passes through a screen which is used to eliminate the large floating objects, such as rags, bottles cans and sticks that may clog pumps, small pipes and downstream processes. The screens range from coarse to fine and are constructed with parallel steel or iron bars with openings of about half an inch, whereas others may be made from mesh screens with much smaller openings. Generally, screens are placed in a chamber or channel and inclined toward the flow of the wastewater. The inclined screen allows the debris to be caught on the upstream surface of the screen, and allows access for manual or mechanical cleaning. Some plants use devices known as comminutors or barminutors, which function as both a screen and a grinder. These devices can catch and then cut or shred the heavy solid and floating material. In the process, the pulverized matter remains in the wastewater flow can be removed later in a primary settling tank. After the wastewater has been screened, it may flow into a grit chamber where sand, grit, cinders and small stones settle in the bottom. It's very important to remove the grit and gravel that washes off streets or land during storms, particularly in the cities with the combined sewer systems. Huge amounts of grit and sand entering a treatment plant can cause serious operating problems, such as excessive wear of pumps and other equipment, clogging of aeration devices, or taking up capacity in tanks that is needed for treatment. In some plants, another finer screen is placed after the grit chamber to remove any additional material that might damage the equipment or interfere with later processes. The grit and screenings can be removed by these processes must be periodically collected and trucked to a landfill for disposal or are incinerated.

12.2.2 Primary Sedimentation

With the screening accomplished and the grit detached, wastewater still contains dissolved inorganic and organic constituents along with suspended solids. The suspended solids contain minute particles of matter that can be removed from the wastewater with additional treatment such as sedimentation or gravity settling, filtration, or chemical coagulation. Gravity settling can't remove effectively the pollutants that are very fine or dissolved and remain suspended in the wastewater. When the wastewater enters a sedimentation tank, it slows down and the suspended solids gradually sink to the bottom. This mass of solids is called primary sludge. Several methods have been devised to remove the primary sludge from the tanks. Newer plants have some type of mechanical equipment to remove the settled solids from sedimentation tanks. Some plants remove the solids continuously whereas others do so at intervals.

12.2.3 Secondary Treatment

After the wastewater has been going through the primary treatment processes, it flows into the next stage of treatment called secondary treatment processes. Using biological treatment processes, secondary treatment processes can remove up to 90% of the organic matter in wastewater. The two most common conventional methods used to achieve the secondary treatment are attached growth processes and suspended growth processes.

12.2.3.1 Attached Growth Processes

Briefly, in attached growth (or fixed film) processes, microbial growth occurs on the surface of stone or plastic media. Wastewater passes over the media along with air to provide oxygen. Attached growth process units include the trickling filters, biotowers and rotating biological contactors. The attached growth processes are very effective at removing biodegradable organic material from wastewater. A trickling filter is only a bed of media (typically rocks or plastic) in which the wastewater passes. The media ranges from 3 to 6 feet deep and allows intensive microorganisms growth. In the older treatment facilities, usually, rocks, stones, or slag are used as media bed material. New facilities might use beds that made of plastic balls, interlocking sheets of furrowed plastic or different types of artificial media. This type of bed material usually offers more surface area and a better environment for promoting and controlling biological treatment than rock. Bacteria, fungi, algae and other microorganisms grow and reproduce, forming a microbial growth or slime layer (biomass) on the media. In this treatment process, the bacteria use oxygen from the air and consume most of the organic matter in the wastewater as food. As the wastewater passes down through the media, oxygen-demanding substances are consumed by the biomass and the water leaving the media is much cleaner. However, portions of the biomass also get rid of the media and must settle out in a secondary treatment tank.

12.2.3.2 Suspended Growth Processes

Similar to the microbial methods in attached growth systems, suspended growth procedures are designed to eliminate decomposable organic and organic nitrogen-containing material by altering ammonia nitrogen to nitrate unless additional treatment is provided. In the suspended growth processes, the microbial growth is suspended in an aerated water mixture where the air is pumped in, or the water is agitated sufficiently to permit oxygen transfer. The units of the suspended growth process include the variations of activated sludge, oxidation ditches and sequencing batch reactors. This process can speed up the work of aerobic bacteria and other microorganisms that decompose the organic matter in the sewage by providing a rich aerobic environment, whereas the microorganisms suspended in the wastewater can work more effectually. In the aeration tank, the wastewater is vigorously combined with air and microorganisms are acclimated to the wastewater in a suspension for several hours. This allows the bacteria and other microorganisms to break down the organic matter in the wastewater. The microorganisms can grow in a countless number and the excess biomass is removed by settling before the effluent is discharged or transferred to further treatment. Now activated with millions of additional aerobic bacteria, some of the biomass can be used again by returning it to an aeration tank for mixing with incoming wastewater.

12.2.3.3 Disinfection

Untreated domestic wastewater contains microorganisms or pathogens that can produce human diseases. Methods used to kill or deactivate these harmful organisms are called disinfection techniques.

Chlorine is the most extensively used disinfectant but ozone and ultraviolet (UV) radiation are also regularly used for the wastewater effluent disinfection.

12.2.3.3.1 Chlorine

Chlorine can kill the microorganisms in the wastewater by destroying cellular material. This chemical can be applied to the wastewater in a liquid, gas or solid form similar to swimming-pool disinfection chemicals. Still, any free (uncombined) chlorine residual in the water, even at low concentrations, is highly toxic to beneficial aquatic life. Hence, removal of even trace amounts of free chlorine by dechlorination is regularly needed to safeguard the fish and aquatic life. Nowadays, due to emergency response and potential safety concerns, chlorine gas is used less commonly than in the past.

12.2.3.3.2 Ozone

Ozone is produced from oxygen exposed to a high voltage current. Ozone is very active at destroying viruses and bacteria and decomposes back to oxygen rapidly without leaving harmful byproducts. Ozone is not very economical due to high energy costs.

12.2.3.3.3 Ultraviolet Radiation (UV)

UV disinfection happens when electromagnetic energy in the form of light in the UV spectrum produced by mercury arc lamps enters the cell wall of exposed microorganisms. The UV radiation delays the capability of the microorganisms to survive by destroying their genetic material. UV disinfection is a physical treatment process that leaves no chemical traces. Organisms can sometimes repair and reverse the negative effects of UV when applied at low doses.

12.2.4 Tertiary Treatment

Advanced treatment technologies can be extensions of conservative secondary biological treatment to more stabilize the oxygen-demanding elements, or to eliminate nitrogen and phosphorus in the wastewater. Additional treatment may also involve physical-chemical separation techniques such as adsorption, flocculation/precipitation, membranes for advanced filtration, reverse osmosis and ion exchange. In numerous combinations, these processes can reach any degree of pollution control desired. As wastewater is purified to a higher and higher degrees by such advanced treatment processes, the treated effluents can be reused for urban, agricultural irrigation, industrial cooling and landscape, processing, recreational uses, water recharge, and even for indirect growth of drinking water supplies.

12.2.5 Land Application

Biosolids are marketed to farmers as fertilizer in many areas of the world. But before land application of municipal biosolids, the minimum necessities for such land application practices should be first defined, including field management, contaminant limits, practices, monitoring, record keeping, treatment requirements and reporting requirements. Appropriately treated and applied biosolids are a good source of organic matter for improving the soil structure and help the farmers to supply nitrogen, phosphorus and micronutrients that are essential for plants. Biosolids have also been used effectively for many years as a soil fertilizer and conditioner and for restoring and revegetating areas with poor soils due to building activities, strip mining or other practices. Under this biosolids management method, treated solids in semiliquid or dewatered form are transported to the soil treatment areas. The slurry or dewatered biosolids, containing nutrients and stabilized organic matter, is spread over the land to give nature a hand in returning grass, trees and flowers to barren land. Restoration of the countryside also helps control the flow of acid drainage from mines that endangers fish and other aquatic life and contaminates the water with acid, salts and excessive quantities of metals (US EPA, 2004).

12.3 Water Quality Assessment Methods

Water quality standards are the scientific and technical information provided for a particular water quality component in the form of numerical data and/or narrative descriptions of its effects on the fitness of water for a particular use or on the health of aquatic ecosystems. Many factors affect the quality of reclaimed wastewater such as the source of wastewater (industrial or domestic), the nature of the wastes added during use and the degree of treatment that wastewater receives (Asano and Pettygrove, 1987) (Figure 12.1). Also, the soil properties that receives wastewater should be taken in consideration. The water quality in WWTPs usually relies on biochemical oxygen demand and suspended solids measurements. In contrary, there are other parameters for water quality that considered more important when used in irrigation. Several water quality guidelines have been developed by many researchers for using water in irrigation under different conditions. However, the classification of US Salinity Laboratory is most commonly used. Parameters such as electrical conductivity (EC), total dissolved solids (TDS), sodium (Na^+), Sodium Adsorption Ratio (SAR), Soluble Sodium Percent, Residual Sodium Carbonate (RSC), Boron (B) and chloride (Cl^-) were used to evaluate the suitability of wastewater for irrigation purposes. Table 12.1 summarizes the different guidelines used for conducting quality of water for irrigation.

However, reclaimed municipal wastewater has other pollutants that weren't considered in these guidelines; hence, there are different indices to classify wastewater pollution and to evaluate its suitability for irrigation purposes. These indices are as follows.

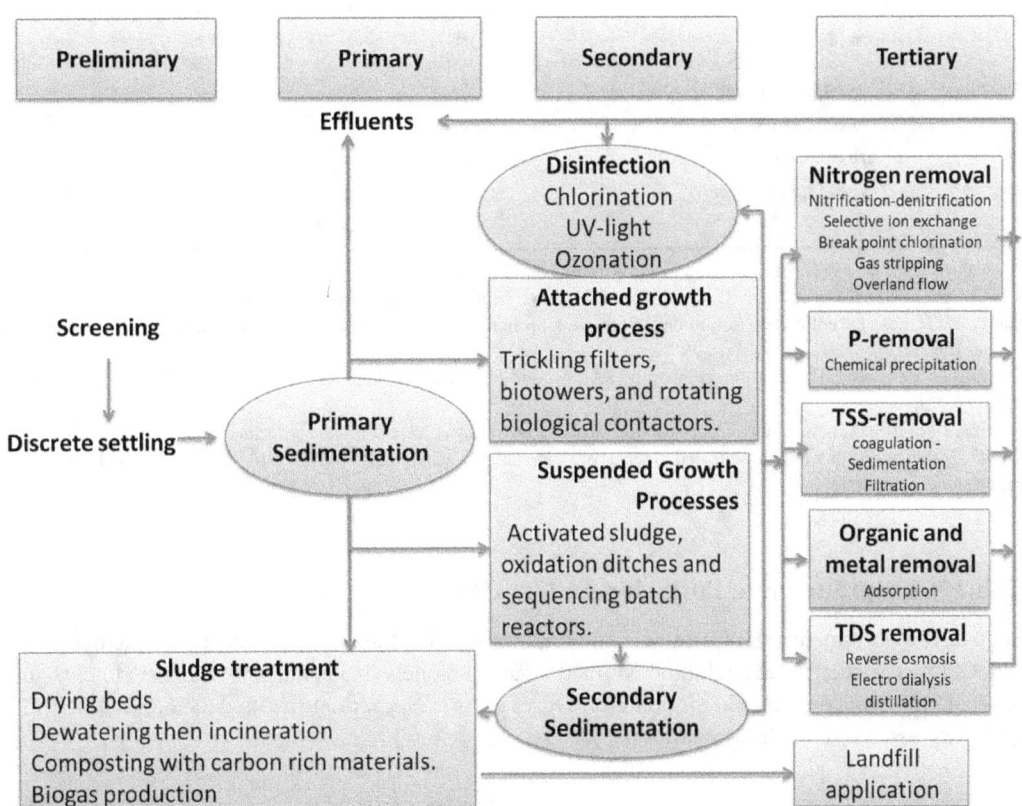

FIGURE 12.1 Schematic diagram of municipal wastewater treatment technologies.

TABLE 12.1   Guidelines for Interpretations of Water Quality for Irrigation[a]

Potential Irrigation Problem			Units	Degree of Restriction on Use		
				None	Slight to Moderate	Severe
Salinity (*Affects crop water availability*)[b]						
EC_w			dS/m	<0.7	0.7–3.0	>3.0
(or)						
TDS			mg/L	<450	450–2,000	>2,000
Infiltration (*Affects infiltration rate of water into the soil. Evaluate using EC_w and SAR together*)[c]						
SAR	=0–3	**and EC_w** =		>0.7	0.7–0.2	<0.2
	=3–6	=		>1.2	1.2–0.3	<0.3
	=6–12	=		>1.9	1.9–0.5	<0.5
	=12–20	=		>2.9	2.9–1.3	<1.3
	=20–40	=		>5.0	5.0–2.9	<2.9
Specific Ion Toxicity (*Affects sensitive crops*)						
Sodium (Na)[d]						
Surface irrigation			SAR	<3	3–9	>9
Sprinkler irrigation			me/L	<3	> 3	
Chloride (Cl)[d]						
Surface irrigation			me/L	<4	4–10	>10
Sprinkler irrigation			me/L	<3	> 3	
Boron (B)			mg/L	<0.7	0.7–3.0	>3.0
Miscellaneous Effects (*Affects susceptible crops*)						
Nitrogen (NO_3-N)[e]			mg/L	<5	5–30	>30
Bicarbonate (HCO_3)						
(*Overhead sprinkling only*)			me/L	<1.5	1.5–8.5	>8.5
pH			**Normal Range 6.5–8.4**			

    [a] Adapted from University of California Committee of Consultants (1974).

    [b] ECw means electrical conductivity, a measure of the water salinity, reported in deciSiemens per meter at 25°C (dS/m) or in units millimhos per centimeter (mmho/cm). Both are equivalent. TDS means total dissolved solids, reported in milligrams per liter (mg/L).

    [c] Evaluate the potential infiltration problem by SAR as modified by ECw. Adapted from Rhoades (1977), and Oster and Schroer (1979).

    [d] For surface irrigation, most tree crops and woody plants are sensitive to sodium and chloride; use the values shown.

    [e] NO_3-N means nitrate nitrogen reported in terms of elemental nitrogen (NH_4-N and Organic-N should be included when wastewater is being tested).

12.3.1 Comprehensive Pollution Index (CPI)

Based on the assessment of Single Factor Index and considering the combined effect of all factors evaluated, CPI was calculated through different mathematical models (Liu et al., 1999; Jiang et al., 1999 and You et al., 2009) and determined the pollution degree using an appropriate method (Guo, 2006). The CPI can be expressed by the following formula:

$$CPI = \frac{1}{n} = \sum_{i=1}^{n} C_i,$$

where n is the pollutants number and C_i is the single factor pollution index and could be calculated according to following equation:

$$C_i = \frac{\text{Measured concentration of individual pollutant}}{\text{Standard permissimble concentration of pollutant}}.$$

The water quality could be summarized according to CPI in following categories (You et al., 2009): CPI<0.8 means qualified (some pollutants are detected but their concentrations accord with the standards), 0.8<CPI< 1.0 basically qualified (concentrations of some pollutants exceed the standard), 1.0<CPI≤2.0 polluted (concentration of a part of pollutants exceeds the standard) and CPI>2.0 seriously polluted (concentration of quite a part of pollutants exceeds the standard many times).

12.3.2 Organic Pollution Index (OPI)

OPI was developed to classify the organic pollution due to organic compounds in the treated wastewater. It was calculated by dividing the values of different parameters indicating organic pollutants, such as dissolved oxygen (DO), dissolved inorganic nitrogen (DIN), biological oxygen demand (BOD), dissolved inorganic phosphate (DIP) and chemical oxygen demand (COD) by their permissible limits in irrigation water as follows:

$$OPI = \frac{BOD_m}{BOD_s} + \frac{DIN_m}{DIN_s} + \frac{DIP_m}{DIP_s} \ldots$$

The water quality is ranked in the following categories: <2: Excellent, 0–1: Good, 1–2: Begin to be contaminated, 2–3: Lightly polluted, 3–4: Moderately polluted and 4–5: Heavily polluted.

12.3.3 Carlson Trophic State Index (C-TSI)

C-TSI is a common method for characterizing a lake's trophic state or overall health. This method uses Secchi's disc transparency, chlorophyll-a and phosphorus measurements Carlson (1977). Trophic state is defined as the total weight of the biomass in a water body at a specific location and time. Trophic state is the biological response for nutrient additions to the water bodies (Naumann, 1929), but these nutrient effects may be modified by the parameters such as seasonal variations, grazing of phytoplankton by zooplankton and mixing depth of the water, etc. C-TSI mainly uses algal biomass involving three variables, namely chlorophyll-a (CA), Secchi disc depth (SD) and total phosphorus (TP). The average values of TSI of these three parameters will be considered in determining the C-TSI. The trophic state index (TSI) of Carlson is calculated using the following formulae:

a. TSI for chlorophyll-a (CA): TSI=9.81ln Chlorophyll-a (μg/L)+30.6
b. TSI for Secchi depth (SD): TSI=60−14.41ln(Secchi depth) (m)
c. TSI for total phosphorus (TP): TSI=14.42ln(total phosphorus) (μg/L)+4.15, where TSI is Carlson's trophic state index and ln is natural logarithm.

$$\text{Carlson's trophic state index (C-TSI)} = \left[TSI(TP) + TSI(CA) + TSI(SD) \right]/3$$

TP and CA are expressed in micrograms per liter and SD transparency in meters. Based on the values of C-TSI, water bodies are classified as follows: <30–40, oligotrophic; 40–50, mesotrophic; 50–80, eutrophic; >80, hyper-eutrophic; and C-TSI < 0.8, qualified.

12.3.4 Eutrophication Index (EI)

The EI is used for initially for assessment of water quality of rivers, lakes and sea, which accounts for the effects of COD, DIN and DIP. The EI greater than or equal to 1 in a water body indicates the condition of eutrophication due to excessive nutrients, while EI values <1 refer to no eutrophication. EI is calculated as follows:

$$EI = \frac{COD \times DIP \times DIN}{4,500} \times 10^{6}.$$

12.3.5 Water Quality Index (WQI)

WQI is a valuable tool to approximate the quality of water and facilitate the work of decision-makers by grouping and analyzing numerous parameters within a single numerical classification system. WQI was established by Horton (1965) in United States by selecting 10 most commonly used water quality variables such as DO, pH, coliforms, specific conductance, alkalinity (Alk) and chloride. Thereafter, a new WQI similar to Horton's index had been developed by Brown et al. (1970), which based on the value of individual parameter. Recently, many modifications have been considered for the WQI concept through various scientists and experts. Several parameters employed to calculate the WQI were pH, EC, DO, TDS, total hardness, Alk, and TP, Cl^-, NO_3, SO_4, Ca, Mg, K, B, As, Cu and Zn.

12.3.5.1 WQI Calculation Method

The WQI was established in three steps, the first of which was to normalize individual values used to design the index. This was done to establish a correspondence of the results obtained for each parameter with a variable scale of 0 to 100 based on the maximum values established in the official standards. A value of 100% indicates the optimal natural conditions, while 50% indicates the significant constraints in water use. The second step in determining the WQI was to assign the numerical weights to the parameters, which were established according to their importance in normal criteria of quality. The third step was calculating the WQI by applying the following equation (Leon-Vizcaino et al., 1991):

$$WQI = \sum Q_n \, W_n \Big/ \sum W_n,$$

where Q_n is the quality rating of nth water quality parameter and W_n is the unit weight of nth water quality parameter.

The quality rating (Q_n) is calculated using the expression given in the following equation:

$$Q_n = \left[(V_n - V_{id})/(S_n - V_{id}) \right] \times 100,$$

where V_n is the estimated value of nth water quality parameter at a given sample location, V_{id} is the ideal value for nth parameter in pure water (V_{id} for pH=7 and 0 for all other parameters) and S_n is the standard permissible value of nth water quality parameter.

The unit weight (W_n) is calculated using the expression given in the following equation:

$$W_n = k/S_n,$$

where S_n is the standard permissible value of nth water quality parameter and k is the constant of proportionality, and it is calculated by using the following expression:

$$k = \left[1 \Big/ \sum 1/S_{n=1,2,...n} \right].$$

The ranges of WQI, the corresponding status of water quality and their possible use are categorized as follows: 0%–39% indicates highly contaminated water, 40%–59% poor water quality, 60%–90% good quality and 90%–100% excellent quality (León-Vizcaíno, 1991).

12.4 Risks of Reuse of Reclaimed Municipal Wastewater in Irrigation

12.4.1 Human Health Risks

Reuse of reclaimed municipal wastewater in crop irrigation seems to have some benefits related to the areas suffering from insufficient water resources, wastewater can be used in poor areas for irrigation since it has high nutrient levels, especially of nitrate and ammonia, and thus, it can improve the productivity of crops (Ebrahimizadeh et al., 2009; Helaly et al., 2018). However, there are also negative effects resulting from pathogens and toxic compounds (organic and non-organic compounds) present in it (Jimeneza, 2006). Irrigation with reclaimed municipal wastewater resulted in a number of potential risks to human health via consumption or exposure to pathogenic microorganisms, harmful organic chemicals and heavy metals. The following are the four groups who could suffer from the harmful effects of irrigation with wastewater: (1) farmers and their families; (2) crop handlers; (3) consumers of crops, meat and milk; and (4) those living near the areas irrigated with wastewater, particularly children and the elderly. Wastewater contains a diversity of the defecated organisms, and their types and concentrations vary according to the background levels of disease in the population and the quality of WWTPs treatments. Many pathogens can survive for long enough periods of time in soil or on crop surfaces and thus be transmitted to humans or animals (WHO, 1989). Consequently, pathogenic microorganisms are considered the greatest threat to human health. Household sewage contains a high percentage of organic materials and pathogenic microorganisms, including bacteria, viruses and protozoa. The diseases associated with such infections are also diverse and include typhoid, dysentery, diarrhea, vomiting and malabsorption. Any human contact with the treated wastewater might be hazardous (Fattal et al., 1991). The process at the treatment plants reduces the level of pathogenic microorganisms, but does not eliminate them completely. This problem can be solved by desalination of the treated wastewater, but this is an expensive process and usually not required from agricultural use aspect. Different crops that are irrigated by the treated wastewater pose various threats to human health. There are some crops with pathogenic microorganisms' contamination that do not seem to pose any health risk, which are:

- Industrial crops such as cotton or fodder.
- Fruits that are dried in the sun for at least 60 days after last irrigation.
- Watermelons grown for edible grains or for seeds that are irrigated only before blooming.
- Groves or flora, without human access (Al-Mahdawi and Al-Kenany, 2009).

Regarding chemical compounds in wastewater, the major health concern is due to metals. Many of them are biologically beneficial in small quantities but become harmful at high levels of exposure.

12.4.2 Environmental Risks

Irrigation with treated wastewater might add specific contaminants, such as chlorides, to the groundwater (Babiker et al., 2004). This risk has an accumulative nature as the contaminant appears in the water supply systems, and flows to the treatment plants and back to the aquifer. Risks in this respect have a long-term influence and are difficult to evaluate. According to the health regulation, treated wastewater irrigation is prohibited in the vicinity of drinking water wells (except for effluent that does not pose any risk). Irrigation crops above the freshwater pipes can be approved, only if the treated wastewater is at the desired quality level, if the water pipe is in good condition and there is no risk of under pressure in the pipe. In addition to microorganisms, household sewage also contains substantial salt additions (Fattal et al., 1991). Irrigation with the treated wastewater causes land salinity, and also causes land sealing and sodium accumulation, which could cause increased runoff and land erosion. One particular concern of the environmental problems is the long-term sustainability issue (e.g., sodium content in soil and salinity increase). High values of soil salinity and SAR cause soil structure deterioration, a decrease in soil permeability, and a reduction of crop yields due to the toxic and osmotic effects (Halliwell et al., 2001).

12.4.3 Miscellaneous Problems

Clogging problems with sprinkler and drip irrigation systems have been reported. Slimes and bacteria in the sprinkler head, emitter orifice, or supply line cause plugging, as do heavy concentrations of algae and suspended solids. The most frequent clogging problems occur with drip irrigation systems. From the standpoint of public health, however, such systems are often considered ideal, because they are totally closed and avoid the problems of worker safety and spray drift. Excessive residual chlorine in treated effluent due to chlorine disinfection causes plant damage when sprinklers are used, and the effluent is sprinkled on foliage. Residual chlorine less than 1 mg/L should not affect the plant foliage, but when it exceeds 5 mg/L, severe plant damage can occur.

12.5 Conclusions

Water is considered as one of the main problems in arid and semi-arid countries, which is becoming an increasingly scarce resource in such areas. To meet the higher demand of water, the planners are obliged to consider any source of water that might be used effectively and economically. Recently, the methodology for managing the reuse of wastewater has shifted from conventional disposal strategies to using value-added products. With the increase of wastewater reuse for different purposes, concerns over the health and environmental effects of this reuse have also increased. WWTPs are considered as a part or whole process for the elimination of different pollutants according to application of effluents, source of wastewater, economic reasons and local regulation for the treatment. This chapter presents the different technologies used in WWTPs and their mechanisms for pollutant elimination and highlighting some effective methods used in the disinfection process.

References

Abdel-Wahab, R, Omar, M. Wastewater Reuse in Egypt: Opportunities and Challenges. Arab World. 2011, Arab Water Council Report.

Al-Mahdawi, G.T., Al-Kenany, K. Efficiency of using treated wastewater in agriculture analytical study for Al-Rustumia station for sewage clarifying – third expansion. In: *Fifth Scientific Conference "Strategies for Spatial Development and Investment in Iraq"*, The Institute of Urban and Regional Planning for Graduate Studies, University of Baghdad, Iraq, 2009.

Asano, T., Pettygrove, G.S. Using reclaimed municipal wastewater for irrigation. *California Agricultural*, 1987, 41 (3), 15–18.

Babiker, I., Mohamed, M., Terao, H., Kato, K., Ohta, K. Assessment of groundwater contamination by nitrate leaching from intensive vegetable cultivation using geographical information system. *Environment International*, 2004, 29 (8), 1009.

Bahri, A. Present and future state of treated wastewaters and sewage sludge in Tunisia. In: *Proceedings of Wastewater Reclamation and Reuse*, Cairo, Egypt, 11–16 December 1988.

Brown, R.M., McClelland, N.I., Deininger, R.A., Tozer, R.G. Water quality index-do we dare? *Water Sewage Works*, 1970, 117 (10), 339–343.

California State Water Resources Control Board (1990). *California Municipal Wastewater Reclamation in 1987*. California State Water Resources Control Board, Sacramento, CA.

Carlson, R.E. A trophic state index for lakes. *Limnology and Oceanography*, 1977, 22(2), 361–369.

Ebrahimizadeh, M.A., Amiri, M.J., Eslamian, S.S., Abedi-Koupai, J., Khozaei, M. The effects of different water qualities and irrigation methods on soil chemical properties. *Research Journal of Environmental Sciences*, 2009, 3 (4), 497–503.

Fattal, S., Shuval, H., Wax, Y., Davies, A. Study of enteric disease transmission associated with wastewater utilization in agricultural communities in Israel. In: *Proceedings of the Water Rescue. Symposium II*, vol. 3, AWWA, Denver, USA, 1991.

Guo, M.M. Application of mark index method in water quality assessment of river. *Environmental Science and Management*, 2006, 31 (7), 175–178.

Halliwell, D., Barlow, K., Nash, D. A review of the effects of wastewater sodium on soil physical properties and their implications for irrigation systems. *Australian Journal of Soil Research*, 2001, 39, 1259–1267.

Helaly, M.N., El-Sheery, N., El-Hoseiny, H., Rastogi, A., Kalaji, H. Impact of treated wastewater and salicylic acid on physiological performance, malformation and yield of two mango cultivars. *Scientia Horticulturae*, 2018, 233, 159–177.

Horton, R.K. An index number system for rating water quality. *Journal of Water Pollution Control Federation*, 1965, 37(3), 300–305.

Jiang, H.H., Zhu, J.P., et al. The relationship between comprehensive pollution index assessment and water quality category decision. *Environmental Monitoring in China*, 1999, 15(6), 46–47.

Jimeneza, B. Special feature on groundwater management and policy, irrigation in developing countries using wastewater. *International Review for Environmental Strategy*, 2006, 6 (2), 229–250.

León-Vizcaíno, R. Water Quality Indices, Ways to Estimate Them and Application in the Lerma Chapala Basin, Mexico; Technological Institute of Water: Morelos, Mexico, 1991; pp. 1–7.

Liu, S., Zhu, J.P, Jiang, H.H. Comparison of several methods of environment quality evaluation using complex indices. *Environmental Monitoring in China*, 1999, 15 (5), 33–37.

MWRI (2010). A summary of the Egypt national strategy for the development and management of water resources until 2050.

Naumann, E. The scope of chief problems of regional limnology. *Internationale Revue der gesamten Hydrobiologie*, 1929, 21, 423.

Oster, J.D., Schroer, F.W. Infiltration as influenced by irrigation water quality. *Soil Science Society of America Journal*, 1979, 43, 444–447.

Pescod, M.B. Wastewater Treatment and Use in Agriculture. F.A.O. Irrigation and Drainage Paper 47, Italy, 1992.

Prasad, A.G.D., Siddaraju, P. Carlson's Trophic State Index for the assessment of trophic status of two Lakes in Mandya district. *Advances in Applied Science Research*, 2012, 3, 2992–2996.

Rhoades, J.D. 1977. Potential for using saline agricultural drainage waters for irrigation. In: *Proceedings of Water Management for Irrigation and Drainage*. ASCE, Reno, Nevada, USA, 20–22 July 1977, pp. 85–116.

Shelef, G. The role of wastewater in water resources management in Israel. *Water Science and Technology*, 1990, 22, 2081–2089.

Shelef, G., Azov, Y. The coming era of intensive wastewater reuse in the Mediterranean region. In: *Proceedings of the 2nd International Symposium on Wastewater Reclamation and Reuse*. International Association on Water Quality, Iraklio, Crete, Greece, 17–20 October 1995, 1, pp. 138–146.

Strauss, M. and Blumenthal, U.J. (1989). Human Waste Use in Agriculture and Aquaculture, Utilization Practices and Health Perspectives. I.R.C.W.D., Report No. 08/09. International Reference Centre for Waste Disposal, Dubendorf, Switzerland.

United States Environmental Protection Agency (US EPA). Primer for Municipal Wastewater Treatment Systems. EPA 832-R-04-001, September 2004.

University of California Committee of Consultants. Guidelines for Interpretation of Water Quality for Agriculture. University of California, Davis, USA, 1974, 13 p.

WHO, World Health Organization. Guidelines for the Safe Use of Wastewater in Agriculture. WHO, Geneva, Switzerland, 1989.

You, Z.J., Lu, J.X., Liu, Y.Y. The improvement of calculation formula of comprehensive pollution index. *Neimenggu Environmental Science*, 2009, 21 (2), 101–102.

Microbiology of Irrigation Water

T. C. Crusberg
Worcester Polytechnic Institute

Saeid Eslamian
Isfahan University of Technology

13.1 Introduction

Today, people worldwide are understandably concerned if not significantly worried about their own health security, as pathogens periodically threaten their daily lives. Those pathogens are designed, by nature through their own genetics, to evolve ways to infect the uninfected and reproduce identical progeny which go on to infect still more uninfected hosts. Most higher organisms have ways to deal effectively with pathogens, through their very complex immune systems. In most cases, infected individuals recover from an illness, often with minor but unfortunately sometimes with major disabilities, and the pathogen plays out its role, replicates into more of its own kind, and goes on to infect others. Water is a well-known and proven vehicle that can deliver pathogens from infected/sick to healthy individuals (Crusberg, 2014). In general, water-borne diseases can be very stressful for an infected person who usually does recover, yet mortality may sometimes be the result.

One of the most serious effects of ingested water-borne pathogens is severe diarrhea, leading to dehydration and corresponding electrolyte imbalances in the blood, thereby affecting kidney function, which can then lead to organ failure and ultimately, mortality. Water used for irrigation of crops as well as for recreational purposes is one such source of those pathogens. Most water-borne diseases affect the intestinal tract of the host, resulting in, as stated above, diarrhea, which in the presence of poor hygienic conditions, can be released into the environment and the pathogens in the fecal matter can be transmitted to and infect other susceptible individuals that consume the water. Special consideration has to be given to those crops whose edible parts come directly in contact with irrigation water, such as lettuce and any plant growing close to the ground surface.

DOI: 10.1201/9780429290114-18

Microbiological characteristics of irrigation water, just as water quantity, availability, dissolved solids, and sediment load are all equally important considerations, and together certainly should be assessed by both water provider and water user. Microbial content of irrigation water is divided into various classes, notably bacterial, algal, protozoan, fungal, and viral, and a few agricultural specialists choose to include some groups of microscopic multicellular organisms. Greatest concerns in most every case center on water-borne pathogens specific to humans, animals, and even crop plants. Identifying pathogens in water in any source can be very difficult, time-consuming and of course costly. A great many scientific studies have been reported on the multitude of microbes found in irrigation waters, with special consideration given to water-borne pathogens.

13.2 Microbiology of Irrigation Water

The world's population living in water-stressed areas is projected to reach 44% by 2050 (Scheierling et al., 2011). As the world faces increasing freshwater scarcity, wastewater use is gaining attention as an option for augmenting available water supplies. Agriculture irrigation is by far the most established application of wastewater reuse in the world (Scheierling et al., 2011). Yet, agriculture still accounts for over 70% of the world's total freshwater withdrawal (FAO, 2012).

This volume presents many practical ideas on irrigation waters and the technologies related to the delivery of those waters as a valuable and strategic product. The origins of irrigation water usually define the microbiological quality of that water. Potable water supplies of high-quality water are usually reserved for use by municipalities, for populations in nearby cities and towns, and only about 10 to sometimes as much as 30% of all available water is in fact used this way, serving domestic needs such as drinking, bathing, and hygiene, retail and commercial needs, and the needs of industry and manufacturing. The lions' share of available freshwater, actually somewhat above 70%, goes to agricultural purposes, growing crops and raising livestock (FAO, 2012).

Although some are fortunate enough to enjoy access to potable water requiring no treatment, mostly from deep well sources, others whose drinking water is from surface and shallow well supplies must realize that their drinking water is susceptible to gross contamination (Crusberg, 2014; Crusberg and Eslamian, 2016). Runoff from fields and pastures carries with it large amounts of animal fecal material. Water from shallow wells has to also be suspect in terms of carrying pathogens, since it is subject to septic leakage from on-site domestic and small-scale sewage treatment systems. As the world's population living in water-stressed environments expands in the near future, more and more dependence on recycled and certainly treated municipal wastewater will be required (Scheierling et al., 2011). Reclaimed municipal wastewater has been shown to provide some nutrient value for crops and a year-round irrigation source in water-stressed areas (Dreschel et al., 2010). Although the modern disinfection technologies used to treat municipal wastewater are quite efficient in removing most pathogens, some pathogens do manage to avoid inactivation and escape into the environment. It is these fugitive pathogens we have to worry about!

13.3 Pathogens in Irrigation Water

The primary sources of any and all water-borne pathogens are those infected organisms which can pass those pathogens on to others, humans and other animals alike. Some pathogens have a very narrow host range limited to a single species or two, while other pathogens have a broad host range able to infect a multitude of hosts. For example, the organism that causes giardiasis *Giardia lamblia*, is found in the feces of most bovine calves at some point in their early life, and runoff from a grazing area, if consumed by a human, can result in the human also acquiring the disease. Even the most aggressive water treatment can fail to inactivate that protozoan parasite (Eslamian, 2016).

Feces derived from agricultural runoff, wildlife, cattle and most other types of livestock, the discharge of domestic wastewater, even that subject to treatment, and septic leakage can pass pathogens directly onto soil and vegetables during irrigation (Bea Clarise et al., 2015; FAO, WHO, 2008; Ijabadeniyi and Buys, 2012; Pachepsky et al., 2011; Tracogna et al., 2013). Preharvest contamination of fresh produce by irrigation waters has been well documented. A strain of *Escherichia coli* (O157:H7) (also known as enterohemorragic *E. coli*) bearing the gene for shiga toxin, causing severe and potentially lethal diarrhea due to acute dehydration, was implicated in sickness from consuming vegetables grown in the Central Valley of California, and *Salmonella saintpaul* was the cause of an outbreak in Mexico probably related to the consumption of Serrano peppers (Benjamin et al., 2013; Solomon et al., 2002).

One important issue for growers is to understand that pathogens can indeed be transmitted to vegetables through their attachment to plant surfaces and internalization through the root, stomata, or through tissue damage, compromising fresh produce and rendering it unsafe for consumption (Berger et al., 2010; Donkor et al., 2010). Irrigation water has been shown in several studies to play an important role in the preharvest contamination of fresh produce (Castro-del Campo et al., 2012; Gerba and Choi, 2006; Park et al., 2013; Van der Linden, et al., 2019). Even plant pathogens, like those of the genus *Phytophthora*, which has at least 98 species, are also of concern, especially for nurseries and greenhouses that utilize the recycled irrigation water. If infested water is applied to crops, the risk of widespread epidemics, crop damage, and lost profits are increased (Gallegly and Hong, 2008; Hong and Moorman, 2005; Hong et al., 2008; Hong et al., 2010).

Rather than losing the treated wastewater effluent by passing it into a receiving body, it is customary in some water-stressed areas to divert the municipal wastewater to local agriculture, often limited to recreational uses such as golf courses and public parks. In arid and semiarid areas around the world, water shortage is one of the most serious environmental problems, which necessitates the search for alternative sources of good-quality water to satisfy this demand. Wastewater reclamation or recycling for the purpose of irrigation provides a readily available alternative water sources when natural water resources are few and is an increasing in many places around the world (Hamilton et al., 2007; Qadir et al., 2007).

Surely, even with advanced wastewater treatment, microorganisms and even human enteric viruses in the discharge can still survive. A number of studies have shown that pathogens can indeed be found in treated wastewater (Santamaria and Toranzos, 2003; Gerba and Smith, 2005). These may include pathogenic bacteria such as *Escherichia coli* O157:H7 (mentioned previously), *Salmonella* spp., *Listeria monocytogenes*, *Campylobacter* spp., *Clostridium botulinum* (agent of botulism, a neurological disease), *Clostridium perfringens* (agent of gas gangrene), *Shigella* spp., various Streptococci, parasites such as *G. lamblia*, and the protozoans *Cryptosporidium parvum* and *Entamoeba histolytica*, as well as several human pathogenic viruses (Ibekwe et al., 2013; Ibekwe et al., 2018; Luo et al, 2015; Scallan et al., 2011). It should be noted that the *Clostridium* spp. mentioned are ubiquitous in nature and their presence in any surface water is expected, and yet they are only pathogenic under special circumstances not normally encountered during normal food preparation or consumption.

Indeed, the human norovirus (HuNoV) is a leading food-borne pathogen often associated with enteric disease outbreaks on cruise ships, and in the United States it is responsible for about 60% of food-borne illnesses (severe diarrhea) annually (Wei et al, 2010; 2011). It should be noted that this virus is quite often the cause of severe outbreaks in confined living environments such as on cruise ships. Recently, there have been increasing outbreaks of HuNoV associated with vegetables and fruits, much of which is consumed raw or with minimal preparation that would help remove or kill this virus (Hamilton et al., 2007; Qadir et al., 2007). Wastewater treatment does not guarantee the removal of HuNoV prior to use of the water for irrigation purposes, for recreational or for food production (Blatchley et al., 2007; Bosch, 2007; El-Senousy et al., 2013; Rusinol et al., 2015). There is one strategy to reduce the likelihood of having HuNoV contaminate a crop and that is to use high-quality water during the last irrigation period prior to harvest and to employ drip irrigation rather than broadcast overhead systems (Hamilton et al., 2006).

13.4 Detecting Pathogens in Irrigation Water

The previous discussion centered on identifying pathogens in irrigation water, without stressing that such assay methods are quite complicated not to mention expensive and certainly not available in most municipal laboratories. Older methods for bacterial detection required the use of culture methods which first involved obtaining a sample of water of interest, and transferring a small volume (1–2 mL) to a 15 cm dia. bacteriological plate (Petrie dish) containing an agar medium (nutrients and mineral salts) known to grow the individual microbial cells of concern into visible "colonies". Sometimes it was necessary to choose colonies which were suspected of being composed of pathogens and using a sterile needle place cells on other media known to support growth. This process could take several days, since overnight growth in an incubator was needed for enough cells to reproduce to be able to visualize a grown colony.

Then, with the advent of molecular biology and the technology that allowed the genetic material of a pathogen to be sequenced, more sophisticated methods were developed. Since the genome of each species of life is unique, a known sequence of genetic material specified exactly, a specific species. All genomes of all microorganisms are composed of DNA, but pathogenic viruses can have either DNA or RNA as genetic material. For RNA viruses, such as the group known as the Noroviruses, which cause enteric infections that result in cases of severe diarrhea, the RNA genome cannot be easily sequenced. However, with studies a half century ago on what are called "retroviruses", an enzyme was coded in those virus genomes that could convert the viral RNA into DNA, which of course can be sequenced. Over the years molecular biologists found numerous enzymes that when added to DNA or RNA with proper, reagents, at specific temperatures, for certain periods of time could make a short sequence of a specific region on a bacterial or viral genome in large enough amounts that could be quantified using sophisticated technologies. A number of biotechnology companies took up the challenge and developed automated methods to determine (mostly) tests for human pathogens. One example of the speed that such a test could be developed is for the presence of the virus that causes covid-19 in the human nasal passage and is the causative virus known as SARS-2 which has a genome of RNA. A short sequence of about 300 contiguous ribonucleotides of the several thousand ribonucleotides that make up the genome is sufficient to allow identification of the virus. It is not necessary to target the whole genome since the 300 ribonucleotide bases are specific only to that virus and are thus employed using polymerase chain reaction technology. Such a test can be done in less than one hour if necessary. This is the same strategy used in developing the assays for several enteric pathogens of concern in agriculture. Fuzawa et al. (2019) used molecular methods to assay for a Norwalk-like virus after treatment with chlorine.

13.5 Detecting Indicators of Pollution in Irrigation Water

Worry over pathogens in water has sparked research that has led to the development of a number of sophisticated testing methods to identify the particular species of concern. Many countries have issued certain criteria for irrigation water. A rapid and rather inexpensive approach is actually to assay, not for any specific pathogen, which may or may not be present in the sample, but for a commonly found type of (usually) harmless resident organism present in most if not all gut tracts (intestines) of warm-blooded animals and indicative of fecal pollution in the water source of concern. Chosen for this test is the bacterium known as *Escherichia coli*, an intestinal resident found in almost all humans and other warm-blooded animals, and eventually excreted in the animal's feces (Crusberg and Eslamian, 2016).

The most used assay (U.S.E.P.A., Office of Water, 2002) involves passing a fixed volume (usually 100 mL) of the water in question through a porous 47 mm dia. plastic filter (always using sterile technique), and then placing the filter into a sterile plastic dish containing a small volume of a special medium (Known as MI) able to support the growth of *E. coli* and other "coliforms" and if present, the individual bacterial cells in the sample will multiply geometrically and yield a small "colony" of cells, after an overnight incubation at 35°C. An *E. coli* colony will appear blue in ambient light, while other coliforms will appear blue under long-wavelength ultraviolet (UV) light (366 nm). Other bacteria also capable of using the same medium and grow form similar colonies, but they do not develop any noticeable color.

Coliforms are a group of bacterial species which can also be found in the intestines of warm-blooded animals and the term really is an "operational" definition, and does not actually identify a specific bacterial genus or species. If more than 200 colonies are present on the filter it is recommended that another water sample, but of lower volume be run; otherwise colony overlap can skew the results. In fact some coliforms may only be derived from soil detritus where they too can grow, and have no animal origin at all. However, it is customary to count all coliforms as having an animal origin. Then why count these colonies? From a purely public health perspective, it is "better to error on the side of safety, than on the side of convince". Therefore, a positive test suggests the presence of fecal contamination in the water sample under consideration. Historically, there is an assay for what are called "fecal coliforms" (FCs) but *Klebsiella* spp., not necessarily of intestinal origins, yield similarly colored colonies in that test, and that test is no longer recommended by many public health agencies. Another assay for a fecal inhabitant, known as "fecal streptococci", is usually reserved for testing saltwater samples for enteric bacteria, certainly not important in a discussion on irrigation water.

Guidelines regarding microbial water quality for irrigation water vary widely depending on the governmental regulatory agency. A limit of 1,000 FCs per 100 mL sample of water has been recommended by the World Health Organization (WHO) for unrestricted irrigation (Shuval, 2007), and the Canadian Water Quality Guidelines for Irrigation is more stringent and suggests a maximum allowable count of 100 FCs per 100 mL and 1,000 total coliforms per 100 mL of irrigation water (Jones and Shortt, 2005). In the United States the U.S Environmental Protection Agency under the Clean Water Act has been regulating water quality for human and environmental health for decades yet there are currently no microbial indicator standards established for irrigation waters (Rock et al., 2019). State and local governments in the United States establish water reuse regulations, including those governing the use of recycled municipal wastewater (Bastian and Murray, 2012; FDA, 2013). Regulations in France specify the several levels of FCs from 200 to 1,000/100 mL depending on the crop, and Spain limits irrigation water to 1,000 FC/100 mL and less than 1 nematode per liter, as examples (Vigneswaran and Sundaravadive, 2004).

The public health impact of the transmission of viruses in water is significant worldwide. Waterborne viruses can be introduced into our recreational and finished drinking water sources through a variety of pathways ultimately resulting in the onset of illness in a portion of the exposed population. Although there have been advances in both drinking water treatment technologies and source water protection strategies, water-borne disease outbreaks (WBDOs) due to viral pathogens still occur each year worldwide. By highlighting the prevalence of viral pathogens in water as well as (1) the dominant viruses of concern, (2) WBDOs due to viruses, and (3) available water treatment technologies, the goal of this review is to provide an insight into the public health impact of viruses in water.

13.6 Making a Case for/Against an "Indicator" Organism

Now, it has been shown in the previous discussion that irrigation water can carry human and even plant pathogens to growing crops. Sometimes irrigation water cannot meet the standards for microbial safety imposed by regulatory bodies. For the most part, most standards rely on a laboratory testing protocol that quantifies the presence of *E. coli* in irrigation streams. *E. coli* serves as an "indicator" of fecal pollution, and *E. coli* has been described as a suitable indicator to predict the prevalence of pathogenic bacteria in irrigation water (Castro-Ibáñez et al., 2015; Ceuppens et al., 2015).

Even though viruses are not the only pathogens present in water that can cause disease, the risk of illness is 10–10,000 times greater for viruses than bacteria at a similar level of exposure (Haas et al., 1993). Certainly, human enteric viruses are clearly a concern for both municipal drinking water and treated municipal wastewater. Viral assays, however, are expensive and require highly trained personnel and expensive equipment. Also, there are many different human enteric viruses carried in water, including hepatitis A virus (Crusberg et al., 1978). If it were desired to carry out a comprehensive assessment of the presence or absence of all of the possible human pathogens in any given water sample, it would be necessary to conduct a multitude of tests for each virus and bacterial species. Certainly, this cannot be done. We still have to rely on a bacterial indicator system to alert us if there is a potential contamination issue. The use of bacterial indicators has been debated for decades, and we are still at a stalemate when it comes to actually predicting the presence of human viral pathogens using either *E. coli*, or the group known as coliforms. Yet, it's the best we have so far.

13.7 Testing for Specific Pathogens in Irrigation Water

If further knowledge of specific pathogens that may be present in a given irrigation water supply is then desired, usually only analysis of their genetic material (DNA or RNA) will provide an accurate accounting of their presence. Often this kind of test is non-quantitative, meaning that it can only determine presence or absence of the pathogen.

Now, it has been shown in the previous discussion that irrigation water can carry human and even the plant pathogens to growing crops. Sometimes irrigation water cannot meet standards for microbial safety imposed by the regulatory bodies. For the most part, most standards rely on a laboratory testing protocol that quantifies the presence of *E. coli* in irrigation streams, as has been previously stated. However, it may be beneficial at times to be able to determine if actual pathogens are present in a specific irrigation water supply, and performing cultural assays are often more difficult and time-consuming. With the technology that allows a genome of a microbe to be determined, by analyzing the actual sequence of the genetic material of an organism or virus, newer testing methods now rely on "genetic methods" for the determination of the presence of pathogenic microbial species.

One example was reported by Fuzawa et al. (2019) studying the action of chlorine disinfection on a Norovirus surrogate, which like Norovirus has a double-stranded RNA genome. Once the virus preparation has been dissociated into its protein and nuclei acid components it has to be first treated at 95°C to separate the two RNA strands, in preparation for conversion of a short sequence on one of the strands to DNA. This reported assay provided a means to determine virus–receptor binding using a cell-free system. In the experiment they quantitatively enumerated virus particles attached to porcine gastric mucin bound to magnetic beads, with the assumption that chlorine treatment of virus particles would/should prevent their adhesion to a possible gut target receptor. At a CT (concentration of chlorine in mg/L x contact time (min.), and discussed in greater detail later in this chapter) of 16 there was a 3 log (1,000 fold) reduction in virus-to-mucin binding, suggesting that chlorine inactivates the virus surrogate by damaging a viral surface protein important in virus attachment to the receptor. El-Senosy et al. (2013) reported an assay for Norovirus comparing two virus recovery methods, one using polyethylene glycol and the other organic flocculation, and then used a molecular approach to quantify virus particles in their samples. They found that in irrigation waters tested almost one-third contained measurable amounts of Norovirus.

Luo et al. (2015) used a combination of standard isolation protocols and molecular identification techniques to identify *Salmonella enterica* strains from 10 irrigation ponds in the southeastern United States. Then they employed molecular DNA sequencing to distinguish the genetic differences between each strain. Using standard cultural methods, they recovered 191 colonies from those ponds and tested each one using DNA sequence analysis. In general, most reagents used were purchased from biotechnology suppliers and DNA sequencing used an automated commercially available machine. They also, but not really surprisingly, found that 20% of the isolates expressed multiple drug resistance, which strongly suggests a human influence in the presence of those strains of the bacterium in the irrigation waters. However, streptomycin resistance, which was prevalent in the strains, could have been naturally acquired since it is a product of the other soil bacteria which co-inhabit the same environment.

Fremaux et al. (2009) were tasked with validating a commercially available and government approved *E. coli* identification system for the examination of irrigation water in Canada, where a limit of 100 coliform/100 mL are permitted. They isolated a number of *E. coli* species using a commercial medium that provides the direct recovery and fluorescent identification of that organism, but wondered what the colonies that gave a similar but not exactly the expected *E. coli* signal actually were. They isolated the ribosomal 16S RNA genes, which are unique for all bacterial species, and subjected those specimens to DNA sequencing. They identified most of those colonies as *Klebsiella planticola* and *Shigella dysente-riae*, which accounted for only 1.5% of all samples. In the end they judged the commercial methodology of interest (the Colilert-18/Quanti-Tray system) as suitable for assessing microbial quality of irrigation water and being a method requiring minimal instrumentation and technical expertise.

Hong et al. (2010) was able to identify a new species of the fungus *Phytophthora hydropathica* found in the irrigation water at two nurseries during warm summer months, by sequencing a segment of its DNA, thereby showing that it was not either of its closest relatives, *Phytophthora parsiana* or *Phytophthora irrigata*. The new species that grew at relatively high temperatures, with an optimum of 30°C and a maximum of 40°C, caused leaf necrosis and shoot blight of the ornamental *Rhododendron catawbiense* and collar rot of *Kalmia latifolia*. Sequencing was performed at the university's advanced genetic technologies center, and not by the investigators. In general, the degree of competence for this kind of effort is not generally available in the labs of principle investigators, but is often available for entire institutions at a "center" with well-trained staff and the appropriate instrumentation.

Certainly one pathogen of great concern is *E. coli* O157:H7 and its presence in soils, irrigation water, and even associated with plant tissue of agricultural crops has been subject to identification using molecular methods (Ibekwe et al., 2004). The designations following the organism's name refer to certain genes carried in its genome. They used real-time polymerase chain reaction or PCR to validate that assay as useful in assessing the presence of that pathogen in the irrigation water and on the plant tissue prior to harvest. In this method, DNA is initially extracted from all bacteria captured in a water sample or plant tissue sample and only the DNA of the specific strain (which is known from previous sequencing studies and found in databases) is artificially replicated using special, buy commercially available enzymes and a device to carry out the synthesis of many strands of only the DNA that the investigator is interested in. The amount of DNA found is related to the population of bacteria of concern in the sample.

Indeed, there are ways to identify microorganisms to the species level using standard brute force cultural methods, but with the advent of molecular technologies over the past 4 decades that can actually provide a DNA (or RNA) sequence, an exact match can be found since almost every organism has had its genome sequenced by now. The unknown sequence of an unknown organism can then be compared to the known sequences found in online databases available to everyone at no cost. Even subtle difference in genome sequences, of one or two nucleotide bases, can be determined, allowing the identification of sub-species. True, sequencing genomes does not come cheaply, but local or national genomic centers are often able to provide assistance and even do most of the work, should identification of a microorganism be needed.

13.8 Abatement of Pathogens in Irrigation Waters

Pathogens are everywhere! Water for human consumption of course is highly regulated by public health authorities, but not so much for water used in irrigation. In all likelihood, almost any water supply used for any purpose would yield evidence of pathogens if appropriate tests were conducted. Without question, irrigation water, as is the case for potable water, should be provided as free of pathogens as is possible. A review of methods available for control of microbes in irrigation water has appeared recently (Raudales et al., 2014). The reality, therefore, is to first choose a water source if at all possible that meets the needs in terms of water quantity and quality. The first to seek out water of quality based upon its clarity and appearance and record it in report form were of course the Romans (Herschel, 1973). If pathogens are of concern, then some form of removal or deactivation will be required. A multitude of technologies are at hand that can accomplish this, yet the goal for any testing regimen is of course to meet regulatory levels which do not really name specific pathogens themselves, but rather name indicator organisms and their numbers per unit of water volume (usually 100 mL) that strongly suggest the level of pollution.

For the most part, technologies cost money. Irrigation volumes are usually large. Treating potable water for human consumption is always cost-effective and many technologies used to provide safe water for human use are quite inexpensive anyway, and many of those approaches lend themselves to disinfecting irrigation water as well. It is important to understand that the term "disinfection" here means only the inactivation of pathogens, not the killing of every single organism and virus in a given water source, which is termed "sterilization". Selection of a certain technology depends on both the costs associated with the chosen process, and the goals needed for providing a defined maximum number of certain measurable organisms that assure a preordained level of safety. This section will review a number of technologies known to be capable of reducing the presence of indicator organisms (non-pathogenic bacteria that are known to inhabit the gut tract of warm-blooded organisms) to satisfactory level in irrigation water.

Considering the yet still unused water available from municipal wastewater treatment plant effluents it is necessary to also be aware of its likelihood of carrying pathogens (Dandie, et al., 2019). Today, secondary wastewater treatment is standard practice as an environmental management strategy worldwide. The typical quality of such wastewater effluents is estimated as containing 10^5 total coliforms/100 mL (Moulin et al., 2010; Levatensi et al., 2010). This level must always be reduced according to the standards set by public health and agricultural agencies that oversee that particular irrigation authority. Many ways are available to accomplish this task, and the most widely used methods of disinfection will be discussed below.

13.8.1 Disinfection with Chlorine and Hypochlorite

Chlorine gas and sodium or calcium hypochlorite provide the inexpensive and well understood methods for disinfection. When added to water chlorine undergoes a disproportionation reaction, forming hypochlorous acid (HOCl) and HCl. The actual disinfection chemical species is the anionic form or OCl⁻. The pKa of the weak acid HOCl requires a pH of the water to be above 7 for HOCl to dissociate into hypochlorite anion and a proton. The process of disinfection is dependent upon several parameters, concentration of the chemical in use, the time that the organisms are in contact with the chemical, pH of course, and temperature. Also of concern is any substance in the irrigation water that can react with the chlorine or hypochlorite and effectively reduce the concentration of disinfectant, such as naturally occurring tannins and humic acids from soil, and some nitrogenous compounds. The product of the concentration of chlorine and time is termed the CT value, the value that gives 90% inactivation of the microbe of concern. Considering *E. coli*, for example at 20°C and at a pH between 6 and 7, the CT has been found to be in the range 0.034–0.05, but for cysts of the protozoan *G. lamblia*, the CT under the same conditions ranges from 15 to 21 (Crusberg, 2014), meaning that it takes 400 to 500 times longer to kill cysts of the protozoan than to kill the so-called indicator (of pollution) bacterium! The bacterial indicators present in water in the presence of small amounts of chlorine (2–4 mg/mL in drinking water) will not be found in the standard assay, but the protozoan pathogen will not be harmed at all! Disinfection still remains perhaps the most important treatment method to prevent the diseases of pathogens found in the water (O'Shea and Field, 1992; Ghernaout and Elboughdiri, 2020a). With respect to enteric virus disinfection, Fuzawa et al. (2019) showed that low levels of chlorine are effective in inactivating a Norwalk-like virus surrogate. Chlorine gas remains the most frequently used disinfectant, which is extensively used in drinking water treatment plants along with other more sophisticated processes that remove not only bacteria but can also remove the protozoan parasites (Ghernaout, 2017).

There are problems, however, when using chlorine as disinfectant, due to the fact that several chemical by-products of serious health concern form from reactions between the disinfectant and naturally occurring organics often found in water (Ghernaout and Elboughdiri, 2020b). Analysis has shown that several chlorinated disinfection by-products can readily form, namely trichloroethane, dichloroacetic acid, trichloroacetic acid, choral hydrate, dichloroacetonitrile, trichloronitromethane, and dichloroacetamide (Chu et al., 2014). Typically, chlorine gas injection for disinfecting potable surface waters and rainwater is usually in the range of 2 mg Cl_2/L (also stated as 2 parts per million or 2 ppm Cl_2), greater disinfectant levels most likely will lead to an increased production of these chemical species (Chu et al., 2016). For this reason, raising chlorine levels in water known to harbor these pathogens in the hope of inactivating the protozoan population is not recommended.

If a goal of attaining 100–200 coliform/100 mL is the case for a typical irrigation water that is ready to be applied to a crop, then in all likelihood, only low chlorine levels for disinfection may be required anyway, unlike water for human consumption which may have to demonstrate 0–1 coliform/100 mL in routine analyses. In the United States the state of Arizona has laid out regulations for the agricultural use of treated domestic wastewater and it is especially noteworthy that strict limits of FCs are expected when that irrigation water is used for food crops, landscape irrigation, and spray irrigation of orchards and vineyards. When reclaimed domestic wastewater is used for farm animal care, however, the levels of FCs permitted are much greater, as would be expected (Rock et al., 2019).

Once chlorine is introduced into the irrigation water it immediately reacts with the chlorine demand substances, if any, also in the water. If any chlorine is left after these reactions, the chlorine is called "free available chlorine" or "residual chlorine". This free chlorine is what actually serves as the disinfectant and inactivates pathogens.

13.8.2 Chlorine Dioxide as Disinfectant

Chlorine is usually supplied as a "bottled" gas made by the electrolysis of a sodium chloride solution, hypochlorite is provided as a solution, usually of the sodium salt, and calcium hypochlorite as a solid. A pH above 7 is required for optimal disinfection. However, another compound, chlorine dioxide ClO_2, is produced economically by reducing chlorate (as the sodium salt) in a sulfuric acid solution, but in situ (on site) and has, according to some critics, better features as a disinfectant than chlorine. It is used by the paper pulp making industry as a bleaching agent, but also plays a role in water disinfection. Chlorine dioxide yields lower levels of organic disinfection by-products in comparison to free chlorine. However, approximately 70% of the chlorine dioxide applied in water treatment is converted to chlorite – a regulated inorganic disinfection by-product with a maximum contaminant level of 1 mg/L in the United States. As a result, the chlorine dioxide dose applied to drinking water is typically limited to 1.4 mg/L

The choice of ClO_2 over chlorine is that the latter results in the formation of fewer types and lesser amounts of organo-halogenated by-products (López-Gálvez et al., 2010; Van Haute et al., 2015, 2017; Veschetti et al., 2003). However, ClO_2 can undergo reactions in water that lead to chlorites (ClO_2^-) and chlorates (ClO_3^-) (AHDB, 2016). Yet studies have shown that ClO2 shows a more favorable bactericidal capability compared to chlorine (Hassenberg et al., 2017). As in the case of chlorine, and most other disinfectants, the bacteriocidal effectiveness of ClO2 depends on dose of the chemical, contact time, water temperature, pH, and organic load (Junli et al., 1997; Ayyildiz et al., 2009). However, as one example of the use of ClO_2 to disinfect and recycle treated municipal wastewater, Decola et al. (2019) found a decrease in the number of coliforms on baby lettuce treated with ClO_2 compared with untreated wastewater alone. Troubling though was the presence of chlorates on the lettuce plants. No difference was observed in quantifying viable but non-culturable *E. coli* comparing disinfectant-treated and untreated wastewater. The possibility that bacterial cells become stressed after disinfection and do not form colonies on the medium that is used to identify coliforms has been of concern. The tact one has to take is to "pre-condition" the cells that were treated with a given disinfectant on a filter first on a richer commercial medium, and then transfer the filter to medium that allows the identification of coliforms. This protocol was not followed in this report, so although such a statement is valid, it was not backed up by experiment. On the other hand, the presence of chlorate anion that forms spontaneously when chlorine dioxide is in an aqueous environment is indeed a serious matter. It is, therefore, essential to balance the potential health effects of chlorate formation from chlorine dioxide vs. the potentially harmful disinfection by-products produced using chlorine gas or hypochlorite anions, when treatment of irrigation water is necessitated using these treatments.

13.8.3 Disinfection by Filtration and Ultraviolet Light

Water purification systems often start with what is called slow sand filtration, a process that will remove most particulates and perhaps, at best, one log (90%) of suspended bacteria, algae, and other microorganisms. However, plastic membranes can be made that when fitted into a proper device and put under pressure can remove even a wide array of microscopic organisms. These so-called ultrafiltration units retain microbes, and allow only water molecules to pass through. The membrane is always subject to a process called "fouling" when a layer of solids (bacteria and other debris) form on the membrane inner surface requiring a maintenance "backwash" that usually removes the material. Membranes have lifetimes and are expensive to replace. Nevertheless, a model study was recently reported using ultrafiltration to prepare municipal wastewater that provided water with a reduced bacterial load for use in irrigation (Verging et al., 2020). The *E. coli* levels in the water were reduced by 3 logs (1,000 fold) using this strategy. The further addition of aUV disinfection unit following ultrafiltration essentially resulted in water free of *E. coli* and with a greatly reduced coliform burden, equal to the ground water found nearby. UV light at 260 nm is absorbed by the purine and pyrimidine monomers that make up the DNA, creating mutations (called sometimes lesions) in the organism's genome, preventing replication. Since UV light is absorbed by standard glass products, flow cells through which the water must pass are made of more expensive silica. There are many cautions that need to be addressed when using these technologies. First, the membranes which have pores of 0.05 μm age and in time can leak, allowing bacteria to enter the effluent. Then the UV system is also subject to formation of a film which can block the light from the effluent stream. UV is used now in many potable water disinfection trains and is effective in killing bacteria and protozoan pathogens alike. For large volume producers of irrigation water, however, these two technologies still must be proven.

Redundancy is an essential when dealing with public health matters. The sand filter used in the above disinfection train coupled with an ultrafiltration system and those together coupled with the UV system provide at the least one composite system that could add some reduction in microbial content should one of the others in the train fail for some reason (Ghernaout et al., 2011; Ghernaout and Naceur, 2011).

13.8.4 Disinfection by Sunlight

Solar disinfection was known to the ancients and we know that sunlight can reduce bacterial numbers given enough time. Actually, the long-wavelength UV component of sunlight is the disinfection agent. In a laboratory study, Bichai et al. (2012) irrigated lettuce crops with solar-treated wastewater effluents from a municipal wastewater treatment plant. Using water treated in this manner, bacterial levels measured as colony forming units or CFUs, were reduced from 10^5 to 10^6 CFUs/100 mL to 200 CFUs/100 mL. Lettuce leaves tested for coliforms 24 hours post-watering were 89% free of *E. coli*. Disinfection of water derived from several sources including wastewater effluents was accomplished by the UVa component of the solar lamps used (natural sunlight was not used), which creates lethal changes in the cell's DNA that prevents replication of the bacterial chromosome. Addition of 5 or 10 mg/L of hydrogen peroxide (H_2O_2) enhanced the efficiency of the disinfection process. Bichai et al. (2012) also suggest that UVa can create reactive oxygen species which can affect cell membrane integrity. The photo-Fenton reaction assisted by endogenous iron in the bacterial cells themselves, is a reaction enhanced by H_2O_2. Low doses of H_2O_2 coupled with natural sunlight have been shown to enhance inactivation of *E. coli* (Gárcia-Fernández et al., 2012), *Salmonella* spp. (Sciacca et al., 2010), and fungal spores (Polo-López et al., 2011).

13.8.5 Ozone Disinfection

Ozone gas O_3 is highly oxidative and its presence when dissolved in water can effectively disinfect even the most resistant organisms (Langlais et al., 1991). It is made in situ using electric discharge in a pure O_2 atmosphere. Most European cities use ozonization as their primary method for disinfecting their drinking water, while most North American cities still use chlorination and sometimes a combination of ozone and chlorine. Ozone, although more effective than chlorine, does not provide a residual inside a water distribution system, as does chlorine, and does not protect water within a system from inadvertent pathogen entry once water passes the ozone generator. However, because ozone disinfects with minimal production of harmful by-products, it is often used in the United States in conjunction with chlorination, first to remove organics from the incoming water stream, but then it is catalytically removed prior to chlorination. Ozone does, however, provide much better CT values for all of the pathogens, especially for protozoans. For example, the CT for *E. coli* using ozone is 0.02, 0.3 for viruses at 5°C, and 0.62 for *G. lamblia* at 5°C.

For the nematode *Meloidogyne enterolobii* and other species of the same genus that cause galls or knots on the roots of valuable crops such as tomato, Fernández et al. (2019) found that ozone was efficient at providing a disinfection efficiency expressed as a CT of 8.25 ((mg-min)/L) at 35°C, a value that actually would kill 99% of the oocytes of the primarily bovine fecal parasite *C. parvum*. They found that the oxidant damaged the nematode eggs, and thereby its viability, and proposed that the use of ozonization in irrigation waters used in tomato, watermelon, and chili peppers production to control this and other species of nematode could improve production.

13.9 Conclusions

Pathogen elimination in drinking (potable) water has been a primary objective of not only individuals with their own needs supplied by wells or clear surface waters, but by public health authorities everywhere. Municipal water treatment technologies have become more and more sophisticated. Initially chlorine gas proved to be a superior option, until it was found that disinfection by-products which were possibly carcinogenic were produced when the disinfectant reacted with mostly naturally occurring organic matter in the same water. Ozone and chlorine dioxide were found to have excellent disinfectant characteristics and produce far less disinfection by-products compared to chlorine gas or hypochlorite. UV light has only been shown to eliminate pathogens in drinking water as well as serve a function in the disinfection of properly treated municipal wastewater. As the need for drinking water increases as population likewise increase, the need for food and therefore additional water for irrigating the additional crops correspondingly is needed. Municipal wastewater has become a commodity initially for irrigating recreational areas such as the public parks and golf courses decades ago, to more recently, agriculturally important issues such as livestock and crop production. These same technologies used for providing safe drinking water for human consumption are being harnessed for producing a safer water for agricultural purposes. Indeed, pathogens that contaminate agriculturally important crops that are consumed as raw products, such as lettuce and other components of salads, have been identified as being present in the irrigation water used to grow the crops. Avoiding situations that negatively affect the perception of safety and food security is most important to the market for those foods. The use of municipal wastewater as an important source of irrigation water will continue into the future, as the need for more portable water sources are diverted from agricultural to municipal uses. It will be essential that irrigation water derived from treated municipal wastewater be monitored regularly, not necessarily for specific pathogens, but for the most reliable microbial indicators of the presence of those potential pathogens. Indeed, public health and agricultural authorities must act together to ensure a reasonable level of safety for the public, both in providing the drinking water and in producing the food supply.

References

AHDB, Agriculture and Horticulture Development Board 2016. Chlorine and its oxides: Chlorate and perchlorate review. Available at: https://horticulture.ahdb.org.uk/sites/default/files/research_papers/CP%20154a_Report_Final_2016.pdf.

Ayyildiz, O., Ileri, B., Sanik, S. 2009. Impacts of water organic load on chlorine dioxide disinfection efficacy. *Journal of Hazardous Materials*, 168, 1092–1097.

Bastian, R., Murray, D. 2012. Guidelines for Water Reuse. US EPA Office of Research and Development, Washington, DC, USA, EPA/600/R-12/618.

Bea Clarise, B., Garcia, M.A., Dimasupil, A.Z., Vital, P.G., Widmer, K.W., Rivera, W.L. 2015. Fecal contamination in irrigation water and microbial quality of vegetable primary production in urban farms of Metro Manila, Philippines. *Journal of Environmental Science and Health, Part B* 50(10), 734–743. DOI: 10.1080/03601234.2015.1048107.

Benjamin, L., Atwill, E.R., Jay-Russell, M., Cooley, M., Carychao, D., Gorski, L., Mandrell, R.E. 2013. Occurrence of generic *Escherichia coli*, E. coli:O157 and *Salmonella spp.* in water and sediment from leafy green produce farms and streams on the Central California coast. *International Journal of Food Microbiology* 165, 65–76.

Berger, C.N.; Sodha, S.V.; Shaw, R.K., Griffin, P.M., Pink, D.; Hand, P., Frankel, G. 2010. Fresh fruit and vegetables as vehicles for the transmission of human pathogens. *Environmental Microbiology* 12, 2385–2397.

Bichai, F., Polo-López, M. I., Iba ñez, P. F. 2012. Solar disinfection of wastewater to reduce contamination of lettuce crops by *Escherichia coli* in reclaimed water irrigation. *Water Research* 46, 6040–6050.

Blatchley III, E.R., Gong, W.L., Alleman, J.E., Rose, J.B., Huffman, D.E., Otaki, M., Lisle, J.T., 2007. Effects of wastewater disinfection on waterborne bacteria and viruses. *Water Environment Research* 79, 81–92.

Bosch, A., 2007. *Human Viruses in Water*, 1st ed. Elsevier, Amsterdam; Boston, MA.

Castro-del Campo, N., Martínez-Rodríguez, C., Chaidez, C. 2012. Occurrence and survival of pathogenic microorganisms in irrigation water. In Garcia-Garizabal, I., and Abrahao, R. (Eds.), *Irrigation—Water Management, Pollution and Alternative Strategies*, InTech, Croatia, pp. 221–234.

Castro-Ibáñez, I., Gil, M.I., Tudela, J.A., Ivanek, R., Allende, A. 2015. Assessment of microbial risk factors and impact of meteorological conditions during production of baby spinach in the Southeast of Spain. *Food Microbiology* 49, 173–181.

Ceuppens, S., Johannessen, G.S., Allende, A., Tondo, E.C., El-Tahan, F., Sampers, I., Jacxsens, L., Uyttendaele, M. 2015. Risk factors for *Salmonella*, shiga toxin-producing *Escherichia coli* and *Campylobacter* occurrence in primary production of leafy greens and strawberries. *International Journal of Environmental Research and Public Health* 12, 9809–9831.

Chu, W.H., Gao, N.Y., Yin, D.Q., Krasner, S.W., Mitch, W.A. 2014. Impact of UV/H_2O_2 pre oxidation on the formation of haloavetamides and other nitrogenous disinfection byproducts during chlorination. *Environmental Science & Technology* 48, 12190–12198. DOI: 10.1021/es502115x.

Chu, W.H., Li, D.M., Deng, Y., Gao, N.Y., Zhang, Y.S., Zhu, Y.P. 2016. Effects of UV/PS and UV/H_2O_2 preoxidations on the formation of trihalomethanes and haloacetonitriles during chlorination and chloramination of free amino acids formation of trihalomethanes and short oligopeptides. *Chemical Engineering Journal* 301, 65–72. DOI: 10.1016/j.cej.2016.04.003.

Crusberg, T.C., Burke, W., Reynolds, J.T., Morse, L.J., Reilly, J., Hoffman, A.H. 1978. The reappearance of a classical epidemic of infectious hepatitis in Worcester, Massachusetts. *American Journal of Epidemiology* 107, 545–551.

Crusberg, T.C. 2014. Chapter 29: Water supply and public health and safety. In Eslamian, S. (Ed.), *Handbook of Engineering Hydrology*, CRC Press, Taylor and Francis Group, Boca Raton FL, pp. 553–575 (ISBN 9781466552357).

Crusberg, T.C., Eslamian, S. 2016. Chapter 42: Choosing indicators of fecal pollution for wastewater reuse opportunities. In Eslamian, S. (Ed.), *Urban Wastewater Handbook*, CRC Press, Taylor and Francis Group, Boca Raton, FL, pp. 513–524. (ISBN 13-978-4823-2915-8).

Dandie, C.E., Ogunniyi, A.D., Ferro, S., Hall, B., Drigo, B., Chow, C.W.K. 2019. Disinfection options for irrigation water: Reducing the risk of fresh produce contamination with human pathogens. *Critical Reviews in Environmental Science and Technology*. DOI: 10.1080/10643389.2019.1704172.

Decola, L.T., López-Gálvezb, F., Truchadob, P., Tondoa, E.C., Gilb, M.I., Allendeb, A. 2019. Suitability of chlorine dioxide as a tertiary treatment for municipal wastewater and use of reclaimed water for overhead irrigation of baby lettuce. *Food Control* 96, 186–193.

Donkor, E.S., Lanyo, R., Kayang, B.B., Quaye, J., Edoh, D.A. 2010. Internalisation of microbes in vegetables: Microbial load of Ghanaian vegetables and the relationship with different water sources of irrigation. *Pakistan Journal of Biological Science* 13, 857–861.

Dreschel, P., Scott, C.A., Raschid-Sally, L., Redwood, M., Bahri, A., 2010. *Wastewater Irrigation and Health-Assessing and Mitigating Risk in Low-Income Countries*, IDRC/IWMI, London.

El-Senousy, W.M., Costafreda, M.I., Pinto, R.M., Bosch, A. 2013. Method validation for norovirus detection in naturally contaminated irrigation water and fresh produce. *International Journal of Food Microbiology* 167, 74–79. DOI:10.1016/j.ijfoodmicro.2013.06.023

Eslamian, S., 2016 (Ed.) *Urban Water Reuse Handbook*, Taylor and Francis, CRC Group, Boca Raton, FL, 1141 Pages.

FAO; WHO. 2008. Microbiological hazards in fresh leafy vegetables and herbs. In Microbial Risk Assessment Series 14, World Health Organization: Rome, Geneva, Italy, Switzerland.

FAO (Food and Agriculture Organization of the United Nations) 2012. Water NEWS: Climate Change & Water. http://www.fao. org/nr/water/news/clim-change.html.

FDA (US Food and Drug Administration) 2013. The New FDA Food Modernization Act (FSMA). Preventive Controls and Produce Safety Rules (online). http://www.fda.gov/Food/GuidanceRegulation/FSMA/ucm242500.html.

Fernández, I.A.L., Monje-Ramirez, I., de Velásquez, M.T.O.L. 2019. Tomato crop improvement using ozone disinfection of irrigation water. *Ozone Science and Engineering* 41, 398–403. DOI: 10.1080/01919512.2018.1549474.

Fremaux, B., Boa, T., Chaykowski, A., Kasichayanula, S., Gritzfeld, J. Braul, L., Yost, C. 2009. Assessment of the microbial quality of irrigation water in a prairie watershed. *Journal of Applied Microbiology* 106, 442–454.

Fuzawa, M., Araud, E., Li, J., Shisler, J.L., Nguyen, T.H. 2019. Free chlorine disinfection mechanisms of rotaviruses and human Norovirus surrogate Tulane virus attached to fresh produce surfaces. *Environmental Science and Technology*, 53, 11999–12006.

Gallegly, M.E., Hong, C.X 2008. Phytophthora: *Identifying Species by Morphology and DNA Fingerprints*. APS Press, St Paul, MN.

Gárcia-Fernández, I., Polo-López, M.I., Oller, I., Fernandez-Ibanez, P., 2012. Bacteria and fungi inactivation using Fe^{3+}/sunlight, H_2O_2/sunlight and near neutral photo-Fenton: A comparative study. *Applied Catalysis B: Environmental* 121–122, 20–29.

Gerba, C. P., Smith J. E. 2005. Sources of pathogenic microorganisms and their fate during land application of wastes. J. Environ. Qual. 42–48. https://doi.org/10.2134/jeq2005.0042a

Gerba, C.P., Choi, C.Y. 2006. Role of irrigation water in crop contamination by viruses. In Goyal, S.M. (Ed.), *Viruses in Foods*, SpringerNew York, pp. 257–263.

Ghernaout, D., Ghernaout, B. and Naceur, M.W. 2011. Embodying the chemical water treatment in the green chemistry: A review. *Desalination* 271, 1–10. DOI: 10.1016/j.desal.2011.01.032.

Ghernaout, D., Naceur, M.W. 2011. Ferrate(VI): *In Situ* generation and water treatment: A review. *Desalination and Water Treatment* 30, 319–332. DOI: 10.5004/dwt.2011.2217.

Ghernaout, D. 2017. Water treatment chlorination: An updated mechanistic insight. *Chemistry Research Journal* 2, 125–138.

Ghernaout, D., Elboughdiri, N. 2020a. Is not it time to stop the use of chlorine for treating water? *Open Access Library Journal* 7, e6007.

Ghernaout, D., Elboughdiri, N. 2020b. Controlling disinfection by-products formation in rainwater: Technologies and trends. *Open Access Library Journal* 7: e6162. DOI: 10.4236/oalib.1106162.

Haas, C.N., Rose, J.B., Gerba, C., Regli, S. 1993. Risk assessment of virus in drinking water. *Risk Analysis* 13, 545–552.

Hamilton, A.J., Stagnitti, F., Premier, R., Boland, A.-M., Hale, G. 2006. Quantitative microbial risk assessment models for consumption of raw vegetables irrigated with reclaimed water. *Applied and Environmental Microbiology* 72, 3284–3290.

Hamilton, A.J., Stagnitti, F., Xiong, X., Kreidl, S.L., Benke, K.K. et al. 2007. Wastewater irrigation: The state of play. *Vadose Zone Journal* 6, 823–840.

Hassenberg, K., Geyer, M., Mauerer, M., Praeger, U., Herppich, W.B. 2017. Influence of temperature and organic matter load on chlorine dioxide efficacy on *Escherichia coli* inactivation. *Lebensmittel-Wissenschaft und -Technologie- Food Science and Technology*, 79, 349–354.

Herschel, C. 1973. *Translation of: Frontinus, S.J. Two Books on the Water Supply of the City of Rome, AD97.* New England Water Works Association, Boston, MA.

Hong, C. X., Moorman, G. W. 2005. Plant pathogens in irrigation water: Challenges and opportunities. *Critical Reviews in Plant Sciences* 24(3), 189–208.

Hong, C.X., Richardson, P.A., Kong P. 2008. Pathogenicity to ornamental plants of some existing species and new taxa of *Phytophthora* from irrigation water. *Plant Disease* 92, 1201–1207.

Hong, C.X., Gallegly, M.E., Richardson, P.A., Kong, P., Moormanc, G.W., Lea-Cox, J.D., Rosse, D.S. 2010. *Phytophthora hydropathica*, a new pathogen identified from irrigation water, *Rhododendron catawbiense* and *Kalmia latifolia*. *Plant Pathology* 59, 913–921. DOI: 10.1111/j.1365–3059.2010.02323.xl.

Ibekwe, A.M., Watt, P. M., Shouse, P. J., Grieve, C. M., 2004. Fate of Escherichia coli O157:H7 in irrigation water on soils and plants as validated by culture method and real-time PCR. *Canadian J. Microbiol.* 50, 1007–1014. https://doi.org/10.1139/w04-097

Ibekwe, A.M., Leddy, M., Murinda, S.E., 2013. Potential human pathogenic bacteria in a mixed urban watershed as revealed by pyrosequencing. *PLoS One* 8 (11), e79490. DOI: 10.1371/journal.pone.0079490.

Ibekwe, A.M., Gonzalez-Rubio, A., Suarez, D.L. 2018. Impact of treated wastewater for irrigation on soil microbial communities. *Science of the Total Environment* 622–623, 1603–1610.

Ijabadeniyi, O.A., Buys, E.M. 2012. Irrigation water and microbiological safety of fresh produce: South Africa as a case study: A review. *African Journal of Agricultural Research* 7, 4848–4857.

Jones, S., Shortt, R. 2005. Improving on-farm food safety through good irrigation practices. OMAFRA Factsheet 05-059.

Junli, H., Li, W., Nanqi, R., Fang, M. 1997. Disinfection effect of chlorine dioxide on bacteria in water. *Water Research* 31, 607–613.

Langlais, B., Reckhow, D.A., Brink, D.R. 1991. *Ozone in Drinking Water Treatment: Application and Engineering.* Lewis Publishers, Chelsea, MI.

Levatensi, C., La Mantia, R., Masciopinto, C., Böckelmann, U., Ayuso-Gabella, M., Salgot, M., Tandoi, V., Van Houtte, E., Wintgens, T., Grohmann, E., 2010. Quantification of pathogenic microorganisms and microbial indicators in three wastewater reclamation and managed aquifer recharge facilities in Europe. *Science of the Total Environment* 408, 4923–4930.

López-Gálvez, F., Allende, A., Martinez-Sanchez, A., Tudela, J.A., Selma, M.V., Gil, M.I. 2010. Suitability of aqueous chlorine dioxide versus sodium hypochlorite as an effective sanitizer for preserving quality of fresh-cut lettuce while avoiding by product formation. *Postharvest Biology and Technology* 55, 53–60.

Luo, G., Gu, G., Ginn, A., Giurcanu, M., C., Adams, P., Vellidis, G., van Bruggen, A.H., Danyluk, M.D., Wright, A.C. 2015. Distribution and characterization of *Salmonella enterica* isolates from irrigation ponds in the southeastern United State. *Applied and Environmental Microbiology* 81, 4376–4387.

Moulin, L., Richard, F., Stefania, S., Goulet, M., Gosselin, S., Goncalves, A., Rocher, V., Paffoni, C., Dume'tre, A., 2010. Contribution of treated wastewater to the microbiological quality of Seine river in Paris. *Water Research* 44, 5222–5231.

O'Shea, M.L., Field, R. 1992. An Evaluation of bacterial standards and disinfection practices used for the assessment and treatment of stormwater. *Advances in Applied Microbiology* 37, 21–40. DOI: 10.1016/S0065-2164(08)70251-X.

Pachepsky, Y., Shelton, D.R., McLain, J.E.T., Patel, J., Mandrell, R.E. 2011. Irrigation waters as a source of pathogenic microorganisms in produce: A review. *Advances in Agronomy* 113, 75–141. DOI: 10.1016/B978-0-12-386473-4.00002-6.

Park, S., Navratil, S., Gregory, A., Bauer, A., Srinath, I., Jun, M., Szonyi, B., Nightingale, K., Anciso, J., Ivanek, R. 2013. Generic *Escherichia coli* contamination of spinach at the preharvest stage: Effects of farm management and environmental factors. *Applied and Environmental. Microbiology* 79, 4347–4358.

Polo-López, M.I., Fernández-Ibá nez, P., García-Fernández, I., Oller, I., Salgado-Tránsito, I., Sichel, C., 2010. Resistance of *Fusarium* spores to solar TiO_2 photocatalysis: Influence of spore type and water (scaling-up results). *Journal of Chemical Technology and Biotechnology* 85, 1038–1048.

Qadir, M., Sharma, B.R., Bruggeman, A., Choukr-Allah, R., Karajeh, F. 2007. Non- conventional water resources and opportunities for water augmentation to achieve food security in water scarce coun-tries. *Agricultural Water Management* 87, 2–22.

Raudales, R.E., Parke, J.L., Guy, C.L., Fisher, P.R. 2014. Control of waterborne microbes in irrigation: A review. *Agricultural Water Management* 143, 9–28.

Rock, C.M., Brassill, N., Deryb, J.L., Carrb, D., McLaina, J.E., Brighta, K.R., Gerba, C.P. 2019. Review of quality criteria for water reuse and risk-based implications for irrigated produce under the FDA Food Safety Modernization Act, produce safety rule. *Environmental Research* 172, 616–629. DOI: 10.1016/j.envres.2018.12.050.

Rusinol, M., Fernandez-Cassi, X., Timoneda, N., Carratala, A., Abril, J.F., Silvera, C., Figueras, M.J., Gelati, E. et al. 2015. Evidence of viral dissemination and seasonality in a Mediterranean river catchment: Implications for water pollution management. *Journal of Environmental Management* 159, 58–67.

Santamaría, J., Toranzos, G.A. 2003. Enteric pathogens and soil: a short review. *Int Microbiol* 6, 5–9 https://doi.org/10.1007/s10123-003-0096-1

Scallan, E., Hoekstra R.M., Angulo, F.J., Tauxe, R.V., Widdowson, M.A., Roy, S.L., Jones, J.L., Griffin, P.M. 2011. Foodborne illness acquired in the United States—Major pathogens. *Emerging and Infectious Diseases* 17, 7–15. DOI: 10.3201/eid1701.P11101.

Scheierling, S.M., Bartone, C.R., Mara, D.D., Drechsel, P., 2011. Towards an agenda for improving waste-water use in agriculture. *Water International* 36, 420–440.

Sciacca, F., Rengifo-Herrera, J.A., Wéthé, J., Pulgarin, J. 2010. Dramatic enhancement of solar disinfection (SODIS) of wild *Salmonella sp.* in PET bottles by H_2O_2 addition on natural water of Burkina Faso containing dissolved iron. *Chemosphere* 78(9), 1186–1191.

Shuval, H. 2007. Evaluating the world new health organization's 2006 guidelines for wastewater. World New Health Organization's 2006 Health Guidelines for Wastewater. In: Zaidi, M.K. (eds) *Wastewater Reuse–Risk Assessment, Decision-Making and Environmental Security*. NATO Science for Peace and Security Series. Springer, Dordrecht. pp. 279-287. https://doi.org/10.1007/978-1-4020-6027-4_27.

Solomon, E.B., Yaron, S., Matthews, K.R. 2002. Transmission of *Escherichia coli* O157:H7 from contami-nated manure and irrigation water to lettuce plant tissue and its subsequent internalization. *Applied and Environmental Microbiology* 68, 397–400.

Tracogna, M.F., Losch, L.S., Alonso, J.M., Merino, L.A. 2013. Detection and characterization of *Salmonella* spp. in recreational aquatic environments in the Northeast of Argentina. *Revista Ambiente & Água* 8, 18–26.

U.S.E.PA. Office of Water 2002. Method 1604. Total Coliforms and *Escherichia coli* in Water by Membrane Filtration Using a Simultaneous Detection Technique (MI Medium), (Method 1604) Publication 10, EPA 821 R 02 024. https://www.epa.gov/sites/default/files/2015-08/documents/method_1604_2002.pdf

Van der Linden, I., Cottyn, B., Uyttendaele, M., Vlaemynck, G., Heyndrickx, M., Maes, M. 2013. Survival of enteric pathogens during butterhead lettuce growth: Crop stage, leaf age, and irrigation. *Foodborne Pathogenic Diseases* 10, 485–491. DOI: 10.1089/fpd.2012.1386.

Van Haute, S., López-Gálvez, F., Gómez-López, V.M., Eriksson, M., Devlieghere, F., Allende, A. 2015. Methodology for modeling the disinfection efficiency of fresh-cut leafy vegetables wash water applied on peracetic acid combined with lactic acid. *International Journal of Food Microbiology*, 208, 102–113.

Van Haute, S., Tryland, I., Escudero, C., Vanneste, M., Sampers, I. 2017. Chlorine dioxide as water disinfectant during fresh-cut iceberg lettuce washing: Disinfectant demand, disinfection efficiency, and chlorite formation. *Lebensmittel-Wissenschaft und -Technologie- Food Science and Technology* 75, 301–304.

Verging, P., Amalfitano, S., Salerno, C., Berardi, G., Pollice, A. 2020. Reuse of ultrafiltered effluents for crop irrigation: On-site flow cytometry unveiled microbial removal patterns across a full-scale tertiary treatment. *Science of the Total Environment* 218, 137298. DOI: 10.1016/j.scitotenv.2020.137298.

Veschetti, E., Cutilli, D., Bonadonna, L., Briancesco, R., Martini, C., & Cecchini, G. 2003. Pilot-plant comparative study of peracetic acid and sodium hypochlorite wastewater disinfection. *Water Research* 37, 78–94.

Vigneswaran, S., Sundaravadive, M. 2004. Recycle and reuse of domestic wastewater. In Vigneswaran, S. V. (Ed.), *Wastewater Recycle, Reuse, and Reclamation, in Encyclopedia of Life Support Systems (EOLSS), Developed under the Auspices of the UNESCO*, EOLSS Publishers, Oxford. http://www.eolss.net.

Wei, J.Y., Sims, J.T., Kniel, K.E. 2010. Manure- and biosolids-resident murine norovirus 1 attachment to and internalization by romaine lettuce. *Applied and Environmental Microbiology* 76, 578–583.

Wei, J., Jin, Y., Sims, T. and Kniel, K.E. 2011. Internalization of Murine Norovirus 1 by *Lactuca sativa* during Irrigation. *Applied and Environmental Microbiology* 79, 2508–2512. DOI: 10.1128/AEM.02701-10.

14

Phosphate and Nitrate Management in Irrigation Water

Alessia Corami
MIUR Ministry of University and Research

Saeid Eslamian
Isfahan University of Technology

14.1 Introduction

Environmental contamination from NO$_3$N and PO$_4$-P has become an issue of great concern, the high levels of nitrate are considered unhealthy for human beings. Furthermore, PO$_4$-P is usually linked with the algal blooms and the eutrophication phenomenon. Even though PO$_4$-P is considered immobile in soils, it could be leached depending on the media (Broschat, 1995). Both P and N could accumulate naturally in soils, their amounts depend on many parameters like parent rocks, grain composition and mainly anthropogenic supply; both could arrive in the saturation zone, thus they could contaminate groundwater. Inter alia, anthropogenic activities are considered the most important sources of phosphate and nitrate pollution in aquatic ecosystems. Unluckily, these non-point sources of nutrients are difficult to measure and regulate because they are derived from activities dispersed over large expanses of land and are time variable because of weather and climate changes (Adesuyi et al., 2015).

Nitrogen is considered a plant nutrient like phosphate and the aim of their copious use is to increase the yield. Generally, it is possible to distinguish between nitrogen in soil, nitrogen in water and/or nitrogen added through fertilizer. Unfortunately, the excess of nitrogen in soil, in water and/or in irrigation water will always have a negative effects, that is, an overstimulation of the crop and consequently a delayed maturity or poor quality of the crop itself (FAO, 2003).

The aim of this chapter is to look over if there is a way to optimize the use of irrigation water and fertilizers to increase the crop, but at the same time avoiding damages to soil, to decrease the amount of wastes in surface water from urban land, if there is a way to clean up water from nitrogen and phosphorus and reuse it for agricultural purposes, also to comprehend some suggestions for the sustainable use of fertilizers, water and soil.

DOI: 10.1201/9780429290114-19

14.2 NO$_3$-N and PO$_4$-P Characteristics

The N available forms are nitrate and ammonium. Nitrate (NO$_3^-$) usually occurs in irrigation waters; ordinarily, ammonium-nitrogen has a low concentration about of 1 mg/L in surface water and the amount is less than 5mg/L in groundwater, unless an excess due to the fertilizers that have been added. It is suggested to monitor the N amount considering fertilizer addition, because of the amount in water supplies. Particularly, in wastewater from domestic source, nitrogen ranges from 10 to 50 mg/L. The NO$_3$ excess is considered toxic because of its reduction to nitrite, which affects the oxidation of normal Haemoglobin to Methaemoglobin, so that it is unable to transport oxygen to the tissues. When metHb concentration reaches 10% of normal Hb concentration, it causes the methaemoglobinaemia; the consequence is cyanosis and, at higher concentrations, asphyxia (WHO, 2016). As a matter of fact, nitrates show a high mobility after the leaching process; it is observed a correlation between land-use, nitrate availability in soil, agriculture and nitrate amount in water therefore a great use of fertilizer could be a huge source of nitrates (Różkowski et al., 2017; Kaushik, 2013; Khurana and Aulakh, 2010). N fertilizers are urea, ammonia and nitrates, and ammonium forms change in nitrates in aerobic soils (Akhavan et al. 2010,).

Phosphorus is a fundamental element for life; it is a component of nucleic acid, intermediary metabolites (sugar phosphates and adenosine phosphate) and they are an important part of the metabolism of life forms (Correll, 1998); it can be also found in earth's crust and soils where phosphorus is usually present in a low amount and it is held tightly, thus leaching is less important than runoff. P compounds are not volatile, the transport is due to dust and aerosol flux, so that P is in surface flows rather than groundwater, moreover, P mostly bound in soils and sediments. Unfortunately, agriculture soils with a high amount of P show a huge phosphorus runoff causing a deep environmental problem. Since P could be desorbed from the soil particles if soil is saturated, the potential for losses of soluble phosphorus in surface runoff will increase (Pierziynski et al., 2005).

In particular, the P pentavalent form occurs in aquatic system, such as orthophosphate, pyrophosphate, longer-chain polyphosphates, organic phosphate esters, phosphodiesters and organic phosphonates, so that P arrives in aquatic system as a mixture of these inputs. Once in the water, P is stored in waters through biological assimilation and sediment deposition. In case of a eutrophic system, water could become anoxic during growing season, thus P in the sediments is released in the waters (Vollenweider model). Moreover, it is observed that N and C are used for algal production if there is an excess of P. Besides, a relationship has been determined between total N and total P in surface waters, the Redfield ratio that is an atomic ratio. Normally, this ratio for N and P is about 15 to 16:1, on the contrary, in the systems that show a bigger ratio than the Redfield ratio, the production of biomass or algal is reduced because of P amount. P results as a key element, controlling the algal or the biomass productivity in streams and water (Correll, 1998). Unfortunately, P in waters leads to an increase of primary production and consequently a high rate of decomposition and depletion of dissolved oxygen, so that it is observed an increasing of dead fish and a shift in species composition. The P is dissolved and P particulate becomes available to phytoplankton and bacteria, that is, a phosphate buffering equilibria. Clavel et al. (2008) have confirmed that the excessive growth of algae is caused by the excessive inputs of P and N in waters; on the contrary, a high concentration of NO$_3^-$ in water can diminish the amount of SO$_4$ reduction and an increase of PO$_4$ amount (Kim and Chung, 2010).

Indeed, the oxidation-reduction reaction among NO$_3$, Fe and SO$_4$ affect P solubility in anoxic flooded soil environments. It is stated that if NO$_3$ is normally added through irrigation water and fertilizers, PO$_4$ solubilisation is observed; this could reduce the amount of PO$_4$ loss through leaching and other ways in the water system. On the other hand, Carter et al. (1971) have observed that if the irrigation water contains PO$_4$-P, most of it will be removed by the irrigation process, and subsurface drainage water will contain an extremely low PO$_4$-P concentration. Unfortunately, the runoff of P from the terrestrial system to the aquatic system could cause a decreasing of water quality and consequently eutrophication could arise with a great amount of algae and aquatic weeds; further nitrogen could also be associated with this phenomenon (Sharpley and Menzel, 1987). Nevertheless, watershed hydrology is dependent on many factors, including land use, climate and soil conditions.

14.3 Land Use Water Quality

Land use and water quality are two useful parameters in case of a diffuse source pollution, contrary to non-point source inputs, which are assumed the most complex source to understand and to regulate like runoff, groundwater and atmospheric deposition. It is a long time that the relationship between the amount of exogenous inputs and algal blooming has been studied. A relationship has been found between agricultural land-use and water quality in case of intensive agricultural activity. Nitrate is the most common water quality factor correlating with agricultural land cover, suspended solids and pesticides, which are also frequently reported to correlate with agricultural land cover. In general, nitrogen, phosphorus and other pollutants in river waters are mainly affected by agricultural land. Tong and Chen (2002) have inferred that the surface runoff is a primary source of non-point pollution and the different land-uses may influenced the runoff with different kinds of contaminants; hence, the quality and quantity of water is mainly influenced by theland use. The hydrogeological system could be altered by changing land use and land management practices, the latter could be one of the most important factors. They have also affirmed that runoff from agricultural activity and from urban land-use produces a huge amount of nitrogen and phosphorus; in case of a change in the land-use, from agriculture to urban land, the management must be considered with care to avoid water quality problems. It is well established that fertilizer, used for agriculture purposes, will get into runoff and later in waters and the consequence is a polluted water (Huang et al., 2013), but vegetation in cultivate land can absorb and retain these pollutants, vegetation can reduce the surface runoff, can avoid the risk of erosion and can ameliorate the quality of water. Thence, Carter et al. (1971) stated that a certain amount of leaching is required for maintaining the salt balance for sustained productivity of any irrigation project, whereas an excessive leaching dissolves and removes the soluble salts than necessary, so that the salt balance could be altered.

As a matter of fact, the amount of fertilizers influence the concentration of nitrogen and phosphorus in surface water and groundwater, this amount must be valued in all systems. At the same time, it is important to value the amount and quality of irrigation return flows; in fact, water from return flow has a low quality and assuring a possible reuse suggests a control of physicochemical characteristics (García-Garizábal and Causapé, 2010).

However, Baker (2006) has pointed out that the relationship between water quality and intensive agricultural land use is less well understood. He has highlighted that, in general, North American and European regions also have moved through the transition from having predominantly point source pollution problems to a situation where diffuse pollution is more important. Studies in the temperate regions have suggested that nitrate export has increased by 3 times to 20 times since industrialization owing to the increased animal and human wastes and increased runoff. Large-scale and land-use changes are occurring in tropical countries; tropical deforestation will lead to a similar phenomenon when tropical rivers are naturally rich nitrate transport environments.

Unfortunately, the eutrophication phenomenon of freshwater turns out as a cost to society owing to the increased water treatment required and the reduction of the health, safety and the amenity value of recreational lakes and reservoirs. Catherine et al. (2010) have written it is widely recognised that phytoplankton biomass is correlated to the nutrient fluxes reaching waterbodies, which is largely dependent on the morphologic parameters of the systems such as catchment size and lake volume and catchment characteristics like land use and drainage network. On the other hand, the increase of urban land means the increase of PO_4-P, BOD_5 and NH_4-N concentration, but also the other variables affect water quality such as climate change, national policy, fertilizers, pollution control, pesticides and other non-point source affect water quality (Chen et al., 2020; Bo et al., 2018; Maberly et al., 2003). In particular, Chen et al. (2020) have studied the quality of the water in a river basin for three years during a period of

17 years, and the results allow to infer that the urban settlements could also strengthen the deterioration of water quality. To avoid the degradation of water, it is suggested to improve the vegetation, but more important the conversion of bare soil into agricultural land and settlement land is also an important reason for the deterioration of water quality. Another highlighted point is the intensive management of urban green, it means watering, fertilizing and insecticide, all have negative effects on water quality. Thence, this anthropogenic management influences water quality. Bo et al. (2018) have suggested choosing plants which need less pesticides and fertilizers. Water system is considered a sensitive environment, and it can be defined as a sink of nutrients like nitrogen and phosphate because of agricultural system, due to this enrichment, it is observed the increasing of eutrophication process and autochthonous organic production of algae.

14.4 Eutrophication

Eutrophication is defined as the nutrients' enrichment in water, causing an increase in autochthonous organic production and the biomass of algae. Generally, two types of eutrophication are distinguished: the natural and the anthropogenic one. The latter is generally due to intensive land-use, misuse of fertilizers, deforestation, growth of urban and rural areas, and also discharge of untreated sewage to surface waters. Researchers have noted that although progresess have been made in curbing the point sources of nutrient loading to surface water in many parts of the world, non-point contributions remained a widespread and often overlooked cause of eutrophication (Taranu and Gregory-Eaves, 2008). Normally, N and P amounts are the main causes of this phenomenon, and the threshold is the N/P ratio, but according to Diatta et al. (2020), P/N has also an important role. The ratio P/N is still a notable criterion where fertilization is still based on phosphorus compounds. So, they have carried out experiments on eutrophication process when physical and chemical factors interact altogether. Particularly, they have observed the ratio N/P and the ratio P/N, which is less studied. If N/P ratio is considered, it is observed that phosphorus is the main control for this process; on the other hand, if P/N ratio is considered, both P and N are interchangeable. Nitrogen seems to be the main regulating factor compared to phosphorus. According to them, it is important to consider other factors such as temperature, type of water, and pH. They have compared three types of water (Tap Water, Double Distilled Water and Lacustrine Water). Eutrophication and the subsequent blooming of biomass have been observed at high temperature, about 23°C. In case of DDW, it has been observed an increase of P concentration and of pH values, whereas it is observed that P/N ratio increases, pH decreases and N concentration increases in solution only at the beginning, whereas a pH increase is observed in TW , considering both ratios: N/P P/N . It is suggested that chemical and biological interactions have been taking place throughout the incubation period, and temperature has supported these interactions. In LW, the temperature is one of the main factors causing algal blooming, pH increases above 7 because ecosystem should buffer any changes, in particular the chemical ones. In conclusion, for DDW, if P and N amounts are low and P is constant, nitrogen controls the algal blooming; in TW, P should be identified as a strictly limiting factor; based on N/P ratio, it was not observed an algal blooming phenomenon, it is suggested an imbalance between N and P and probably N is the key factor. In LW, where the N/P ratio was high, P is considered the controlling factor, but P and N are not the only factors to consider, water pH and temperature have a fundamental role, too.

Heisler et al. (2008) have anticipated in a round table about Harmful Algal Blooming (both autotrophic and heterotrophic) that many HABs may be promoted by other factors such as physical, biological and other chemical factors, not only nutrients. Potentially, toxic species and high-biomass producers were defined harmful because they could cause hypoxia and anoxia in marine life if the concentration is very high whether, or not the toxins were produced. Not only the high concentration of nutrients in water, other factors may alter the water quality, so macronutrients are considered as the main factors; but also micronutrients like trace metals are assumed to be fundamental for the growth of phytoplankton and have an important role in assimilating macronutrients. Moreover, allochthonous sources such as fuel biomass are considered a cause for HAB. It has been stated that anthropogenic nutrients do not always stimulate the presence of HABs, they could be assumed one of the factors; later, these nutrients influence HABs growth and biogeochemical processes and/or stimulation of the other components of the food web. It is inferred that the flux of nutrients and not their concentration is the term to consider defining the non-point source inputs like runoff and/or groundwater, therefore, reducing nutrients does not always mean a decrease in algal blooming in estuarine and marine waters. Unfortunately, this is not valid for all the species, in some cases this relationship seems to be more complex, thus, it is suggested that to decrease HABs it is not only a question of time whether the alteration of the ecosystem is very deep; since other systems result permanently altered so that the nutrients reduction is not the only solution to restore the previous conditions.

Moreover, Smolders et al. (2010) have stated that sulphate could induce phosphate mobilization in wetlands because of the internal eutrophication, even though phosphate is quite immobile in soil. In addition, nitrate could interfere with sulphur and iron biogeochemistry in the subsoil by the oxidation of iron sulphides (including pyrite). The products are insoluble iron sulphide minerals and iron phosphate which are also mobilized. If the sulphate amount increases, the consequence is a strong eutrophication of phosphate in wetlands via surface waters. In particular, nitrate is considered an electron acceptor, but it could be reduced by ferrous iron influencing the iron cycle, so that ferrous iron-bearing carbonate (siderite) or sulphides (pyrite) become an electron donor for denitrification in aquifer soils. At these conditions, P could be bound to oxidize iron, so that P amount decreases in groundwater and in the wetlands too. At the same time, in anaerobic conditions, it is observed that sulphate and nitrates interfere with iron phosphorus chemistry stimulating anaerobic decomposition of organic matter. In case of anaerobic decomposition, nitrate and sulphate regulate the availability of organic matter if electron acceptors are available. Moreover, it is observed that increasing the concentration of sulphate and decreasing the input of iron in groundwater, there is a release of phosphate because of the interference of sulphide itself in the iron–phosphorus cycle and an increase of anaerobic decomposition of organic matter. It is observed that iron is bound to reduced sulphur, unfortunately, P is not retained from the sediments showing a high mobility in the soils and unluckily it is observed a strong eutrophication phenomenon in the system. Consequently, iron input is decreased because of the nitrate immobilization.

Sansfica et al., (2010) also studied the chemistry of surface waters and groundwater draining agricultural catchments in paddy soil in Sri Lanka to avoid the eutrophication phenomenon. They noted that the concentration of nitrate in surface waters is not related to the amount of the same nutrients in groundwater. In particular, nitrate and phosphate concentration show a positive correlation with different clustering between groundwater and surface water. The low amount of nitrate in surface waters, notwithstanding the great amount of fertilizers, is due to the dilution, whereas the high values of P are caused by the agricultural use. They also noted that in some lakes, P values are low because it is leached. On the other hand, fertilizers used for agriculture purposes plays an important role in groundwater hydrochemistry. It is inferred that the source of high value of nitrate in groundwater is anthropogenic one, whereas the phosphate source in surface water might be due to the dissolution of rock phosphate. They stated that in groundwater, both nitrate and phosphate undergo the processes of reduction, precipitation, transportation, and adsorption depending on the condition of the aquifers and their concentration. In particular, in a reducing environment, a decreasing of nutrients is observed. There is nitrate reduction because the aquifer is shallow and there might be a high amount of dissolved organic carbon below the oxicline. It is suggested that organic matter is a key in controlling nitrate reduction, this reduction could be increased from the presence of Fe-silicates, such as pyroxenes and amphiboles, which are commonly found in the basements of the studied area.

Phosphate may be removed through precipitation with cation or through sorption on clay layers, if the retention time is long and if there is an ionic interaction between clay and phosphates. Groundwater quality depends on the surface water, on the process of dilution and on the reducing condition. Concerning surface waters, the source of nutrients is an anthropogenic one; the excessive use of fertilizers for agriculture purposes will increase the amount of nitrogen and phosphate in surface water by leachates and consequently in the groundwater too.

14.5 Water Quality Index (WQI)

In the world, it has been observed an intensification in the use of the irrigation water because of the increase in population and food production. It is necessary a certainty in water quality, but most important it is to define parameters for water quality. A classification of water quality that considers the interaction between salinity and soil sodicity with the toxicity risk is not still available. The water quality is defined in terms of its physical, chemical and biological parameters, and ascertaining its quality is important before its use to minimize the impact on agriculture and/or to avoid the utilization of degraded quality of water. Water quality indices (WQI) (Meireles, 2010) are considered a very simple and useful tool to manage water, in particular irrigation water. Almeida et al., (2008) suggested not considering only one index, as it might not be enough to better evaluate the quality of the water, and. in some cases, results may not be correct. Concerning the quality of irrigation water, this could affect crop and soil; hence, it must be determined more than one quality index for this purpose, about quality indices, the aim is to value the alterations in time and space. It is observed that these indices vary from time to time and region to region, so, it is extremely hard to develop a single general water quality index. One of the advantages in summing up the numerous data in a single value is an impartial, rapid and logical way, as it evaluates different areas and detects the changes in water quality and also an index value can relate to a potential use of that source of water (Abbasnia et al., 2018). Therefore, it is fundamental to bring many values to get a better understanding about the water. Candela et al., (2009) proposed water from urban zone as a principal pollution source and agriculture as a non-point source of nutrients influenced by climatic conditions as well as geomorphological, lithological and pedological properties. This proposed model can be adjustable in adding or removing the sub-models like Storm Water Tank (SWT) or the Combined Sewer Overflow (CSO). With this model, it is possible to evaluate the net rainfall and simulate the net runoff, the surface storage and the infiltration, and the solid transfer module. The CSO structures describe the hydraulic behaviour of the overflow. Candela et al., (2009) proposed, according to the Water Framework Directive (Europe), either point and non-point sources as fundamental to have an integrated view, to better understand the roles of all the components, both natural and anthropic drainage systems. This kind of modelling is complex, but it confirms the interplay of each contribution to the pollution from urban and agriculture areas. It is determined that metals and Biochemical Oxigen Demand BOD are mainly from urban areas; nitrogen is from urban area, and mainly from agriculture like phosphate, whereas phosphate from urban areas is really limited.

These results confirm the necessity to use an integrated model approach to better understand every component and source of the system. Even though these results depend on the specific case study, an important view into pollution propagation in water systems is provided. It is important to assess the health of water bodies to develop strategies to improve water resource and watershed management.

14.6 Irrigation: Leaching and Management

The negative environmental effects of intensive agriculture threaten the expansion of irrigation in many areas of the world. The sustainable use of irrigation water is a priority for agriculture. Nitrogen is the nutrient that requires a better management through runoff, leaching, denitrification, and volatilization, and particularly leaching is frequently the most important loss processes in agriculture. A suitable, adequate management of irrigation water is required for an efficient use of nitrogen and phosphate too. It is fundamental to calculate these losses and to establish a best management practices aimed at their reduction.

Lee et al., (2010) have written that the relationship between land use and water quality is bidirectional. In fact, land-use activities have a direct impact on river water quality, while water quality influences the siting of land-use activities. As a matter of fact, irrigation is needed for increasing, securing, and diversifying agricultural production. A sustainable irrigation means to follow the concept highlight by the

Brutland Commission Report (WCED, 1987), which is to use resources without compromising the ability of future generations to meet their own needs. Modernization in irrigation is a key strategy to secure a profitable crop yield, meeting adequate living standards for farmers and conserving water resources in terms of quantity and quality.

Speaking of sustainable irrigation, two goals must be kept in mind: irrigation for agriculture and preserving the environment. The most important problem related to irrigation is the huge amount of water and the consequences are the depletion of the water resources, the quality reduction and the salinization. Indeed, inappropriate irrigation practices, accompanied by inadequate drainage, have often damaged the soils through oversaturation and salt buildup (Cai et al., 2001). It is suggested a sustainable irrigation water management to shorten the interference of the irrigation system within the environmental system, including the effects on the water bodies. Water bodies receive irrigation water through wind-drift, surface runoff, or drainage to groundwater. It is suggested an improving of soil and plant practices, water pricing, reuse of treated wastewater, fertilizer application and disease and pest control (Chartzoulakis and Bertaki, 2015).

At present, about 90% of freshwater is for irrigation purposes, to prevent water scarcity or salinization, hence, it is necessary to model irrigation water requirements. According to Döll and Siebert (2002), the irrigation model will be used to assess the impact of climate change on irrigation water requirements regarding the differences among world regions. The quality of water is also important. It ought to be identified in terms of its physical, chemical and biological parameters. River water quality is characterized by a high level of heterogeneity in time and space, the consequence is the difficulty to identify water conditions and pollution sources. It is necessary to effectively control pollution, in addition to building up a successful strategies for minimizing the contamination of resources. The variation of water quality depends on the change in land-use. thence, the land-use activities in the basin must be carefully planned and controlled on account of protecting the water resource and quality status.

Fawaz et al. (2013) stated that in Semenyih River, nitrogen occurs as nitrate, which is very mobile in water; luckily nitrate is much less toxic than ammonia and nitrite. On the other hand, NH_3-N could be considered as nutrients in small concentrations for excessive growth of algae (Corwin et al., 1999), whereas high concentrations of phosphates are generally the indication of pollution associated with eutrophication conditions and the sources are anthropogenic and agricultural ones. They have determined that the responsible parameters for variations of water quality are principally associated with soluble minerals and temperature as natural sources as well as agricultural activities, domestic wastewater and surface runoff from roads and villages as anthropogenic activities. However, water quality of the Semenyih River varies according to the seasons and location of the sampling stations. At the same time, the amelioration of irrigation for agricultural productions allows a stability in the supply of food. In fact, it has such an important impact that only 20% of the world's agrarian surface generates 40% of agricultural production, meanwhile using 70% of water resources (FAO, 2003). Therefore, it is very necessary to improve the use of water and fertilizer, and their management to avoid the contamination of the water, this could be implemented monitoring and controlling the drainage networks, in particular, the river water quality whose chemical concentration is due to the climatic, hydrogeological, and agronomic characteristics of the territory.

García-Garizábal and Causapé (2010) suggested that controlling the amount of water and fertilizers, the environmental impact on soil is minimized. A nitrogen fertilization with higher efficiency in the application is possible, not always higher irrigation efficiency was sufficient to compensate a higher evapo-concentration, and it was observed an increase in salinity and nitrate concentration in the drainage water. Moreover, Klocke et al., (1999) have confirmed that in a semiarid climate, the leaching in non-irrigated periods depends on the nitrate present in the soil and on the precipitation. The absence of crops that takes the nitrate available in the soil and decreases soil water content is a contribution to leaching. Soil must be also considered as a key factor such as the type of crop, the irrigation and fertilizer rate amount. Alternatives in the irrigation management allow efficiency in irrigation and improvement in water quality in the river and return flows.

The intensification of agriculture, the extensive use of fertilizers and water have caused many environmental problems, Stamatis et al., (2011) observed in Thessaly valley that water of shallow aquifers are considered unsuitable for human use because of the amount of pollutants, whereas waters from deeper aquifers are suitable. The scarce quality of water from shallow aquifers is due to the intense agriculture (large use of nitrogen and phosphate), and unfortunately the disposal of urban wastes, livestock, fertilizers and pesticides; the consequences are the decreasing of organic matter in soil and the degradation of soil too. Nitrogen is a nutriment for the crop and generally it is added to increase the productivity, on the other hand, it is of great concern because of the percolation phenomenon, as deep percolation contributes to an increase in nitrogen amount in the aquifers and depletion of freshwater within the aquifers. Generally, when an element is applied to soil, it may be leached, retained by the soil particles, incorporated by soil microorganisms or taken up by plant. If the amount in plants is high, it could damage the crop or reduce the yields. Intense precipitations, irrigation methods and floods must be also taken into account to avoid high nitrogen concentration in water. Phosphates give the same issues. an increase pf algae growth blooms and consequently deterioration in the aquatic biotope of fish and of other aquatic organisms are observed, this is the reason why high amounts of NO_3 and PO_4 in drinking water are not allowed. It was suggested that this pollution is caused by anthropogenic activities, in particular runoff from nitrate fertilizer and leaching in deep soil layers from the intense use of these substances, but it is also partially due to the atmospheric inputs. The great problem is the accumulation in plants causing a damage or a reduction to the crop.

Islam et al., (2008) observed also that NO_3 and PO_4 retention inversely depends on the clay content, and a little amount of them are uptaken from the growing plants; in case of high pH values, a lower retention of these two ions is observed. Nitrate and phosphate leaching increases with the increase of their concentrations in solution. Unfortunately their accumulation in soil means a reduction in soil fertility, soil microbial activity, plant growth and the agriculture products, too. Anderson and Siman (1991) have written that a high amount of phosphorus in a freely draining soil and afterwards a strong rainfall caused a P loss through volatilization and leaching processes, and the consequences are the pollution and freshwater eutrophication. Islam et al. (2008) have stated that the leaching of NO_3 and PO_4 increases with the amount of nitrate and phosphate in solution but clay fraction is the fundamental parameter that controls this phenomenon. This is already stated by Conway and Pretty (1991),. they have also determined that nitrate leaching from a high nitrate amount is higher than an applied low dose. Likewise, Greenland and Hayes (1981) have determined that phosphate leaching from high doses of applied phosphate was significantly higher than that of the low doses.

Khurana and Aulakh (2010) have confirmed that leaching is due to the kind of soil; hence, leaching and nitrification in permeable or porous soil within a high amount of N fertilizer make nitrate leaching a serious problem in irrigated soils. It was determined that the presence of vegetation retards NO_3 leaching from the root zone absorbing nitrate and water. It was also determined that a high amount of residual P in soil may enhance a downward movement, thence P could reach groundwater; at the same time, P leached from agricultural soil may be intercepted by artificial drainage or subsurface flow. P could be transported to water bodies and the water quality could decline. Again, soil characteristics are fundamental for the interplay between P from fertilizer and labile P in soil profile. Not only soil characteristics play a role in N and P concentration, also the water chemistry plays an important role as in Lorite-Herrera and Jiménez-Espinosa (2008). They observed that the application of N fertilizers is a common practice during the growing season. They also reported that a reduction in the NO_3 proportion in applied fertilizers leads to a decrease in the amount of NO_3 leached. In case of obsolete irrigation (flood irrigation), the amount of water is high and it is also high the irrigation return flow which could leachate nitrogen and consequently high Nitrogen level in irrigation springs. They have found a high value of nitrate and NO_3 (> 50 mg/L) in drinking water, irrigation wells and spring; on the other hand, phosphorus concentration in fertilizer is between 15% and 18% as P_2O_5, but phosphate is not detected. In aerobic conditions, concentration decreases rapidly and later the amount continues to reduce, an adsorption phenomenon on the surface of iron and aluminium hydroxides is observed, coating clay particles control the behaviour of phosphate in solution. The aerobic conditions are inferred from the amount of oxygen in groundwater; it seems this adsorption process might be the key factor controlling phosphate concentration. Because of the aerobic condition, the active decay of nitrate is unlikely, and it is not determined as a denitrification process in groundwater. The oxygen concentration is fundamental to determine the freshness of water; when wastes are dumped in water, it is observed that nitrate and phosphate increase and subsequently a great amount of plants and algae grow in these waters and oxygen is depleted. Runoff and organic matter decomposition in surface water also produced inorganic nutrients (ammonia, nitrate, and phosphates) with eutrophication phenomenon and other serious ecological impairments of the water body (Adesuyi et al., 2015).

It is important to adjust the rate and time of N fertilizer to get the maximum yield and, at the same time, save the nitrogen rates. Therefore, the management of fertilizers is fundamental for the increasing of the production and at the same time also to get a better environment; so the use of fertilizers must be done according the climatic and soil physico-chemical conditions. In Gadalla et al. (2010), it is underscored that an excessive use of nitrogen fertilizers could be considered by the farmers as a reasonable insurance against yield losses, without considering the leaching, volatilization, denitrification and surface runoff to avoid the accumulation of nitrate in vegetables. In particular, it has been observed that the movement and uptake of nitrate was negatively affected by the reduction in water quantities, minimizing the amount of water means minimizing the amount of fertilizers. Isidoro and Aragüés (2007) noted in Ebro river (Spain) a decreasing trend of PO_4 after 1993, thanks to the less amount of P in cleansing detergent. However, they suggested a common source for the presence of NO_3 and PO_4 in the river, that is an urban waste discharge, although PO_4 influence in irrigation water is less apparent than NO_3. High NO_3 concentration is due to the irrigation return flows from large irrigation schemes, whereas high NH_4 and PO_4 concentrations are normally associated with the return flows from urban areas, and a correlation between NH_4 and PO_4 is also suggested. The management of urban discharge is also important to avoid an high amount of these minor ions (Kuyeli et al., 2009).

Kim and Chung (2010) have suggested that the introduction of high level NO_3 with irrigation water in rice paddy can limit the P solubilisation in root zone soil layer in addition to the excessive supply of N to the rice plants. Commonly, submergence enhances available PO_4 levels in flooded anoxic rice paddy soils, which is attributable to reduction of hydrous ferric compounds, particularly Fe-oxides, and involving NO_3 and SO_4 in anoxic soil. The presence of NO_3 in water acts as an effective redox buffer and substantially delays the release of PO_4 in various anoxic flooded ecosystems including rice paddies. It is determined that the presence of NO_3 in anoxic layer of rice paddy soil with fertilization and irrigation can inhibit the reduction of Fe and P solubilisation. High concentration of NO_3 in the water can suppress the SO_4 reduction and PO_4 accumulation is also observed. This high NO_3 concentration inhibits the eutrophication in river water and ground water. it limits the availability of PO_4 in fresh water wetlands. It seems that PO_4 loss is reduced through leaching. The amount of P, sorbed and/or coprecipitated, increases the amount of PO_4 solution or extractable in anoxic flooded soils.

Różkowski et al., (2017) have noted that the nitrates leaching from the vadose zone means a complex interaction of many factors (land use, dynamics of soil nitrogen, and depth of the groundwater table), whereas the leaching of phosphorus is considered less important, more important is the transport of particulate and soluble phosphorus increasing the concentration of the element in surface waters. In case of soil saturation, the potential for losses of soluble phosphorus in surface runoff will deeply increase. They carried out a study on the water in a cave and found that the reduced forms of nitrogen (NH_4^+, NH_3 and NH_4SO_4) prevail in surface waters, and phosphates were present as dissolved forms. It is observed the loss of phosphorus is due to the precipitation of stable mineral phases and at the same time is due to the sorption processes with the ions of Fe, Al and Mg. If temperatures conditions of and redox conditions change these ions are able to cause a phosphorous recovery from sediments. They assumed that one of the most important factors affecting the quality of groundwater with phosphorus and nitrogen compounds is the improper use of land: runoff from fields, crop residues and manure, discharge of urban and industrial waste and the most important fertilizers for agricultural purposes.

14.7 Remediation

Water pollution due to the excessive presence of nitrogen and phosphorus compounds is one of the most frequently faced environmental problems around the world. Developing an efficient low-cost removal method is important to protect the aquatic environment.

Many biological, chemical and physical processes have been applied to remove NO_3 and PO_4 from water. Adsorption process by the ion exchange mechanism is evaluated as an efficient and cost-effective method. Nguyen et al., (2020) have used ferric oxide as a support to synthesize ferric-amine oxide to remove NO_3 and PO_4. It is observed (by FTIR), after the adsorption, some peaks are considered to be of nitrate and phosphate, confirming that ferric oxide-amine can adsorb both anions. By thermogravimetric analysis, it is observed that at 600°C the ferric oxide as γ-hematite (cubic structure) is converted to α-hematite (hexagonal structure); between 300°C and 500°C the weight of ferric oxide-amine decreases, for a loss of water and it is observed a restructuring of the oxide increasing the temperature. The adsorption equilibrium of nitrate is reached after 30 minutes and after 60 minutes for phosphate. For the latter, it is a three-stage process, that is, a rapid increase for the first 30 minutes by the externally exposed surface of the amine group, secondly a slow increase for 30 minutes by the internal ferric oxide surface and after 60 minutes the equilibrium is reached, and the saturation of internal and external surfaces is completed. The pH values, to get the best results, are in the range of 5–6 for nitrate and 6 for phosphate. Adsorption capacity increases with temperature and concentration of nitrate and phosphate, but it is observed a decreases with the increase of adsorbent dosage.

Bektaş et al., (2004) have proposed Electro-Coagulation (EC) to remove phosphate. EC is considered a reliable and cost-effective method, advantages are a short operation time and fewer amounts of chemicals. Some suitable values of EC duration (15 minutes) are determined, whereas increasing EC duration is not so effective in removing a great amount of phosphate. The initial concentration of the solution influences the time in removing phosphates, increasing the concentration the percentage phosphate removal decreases. According to the results, they have supposed that the amount of Al^{3+} ions at the first stages of EC are not enough to bind all of the phosphate ions, whereas increasing the duration of EC process, the amount of Al^{3+} ions increase and therefore an higher percentage of phosphate is bound. It is demonstrated that EC processes provide stable effects and it is possible to remove phosphate from aqueous solution in spite of the high concentration.

Berkessa et al. (2019) studied the feasibility of using Solid Waste Residual (SWR) from manufacturing of aluminium sulphate and sulphuric acid for the simultaneous removal of nitrate and phosphate ions from aqueous solution. It is observed that the adsorption on SWR of both nutrients is dependent on pH solution. In particular, nitrate is better removed at pH values ranging from 2 to 7, and if pH increases, the efficiency decreases. The pH value, for the suitable efficiency, ranges from 5 to 9. The decrease in nitrate removal seems to be due to internal competitions between hydroxide ions, while the low removal is due to the instability of oxides from SWR. On the contrary, the low phosphate removal at pH < 5 seems to be caused by H_3PO_4, which is weakly attached to the sites of the SWR. The adsorbent shows a low tendency to adsorb nitrate, whereas phosphate shows a good affinity for SWR. Reached the equilibrium, the removal mechanism for nitrate was physical adsorption, the maximum adsorption is reached after 60 minutes. Concerning phosphate, the equilibrium is reached in 90 minutes, after the first 60 minutes, the adsorption efficiency is high and remained unchanged afterwards.

Mikhak et al., (2017) have studied if $Fe(OH)_3/CpNPs$ is suitable for removing PO_3 and NO_4 from water. They have chosen these two nutrients for their role in the eutrophication process. It is observed that nitrate was initially fast removed in the first 8 hours, later removal is constant for the following 4 hours. Phosphate behaviour is quite similar, the difference is after 8 hours, and removal gradually increases. Furthermore, increasing the adsorbent concentration, the contact surface increases; hence it is observed more elimination of nutrients. At the same time, it is observed t increasing the amount of FeNPs, the phosphate removal increases. This is probably due to the better phosphate tendency to bond with Fe than NO_3. In this case, the relation between the removal and initial pH is not positive. As a matter of fact, the phosphate removal decreases with the increase of pH, this effect is more intense on nitrate because nitrate results are more sensitive to the pH variation. Neutral pH values show the best efficiency in removing both nitrate and phosphate. It is supposed that the presence of both nutrients slightly decreases the removal efficiency of nitrate.

Rumhayati et al. (2012) studied acrylamide-ferrihydrite gel to remove phosphate and nitrate ions from drinking water. Ferrihydrite has been chosen because of its reactivity and the large specific area, the disadvantage is the reduction Fe^{3+} into Fe^{2+}, to avoid this drawback, the adsorbent must be stored in a dark bottle. It is obvious pH has a key role in the adsorption process, as a matter of fact, the nature of surface site groups of acrylamide-ferrihydrite gel is changed. The suitable pH value is 5, unfortunately, if pH is less or more than this value, the percentage of removal is less for both nitrate and phosphates. These low values seems due to the ferric iron dissolution at pH less than 4; at the same time, it is supposed the reduction of ferric iron to ferrous iron which is soluble up at pH 8. Hence, in presence of ferrous iron, it is observed the reduction of nitrate or ammonia and consequently, a decreasing in removing nitrate and phosphate at pH more than 5. The nutrients' concentration is also an important factor in controlling the percentage of removal, if the concentration is low, the removal is higher; increasing their concentrations, the removal decreases. Nevertheless, the acrylamide-ferrihydrite gel could remove phosphate and nitrate in the drinking water as high as 70%.

Rahmani et al., (2015) have carried out tests on anionic compounds adsorbent based on Polypropylene nonwoven. It is a convenient method without affecting the bulk properties of the fibres at atmospheric pressure. The results show that the adsorption process of nitrate and phosphate depends on the amination degree of the PP fibres. In particular, the adsorption capacity increases up to 120%, but increasing the amination degree the removal capacity decreases, this is probably due tothe thickness of the fibre decreases and the effective surface too, and consequently the adsorption functionality. They suggested an increasing of the contact time to reach the equilibrium. Moreover, the degree of grafting depends on the monomer amount and the time of plasma treatment.

Developing an efficient low-cost removal method is important to protect the aquatic environment. Further studies of quantitative indicator analysis are needed to carry out a sustainability appraisal in terms of social, economic and environmental balance. Studying the economics of different water treatments is an essential pre-requisite for the identification of cost-effective solutions. In addition, the effectiveness and sustainability of a remediation project is affected by many qualitative and quantitative indicators, in particular the water and land use after remediation and available information of the site.

Policies require frequent reviews since the environmental situation is constantly changing. Although industrial and technological developments have made life easier in many aspects, they have also caused some losses. Conserving precious water resources is absolutely essential to protect the environment during these times of climate change.

14.8 Conclusions

It is well established that the excess of nutrients like phosphorus and nitrogen has a negative environmental effect in waters, such as eutrophication reducing the biodiversity in aquatic ecosystem and the quality of the water. The limiting nutrients for N and P are well determined, but the bioavailability could be a better parameter to consider because it could change according to the redox process, pH changes and enzymatically mediated hydrolysis processes, in particular for P nutrient according to Reynolds and Davies (2001). Generally, these nutrients are lost from the agriculture system and livestock production, due to their high availability, they can reach the water system because of runoff. Their concentration in water could also be increased because of the soil erosion and leaching process too (Chapman et al., 2005). Some soil managements have been suggested such as tillage strategies, crop rotation and management (Newell Price et al., 2011).

To avoid eutrophication process, it is suggested that N application ought to be balanced between crop requirement, manure and fertilizer keeping in mind the N amount released from the soil; similarly, it is fundamental considering the amount of P available in soil so that to best manage the amount of nutrients at farm and fields scales, as a matter of fact, the P surplus accumulates in the soil (Lemercier et al., 2008). It is stated that integrated fertilizer and manure nutrient strategies are important and these strategies could improve the water quality reducing the annual rate of fertilizers (Newell Price et al., 2011). It has been observed that the placement, the timing of manure and fertilizer are important. It is also important the rates and the forms of applied nutrients, they should be suitable according the crop requirements and the soil. It has been suggested that splitting nutrient applications and the timing application increase the efficiency in the uptake from plants. Schoumans et al., (2014) have asserted that separating the manure into a liquid fraction (with high N and low P content) and a solid one (high P and low N) could be a better solution. Moreover, the liquid fraction could be applied locally, while solid fraction could be transported for an incineration process and later for the energy production. This can contribute to sustainably closing the terrestrial P cycle and to reduce the risk of surface water pollution.

As soil and crop management is suggested the direct drilling avoiding the ploughing. Further, it is suitable to increase the aggregate stability through planting material, hence the biological activity is improved and the risk of erosion is reduced (Ulén et al., 2012). Mainly to reduce the erosion risk, it is suggested to increase water infiltration, to reduce runoff volume and erosivity and to protect soil surface from erosion with plants or residue cover (Govers et al., 2004). It must be considered the timing to immobilize N between the growing season of two crops by harvest. On the contrary, the amount of P in soil solution is buffered by the amount of P in solid fraction, moreover, dissolved P in soil solution is not dynamic as N, so that P amount can decrease in soil solution.

The management strategies for water are the field water management, the land-use changes and the landscape management. The first strategy has important consequences. If the water flow l changes, the soil moisture will change and the effects are on chemical and biological processes. Water management means to reduce nutrients in overland and subsurface pathways' water flow. As land-use measure, it is suggested to change the agricultural land use (e.g. agro-forestry) to a non-agricultural use. As water management strategy, it is suggested to regulate the surface flow connectivity between surface water and farm infrastructure, which means waste. The surface water management includes the river maintenance, lake re-establishment and wetland restoration. About the river maintenance, it is necessary a physical improvement of the channel and the restoration of riparian areas, it is also important to consider the residence time of the nutrients in surface water (De Klein and Koelmans, 2011).

Finally, it results that controlling the water flow from fields to surface water is the first key factor to reduce nutrients from agriculture system. Nutrients from agricultural land can be changed by changing the drainage system. Surface water management to increase the removal efficiency is cost-effective for both N and P. To understand which water system is more susceptible to eutrophication is fundamental for the most suitable management, preventing this phenomenon.

References

Abbasnia A., Radfard M., Hossein Mahvi A., Nabizadeh R., Yousefi M., Soleimani H., Alimohammadi M., 2018 "Groundwater quality assessment for irrigation purposes based on irrigation water quality index and its zoning with GIS in the villages of Chabahar, Sistan and Baluchistan, Iran" *Data in Brief* Vol. 19, pp. 623–631.

Adesuyi A. A., Nnodu V. C., Njoku K. L., Jolaoso A., 2015 "Nitrate and phosphate pollution in surface water of Nwaja Creek, Port Harcourt, Niger Delta, Nigeria" *International Journal of Geology, Agriculture and Environmental Sciences*, Vol. 3 (5), pp. 14–19.

Akhavan S., Abedi-Koupai, J, Mousavi, S. F., Afyuni, M., Eslamian, S. S., Abbaspour, K. C., 2010 "Application of SWAT model to investigate nitrate leaching in Hamadan–Bahar Watershed, Iran" *Agriculture, Ecosystems and Environment*, Vol. 139, pp. 675–688.

Almeida C., Quintar S., González P., Mallea M., 2008 "Assessment of irrigation water quality. A proposal of a quality profile" *Environmental Monitoring and Assessment*, Vol. 142, pp. 149–152.

Anderson A., Siman G., 1991 "Levels of Cd and some other trace elements in soils and crops as influenced by lime and fertilizer level." *Acta - Agriculture - Scandinavica*, Vol. 41, pp. 3–11.

Baker A. 2006 "Land use and water quality" *Encyclopedia of Hydrological Sciences*, Vol. 10, pp. 1–6.

Bektaş N., Akbulut H., Inan H., Dimoglo A., 2004 "Removal of phosphate from aqueous solutions by electro-coagulation" *Journal of Hazardous Materials*, 106B, pp. 101–105.

Berkessa Y. W., Mereta S. T., Feyisa F. F., 2019 "Simultaneous removal of nitrate and phosphate from waste-water using solid waste from factory" *Applied Water Science*, Vol. 9 (2), pp. 9–28.

Bo W., Wang X., Zhang Q., Xiao Y., Ouyang Z., 2018 "Influence of land use and point source pollution on water quality in a developed region: A case study in Shunde, China" *International Journal of Environmental Research and Public Health*, Vol. 15, pp. 51–59.

Broschat T. K., 1995 "Nitrate, phosphate, and potassium leaching from container-grown plants fertilized by several methods" *Hortscience*, Vol. 30 (1), pp. 74–77.

Cai X., Mckinney D. C., Rosegrant M.W. 2001 "Sustainability analysis for irrigation water management: Concepts, methodology, and application to the Aral sea region" Eptd Discussion Paper No. 86, pp. 1–60.

Candela A., Freni G., Mannina G., Viviani G., 2009 "Quantification of diffuse and concentrated pollutant loads at the watershed-scale: An Italian case study" *Water Science and Technology* Vol. 59 (11) pp. 2125–2135.

Carter D. L., Bondurant S. A., Robbins C. W., 1971 "Water-soluble NO_3-nitrogen, PO_4-phosphorus, and total salt balances on a large irrigation tract" *The Soil Science Society of America Proceedings*, Vol. 35 (2), pp. 331–335.

Catherine A., Mouillot D., Escoffier N., Bernard C., Troussellier M., 2010 "Cost effective prediction of the eutrophication status of lakes and reservoirs" *Freshwater Biology*, Vol. 55 (15), pp. 1–11.

Chapman A. S., Foster I. D. L., Lees J. A., Hodgkinson R. A., 2005 "Sediment delivery from agricultural land to rivers via subsurface drainage" *Hydrological Processes*, Vol. 19 (15), pp. 2875–2897.

Chartzoulakis K., Bertaki M., 2015 "Sustainable water management in agriculture under climate change" *Agriculture and Agricultural Science Procedia*, Vol. 4, pp. 88–98.

Clavel, L., Ducrot, R., Sendacz, S., (2008). *"Gaming with Eutrophication: Contribution to Integrating Water Quantity and Quality Management at Catchment Level."* IWRA.

Conway G. R., Pretty J. N., 1991 *"Unwelcome Harvest Agriculture and Pollution"*, Earthcan Publications Ltd. pp. 445–504. Island Press, London UK.

Correll D. L., 1998 "The role of phosphorus in the eutrophication of receiving waters: A review" *Environmental Quality*, Vol. 27, pp. 261–266.

Corwin D. L., Loague K., Ellsworth T. R., 1999 "Advanced information technologies for assessing nonpoint source pollution in the Vadose Zone: Conference overview" *Journal of Environmental Quality*, Vol. 28 (2), pp. 357–365.

Chen D., Elhadj A., Xu H., Xu X., Qiao Z., 2020 "A study on the relationship between land use change and water quality of the Mitidja watershed in Algeria based on GIS and RS" *Sustainability*, Vol. 12, pp. 1–20.

De Klein J. J. M., Koelmans A. A., 2011 "Quantifying seasonal export and retention of nutrients in West European lowland rivers at catchment scale" *Hydrological Processes*, Vol. 25, pp. 2102–2111.

Diatta J., Waraczewska Z., Grzebisz Z., Niewiadomska A., Tatuśko-Krygier N., 2020 "Eutrophication induction via N/P and P/N ratios under controlled conditions—Effects of temperature and water sources" *Water Air & Soil Pollution*, Vol. 231, p. 149.

Döll P., Siebert, S., 2002 "Global modeling of irrigation water requirements" *Water Resources Research*, Vol. 38 (4), pp. 1–10.

FAO, 2003, *"Unlocking the Water Potential of Agriculture"*, Food and Agriculture Organization, United Nations. Rome, Italy.

Fawaz A.-B., Shuhaimi-Othman M., Barzani Gasim M., 2013 "Water quality assessment of the Semenyih River, Selangor, Malaysia" *Hindawi Publishing Corporation Journal of Chemistry*, Vol. 2013, ID 871056, pp. 1–10.

Gadalla A. M., Galal Y. G. M., Hamdy A. Ismail M. M., Sherin A. El Degwy, Hamed L. M. M., 2010 "Proper management of irrigation water and nitrogen fertilizer to improve spinach yield and reserve environment using 15N tracer technique" *Radiation Research*, Vol. 42 (2), pp. 3–17.

García-Garizábal I., Causapé J., 2010 "Influence of irrigation water management on the quantity and quality of irrigation return flows" *Journal of Hydrology*, Vol. 385, pp. 36–43.

Greenland D.J., Hayes M.H.B., 1981 "*The Chemistry of Soil Processes*", John Wiley & Sons, London.

Govers G., Poesen J., Goosens D., 2004 "Soil erosion — Processes, damages and counter measures" In: Schjønning P, Elmholt S, Christensen BT, Editors. *Managing Soil Quality: Challenges in Modern Agriculture*, CABI Publishing, Wallingford. pp. 199–217.

Heisler J., Glibert P. M., Burkholder J. M., Anderson D. M., Cochlan W., Dennison W. C., Dortch Q., Gobler C. J., Heil C. A., Humphries E., Lewitus A., Magnien R., Marshallm H. G., Sellner K., Stockwell D. A., Stoecker D. K., Suddleson M., 2008 "Eutrophication and harmful algal blooms: A scientific consensus" *Harmful Algae*, Vol. 8, pp. 3–13.

Huang J., Zhan J., Yan H., Wu F., Deng X., 2013 "Evaluation of the impacts of land use on water quality: A case study in The Chaohu Lake Basin" *The Scientific World Journal*, Vol. 2013, pp. 1–8.

Isidoro D., Aragüés R., 2007, "River water quality and irrigated agriculture in The Ebro Basin: An overview" *Water Resources Development*, Vol. 23 (1) pp. 91–106.

Islam M. S., Ullah S. M., Khan T. H., Imamul Huq S. M., 2008 "Retention of nitrate and phosphate in soil and their subsequent uptake by plants" *Bangladesh Journal of Scientific and Industrial Research*, Vol. 43 (1), pp. 67–76.

Kaushik S. C., 2013 "Nitrate Pollution in Ground" Water. https://www.biotecharticles.com/Biology-Article/Nitrate-Pollution-in-Ground-Water-2998.html.

Khurana M. P. S., Aulakh M. S., 2010 "Influence of wastewater application and fertilizer use on the quality of irrigation water, soil and food crops: Case studies from Northwestern India" *Conference Paper August 19th, World Congress of Soil Science, Australia (Brisbane) Soil Solutions for a Changing World*.

Kim B-H., Chung J-B., 2010 "Effect of nitrate in irrigation water on iron reduction and phosphate release in anoxic paddy soil condition" *Korean Journal of Soil Science and Fertilizer*, Vol. 43 (1), pp. 68–74.

Klocke N. L., Watts D. G., Schneekloth J. P., Davison D. R., Todd R. W., Parkhurst A. M., 1999 "Nitrate leaching in irrigated corn and soybean in a semi-arid climate" *Transactions of the ASAE 42*, Vol. 6, pp. 1621–1630.

Kuyeli S. M., Masamba W. R. L, Fabiano E., Sajidu S. M., Henry E. M. T., 2009 "Temporal And Spatial Physicochemical Water Quality In Blantyre Urban Streams" *Malawi Journal of Science and Technology*, Vol. 9 (1), pp. 5–10.

Lee J. Y., Yang J. S., Kim D. K., Han M. Y., 2010 "Relationship between land use and water quality in a small watershed in South Korea" *Water Science & Technology*, Vol. 62 (11), pp. 2607–2615.

Lemercier B., Gaudin L., Walter C., Aurousseau P., Arrouays D., Schwartz C., Saby N. P. A., Follain S., Abrassart J., 2008 "Soil phosphorus monitoring at regional level by means of soil test database" *Soil Use Manage*, Vol. 24, pp. 131–138.

Lorite-Herrera M., Jiménez-Espinosa R., 2008 "Impact of agricultural activity and geologic controls on groundwater quality of the alluvial aquifer of the Guadalquivir River (province of Jaén, Spain): A case study" *Environmental Geology*, Vol. 54, pp. 1391–1402.

Maberly S. C., King L., Gibson C. E., May L., Jones R. I., Dent M. M., Jordan C., 2003 "Linking nutrient limitation and water chemistry in upland lakes to catchment characteristics" *Hydrobiologia*, Vol. 506–509, pp. 83–91.

Meireles A. C. M., Maia de Andrade E., Guerreiro Chaves L. C., Frischkorn H., Crisostomo L. A., 2010 "A new proposal of the classification of irrigation water Uma nova proposta de classificacao da agua para fins de irrigacao" *Revista Ciencia Agronomica*, Vol. 41 (3), pp. 349–357.

Mikhak A., Sohrabi A., Kassaee M. Z., Feizian M., Disfani M. N., 2017 "Removal of nitrate and phosphate from water by clinoptilolite-supported iron hydroxide nanoparticle" *Arabian Journal for Science and Engineering*, Vol. 42, pp. 2433–2439.

Newell Price J. P., Harris D., Taylor M., Williams J. R., Anthony S. G., Duethmann D., Gooday R. D., Lord E. I., Chambers B. J., Chadwick D. R., Misselbrook T. H., 2011 "An inventory of mitigation methods and guide to their effects on diffuse water pollution, greenhouse gas emissions and ammonia emissions from agriculture" Defra Project WQ0106 ADAS and Rothamsted Research.

Nguyen T. T., Thich Le T., Phan P. T., Nguyen N. H., 2020 "Preparation, characterization, and application of novel ferric oxide-amine material for removal of nitrate and phosphate" *Water Journal of Chemistry*, Vol. 4 (24), pp. 1–12.

Pierziynski G. M., Mcdowell R. W., Sims J. T., 2005 "Chemistry, cycling, and potential movement of inorganic phosphorus in soils phosphorus reactions and cycling in soils section" In: Thomas Sims J, Andrew N, Editors. *Phosphorus: Agriculture and the Environment*, Volume 46, Sharpley American Society of Agronomy-Crop Science. pp. 51-86. Wiley Online Library.

Rahmani S., Abdouss M., Kowsari E., Shoushtari A. M., Haji A., 2015 "Adsorption of phosphate and nitrate ions on surface modified polypropylene nonwoven" *Materiale Plastice*, Vol. 52 (2) pp. 218–220.

Reynolds C. S., Davies P. S., 2001 "Sources and bioavailability of phosphorus fractions in freshwaters: A British perspective" *Biological reviews of the Cambridge Philosophical Society*, Vol. 76 (1) pp. 27–64.

Różkowski J., Różkowski K., Rahmonov O., Rubin H., 2017 "Nitrates and phosphates in cave waters of Kraków-Częstochowa Upland, southern Poland" *Environmental Science and Pollution Research International*, Vol. 24 (33), pp. 25870–25880.

Rumhayati B., Bisri C., Kusumawati H., Yasmin F., 2012 "Phosphate and nitrate removal from drinking water sources using Acrylamide-Ferrihydrite gel" *Indonesian Journal of Chemistry*, Vol. 12 (3), pp. 287–290.

Sansfica M. Y., Pitawala A., Gunatilake J., 2010 "Fate of phosphate and nitrate in waters of an intensive agricultural area in the dry zone of Sri Lanka" *Paddy Water Environment*, Vol. 8, pp. 71–79.

Schoumans O.F., Chardon W.J., Bechmann M.E., Gascuel-Odoux C., Hofman G., Kronvang B., Rubæk G.H., Uléng B., Dorioz J.-M., 2014 "Mitigation options to reduce phosphorus losses from the agricultural sector and improve surface water quality: A review" *Science of The Total Environment*, Vol. 468–469, pp. 1255–1266

Sharpley N.L., Menzel A. G., 1987 "The impact of soil and fertilizer phosphorus on the environment" *Advances in Agronomy*, Vol. 41, pp. 297–324.

Smolders A. J. P., Lucassen E. C. H. E. T., Bobbink R., Roelofs J. G. M., Lamers L. P. M., 2010 "How nitrate leaching from agricultural lands provokes phosphate eutrophication in groundwater fed wetlands: The sulphur bridge" *Biogeochemistry*, Vol. 98, pp. 1–7.

Stamatis G., Parpodis K., Filintas A., Zagana E., 2011 "Groundwater quality, nitrate pollution and irrigation environmental management in the Neogene sediments of an agricultural region in central Thessaly (Greece)" *Environmental Earth Sciences*, Vol. 64, pp. 1081–1105.

Taranu Z. E., Gregory-Eaves I., 2008 "Quantifying relationships among phosphorus, agriculture, and lake depth at an inter-regional scale" *Ecosystems*, Vol. 11, pp. 715–725.

Tong S. T. Y., Chen W., 2002 "Modeling the relationship between land use and surface water quality" *Journal of Environmental Management*, Vol. 66, pp. 377–393.

Ulén B., Alex G., Kreuger J. A. S., Etana A., 2012 "Particulate-facilitated leaching of glyphosate and phosphorus from amarine clay soil via tile drains", *Acta Agriculturae Scandinavica. Section B*, Vol. 62, pp. 241–251.

WHO/FWC/WSH/16.52, 2016, Guidelines for Drinking-water Quality, Nitrate and Nitrite in drinking-Water.

World Commission on Environment and Development, 1987, Our Common Future, Brutland Commision Report, pp. 1–300.

15

Removal of Organic Pollutants and Enteric Pathogens by *Typha Latifolia* and Sand Filter from Domestic Irrigation Wastewater Reuse

Nedjma Mamine
and Fadila Khaldi
Mohamed-Cherif
Messaadia University

Nedjoud Grara
Faculté des Sciences de
la Nature et de la Vie et
Sciences de la Terre et de
l'Univers, Université

Saeid Eslamian
Isfahan University
of Technology

15.1 Introduction

Nowadays, most developing countries face enormous environmental problems, especially those related to the treatment of raw sewage. The discharge of these effluents into surface waters without adequate treatment causes environmental degradation (eutrophication, loss of biodiversity, reservoirs of pathogenic microorganisms) and renders them unfit for human consumption (Fagrouch et al., 2011; Youbi et al., 2018).

On the other hand, despite the efforts made in the implementation of conventional treatment systems (activated sludge, bacterial beds), raw wastewater still poses a threat to the environment and the populations of these countries (Kone et al., 2011). The water is mainly required for population drinking and plant irrigation, in addition, that the use of sewage wastewaters provide a nutrient source for irrigation purpose (Shingare et al., 2017), meanwhile the untreated wastewaters are responsible for serious human and environmental health problems (Scott et al., 2004). Accordingly, human health can be affected by wastewaters containing various pathogens, like viruses and protozoa. As previously reported, nearly 2 million people, including especially children, die by diarrheal illness every year (43% of deaths) (WHO, 2010).

DOI: 10.1201/9780429290114-20

Furthermore, the reuse of wastewater is one of the alternatives that could be reliable and very beneficial for irrigation and, at the same time, for agriculture (Hannachi et al., 2016). The wastewater can be an alternative to using clean water in agriculture, leaving fresh water for other uses, including providing drinking water. In reality, the environmental and socio-economic benefits of this reuse can only materialize if this water passes through a treatment plant that ensures the elimination of all elements that can harm the environment and public health (Mamine et al., 2020).

The agricultural vocation and surface water resources of Souk Ahras city (North-East of Algeria) contribute mainly to the reservoir water quality of the Medjerda River, one of the most important rivers in eastern Algeria (Athmani, 2008). This city faces many environmental problems by raw wastewaters that are becoming a risk to the health of the populations of Souk Ahras. However, their reuse in agriculture constitutes attractive inputs for agriculture (Keffala et al., 2012) and increasing danger for human health due mainly to harmful pathogens, such as helminths and bacteria (El Ouali Lalami et al., 2014).

Another point worth noting is that this region gets worse due to the multiplicity of the urban installations, often unaccomplished and to the lack of the appropriate sanitation facilities for wastewater discharge (Fonkou et al., 2010).

Therefore, it seems necessary to explore new reliable wastewater treatment technologies adapted to the realities of countries around the world. Phytopurification of wastewater is presented as an alternative technique for wastewater treatment (Kadaverugu et al., 2016; Ouatrra et al., 2008; Grara et al., 2017; Mamine et al., 2019). Its application now seems the most appropriate, given the diversity of plant species that can be used, the low installation costs and the good performance (Czudar et al., 2011).

Constructed wetlands (CWs) are artificial ecosystems for the treatment of wastewater and represent an alternative to conventional biological systems (Licata et al., 2019). Plants, substrate and microorganisms are the main components of CWs and their interaction is fundamental for the optimal functioning of the system (Leto et al., 2013).

It is worth considering that the use of macrophytes as bioindicators in domestic and industrial contexts is also appropriate, because, on the one hand, plants integrate temporal, spatial, chemical, physical and biological data from the environment (Kaldy et al., 2017; Yu et al., 2020), and on the other hand, simple measurements based on observations, morphological and/or physiological, can transcribe the deleterious effects linked to exposure to pollutants (El-Khatib et al., 2014).

Several studies have also shown the effective role of this technique in cleaning up raw wastewaters around the world (Grara et al., 2017; Rana et al., 2018; Mamine et al., 2019). Up to now, this technique is applied as a new technology, promoting the diversity of macrophyte species, the low installation costs and the good performance of the pollutants (Osei et al., 2019). The aquatic tested plants are either free or floating, like *Lemna gibba* (Chaudhary and Sharma, 2019) and helophytes such as *Phragmites australis* (Khaldi et al., 2019) and *Typha latifolia* (Yalçuk and Ugurlu, 2019).

The aim of the study was to evaluate the comparative performances of the two treatment systems (a planted filter and a sand filter) to purify urban wastewaters not only by following the evolution of the physico-chemical parameters and the elimination of organic pollution but also by the reduction of the pathogenic bacteriological load.

15.2 Materials and Methods

15.2.1 Plant Material

The plant model used in this study was monocotyledonous angiosperm, which is a cattail plant (*T. latifolia*) as defined by Bergner and Jensen (1989). It was collected from the Souk Ahras city. It's worth noting that it is a vertical and horizontal rhizome plant growing spontaneously in humid environments and forming monospecific fields. In addition, it has a height ranging from 1.5 to 3 m, and large leaves (2–4 cm), making it a helpful plant for humans, an important nutrient for semi-aquatic mammals (Julve, 1998) and a very productive species that easily adapts to adverse environmental conditions (Zaimoglu, 2006).

FIGURE 15.1 Experimental Design

15.2.2 The Experimental Description

The experimental design (fields) includes two serial systems (Figure 15.1) composed of two trays of 40 L capacity and 50 cm height, filled successively from top to bottom by three layers. The first two layers are composed of gravel of decreasing diameter (4–25 mm) and thickness 8 cm, and the third layer is composed of 18 cm thick sand. Each tray has a slope of 10% toward the downstream and equipped with an outlet valve to evacuate the percolation water (treated water). The first tray is planted with young stems of *T. latifolia* (with a density of 10 plants /m²), but the second tray remained empty and was kept as a control for the effectiveness of the planted filter. Both were retained for a 14-day treatment period. The systems are supplied with wastewater from the city of Souk Ahras flowing into a Medjerda River, and the filtration water is recovered by means of a tap placed at the base of each tray.

15.2.3 Raw Wastewater Sampling Site

Our sampling station was chosen on the Medjerda River, because it is the only receiving environment for the domestic wastewater effluents of the city. This point of sampling is located downstream of a first outlet of wastewater and a significant amount of discharge (Figure 15.2). It should be noted that this river is an important source of irrigation water in this region, exhibiting an agricultural vocation (Guasmi et al., 2006).

The city of Souk Ahras is characterized by subhumid climate and significant agricultural activities (Mamine et al., 2019). In addition, a wide variety of crops, including grains, vegetables and arboriculture, are grown by people in this region (Athmani, 2008). The experiment was conducted from April to July 2018. Throughout this study period, the operation of the experimental pilot plant was controlled by measuring the physico-chemical and bacteriological parameters with a retention time of 14 days. Samples were taken from the raw wastewater applied to the treatment in trays and at the outlet of these trays, under strict aseptic conditions to avoid accidental contamination during handling. Then, collected samples were carefully labeled, placed in sterilized vials and stored in a laboratory at low temperature (4°C) until the day of analysis (Rodier, 2009).

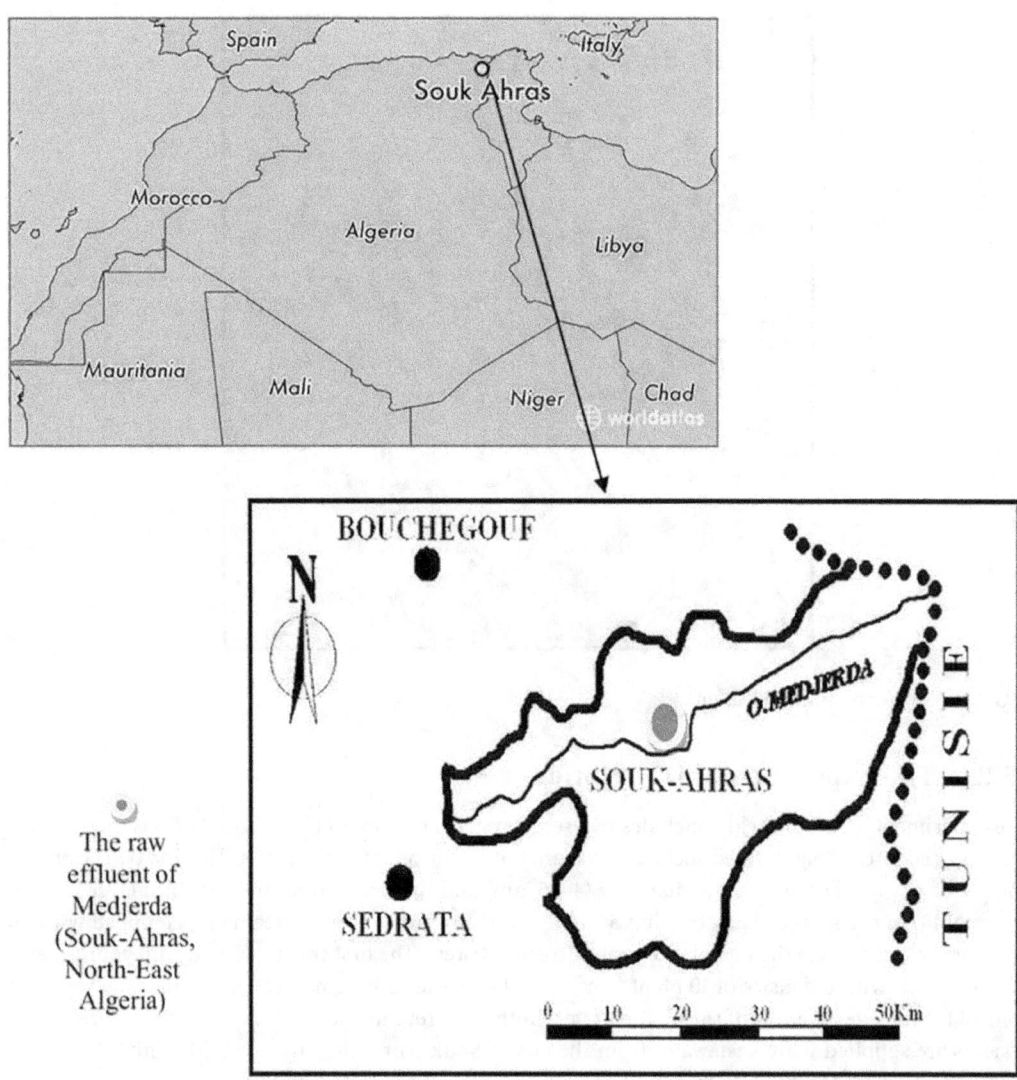

FIGURE 15.2   Location of the studied site (Guasmi et al., 2006).

15.2.4 Determination of Pollution Indicators

Table 15.1 summarizes the different methods of analysis of physico-chemical and bacteriological parameters.

15.2.5 Purification Yields

The abatement rates of physico-chemical and bacteriological parameters are calculated according to the following equation:

$$\text{Yield}(\%) = \left(X_{\text{RW}} - X_f / X_{\text{RW}}\right) * 100,$$

TABLE 15.1 Methods of Physico-Chemical and Bacteriological Analyses

Parameters	Determination Methods	Standards
pH	Multi-parameter probe	HQ40d multi/NACH
TSS	Spectrometric (HACH : DR5000)	NF, T 90-105
BOD_5	Manometric method	NF, T90-103
COD	Oxidation of potassium dichromate	NF, T90-101
Cadmium	Spectrometric method	NF EN ISO 5961
Total germs	General method by seeding in a nutrient agar culture medium	ISO 6222:1999
E. coli	The miniaturized method by seeding in liquid medium	AFNOR T90-433

AFNOR: French Standards Association; FD: Documentation fascicle; ISO: International Organization for Standardization; NF: French standard

TABLE 15.2 Physico-Chemical and Bacteriological Characteristics of the Raw Wastewater

Parameters	Minimum	Maximum	Mean±SD	References
pH	7.74	8.12	7.95±1.12	(6.5–8.4) (J.O.R.A, 2012) (FAO, 1985)
TSS (mg/L)	356.32	367.21	359.11±56.3	(30 mg/L of O_2) (WHO, 1989)
COD (mg/L of O_2)	242	265	258.23±78.61	(90 mg/L of O_2) (J.O.R.A, 2012)
BOD_5 (mg/L of O_2)	155	162	139.32±34.2	(30 mg/L of O_2) (J.O.R.A, 2012)
COD/BOD_5	1.57	1.63	1.6±0.15	/
Cadmium (mg/L)	0.034	0.046	0.039	0.01 mg/L (OMS, 2012)
Total germs (germs/100 mL) ($\times 10^5$)	17	21	18.5±4.56	Absolute absence (WHO, 1989)
Escherichia coli (germs/100 mL) ($\times 10^2$)	10	22	16.1±5.8	Absolute absence (WHO, 1989)

SD: Standard deviation

where:

X_{RW} ((mg/L) (germs/100 mL)) represents the concentration of the parameter selected in the raw wastewater and applied on the tray.

X_f ((mg/L) (germs/100 mL)) represents the concentration of the parameter selected in the filtrate.

15.2.6 Statistical Analysis

The collected data were analyzed by utilizing Statistica (version 7.1, XLSTAT, 2014) software, and the findings are displayed as arithmetic means±standard deviations. On the other side, the difference of performance between the trays was tested at a significance level of 0.05, after the normality check by the Kolmogorov–Smirnov test, since the analysis of variance was used to compare the mean leads subsequently to the determination of the significant difference between trays (Dagnelie, 2011), XLSTAT version 2014.5.03 has been used for this analysis.

15.3 Results and Discussion

15.3.1 Typology of Feed Wastewater

The physico-chemical characteristics of wastewater collected were evaluated (Table 15.2). The maximum pH values recorded for wastewater reused for irrigation purification trays (planted and unplanted) are slightly alkaline. These values comply with Algerian standards for the quality of wastewater used for irrigation (J.O.R.A, 2012) and are also within the range recommended by the Food and Agriculture Organization (FAO) (6.5–8.4) (FAO, 1985). Concerning the analysis of organic pollution indicator parameters, we noted that the recorded chemical oxygen demand (COD) and biochemical oxygen demand (BOD5) values exceeded the standard for irrigation water and the standard of the *Official Journal of the Algerian Republic* for discharges into the receiving environment (J.O.R.A, 2012). Grara et al. (2017) have analyzed these indicators and found that pollutant contamination was due to toxic organic compounds. Its biodegradability index is less than 3 conforms according to (Rodier, 2009) to that of predominantly domestic urban wastewater. In reality, this environment almost receives domestic effluents because wastewater channels pass through residential areas with high human concentration (Mamine et al., 2020). This situation can be explained by the level of the wastewater collection, treatment and disposal system and public hygiene at the study site. Bougherira et al. (2017) indicated that high organic loading of effluents could increase toxic organic contamination of the aquatic environment, which will increasingly lose its self-purification capacity.

The presence of suspended matter, in quantities exceeding the standard recommended by the World Health Organization (WHO) (30 mg/L), may result in soil clogging, with damaging effects on the agriculture (Abouelouafa et al., 2002).

Regarding the cadmium concentration, the recorded values are superior to the standard 0.01 mg/L, considered a limit value, for wastewater used in irrigation (WHO, 2012). The presence of heavy metals (Cd) reflects the chemical contamination of the waters of the Medjerda River reused for irrigation. The presence of these metals in water, even in low concentrations, induces harmful effects on the environment and health (WHO, 2013). However, other studies, such as Kinuthia et al. (2020), have shown that wastewater is generally blamed for the presence of heavy metals in soils and irrigated crops, in addition to the fact that this heavy metal absorbed by vegetables enters the food chain and leads to a phenomenon of bioconcentration at each step in the upper trophic link.

The results related to the search for pathogenic germs in raw water intended for use in purification systems show the presence of total germs (TG) and *Escherichia coli*, evidently due to the contamination of fecal origin or organic matter present in water (Barthe et al., 1998). These results do not correspond to the WHO standards that require an absolute absence of these germs in the water intended for the irrigation of crops or orchards (WHO, 1989).

15.3.2 Performance of Purification Systems

15.3.2.1 The Behavior of the Hydrogen Potential (pH)

The pH behavior of the raw wastewater and filtrates from both trays is shown in Figure 15.3. From our findings, a slight pH variation was noticed at the outlet of both trays compared to that of the raw

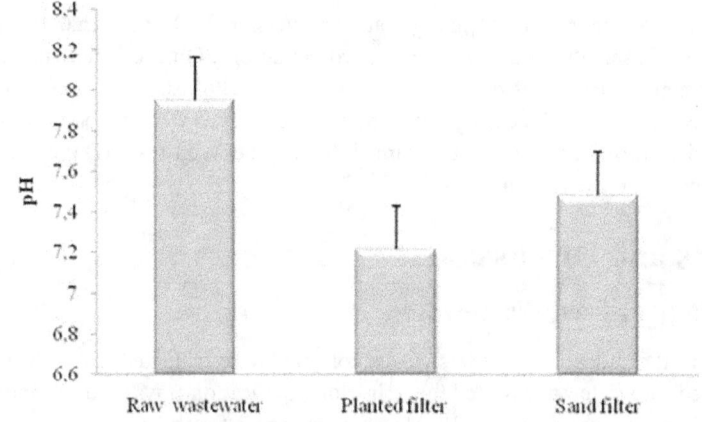

FIGURE 15.3 The behavior of the hydrogen potential (pH) of the raw feed wastewater and the filtrates of the tray planted with *Typha latifolia* and of the sand filer.

wastewater. The pH of the used wastewater ranges between 7.74 and 8.12, while in the filtrates for the planted tray, the maximum values vary between 7.19 and 7.25 with a mean value of 7.22 ± 2.41, and between 7.42 and 7.54 with a mean value of 7.49 ± 1.11 for the unplanted tray. These values meet the Algerian standards for the quality of wastewater used for irrigation (J.O.R.A, 2012) and are also within the range recommended by the FAO (1985) (6.5–8.4). The difference in the equality of the averages is no significant ($p=0.061$) between the filter trays compared to the raw wastewater. The decrease in pH at the outlet of the purification systems can be explained by the acidification of the waste during the purification process (Ouattara et al., 2008), which could result from the oxidation of COD and NH4+. According to Franck (2002), the biomass needs a pH in the range of 6.5–8.0 for the aerobic purification process. A very low pH inhibits the growth of bacteria required for the biological degradation of organic matter. Indeed, the helophytes release root exudates, which are mainly tannic and gallic acids that can lead to acidification of the environment (Boutin, 2006). These results are in agreement with those reported by Ouattara and Coulibaly (2019) for systems planted with *Panicum maximum*.

15.3.2.2 Total Suspended Matter (TSS)

The TSS concentrations in the feed affluent oscillates from 356.32 to 367.21 mg/L with an average value of 359.11 ± 56.3 mg/L, while the planted filter provides a 92.1% abatement rate, showing contents ranging from 19.58 to 39.14 mg/L with an average value of 28.51 ± 8.42 mg/L. At the same time, the sand filter gives contents between 137 and 201.11 mg/L with a mean value of 192.23 ± 59.22 mg/L and a reduction rate of 46.47%. The difference in equality of the averages is highly significant ($p=0.008$) between the trays compared to the raw wastewater. The removal of TSS in both treatment systems can be explained by physical processes (sedimentation and filtration). Vertical flow systems are fed from the surface and the effluent percolates vertically through the substrate. The effluent then undergoes a first stage of filtration allowing physical retention of suspended solids at the surface (Achak et al., 2011). The average TSS concentration at the outlet of the system planted with *T. latifolia* is constantly below the standard for discharge into the receiving environment recommended by the WHO (30 mg/L) (1989). The work of Vymazal and Dvořáková Březinová (2018) has shown a decrease in the concentrations of suspended solids in water used for agricultural irrigation in wetland dominated by *Phalaris arundinacea*, *Carex nigra* and *Scirpus sylvaticus*. According to Stang et al. (2016), this behavior is explained by biological processes that include aerobic and anaerobic bacterial degradation and adsorption by vegetation (Figure 15.4).

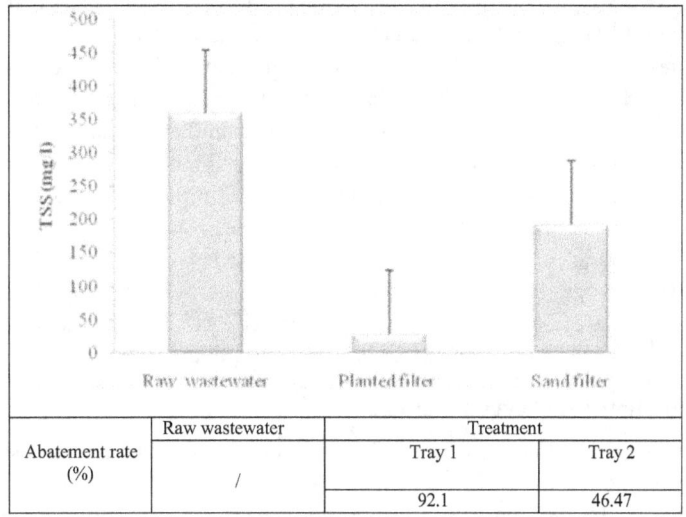

FIGURE 15.4 Evolution of the TSS of the raw feed wastewater and the filtrates from the planter planted with *Typha latifolia* and the sand filer.

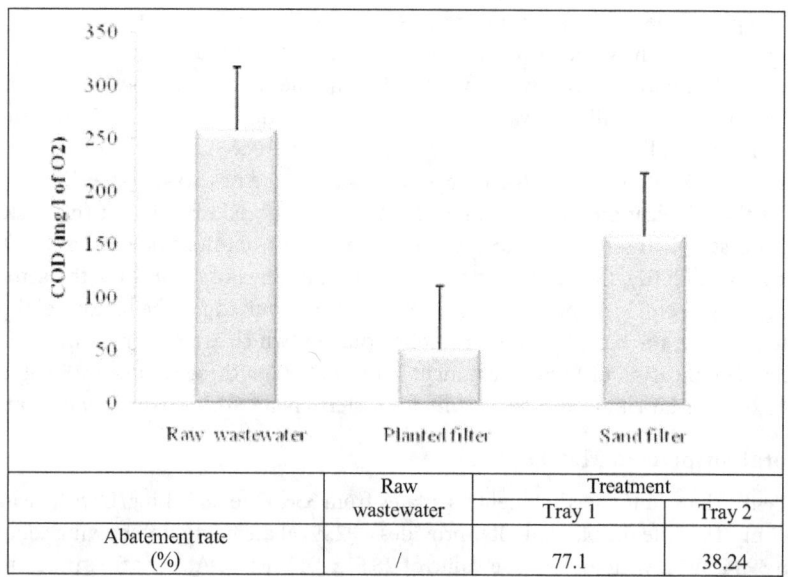

| | Raw | Treatment | |
	wastewater	Tray 1	Tray 2
Abatement rate (%)	/	77.1	38.24

FIGURE 15.5 Evolution of the COD of the raw feed water and the filtrates from the tray planted with *Typha latifolia* and the sand filer.

15.3.3 Carbonaceous Pollution

15.3.3.1 Chemical Oxygen Demand

Figure 15.5 illustrates the evolution of the COD content at the entry and exit of the two systems: planted and non-planted. The analysis of the COD shows that the organic matter content of the decreases after passing through the planted filter. Indeed, the COD decreases from 258.23 ± 78.61 mg/L of O_2 to 52.1 ± 11.1 mg/L of O_2 a reduction. The average COD value obtained at the level of the planted system is lower than the standard (90 mg/L of O_2) limited to irrigation water considered as a limit value, recommended by the Algerian standards (J.O.R.A, 2012). The strong reduction in COD (77.1%) indicates better oxygenation of the substrate in the tank planted with *T. latifolia*, which allows aerobic bacteria to proliferate and consequently ensure better mineralization and oxidation of organic matter (Kaverugu et al., 2016). On leaving the unplanted system, the COD analysis shows that the organic matter content of the treated wastewater decreases after it has passed through the sand with hovers around an average of 159.47 ± 25.4 mg/L of O_2 with a 38.24% reduction, compared to the raw wastewater (Figure 15.4). The elimination of COD involves physical phenomena of sedimentation and filtration of the particulate form (Achak et al., 2011) and the biofiltration process linked to the presence of microorganisms, bacteria and/or fungi) which makes it possible to degrade compounds of these pollutants (Kiedrzyńska et al., 2017). The difference in the equality of the averages is highly significant ($p=0.003$) between the trays compared to the raw wastewater. These results are in agreement with those reported by Mello et al. (2019), who achieved 80% removal efficiency of COD from raw wastewater for the macrophyte *Eichhornia crassipes*.

15.3.3.2 The Biochemical Oxygen Demand

Figure 15.6 reveals a significant decrease in the BOD_5 load at the outlet by both trays compared to raw wastewater. This reduction in the pollutant load of planted filter (28.41 ± 8.2 mg/L of O_2) is lower than that of the sand filter (87.44 ± 18.7 mg/L of O_2) and raw wastewater (139.32 ± 34.2 mg/L of O_2). Moreover, the average treatment efficiencies provided by the two treatment trays are in the following order: planted filter=79.61%>sand filter=37.24%. The results of the value obtained in the planted purification system

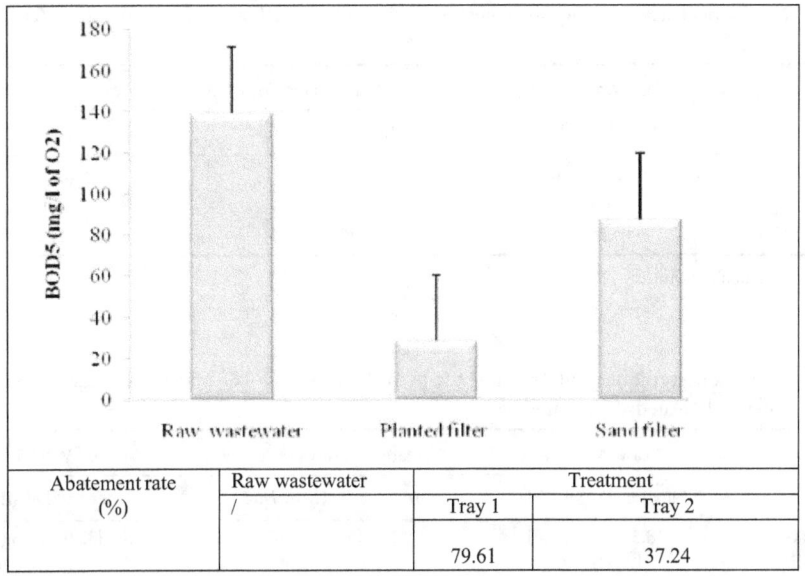

Abatement rate	Raw wastewater	Treatment	
(%)	/	Tray 1	Tray 2
		79.61	37.24

FIGURE 15.6 Evolution of the BOD$_5$ of the raw feed wastewater and the filtrates from the planter planted with *Typha latifolia* and the sand filer.

level do not exceed the standard (30 mg/L of O$_2$) considered as the value limited to irrigation water recommended by Algerian standards (J.O.R.A, 2012). The difference in the equality of the averages is highly significant ($p=0.000$) between the trays compared to the raw wastewater. The lower biochemical oxygen demand (BOD$_5$) concentrations in treated wastewater by the plant system is associated with the effect of aerobic conditions resulting in the release of oxygen at the substrate level by the root system that could be responsible for better oxygenation of the latter allowing aerobic exercise. Bacteria multiply and thus ensure the oxidation of organic matter (Abissy and Mandi, 1999).

Queiroz et al. (2017) reported that the use of aquatic macrophytes (*Polygonum* sp. and *Eichhornia paniculata*) can ensure the reduction of organic matter in dairy effluents. Roé-Sosa et al. (2019) showed that the purifying plant *T. latifolia* is highly hydrophilic and its growth requires significant amounts of organic matter and nutrients.

15.3.3.3 Wastewater Cadmium (Cd) Concentrations

During the treatment trial, the concentration of cadmium (Cd) decreases significantly ($p=0.000$) in the planted filter compared to the sand filter and raw wastewater. The maximum concentrations of cadmium in sand filer oscillate between 0.029 and 0.038 mg/L with a mean value of 0.032 mg/L. These values remain above the standard (0.01 mg/L), considered as a limit value, for wastewater used in irrigation (WHO, 2012). The expression of the results obtained by the planted system shows an almost total elimination of cadmium (100%) (Table 15.3). According to Ayaz et al. (2019), the halophytes play an important role in the removal of metals of wastewater by either the process of phytostabilization or phytoaccumulation. The work by Hejna et al. (2020) showed that halophytes such as *T. latifolia* and *Thelypteris palustris* have the capacity to bioaccumulate heavy metals. The metals were taken up into the (stem, leaves and roots) parts of the plants, with the roots being the most significant. At the same time, the bioavailability of metals also influences their removal by plants in the constructed wetland (Usharani and Vasudevan, 2016). Our results indicate that the sand filter has a low removal rate of this metal (12.82%). Removal of this metal in sand filter is a complicated process including abiotic reaction (precipitation, flocculation, sedimentation, exchanges of cations and anions, reduction and oxidation)

TABLE 15.3 Mean Concentrations and Abatement Rates of Cadmium Obtained for Raw wastewater and Treated Wastewater by Trays

	Raw Wastewater	Treated Wastewater (Tray 1)	Treated Wastewater (Tray 2)
	(Min-Max) Mean	(Min-Max) Mean	(Min-Max) Mean
Cadmium (Cd) mg/L	(0.034 to 0.046)	0	(0.029 to 0.038)
	0.039	0	0.032
Abatement rate (%)	/	100	12.82

Min: minimum; Max: maximum

TABLE 15.4 Mean Concentrations and Abatement Rates of Total Germs (TG) and *E. coli* Concentrations Obtained for Raw Wastewater and Treated Wastewater by Trays

	Raw Wastewater	Treated Wastewater (Tray 1)	Treated Wastewater (Tray 2)
	Mean±SD (Min-Max)	Mean±SD (Min-Max)	Mean±SD (Min-Max)
Total Germ (TG)	18.5±4.56 ($\times10^5$)	12.83±6.12 ($\times10^4$)	13.59±2.46 ($\times10^3$)
(germs/100 mL)	(17–21) ($\times10^5$)	(11–14) ($\times10^4$)	(9–16) ($\times10^3$)
Abatement rate (%)	/	93.1	99.26
E. coli (germs/100 mL)	19.13±9.2 ($\times10^2$)	10.43±2.31 ($\times10^2$)	0.62±0.19 ($\times10^2$)
	(10–22) ($\times10^2$)	(8–11) ($\times10^2$)	(0.52–1.21) ($\times10^2$)
Abatement rate (%)	/	45.47	96.76

SD: Standard deviation; Min: minimum; Max: maximum

(Chai et al., 2018; Yang et al., 2016). Even so, the presence of this metal in wastewater treated by sand filter with low concentrations induces adverse ecological and health effects (OMS, 2013).

15.3.3.4 Evolution of the Bacteriological Load

As indicated in Table 15.4, the number of TG colonies in the raw wastewater is generally higher than that in the filtrates from the different trays. It ranges between 18.5±4.56 ($\times10^5$) germs/100 mL in the raw wastewater and 13.59±2.46 ($\times10^3$) germs/100 mL and 12.83±6.12 ($\times10^4$) germs/100 mL in the filtrates from the planted and control trays, respectively. The TG count removal rates obtained in the planted and control trays are 99.26% and 93.1%, respectively. Statistical analysis shows a very highly significant difference ($p=0.000$) between the TG count of the *T. latifolia* planter and the control. For *E. coli*, the number of colonies in the waste feed water is highly variable (10–22) $\times10^2$ germs/ 100 mL), but it remains higher than that in different filtrates. This number averages 0.62±0.19 $\times10^2$ germs/100 mL in the planted bed filtrates and 10.43±2.31 $\times10^2$ germs/100 mL in the control filtrates. Statistical analysis shows also a highly significant difference ($p=0.002$) between the removal of *E. coli* from the tray planted with *T. latifolia* and that of the control. The number of colonies decreases in the filtrates of the trays compared to the raw wastewater, and this could be related to adsorption of pathogens in the trays and/or physical retention (Ouattara et al., 2008). It could also be favored by the biological antagonism relationships (amensalism, competition, parasitism and predation) as described by Dommergues and Mangenot (1970) between the microorganisms constituting the microbial flora that develop in the trays. The planted tray provides the best yield of the elimination of pathogenic germs (TG=99.26% and *E. coli*=96.76%). This result is probably due to the action of certain exudates secreted by *T. latifolia* contributing to removing the TG and *E. coli* (Von Rein et al., 2016).

15.4 Conclusion

The physico-chemical and bacteriological characterization of the raw wastewater of Souk Ahras city before treatment showed that the quality of this water can lead to environmental problems and potential health risks when reused in irrigation.

The treatment of the raw wastewater by the filter planted with *T. latifolia* caused a significant decrease in pH, and removal of suspended solids (TSS: 92.1%) and cadmium (Cd): (100%). It provides treated water with a reduced organic load (BOD5: 79.61% and COD: 77.1%) and significant elimination of bacteriological (TG: 99.26% and *E. coli*: 96.76%) germs compared to that of the raw wastewater and the control. Thus, it is concluded that the vertically draining artificial marshland system planted with *T. latifolia* is a robust domestic sewage treatment system in Souk Ahras city.

References

Abissy, M. and Mandi, L. (1999) 'The use of rooted aquatic plants for urban wastewater treatment: case of Arundo', *Revue des sciences de l'eau*, Vol. 12, No. 2, pp. 285–315.

Abouelouafa, M., El Halouani, H., Kharboua, M. and Berrichi, A. (2002) 'Caractérisation physico-chimique et bactériologique des eaux usées brutes de la ville d'Oujda: canal principal et Oued Bounaïm', *Revue Marocaine des Sciences Agronomiques et Vétérinaires*, Vol. 22, No. 3, pp. 143–150.

Achak, M., Ouazzani, N. and Mandi, L. (2011) 'Élimination des polluants organiques des effluents de l'industrie oléicole par combinaison d'un filtre à sable et un lit planté', *Revue des Sciences de l'Eau*, Vol. 24, No. 1, pp. 35–51.

Athmani, A. (2008) *Evolution de la qualité des eaux de surface cas du sous bassin versant Medjerda région de souk Ahras (Algérie)*. Mémoire de magister, Université de Souk Ahras, Algérie.

Ayaz, T., Khan, S., Khan, A.Z, Lei, M. and Alam, M. (2019) 'Remediation of industrial wastewater using four hydrophyte species: A comparison of individual (pot experiments) and mix plants (constructed wetland)', *Journal of Environmental Management*. http://www.researchgate.net/publication/337424691_Remediation_of_industrial_wastewater_using_four_hydrophyte_species_A_comparison_of_individual_pot_experiments_and_mix_plants_constructed_wetland.html.

Barthe, C., Perron, J. and Perron, J.M.R. (1998) *Guide d'interprétation des paramètres microbiologiques d'intérêt dans le domaine de l'eau potable, Document de travail (version préliminaire)155*. Ministère de l'Environnement du Québec, Canada.

Bergner, I. and Jensen, U. (1989) 'Phytoserological contribution to the systematic placement of the Typhales', *Nordic Journal of Botany Marketing*, Vol. 8, pp. 447–456.

Boutin, C. (2006) *Usage de filtres plantés de roseaux dans le traitement des eaux usées du« petit collectif »*. Ministère de l'Agriculture, Paris, France.

Bougherira, N., Hani, A., Toumi, F., Haied, N. and Djabri, L. (2017) 'Impact des rejets urbains et industriels sur la qualité des eaux de la plaine de la Meboudja (Algérie)'. *Hydrological Sciences Journal*, Vol. 62, No. 8, pp. 1290–1300.

Chai, L., Li, Q., Wang, Q and Yan, X. (2018) 'Solid-liquid separation: an emerging issue in heavy metal wastewater treatment', *Environmental Science and Pollution Research International*, Vol. 25, No. 18, pp. 7250–17267

Czudar, A., Gyulai, I., Keresztúri, P., Csatári, I., Serra-Páka, S and Lakatos, G. (2011) 'Removal of organic materials and plant nutrients in a constructed wetland for petrochemical wastewater treatment'. *Studia Universitatis Vasile Goldis*, Vol. 21 No. 1, pp. 109–114.

Dagnelie, P. (2011) *Statistique théorique et appliquée: Tome 2, Inférence statistique à une et à deux dimensions*, 736. Univ De boeck, Bruxelles.

Dommergues, Y. and Mangenot, F. (1970) *Ecologie microbienne du sol*. Masson ad Cie edi, Paris, France.

El-Khatib, A.A., Hegazy, A.K. and Abo-El-Kassem, A.M. (2014) 'Bioaccumulation potential and physiological responses of aquatic macrophytes to Pb pollution', *International Journal of Phytoremediation*, Vol. 16, No. 1, pp. 29–45.

El Ouali Lalami, A., Zanibou, A., Bekhti, K.F., Zerrouq, F. and Merzouki, M. (2014) 'Contrôle de la qualité microbiologique des eaux usées domestiques et industrielles de la ville de Fès au Maroc', *Journal of Materials and Environmental Science*, Vol. 5, pp. 2325–2332.

Fagrouch, A., Berrahou, A. and El Halouani, H. (2011) 'Impact d'un effluent urbain de la ville de Taourirt sur la structure des communautés de macroinvertébrés de l'oued Za (Maroc oriental) ', *Revue des Sciences de l'Eau*, Vol. 24, pp. 87–101.

FAO (1985) Water quality for agriculture. In R.S. Ayers and D.W. Westcot. *Irrigation and Drainage*-Paper 29 Rev. 1. FAO, Rome.

Fonkou, T., Fonteh, M.F., Djousse Kanouo M. and Akoa A. (2010) 'Performances des filtres plantes de Echinochloa pyramidalis dans l'épuration des eaux usées de distillerie en Afrique subsaharienne', *Tropicultura*, Vol. 28, No. 2, pp. 69–76.

Franck, R. (2002). *Analyse de l'eau (Aspects réglementaires et techniques) Ed*, Collection biologie technique, 360 p.

Julve, P. (1998) *Baseflor. Index botanique, écologique et chorologique de la flore de France*, Ed, Morin.

J.O.R.A. (2012) 'ANNEXE, spécifications des eaux usées épurées utilisées à des fins d'irrigation', *Journal Officiel de la République Algérienne*, N. 41, pp. 18–21.

Hannachi, A., Gharzouli, R., Djellouli Tabet, Y. and Daoud, A. (2016) 'Waste water reuse in agriculture in the outskirts of the city Batna (Algeria)', *Journal of Fundamental and Applied Sciences*, Vol. 8, No. 3, pp. 919–944.

Hejna, M., Moscatelli, A., Stroppa, N., Onelli, E., Pilu, S., Baldi, A and Rossi, L. (2020) 'Bioaccumulation of heavy metals from wastewater through a *Typha latifolia* and *Thelypteris palustris* phytoremediation system', *Chemosphere*, Vol. 241, p. 125018. http://www.sciencedirect.com/science/article/pii/S004565351932257X.

Grara, N., Boucheleghm, A. and Bouregaa, M. (2017) *Détoxification des eaux usées par différent procédés: Purification biologiques (lentilles d'eau) et Catalytique (les Nanoparticules) des eaux usées*, édition université européenne, France.

Guasmi, I., Djabri, L., Hani, A. and Lamouroux, C. (2006) 'Pollution des eaux de l'Oued Medjerda par les nutriments', *Larhyss Journal*, Vol. 05, pp. 113–119.

Kadaverugu, R., Shingare, R.P, Raghunathan, K., Juwarkar, A.A, Thawale, P.R. and Singh, S.K. (2016) 'The role of sand, marble chips and *Typhalatifolia* in domestic wastewater treatment. A column study on constructed wetlands', *Environmental Technology*, Vol. 37, No. 19, pp. 1–26.

Kaldy, J.E., Brown, C., Nelson, W. and Frazier. M. (2017) 'Macrophyte community response to nitrogen loading and thermal stressors in rapidly flushed mesocosm systems', *Journal of Experimental Marine Biology and Ecology*, Vol. 497, pp. 107–119.

Keffala, C., Harerimana, C. and Vasel, J.L. (2012) 'OEufs d'helminthes dans les eaux usées et les boues de station d'épuration: enjeux sanitaires et intérêt du traitement par lagunage', *Environ Risque Sante*, Vol. 11, pp. 511–520.

Khaldi, F., Menaiaia, K., Ouartane, N. and Grara, N. (2019) 'Biochemical and enzymatic characterization of macrophyte plant *Phragmites australis* affected by zinc oxide', *Scientific Study & Research*, Vol. 20, No. 2, pp. 237–252.

Kiedrzyńska, E., Urbaniak, M., Kiedrzyński, M., Jóźwik, A., Bednarek, A., Gągała, I. and Zalewski, M. (2017) 'The use of a hybrid Sequential Biofiltration System for the improvement of nutrient removal and PCB control in municipal wastewater', *Scientific Reports*, Vol. 7, No. 1. http://doi.org/10.1038/s41598-017-05555-y.

Kinuthia, G.K., Ngure, V., Beti, D., Lugalia, R., Wangila, A. and Kamau, L. (2020) 'Levels of heavy metals in wastewater and soil samples from open drainage channels in Nairobi, Kenya: community health implication', *Scientific Reports*, Vol. 10, No. 1, p. 8434. https://doi:10.1038/s41598-020-65359-5.

Kone, M., Zongo, I., Bonou, L., Koulidiati, J., Joly, P., Bouvet, Y. and Sodre, S. (2011) 'Traitement d'eaux résiduaires urbaines par filtres plantés à flux vertical sous climat Soudano-Sahélien Int', *International Journal of Biological and Chemical Sciences*, Vol. 5, No. 1, pp. 217–231.

Leto, C., Tuttolomondo, T., La Bella, S., Leone, R. and Licata, M. (2013) 'Effects of plant species in a horizontal subsurface flow constructed wetland–phytoremediation of treated urban wastewater with *Cyperus alternifolius* L. and *Typha latifolia* L. in the West of Sicily (Italy)', *Ecological Engineering*, Vol. 61, pp. 282–291.

Licata, M., Gennaro, M.C., Tuttolomondo, T., Leto, C. and La Bella, S. (2019) 'Research focusing on plant performance in constructed wetlands and agronomic application of treated wastewater – A set of experimental studies in Sicily (Italy)', *PLoS One*, Vol. 14, No. 7. https://doi.org/10.1371/journal. pone.0219445.

Mamine, N., Grara, N. and Khaldi, F. (2019) 'The use of macrophyte *Typha latifolia* filters in the treatment of wastewaters of Medjerda river, in Souk-Ahras city (North-East Algeria)', *Studia Universitatis "Vasile Goldiş", Seria Ştiinţele Vieţii*, Vol. 29, No. 2, pp. 70–81.

Mamine, N., Grara, N and Khaldi, F. (2020) 'Survey of the physico-chemical and parasitological quality of the wastewaters used in irrigation (Souk Ahras, North-East of Algeria)', *Iranian (Iranica) Journal of Energy and Environment*, Vol. 11, No. 1, pp. 79–88.

Mello, D., Carvalho, K.Q., Passig, F.H., Freire, F.B., Borges, A.C., Lima, M.X. and Marcelino, G.R. (2019) 'Nutrient and organic matter removal from low strength sewage treated with constructed wetlands', *Environmental Technology*, Vol. 40, No. 1, pp. 11–18.

Osei, A.R, Konate, Y. and Abagale, F.K. (2019) 'Pollutant removal and growth dynamics of macrophyte species for faecal sludge treatment with constructed wetland technology', *Water Science & Technology*, Vol. 80, No. 6, pp. 1145–1154.

Organisation mondiale de la Santé(OMS) (2012) *L'utilisation sans risque des eaux usées, des excrétas et des eaux ménagères. Volume II. Utilisation des eaux usées en agriculture*, Genève, OMS.

Organisation mondiale de la Santé(OMS) (2013) *Directives OMS pour l'utilisation sans risque des eaux usées, des excreta et des eaux ménagères, Volume 2, chapitre 8*, p. 125.

Ouattara, J.P., Coulibaly, L., Manizan, P.N. and Gourene, G. (2008) 'Traitement des Eaux Résiduaires Urbaines par un Marais Artificiel à Drainage Vertical Planté Avec *Panicum maximum* sous Climat Tropical', *European Journal of Scientific Research*, Vol. 23, No. 1, pp. 25–40.

Ouattara, J.P. and Coulibaly, L. (2019) 'Effet de la charge hydraulique appliquéesur le fonctionnement d'un marais artificiel à drainage vertical planté avec *Panicum maximum* traitant des eaux domestiques', *International Journal of Biological and Chemical Sciences*, Vol. 13, No. 5, pp. 24–38.

Queiroz, R.C.S., Andrade, R.S., Dantas, I.R., Rodrigues, L.B. and Almeida Neto, J.A. (2017) 'Constructed wetland', *Environmental Monitoring and Assessment*, Vol. 190, p. 328. http://doi.org/10.1007/s10661-018-6705-4.

Rana, V. and Maiti, S.K. (2018) Municipal wastewater treatment potential and metal accumulation strategies of *Colocasia esculenta* (L) Schott and *Typha latifolia* L. in a constructed wetland. Vol. 190, p. 328, http://doi.org/10.1007/s10661-018-6705-4.

Rodier, J. (2009) *L'analyse de l'eau.*, 9ème édition, Dunod, Paris.

Roé-Sosa, A., Rangel-Peraza, J.G, Rodríguez-Mata, A.E, Pat-Espadas, A., Bustos-Terrones Y., Diaz-Peña, I., Vu, C.M. and Amabilis-Sosa, L.E. (2019) 'Emulating natural wetlands oxygen conditions for the removal of N and P in agricultural wastewaters', *Journal of Environmental Management*, Vol. 236, pp. 351–357.

Scott, C.A., Faruqui, N.I. and Raschid-Sally, L. (2004) Wastewater use in irrigated agriculture: management challenges in developing countries. In *Wastewater Use in Irrigated Agriculture: Confronting the Livelihood and Environmental Realities*, CABI Publishing, Wallingford, UK, pp. 1–10.

Shingare, R.P., Nanekar, S.V., Thawale, P.R., Karthik, R. and Juwarkar, A.A. (2017) 'Comparative study on removal of enteric pathogens from domestic wastewater using *Typha latifolia* and *Cyperus rotundus* along with different substrates', *International Journal of Phytoremediation*, Vol. 19, No. 10, pp. 899–908.

Stang, C., Bakanov, N. and Schulz, R. (2016) 'Experiments in water-macrophyte systems to uncover the dynamics of pesticide mitigation processes in vegetated surface waters/streams', *Environmental Science and Pollution Research International*, Vol. 23, No. 1, pp. 673–682.

Usharani, B. and Vasudevan, N. (2016) 'Impact of heavy metal toxicity and constructed wetland system as a tool in remediation', *Archives of Environmental and Occupational Health*, Vol. 71, No. 2, pp. 102–110.

Von Rein, I., Kayler Z.E, Premke, K. and Gessler, A. (2016) 'Desiccation of sediments affects assimilate transport within aquatic plants and carbon transfer to microorganisms', *Plant Biology (Stuttgart, Germany)*, Vol. 18, No. 6, pp. 947–961.

Vymazal, J. and Dvořáková Březinová, T. (2018) 'Treatment of a small stream impacted by agricultural drainage in a semi-constructed wetland', *The Science of the Total Environment*, Vol. 643, pp. 52–62.

Wang, R., Baldy, V., Périssol, C. and Korboulewskyac, N. (2012) 'Influence of plants on microbial activity in a vertical-downflow wetland system treating waste activated sludge with high organic matter concentrations', *Journal of Environmental Management*, Vol. 95, pp. 158–164.

World Health Organization (WHO). (1989) Health Guidelines for the Use of Wastewater in Agriculture and Aquaculture. Report of a Scientific Group, Technical Report Series 778, World Health Organization, Geneva.

World Health Organization (WHO). (2010) *Enterotoxigenic Escherichia coli (ETEC) diarrhoeal diseases.* http://www.who.int/vaccine_research/diseases/diarrhoeal/en/index4.htm.

Yalçuk, A. and Ugurlu, A. (2019) 'Treatment of landfill leachate with laboratory scale vertical flow constructed wetlands: plant growth modeling', *International Journal of Phytoremediation*, Vol. 12, pp. 1–10.

Yang, H., Hu, Y. and Cheng, H. (2016) 'Sorption of chlorophenols on microporous minerals: mechanism and influence of metal cations, solution pH, and humic acid', *Environmental Science and Pollution Research International*, Vol. 23, No. 19, pp. 19266–19280.

Youbi, A., Houilia, A., Soumati, B., Berrebbah, H., Djebar, M.R. and Souiki, L. (2018) 'Assessment of the physico-chemical and bacteriological quality of the surface waters of wadis "Boukhmira, Meboudja and Seybouse" used in irrigation in the North-East of Algeria', *Studia Universitatis "Vasile Goldiş", Seria Ştiinţele Vieţii*, Vol. 28, No. 2, pp. 95–106.

Yu, H., Zhang, X., Hu, J., Peng, J., Qu, J. (2020) 'Ecotoxicity of polystyrene microplastics to submerged carnivorous *Utricularia vulgaris* plants in freshwater ecosystems', *Environmental pollution*, Vol. 265, p. 114830. https://www.sciencedirect.com/science/article/abs/pii/S026974911936823X.

Zaimoglu, Z. (2006) 'Treatment of campus wastewater by a pilot-scale constructed wetland utilizing *Typha latifolia, Juncus acutus* and *Iris versicolor*', *Journal of Environmental Biology*, Vol. 27, No. 2, pp. 293–298.

VI

Water Harvesting for Irrigation

16

Surface Runoff Water Harvesting for Irrigation

Saeid Eslamian and
Hosein Roknizadeh
*Isfahan University
of Technology*

Mousa Maleki
*Illinois Institute
of Technology*

Jahangir Abedi
Koupai
*Isfahan University
of Technology*

16.1 Introduction

In the 20th century, the world's population has tripled, but water consumption has increased sixfold, exacerbating the problems of water scarcity (Cosgrove and Rijsberman, 2000). Water scarcity is the first and most challenging crisis worldwide. By growing the world's population, it is inevitable to find the alternative water resources to provide demanding water. Atmospheric water also known as air humidity is one of the most accessible resources and therefore could be used as a sustainable resource for water harvesting (Maleki et al., 2021). In view of high population growth and water resources deficit in arid and semi-arid area and groundwater resources shortage in mountainous area, there is an urgent need to identify the alternative sources of potable water (Zamani et al., 2021). Predicting the future of population is important to direct and adjust the management to achieve the goals of sustainable development. In this regard, the United Nations has published a report in which about 83 million people are added to the world population annually; it predicts that the present world population of 7.6 billion is expected to reach 8.6 billion in 2030, and 9.8 billion in 2050 (UN, 2017). Therefore, because of the increasing population, urbanization, water needs, and lack of sufficient water resources, there is need to use the renewable resources. Nowadays, one of the renewable resources that has a high potential to meet some of the water needs is the use of runoff, and water engineers and designers collect and treat the runoff in different ways and store them in various ways. The fraction of the rainfall flowing on the landscape which is higher than the lower height is known as runoff. Runoff is usually associated with negative consequences such as erosion and loss of water. However, it can be used for surface irrigation of agricultural products in rainfall events. This is done by diverting runoff water into surrounding areas that grow agricultural products. The collection of runoff provides water for the agricultural activities in areas. This technique is called farming or harvesting water runoff. The

DOI: 10.1201/9780429290114-22

latter is usually used to demonstrate water collection for domestic use. The amount of water that can be collected during a rain event depends on the rainfall characteristics such as the amount, intensity and distribution, and on the production area, such as size, geomorphology and surface characteristics. The main difference between the irrigation of the runoff and conventional irrigation is that timing and the amount of demand cannot be determined. In order to model in such a system, it is necessary to make the different models involved, such as rainfall, penetration, surface flow and consumer water consumption (Ben-Asher and Berliner, 1994). Runoff farming uses the surface currents such as stormwater.

Runoff farming and related approaches covered in this book chapter have a special value for remote environments in which other strategies are technically impossible, too expensive or bad. In addition, runoff farming often uses the local materials and work, so it can be adopted at a little risk of dependence, maybe expanded with limited foreign aid. With food production in many developing countries, reducing the land degradation and uncertainty about future climate change as a result of global warming, runoff farming will be much more important (Barrow, 2014).

Runoff harvesting is particularly valuable in drylands but can also help better to irrigate the areas, even for the environments as wet as Amazonia that can have many rainy periods of under stress and damaged products. Where practiced, removing the runoff should achieve one or more items: improved security harvest, increased production, diversity in products, protected soil, and improved stability that in the dryer environments, runoff harvesting, due to "multiplicity of rainfall", can provide planting, the increased livestock, livestock, or forestry, even if precipitation is technically very low. But runoff harvesting may only provide the insufficient support for drought products in unusual dry periods. Where the drought is a risk, runoff harvest should be combined with water storage (as wet soil or water in reservoirs) – if the economics, geology, and land allowed (Verma and Sarma, 1990). For example, a 1.2 ha basin in Negev, where only 100 mm of rain per year, can provide a 440 m^3 tank that can be reliably from 300 to 500 sheep (Pacey and Cullis, 1986) support.

Not only does runoff harvesting increase water efficiency, but it also prevents the contaminated surface water from entering streams, soil erosion, and ecosystem degradation. In some countries, the most common method of flood control is to collect a flood and remove it from the urban areas. Although this method is effective in reducing the volume of floods and preventing the flooding of roads, in addition to the high costs that must be spent on the construction of water conveyance structures, these structures have the destructive effects such as soil erosion and pollution transfer (Hernández-Crespo et al., 2019; Kamali et al., 2017). Problems related to the surface water collection and increasing the development of urban areas with human encroachment on nature have reduced the possibility of water infiltration into the ground, changes in the hydrological cycle, and increased runoff volume (Jackson et al., 2001; Li et al., 2019). Increasing the volume and velocity of rainwater runoff leads to the increased destructive power and the transfer of pollutants such as petroleum products and oils, toxins, insecticides, sewage, and fine-grained and coarse-grained sediments, which endangers lives of many aquatic animals. In such conditions, if it is possible to increase the density of vegetation, this will reduce the speed of runoff, increase the possibility of water percolation into the soil, and eventually prevent the waterways from joining and causing floods, to a large extent. One of the surface runoff management models is the stormwater management model (SWMM), which is widely used to simulate the continuous and event-based phenomena. The focus of this model is on rainfall-runoff simulation of the urban catchment area and can simulate and predict the performance of this system in runoff drainage by simulating the quantitative and qualitative runoff from rainfall and modeling the runoff drainage system. Predicting system performance after climate change and urban development can also be used in the future. Modeling in SWMM software is done in three parts: hydraulic, hydrological, and qualitative ones. The hydrological section of the software calculates the volume of runoff as a model output by taking precipitation data from the user and applying it to the sub-basins. Using the results obtained in the hydraulic section, parameters such as flood volume, basin outlet's hydrograph, runoff depth, and velocity are calculated and users can have an accurate planning for future water management in cities by predicting and evaluating the performance of the urban drainage system (Elliott and Trowsdale, 2007; Gironás et al., 2010). The aim of this chapter is the investigation of surface runoff harvesting for irrigation purpose.

16.2 New Methods of Runoff Management

Collection, conveyance, and storage of runoff, flood control, transport of pollutants, and soil erosion, are of particular importance. Using water harvesting methods in modern ways, in addition to the required water, the agricultural sector could also provide the water for domestic, drinking, and non-drinking purposes (Jones and Hunt, 2010; Li et al., 2021; Villarreal and Dixon, 2005). The transfer of runoff to the treatment plant and their treatment, in addition to causing destructive phenomena such as soil erosion and water pollution, are costly and not economic. As in most of the developed countries have been considered, controlling runoff created in the same place where they are originated (Xu et al., 2019).

16.3 Low Impact Developments (LID)

In this method, the conditions of the study area are simulated before urban development, that are used to manage the urban floods (Gilroy and McCuen, 2009; Li et al., 2021; Liu et al., 2017; Liu et al., 2018). At LID, the floods are attempted to manage by using the storage, infiltration, and treatment of runoff at the source of their production instead of the disposal of urban floods and treating them at the great costs. These methods, which are highly compatible with the environment and urban structures, have a vital role in reducing the network operation and maintenance costs, runoff volume, improving the physical and chemical quality of water, and ecosystem protection (Jia et al., 2012).

16.4 Best Management Practices

Best management practices (BMP) for water harvesting is called a set of solutions that cause the quantitative and qualitative controls of runoff. Most of these solutions work to delay the runoff accumulation in order to reduce the runoff volume and feed aquifers. Designers offer the best solution among the available solutions according to the experience, climate, and conditions of a region. Today, effects of urbanization and increasing water productivity are reduced by giving an importance to the sustainable development, using BMPs and LIDs, which is discussed below (Jia et al., 2012).

16.5 Green Roofs

One method to control the water efficiency and runoff from rainfall is to use a green roof system. The roof or balcony of a house is covered by the plants that are resistant to adverse conditions such as drought and cold weather, and the types of plants depend on the climate of the region. Important advantages of this system are improving the quality of the climate, improving the ecological conditions, the beauty of the area, preventing global warming, and preventing the loss of energy of the building (Nazif et al., 2021). Disadvantages include the inadequacy of the system for sloping roofs, plant care, and the relatively high cost of insulation. It is worth mentioning that this method will be more effective in dense urban, commercial, and industrial areas where the percentage of green roof coverage is higher (Ahmed and Alibaba, 2016) (Figure 16.1).

16.6 Permeable Surfaces

Another method of controlling surface water and water efficiency, which is used in many developed countries, is the surface permeability, which can be applied from the porous asphalt, permeable concrete, bricks, and coarse grains that are used on the streets and passages. Infiltrated water can either

FIGURE 16.1   Examples of green roofs in Colombia (Image: Santalaia, via Greenroofs.com, courtesy of Groncol).

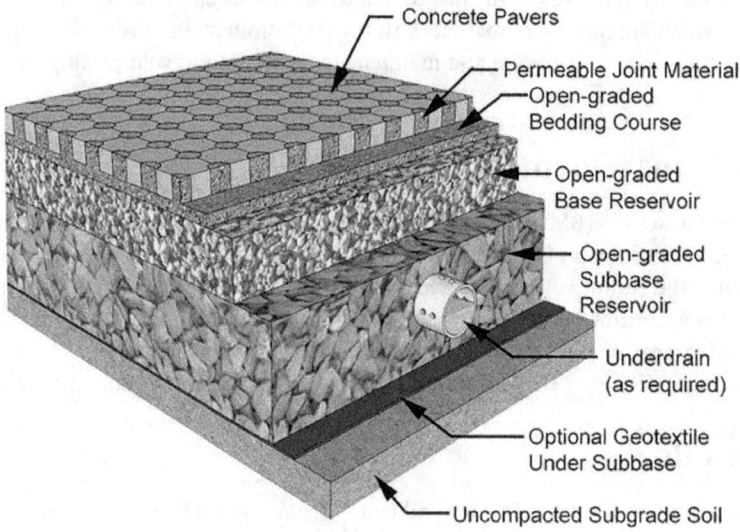

FIGURE 16.2   Cross section of typical permeable pavement (Smith, 2006).

be stored temporarily and then be used to feed canals or rivers, or be significantly filtered by penetrating deep into the soil and passing through filters, and then joining the aquifers. The advantages of this method include the groundwater recharge, reduction of urban surface runoff, and peak flow (Hernández-Crespo et al., 2019; Tedoldi et al., 2016) (Figure 16.2).

16.7 Mulched Area

Mulched area is a layer of organic matter that is applied to the soil surface, the slope of which is usually proportional to the slope of the ground, and the lower the slope, the greater the degree of permeability and removal of pollution from water. In general, the mulched areas are highly efficient for small areas

FIGURE 16.3 High turf density and mulched areas to manage the runoff in a lawn setting (Freeborn et al., 2012).

and low slopes, but in addition to the above, they also depend on land use and soil type. Its benefits include reducing erosion, increasing permeability, maintaining soil moisture, and regulating temperature in the root zone of the plant (Freeborn et al., 2012; Relf, 2009) (Figure 16.3).

16.8 Stormwater Ponds

Downstream of some structures, such as the mulched areas, which perform sedimentation and primary treatment, are located at the stormwater ponds. The water collected in this system penetrates into the soil from the surface layers and finally feeds the downstream tank by directing it to the channel outlet. Plants and microorganisms in the soil treat pollutants in runoff through physical processes (such as filtration, infiltration, or adsorption) and biological processes (such as biological adsorption or microbial decomposition) (Freeborn et al., 2012).

16.9 Infiltration Trenches

By digging trenches and filling them with coarse-grained soils, the existing runoff is drained and in addition to the feeding the aquifers, it reduces their salinity. It should be noted that this method requires the maintenance and care because there is a possibility of clogging in this system and it is better to be used for the low slope lands that have a greater chance of infiltration. If the slope is high or the runoff volume in that area is high, it is better to place a pipe inside the trench for better drainage, the example of which can be seen in Figure 16.4. Although the application of runoff infiltration in urban environments is limited due to built environments, groundwater level, economic aspects, soil pollution, and runoff quality, one of the important advantages of infiltration trenches is that they are at ground level. It should not be forgotten that the advantage is especially important in urban areas (Göbel et al., 2004; Locatelli et al., 2015; Mikkelsen et al., 1994; Revitt et al., 2003).

16.10 Use of Rainwater for Irrigation

In order to make an optimal use of rainwater, it should first be stored in a suitable place. To achieve this, the method of direct storage in soil profiles or storage in facilities such as reservoirs is used (Oweis et al., 2012; Zhang et al., 2021). Rainwater harvesting has always been a great option to deal with the problem of water shortage in agriculture and home consumption. By creating a small rainwater collection system in houses and using a simple sand filter, a significant amount of water can be recovered to meet the

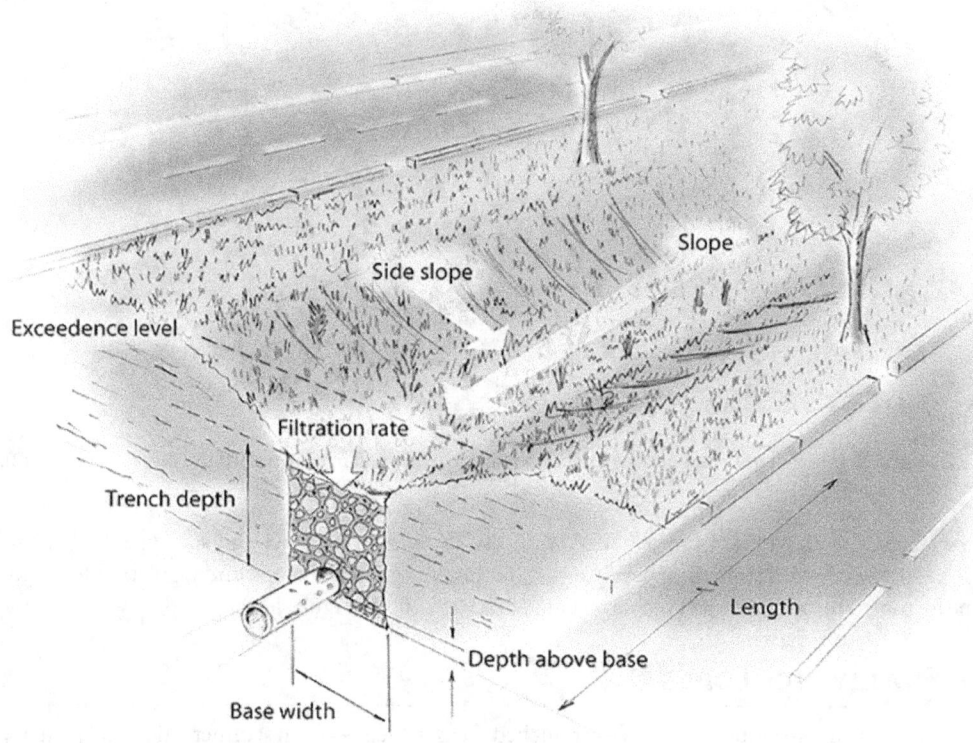

FIGURE 16.4   Outline of infiltration trenches.

collected household water demands. The ultimate solution to water scarcity is to take the effective measures to conserve the existing water resources and store the rainwater for future use. It is also the most cost-effective system for using an alternative water source in farm irrigation (Roshan Akther, 2014). One of the advantages of a rainwater collection system, in addition to reduce the peak flow, is the reduction and saving of monthly bills and municipal tolls (Campisano et al., 2017; Deng, 2021; Kim et al., 2021; Reidy, 2009; زخقفدع.com.au, 2020). These tanks are also easy to fill so that one millimeter of rain per square meter of the roof is equivalent to one liter of water in the tank (sydneywatertanks.com.au, 2019; abc.net.au, 2020). Collecting rainwater from the roof of houses and directing them by

gutters into the tank is the most common, convenient, and least expensive method of extracting rainwater, which is widely used for home use, garden irrigation, and green space (Kim et al., 2021; Ward, 2007). Although rainwater harvesting is widely used in agriculture and irrigation, it can contain the contaminants such as toxic metals (such as Cu, Cd, and Zn) and synthetic organic chemicals (such as agricultural chemicals) that threatens the crops productivity, soil health, and consumers (Deng, 2021). In arid regions, water is the most limited source for improving agricultural production. Therefore, maximizing water efficiency is the best strategy for dry farming systems, which can be achieved by using the efficient water management techniques. By collecting and directing the rainwater to agricultural lands, more than 50% of lost water can be restored at a very low cost, which increases agricultural water productivity and also has significant socio-economic and environmental benefits. In West Asia and North Africa (WANA), agriculture accounts for 75% of total water consumption. Taking into account with the development of urbanization, industrialization, population growth, and increasing demand for food, it is predicted that the share of water allocated to the agricultural sector will decrease in the future, which should be an alternative water source for it to avoid losses to this sector. One of the methods of obtaining

FIGURE 16.5 The six cisterns are interconnected in series with a 4″ pipe which allows the cisterns to all share volume and fill equally (www.watercache.com, 2020).

water is groundwater extraction, but it should be noted that the use of this method, in addition to cause the subsidence, both reduces the groundwater reserves and endangers their quality. Many of the WANA lands are exposed to desertification and their annual rainfall is less than 250 mm. Given that the average annual rainfall in these areas is low, when it affects the large areas, the volume of runoff will be very high. Although rainwater runoff is a major source, agricultural productivity from rainwater in these areas is very low because runoff produced from rainfall is almost completely eliminated by evaporation or uncontrolled runoff (Oweis and Hachum, 2006) (Figure 16.5).

The central part of Gansu Province in China is one of the poorest parts of China due to its unique physical geography and lack of rainfall. For thousands of years, water needs in the domestic sector have been met in this region by collecting rainwater, but recently, the collection of rainwater for agricultural purposes has taken a new direction (Yuan et al., 2003). The number of people living in poverty in China in 1978 was 250 million (30.7% of the total population), which increased to 30 million in 2000 (3% of the total population). This indicates that China has taken the appropriate measures in the last few decades, but still faces challenges in the 21st century to supply the water to mountainous areas with natural limitations. From 1995 to 1996, with the support of the Gansu Provincial Government and social financial assistance, a project was implemented using the RWH (Rain Water Harvesting) system, during which each household received a grant to build two reservoirs to collect rainwater. One of these tanks was used for domestic purposes and the other tank was used for irrigating farms. Water storage in this way caused problems in water quality, which was that the number of bacteria in it was much higher than the national standard and the water stored in newly built cement tanks had a very high pH. To deal with these problems, it is recommended not to use the first water stored in the home storage tank and to boil it before drinking. Regarding irrigation of agricultural lands, it is better for farmers to pass the water stored in the reservoir through the filter and use the solar heaters in cold climates to protect against frost. It should be noted that due to the limited water stored in this method, efficient management should be used to use it because this water can prevent the plant death only in critical periods of crop growth when there are the severe water shortages and keeping it alive until the rainy season arrives so that the farmer can see the satisfactory harvests (Qiang, 2003; Sun et al., 2022). In areas with very dry climates, especially those with an altitude of more than 2000 meters above the sea level, the cultivation of vegetables such as cucumbers, tomatoes, and eggplants due to lack of water and frost is limited by the use of the RWH system; in addition to compensate for water shortages, allows the farmers in these areas to protect crops from freezing by building greenhouses. Although implementation of the RWH system has low costs, if they are properly designed, they can compensate for it in 1–2 years. The RWH system has no negative impact on the environment and has been proven to be effective for semi-humid and humid mountainous areas including southwestern China with an annual rainfall of more than 1000 mm; because of periods of water drought and monsoon weather, drought prevails in these areas (Qiang, 2003; Xiaoyan et al., 2002). To prevent mud from entering the tank, filters can be placed in the water passage before entering the tank, and also for easier access, it is better to place the tank near the farm. The important point about reservoirs that are placed beneath the ground is that they should be at least 5 m away from the large trees to prevent their roots from entering the reservoirs and damaging them. Today, rainwater collection has developed so much that water collection tanks can be seen in greenhouses. The rainwater stored in these tanks is not enough to irrigate the vegetables grown in the greenhouse, and complementary methods must be used. However, building a tank inside a greenhouse is a great idea because it solves the problems related to the water transportation and is effective in organizing agricultural work. On the other hand, with this work, farmers no longer only look at the sky when it rains and continues their work during the dry season. With the help of such reservoirs, the vegetables and other economic products can be grown in the mountainous areas. Due to the fact that the water collected in this method is limited, the use of specific irrigation methods for optimal consumption is essential; among the existing methods, subsurface drip irrigation is the most effective method (Consoli et al., 2014; Zhang et al., 2017; Zhang et al., 2021). In this method, water is pumped with the help of a pump into pipes whose walls are perforated in certain places, and according to the need, water seeps out of them and reaches the roots of the plant slowly and evenly.

Various cultivation techniques have been developed to save water. One of these techniques, which has gradually become a promising cultivation method in China's semi-arid ecosystems, is the use of plastic film mulch based on the rainwater harvesting (Li et al., 2016). Some farmers believe that water should not be wasted in any amount, so during the growing season, they spread the plastic layer on a farm and plant the plants on both sides of the plastic. By doing this, when it rains, the water moves on the plastic and reaches the roots of the plants from both sides. The rate of crop growth in this farm is higher than in the farm in which plastic is not used and more harvest will be done. The other local farmers also use the polyethylene plastic to collect the water from non-cultivated land and send it into the reservoir. Of course, before spreading the plastic, the weeds can be removed, the soil completely can be spread, and pound it so that it does not sink under the plastic. Several tires are placed around the plastic so that strong winds do not lift them (aparat.com, 2018) (Figures 16.6 and 16.7).

FIGURE 16.6   Rainwater collection using plastic membrane.

FIGURE 16.7   Overview of rainwater collection for field irrigation (abc.net.au, 2020).

16.11 Conclusions

In addition to the availability of water and the associated price, the amount that is required in the domestic, agricultural, and industrial sectors is very important and has the influential factors on the rate of water demand, which is approximately 70% of water demand in the agricultural sector. As the population grows, so does the demand for water, but it has a higher growth rate, with the world's population doubling between 1900 and 1995, but water demand around the world has grown by 600%. In the prediction of population growth and subsequent increment in water demand, water scarcity is one of the major concerns in the future, so we must optimally manage the existing water resources that are very limited and look for the ways to increase the resources to avoid the effects of migration, prevent the habitat destruction, disease, poverty, and other water shortage problems (Appan, 1999). Following population growth and urban development, the volume of runoff will also increase, which should be considered as one of the most important sources of water supply, because with population growth, global warming, and climate change, in the not too distant future, especially in arid and semi-arid regions, we will face water shortage (Baek et al., 2015; Xu et al., 2019).

The integration of rainwater harvesting and drip irrigation techniques, in addition to improving water productivity and preventing desertification, helps to alleviate poverty and ensures the food security for the occupants through agricultural sustainability in arid areas (Qiang, 2003; Zhang et al., 2021). In this case, it will lead to the great achievements such as creating job opportunities, reducing food prices, beautifying society, and lowering air pollution (Deng, 2021; Lawson, 2016; McDougall et al., 2019; Palmer, 2018; Reidy, 2009; Whitfield, 2009). To prevent or minimize these problems, we must now reduce the water losses and increase the water productivity by using the appropriate management strategies, with the help of modern irrigation methods, adjusting irrigation planning, and proper cultivation pattern. Considering the climate of the area and the available water resources, irrigation is possible when the intensity of evapotranspiration and deficit irrigation technique are feasible.

References

abc.net.au. (2020). https://www.abc.net.au/news/rural/2020-06-20/water-harvesting-dam-helps-cleve-farmer-get-through-hard-times/12355538.

Ahmed, R. M., and Alibaba, H. Z. (2016). An evaluation of green roofing in buildings. *International Journal of Scientific and Research Publications*, 6(1), 366–373.

aparat.com. (2018). (in Persian). https://www.aparat.com/v/oyrpz.

Appan, A. (1999). Trends in water demands and the role of rainwater catchment systems in the next millennium. In *9ª Conferência Internacional sobre Sistemas de Captação de Água de Chuva. Anais.... (Cd-rom)*. Petrolina-PE., Brazil.

Baek, S. S., Choi, D. H., Jung, J. W., Lee, H. J., Lee, H., Yoon, K. S., and Cho, K. H. (2015). Optimizing low impact development (LID) for stormwater runoff treatment in urban area, Korea: Experimental and modeling approach. *Water Research*, 86, 122–131.

Barrow, C. J., 2014. *Alternative Irrigation: The Promise of Runoff Agriculture*. Routledge, New York.

Ben-Asher, J. and Berliner, P. R. (eds.), 1994. Runoff irrigation. In: *Management of Water Use in Agriculture* (pp. 126–154). Springer, Berlin, Heidelberg, Germany.

Campisano, A., Butler, D., Ward, S., Burns, M. J., Friedler, E., DeBusk, K., ... Han, M. (2017). Urban rainwater harvesting systems: Research, implementation and future perspectives. *Water Research*, 115, 195–209.

Consoli, S., Stagno, F., Roccuzzo, G., Cirelli, G. L., and Intrigliolo, F. (2014). Sustainable management of limited water resources in a young orange orchard. *Agricultural Water Management*, 132, 60–68.

Cosgrove, W. J. and Rijsberman, F. R. (2000). *World Water Vision: Making Water Everybody's Business*. Earthscan Publications, 2(15): (2000). London, UK.

Deng, Y. (2021). Pollution in rainwater harvesting: A challenge for sustainability and resilience of urban agriculture. *Journal of Hazardous Materials Letters*, 2, 100037.

Elliott, A. H. and Trowsdale, S. A. (2007). A review of models of low impact urban stormwater drainage. *Environmental Modelling and Software*, 22, 394–405.

Freeborn, J., Sample, D., and Fox, L. (2012). Residential stormwater methods for decreasing runoff and increasing stormwater infiltration. *Journal of Green Building*, 7, 15–30. Doi: 10.3992/jgb.7.2.15.

Gilroy, K. L. and McCuen, R. H. (2009). Spatio-temporal effects of low impact development practices. *Journal of Hydrology*, 367(3–4), 228–236.

Gironás, J., Roesner, L., Rossman, L., and Davis, J. (2010). A new applications manual for the Storm Water Management Model (SWMM). *Environmental Modelling and Software*, 25, 813–814.

Göbel, P., Stubbe, H., Weinert, M., Zimmermann, J., Fach, S., Dierkes, C., ... Coldewey, W. G. (2004). Near-natural stormwater management and its effects on the water budget and groundwater surface in urban areas taking account of the hydrogeological conditions. *Journal of Hydrology*, 299(3–4), 267–283.

greenroofs.com. (2022). https://www.greenroofs.com/2022/02/15/happy-international-greenwallday-2022/.

Hernández-Crespo, C., Fernández-Gonzalvo, M., Martín, M., and Andrés-Doménech, I. (2019). Influence of rainfall intensity and pollution build-up levels on water quality and quantity response of permeable pavements. *Science of the Total Environment*, 684, 303–313.

Jackson, R. B., Carpenter, S. R., Dahm, C. N., Mcknight, D. M., Naiman, R. J., Postel, S. L., and Running, S. W. (2001). Water in a changing world. *Ecological Applications*, 11, 1027–1045.

Jia, H., Lu, Y., Yu, S., and Chen, Y. (2012). Planning of LID–BMPs for urban runoff control: The case of Beijing Olympic Village. *Separation and Purification Technology*, 84, 112–119.

Jones, M. P. and Hunt, W. F. (2010). Performance of rainwater harvesting systems in the southeastern United States. *Resources, Conservation and Recycling*, 54(10), 623–629.

Kamali, M., Delkash, M., and Tajrishy, M. (2017). Evaluation of permeable pavement responses to urban surface runoff. *Journal of Environmental Management*, 187, 43–53.

Kim, J. E., Teh, E. X., Humphrey, D., and Hofman, J. (2021). Optimal storage sizing for indoor arena rainwater harvesting: Hydraulic simulation and economic assessment. *Journal of Environmental Management*, 280, 111847.

Lawson, L. (2016). Agriculture: Sowing the city. *Nature*, 540(7634), 522–523.

Li, C., Wen, X., Wan, X., Liu, Y., Han, J., Liao, Y., and Wu, W. (2016). Towards the highly effective use of precipitation by ridge-furrow with plastic film mulching instead of relying on irrigation resources in a dry semi-humid area. *Field Crops Research*, 188, 62–73. Doi: 10.1016/j.fcr.2016.01.013.

Li, F., Liu, Y., Engel, B. A., Chen, J., and Sun, H. (2019). Green infrastructure practices simulation of the impacts of land use on surface runoff: Case study in Ecorse River watershed, Michigan. *Journal of Environmental Management*, 233, 603–611.

Li, S., Liu, Y., Her, Y., Chen, J., Guo, T., and Shao, G. (2021). Improvement of simulating sub-daily hydrological impacts of rainwater harvesting for landscape irrigation with rain barrels/cisterns in the SWAT model. *Science of The Total Environment*, 798, 149336.

Liu, Y., Engel, B. A., Flanagan, D. C., Gitau, M. W., McMillan, S. K., and Chaubey, I. (2017). A review on effectiveness of best management practices in improving hydrology and water quality: Needs and opportunities. *Science of the Total Environment*, 601, 580–593.

Liu, Y., Engel, B. A., Flanagan, D. C., Gitau, M. W., McMillan, S. K., Chaubey, I., and Singh, S. (2018). Modeling framework for representing long-term effectiveness of best management practices in addressing hydrology and water quality problems: Framework development and demonstration using a Bayesian method. *Journal of Hydrology*, 560, 530–545.

Locatelli, L., Mark, O., Mikkelsen, P. S., Arnbjerg-Nielsen, K., Wong, T., and Binning, P. J. (2015). Determining the extent of groundwater interference on the performance of infiltration trenches. *Journal of Hydrology*, 529, 1360–1372.

Maleki, M., Eslamian, S., and Hamouda, B., 2021. Principles and applications of atmospheric water harvesting. *Handbook of Water Harvesting and Conservation: Basic Concepts and Fundamentals* (pp. 243–259) Edited by Saeid Eslamian and Faezeh Eslamian., Published 2021 by John Wiley & Sons Ltd., https://doi.org/10.1002/9781119478911.ch16.

McDougall, R., Kristiansen, P., and Rader, R. (2019). Small-scale urban agriculture results in high yields but requires judicious management of inputs to achieve sustainability. *Proceedings of the National Academy of Sciences*, 116(1), 129–134.

Mikkelsen, P. S., Weyer, G., Berry, C., Waldent, Y., Colandini, V., Poulsen, S., ... Rohlfing, R. (1994). Pollution from urban stormwater infiltration. *Water Science and Technology*, 29(1–2), 293–302.

Nazif, S., Razavi, S. G., Soleimani, P., and Eslamian, S. (2021). Rainwater and green roofs, 355–372. Doi: 10.1002/9781119478911.ch23.

Oweis, T. and Hachum, A. (2006). Water harvesting and supplemental irrigation for improved water productivity of dry farming systems in West Asia and North Africa. *Agricultural Water Management*, 80(1–3), 57–73. https://doi.org/10.1016/j.agwat.2005.07.004.

Oweis, T. Y., Prinz, D., and Hachum, A. Y. (2012). *Rainwater Harvesting for Agriculture in the Dry Areas*. CRC Press, Boca Raton, FL.

Pacey, A. and Cullis, A. (1986). *Rainwater Harvesting. The Collection of Rainfall and Runoff in Rural Areas*. Intermediate Technology Publ., London.

Palmer, L. (2018). Urban agriculture growth in US cities. *Nature Sustainability*, 1(1), 5–7.

Qiang, Z. (2003). Rainwater harvesting and poverty alleviation: A case study in Gansu, China. *Water Resources Development*, 19(4), 569–578.

Reidy, P. C. (2008). Integrating rainwater harvesting and stormwater management infrastructure: Double benefit-single cost. In *Low Impact Development for Urban Ecosystem and Habitat Protection* (pp. 1–7)., 2008 International Low Impact Development Conference., https://doi.org/10.1061/41009(333)28.

Relf, D. (2009). Mulching for a healthy landscape. http://hdl.handle.net/10919/48323

Revitt, M., Ellis, B., and Scholes, L. (2003). Report 5.1 Review of the use of stormwater BMPs in Europe. Project under EU RTD 5th Framework Programme. WP5, 5.

Roshan Akther, M. S. (2014). Rainwater Harvesting. https://www.researchgate.net/publication/321329

Smith, D. (2006). *Permeable Interlocking Concrete Pavement? Selection Design, Construction and Maintenance*, Third Edition. Interlocking Concrete Pavement Institute, Herndon, VA.

Sun, M., Gao, X., Zhang, Y., Song, X., and Zhao, X. (2022). A new solution of high-efficiency rainwater irrigation mode for water management in apple plantation: Design and application. *Agricultural Water Management*, 259, 107243.

sydneywatertanks.com.au. (2019). https://www.sydneywatertanks.com.au/save-space-water-and-money-with-underground-water-tanks/.

sydneywatertanks.com.au. (2020). https://www.sydneywatertanks.com.au/save-space-water-under ground-concrete-tank/.

Tedoldi, D., Chebbo, G., Pierlot, D., Kovacs, Y., and Gromaire, M. C. (2016). Impact of runoff infiltration on contaminant accumulation and transport in the soil/filter media of sustainable urban drainage systems: A literature review. *Science of the Total Environment*, 569, 904–926.

UN. (2017). United Nations. https://www.un.org/development/desa/en/news/population/world-population-prospects-2017.html accessed June 21th 2021.

Verma, H. N. and Sarma, P. B. S., 1990. Design of storage tanks for water harvesting in rainfed areas. *Agricultural Water Management*, 18(3), pp. 195–207.

Villarreal, E. L. and Dixon, A. (2005). Analysis of a rainwater collection system for domestic water supply in Ringdansen, Norrköping, Sweden. *Building and Environment*, 40(9), 1174–1184.

Ward, S. (2007). *Rainwater Harvesting in the UK-Current Practice and Future Trends* (p. 1). Centre for Water Systems, University of Exeter, Exeter.

watercache.com. (2020). https://www.watercache.com/faqs/anodamine-rain-catcher-award.

Whitfield, J. (2009). Seeds of an edible city architecture. *Nature*, 459(7249), 914–915.

Xiaoyan, L., Ruiling, Z., Jiadong, G., and Zhongkui, X. (2002). Effects of rainwater harvesting on the regional development and environmental conservation in the semiarid loess region of Northwest China. In *12th ISCO Conference*, Beijing.

Xu, C., Jia, M., Xu, M., Long, Y., and Jia, H. (2019). Progress on environmental and economic evaluation of low-impact development type of best management practices through a life cycle perspective. *Journal of Cleaner Production*, 213, 1103–1114.

Yuan, T., Fengmin, L., and Puhai, L. (2003). Economic analysis of rainwater harvesting and irrigation methods, with an example from China. *Agricultural Water Management*, 60(3), 217–226. Doi: 10.1016/S0378-3774(02)00171-3.

Zamani, N., Maleki, M., and Eslamian, F., 2021. Fog water harvesting investigation as a water supply resource in arid and semi-arid areas. *Water Productivity Journal*, 1(4), pp. 43–52.

Zhang, H., Wang, D., Ayars, J. E., and Phene, C. J. (2017). Biophysical response of young pomegranate trees to surface and sub-surface drip irrigation and deficit irrigation. *Irrigation Science*, 35(5), 425–435.

Zhang, W., Sheng, J., Li, Z., Weindorf, D. C., Hu, G., Xuan, J., and Zhao, H. (2021). Integrating rainwater harvesting and drip irrigation for water use efficiency improvements in apple orchards of northwest China. *Scientia Horticulturae*, 275, 109728.

17

Water Harvesting for Rain-Fed Farming

Adebayo Oluwole Eludoyin, Adedayo Oreoluwa Adewole, and Mayowa Emmanuel Oyinloye
Obafemi Awolowo University

Saeid Eslamian
Isfahan University of Technology

17.1 Introduction

The problem of water scarcity has become more severe, owing to increased urbanization, occurrence of drought and changing climatic patterns, among others (Güneralp et al., 2015; Omer et al., 2020; Eludoyin & Olanrewaju, 2020; Eludoyin et al., 2021). Studies revealed that human water requirements would rise by 40%, and additional 17% of water will be required to increase food production to meet up with increasing population (Gleick, 1996; Sharma, 2007; Di Baldassarre et al., 2018). Most of the studies have argued that economic and urbanization forces will place increasing demands on scarce water resource availability, and therefore recognized the urgent need to adopt measures to ensure water sustainability at different (households, catchment, regional and global) scales (Hoekstra et al, 2018; Chen, et al., 2020; Eludoyin, 2020; Eludoyin et al., 2021; Onanuga et al., 2022). Sustainability is an act or measure that involves improving the quality of human life while living within the carrying capacity of supporting ecosystems. Water sustainability has in earlier work (Eludoyin, 2016) been used to mean 'supplying or being supplied with water for life', and an important way to tackle acute water shortage in many tropical countries is through the application of rainwater-harvesting systems (Lye, 2002; Eludoyin et al., 2021). Rainwater harvesting is an important source of water for domestic and agricultural uses (Velasco-Muñoz et al., 2019; Eludoyin et al., 2021).

Rain-fed farming is a type of farming which relies heavily on rainfall, and is also vulnerable to potential and real climate extremes. According to Molden et al. (2011), rain-fed farming often involves

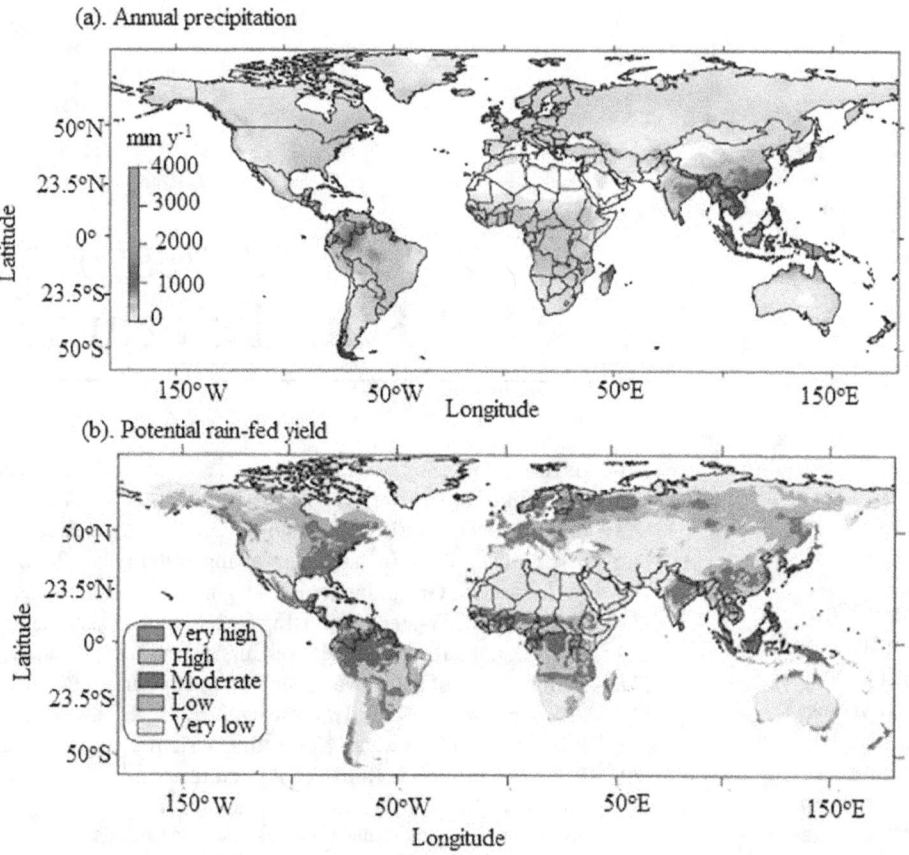

FIGURE 17.1 (a and b) Global distribution of annual rainfall and potential rain-fed yields (modified from Droogers et al., 2001).

cultivation of both short (annual, biannual and biennial) and long-term crops, some of which have not been influenced by technological developments. Rain-fed farming constitutes 80% of the world's cropland and produces more than 60% of the world's cereal grains, generating livelihoods in rural areas while producing food for cities. In temperate regions with relatively reliable rainfall and good soils, rain-fed agriculture generates high yields and supplemental irrigation practices boost yields even higher (Molden et al., 2011). A major problem of rain-fed agriculture is the recently climatic extremes and variability, poor weather/climate prediction and poor farm technology (Eludoyin et al., 2017; Idowu et al., 2022; Adewoyin et al., *under review*). According to Molden et al. (2011), there appears to be a strong relationship between poverty, hunger and water stress in part because of the dependencies on rain-fed agriculture in developing economies. Figure 17.1 compares the global distribution of annual rainfall and potential for rain-fed yields. Areas within the tropics ($23\frac{1}{2}°$N – S) with more annual rainfall distribution tend to possess more potential for rain-fed yield than other parts of the world (Figure 17.1a and b). Yields from rain-fed farming are known to exhibit high uncertainty in terms of the start and end of the rainy season, rainfall amount and duration (Mackill et al., 1999). FAOSTAT (2005) reported that rain-fed farming is practiced in 80% of the world's physical agricultural area and generates 62% of the world's staple food.

This chapter discusses the concept of water harvesting in rain-fed farmlands, especially in Africa and Asia. Its goal is to explain the concept of rainwater harvesting, the prospect and challenges involved. The chapter is subdivided into seven subsections, consisting of the introduction, concept of rainwater water harvesting, rainwater harvesting technology, agricultural water harvesting, rainwater for groundwater recharging, advantages of water-harvesting systems and reasons to upgrade rain-fed agriculture. The discussion is entirely based on literature review.

17.2 Concept of Rainwater Water Harvesting

Water harvesting, collection and storage water from runoff, precipitation or creek flow for irrigation use, is essential for rain-fed farming activities. Rainwater-harvesting involves stages that include rainwater-interception, diversion, storage and distribution. Various rainwater-harvesting systems are currently applied in different parts of the world, including contour terracing (Sánchez, 2019), floodwater farming (Bryan, 1929; Brunner & Haefner, 1986) and *khadin* systems (Kolarkar et al., 1983). Some other systems include rock bunds and stone terraces (Reddy, 2016), basin systems in Mali and the *caag* system in the Hiraan region of central Somalia (Wani et al., 2009) (Figure 17.2). In northern coastal areas of Egypt, the tradition of rainwater-harvesting system exists in the form of *wadi* terracing. In Ethiopia rainwater harvesting dated back to 560 BC with evidence of a water-harvesting setup in the palace of the legendary Queen of Sheba (Yosef and Asmamaw, 2015). In recent decades, scientists in the Sub-Saharan Africa, the Middle East and Southeast Asia have made efforts to develop and test a wide variety of techniques for collecting, storing and using natural precipitation for agricultural purposes (Biazin et al., 2012).

FIGURE 17.2 (a–d) Some examples of rainfall harvesting procedure in rain-fed farming systems (see Bryan, 1929; Brunner & Haefner, 1986; Reddy, 2016; Sánchez, 2019, for further information).

Harvested rainwater has served as supplemental water for irrigation of crops, livestock, fodder and tree production and, in some cases, supply for fish and duck ponds.

The practice of rainwater harvesting varies across different regions and the focus is to reduce costs incurred when purchasing water from the centralized system as operating costs (Słyś and Stec, 2020). It is also an approach aimed at saving energy by directly reducing water use and energy requirements needed to pump water from water treatment plants to individual homes.

Rainwater harvesting is being promoted by non-governmental organizations, national agricultural extension services and government agencies in African countries owing to the fact that it is arguably the most common soil and water conservation techniques (Stroosnijder, 2003). Adoption of rainwater harvesting can increase water resources (Rohitashw et al., 2011).

17.3 Rainwater-Harvesting Technology

Martínez-Acosta et al. (2019) in a review of rainwater structure indicated four groups of rainwater harvesting: micro- and macro-basins, aquifer recharge and reserve tillage (Figure 17.3).

Rainwater harvesting is a subject of science and involves technology for its design, installation, operation and maintenance. Rainwater harvesting in a building deals with managing rainwater and is based on hydrological philosophy as well as sanitation and plumbing technology (Haq, 2017). This technology may be adopted as a household project or in hospitals, schools and housing complexes. Components of typical rainwater-harvesting facility system are conveyance, storage, overflow, outlet and delivery. A first-flush diverter may be installed for improved water quality. Important factors are engineering specifications for storage contributing area, rainfall patterns, anticipated usage, cost and aesthetics. Collected water can be used for non-potable uses or for potable supply with appropriate treatment. The technology requires a little/ or no energy because capture systems often use low-volume, non-pressurized, gravity-fed systems or low power pumps (Biswas et al., 2021).

Rainwater harvesting is the only sustainable alternative for ensuring continued access to safe drinking water in drought-prone areas or where the surface water/groundwater is saline or polluted, rooftop rainwater harvesting. It is therefore the best approach for communities potentially vulnerable to climate change and for rainwater conservation (Biswas et al., 2021). The construction of rainwater-harvesting systems is a source of employment to persons having required skills because local people can be trained and mobilized to implement the technology. Figure 17.4 shows the different types of rainwater-harvesting approaches for

Type 1	Type 2	Type 3
Macro - basins	Micro-watersheds	Acquifer recharge
Dams	Contemplating jessour Tabias Excavation with a dyke (pit contours) Contour ridges	Gully plugs Aquifer recharge Subsurface dams Percolation tanks
Small dams Control dams Reservoirs Cisterns nala bunds	Contour strips (runoff strips) Negarim Dykes Terraces	

Type 4

Reserve Tillage

In situ uptake of rainwater with mechanical devce

FIGURE 17.3 Rainfall harvesting structure.

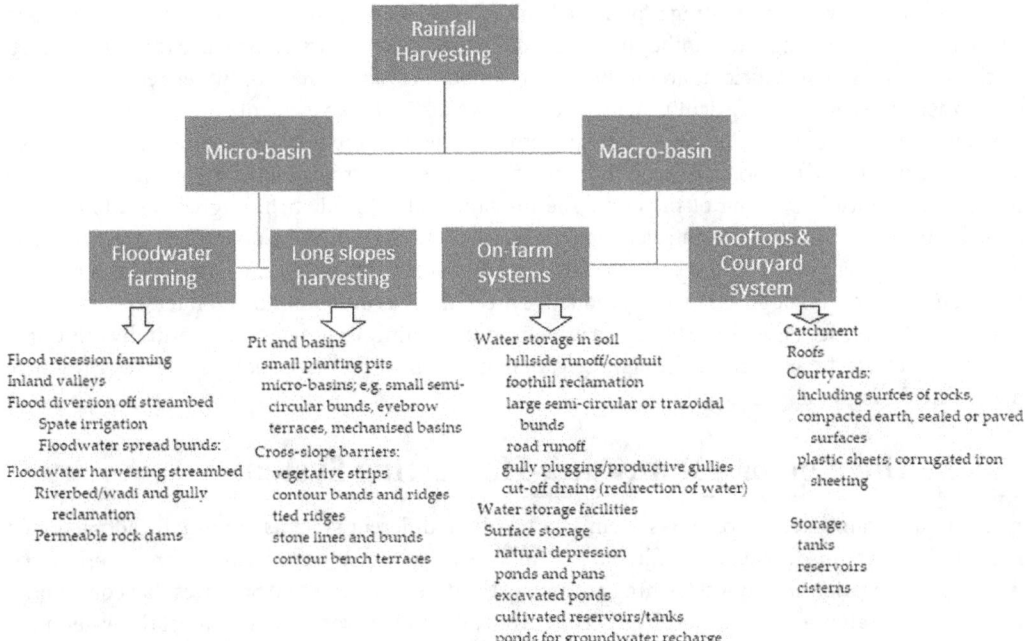

FIGURE 17.4 Characteristics of different rainwater-harvesting techniques.

rain-fed agriculture and farmers in most developing countries. Construction materials are readily available and system provides water at the point of consumption, and family members have full control of their own system. The technology also enhances farmers' capacities because it provides water, which is otherwise brought from distanced water sources. It also reduces the physical hardship and mental stress of the as well as time required to fetch water from other water sources. The saved time will serve other productive purposes such as domestic work, agriculture and livestock activities and childcare (Biswas et al., 2021). Rainwater harvesting from the rooftops would reduce the total volume of runoff from the roofs.

In general, Figure 17.4 includes micro-catchment techniques, which directly supply runoff water from a small catchment to the crop, and thus water accumulates around the plant, infiltrates into the soil and is stored in the crop root zone. It also includes macro-catchment techniques, which concentrate rainwater runoff flowing in an ephemeral *wadi* (natural channel) and store it in a prepared storage facility (such as a reservoir) for subsequent productive use. Water may also be diverted out of the *wadi* course to inundate nearby lands through proper damming or cross-structure. The macro-catchment systems involve having runoff water collected from relatively large catchments. Often the catchments are natural rangeland or a mountainous area. In most scenarios, these catchments are located outside farm boundaries, where farmers have little or no control over them. Harvested runoff can be stored in the soil profile for direct use by the crop, in aquifers as a recharge system or in a storage facility ranging from an on-farm pond or tank to a small dam constructed across the *wadi*, and used later for domestic purpose, livestock and supplemental irrigation. Generally, designed rainwater-harvesting systems will have minimal maintenance costs associated with its upkeep and therefore will show the best long-term relationship between cost and financial benefit. Rainwater-harvesting systems with a connected vaporization system can raise site humidity and create a healthier microclimate, which is ideal for city areas dealing with air pollution.

Agricultural rainwater harvesting is premised on the concept of depriving part of the land of its share of precipitation and giving it to another part to increase the amount of water available to the latter part, which originally was insufficient, and to bring this amount closer to the crop water requirements in order to achieve economical agricultural production. Two distinct management periods are involved in maximizing the use of precipitation for dryland crop production: the first period of rain storage, lasting from harvesting of the previous crop until planting of the next crop; and the second period, lasting from planting until harvesting of the crop. The fundamental principle behind green and blue water resources in agriculture is that plants take up water from the root zone in the uppermost part of the soil profile, i.e. the green water resource, which subsequently leaves the plant as transpiration, i.e. a productive green waterflow (as opposed to non-productive green flows as evaporation and interception) (Wani et al., 2009). In rain-fed agriculture the green water resource mainly originates from naturally infiltrated rainwater but it can be augmented through irrigation by allowing the application of blue water resource to infiltrate the land.

17.4 Setting Up for a Rainwater-Harvesting System

In setting up a rainwater-harvesting system in a building, different systems are usually adopted, and these include direct use system, non-filtered system, filtered system and complete systems. Runoff irrigation, spate irrigation and runoff farming are among the different forms and practices that come under the umbrella of rainwater harvesting. The effective roof area and the roof material affect the water quality and chemically inert, non-toxic materials should construct efficiency of collection: drainpipes, roof surfaces and the storage tank in order to avoid adverse effects on water quality (Biswas et al., 2021). Limitations associated with the implementation of rainwater harvesting are increased soil erosion when slopes are cleared for higher runoff rates, loss of habitat of flora and fauna on those slopes, loss of habitat of flora and fauna in depressions, upstream–downstream conflicts and competition among farmers and herders. Specific components are as follows:

i. Catchment area/runoff area, varying from a few square metres (micro-catchment) to as large as several square kilometres (macro catchment)
ii. Storage facility i.e. the place where the runoff water is held from the time it occurs until it is utilized. Storage can be above the soil surface in reservoirs or ponds, in the soil profile as soil moisture, and/or underground in cisterns or as groundwater in aquifers
iii. Target or use: i.e. the beneficiary of the stored water. In agricultural production, the target is the plant or the animal and in domestic use, it refers to human beings and their needs (Table 17.1).

TABLE 17.1　Methods of Setting Up a Rainwater Harvesting

Methods	Description
Direct use system	This involves direct collection of rainwater from catchment for immediate use. It is the simplest and most basic form of rainwater harvesting. It requires only a clean catchment area and a well-maintained storage tank.
Non-filtered system	Sedimentation using this method is facilitated in the storage tank; no extra settling tank. The storage tank is designed to play dual role of settling and storage.
Filtered system	In this system, a filtration system is employed, which is installed after the sedimentation tank. An extra tank is required to collect the filtered rainwater from where the rainwater is supplied to various locations in the buildings where the rainwater is needed.
Complete system	All the conditioning systems of rainwater harvesting are incorporated in this system. This complete system should be developed for consuming rainwater as potable water. The success of rainwater harvesting in buildings fully depends on the effectiveness of the system in place. Ignorance and/or failure of any of these aspects can stop the establishment of this system in a building.

17.5 Rainwater for Groundwater Recharging

In areas where the source of water is predominantly groundwater, over-extraction of groundwater could cause various hydrogeological problems. Groundwater recharging is the most effective approach for improving the groundwater situation, and the use of rainwater is highly suitable for this purpose. Groundwater recharging involves collection of rainwater from suitable catchments, conveyance of the collected water to the recharge structure, qualifying rainwater based on the method of recharging adopted and draining of excess rainwater.

The methods for increasing groundwater recharge include pumping surface water directly into an aquifer and enhancing infiltration by spreading water into infiltration basins. Runoff could be stored on the surface for future use or recharged to groundwater and shallow aquifer. Rainwater harvesting has a positive impact on soil conservation, erosion prevention, replenishment of groundwater and ecosystem restoration (Yosef & Asmamaw, 2015). Contour trenching, construction of bunds (small dams) or sub-surface dams, tree planting and terracing are some of the options for increasing groundwater recharge.

17.6 Advantages of Water-Harvesting Systems

In arid and semi-arid areas, rainwater harvesting makes farming possible on part of the land, provided other production factors such as climate, soils and crops are favourable. The economy of arid lands highly depends upon livestock; consequently, rainwater harvesting has been aimed at providing water for livestock. In rain-fed areas, rainwater-harvesting systems can provide additional water to supplement rainfall to increase and stabilize production. With rainwater harvesting, farmers have diversified to include horticultural cash crops and the keeping of diary animals, which has contributed to food security, better nutrition and family income. It can also alleviate the risk associated with the unpredictability of rainfall in these areas. Rainwater-harvesting system is usually equipped with a facility (above- or underground type) to store the harvested water for later use in supplemental irrigation during drought. In areas where there is lack of public water supply for domestic and animal production, inducing runoff from a treated area and storing it in a cistern or other type of reservoir for later use is a common practice in remote areas where no other water resources are available. In arid lands suffering from desertification, rainwater harvesting would improve the vegetative cover and can stop environmental degradation. Water harvesting is also effective in recharging groundwater aquifers. It is important to note that rainwater harvesting can be used to rehabilitate degraded land and retain moisture in situ through structures that reduce runoff from fields and hold water long enough to allow infiltration (Yosef & Asmamaw, 2015). An improved in-field water harvesting can increase the required time for crop moisture stress to set in and can result in improved crop yield as well as soil fertility, moisture conservation and agricultural productivity. Indirect benefits of rainwater harvesting in dry environments apart from food and feed production include substantial environmental and social returns such as combating land degradation and migration from rural to urban areas and employment.

17.7 Planning of a Rainwater-Harvesting System

A reinforced concrete building is supposed to have a maximum useful structural life of 75 years (Haq, 2017). During its life span, the water–environment scenario of the area might be changed; consequently, alternative options for water management might be required for the building. After a building has been constructed, the incorporation of a rainwater-harvesting system, as a supplement will prove difficult majorly in terms of accommodating the storage and conditioning structures of appropriate size at the proper locations. Therefore, during the preparation of the architectural drawing of a building, planning for the incorporation of rainwater-harvesting elements must be performed based on the following guidelines.

  i. One must synchronize the harvesting system with the normal water-supply and drainage system.

  ii. The components of the rainwater-harvesting system should be well positioned.

  iii. The provisions for maximizing collection in case of shortage should be highly incorporated.

17.8 Maintenance of a Rainwater-Harvesting System

Planning a well-planned rainwater-harvesting system involves identification of the purposes requiring the use of rainwater, planning of the catchments, collecting system, the manner of storing and assigning the storage system a location, conditioning system, supplying system and location of the recharge structure. The components of a rainwater-harvesting system are required to be regularly maintained at certain intervals. The following maintenance requirements of the major components of a rainwater-harvesting system are suggested (Haq, 2017).

  i. *Catchments*: The areas from which the rainwater will be collected must be swept every month to remove leaves, litter and dirt. Catchments must be washed off with water when dust or dirt accumulates, thus diverting runoff away from the filter or storage tank before rain starts. Tree branches hanging over the catchments must be trimmed when required.

  ii. *Gutters*: Gutters must be kept clean by washing out bird droppings, leaves and other dirt with water before it rains or after heavy winds. The cleanliness, stability, alignment and slope of the gutters must be checked every 3 months, as well as after a storm to ensure their proper functioning. After a heavy shower, the slope of gutter should be checked specially and corrected if any deviation is observed.

  iii. *Pipes*: All of the pipes—including inlet screens installed for rainwater harvesting—must be maintained and monitored in the way gutters should be addressed. The pipes must be checked regularly for leakages and must be repaired when required.

  iv. *First-flush device*: The first-flush device must be checked for cleanliness. It should be cleaned before and after the monsoon rain starts and after every rooftop-cleaning operation.

  v. *Treatment system*: Filters should be cleaned at least once in a year before the monsoon rain begins and backwashed after the monsoon rain is over. A chlorine level at or slightly greater than 0.5 mg/L and a pH level of 6.5–8.5 must be ensured by testing weekly and after heavy rains. The UV light must be inspected weekly to keep it free from scum.

  vi. *Storage tank*: Extra care for the storage tank should be taken in the regular maintenance programme of the rainwater-harvesting system. The tank must be cleaned before the monsoon rain starts. Occasionally the tank must be checked for the development of any cracks or leakages. If any crack or leakage is noticed, it must be repaired at the earliest possible time. In case of underground storage tanks, the growth of trees on or nearby the tank must be checked all of the time, and roots that might affect the tank must be cut when necessary. The lid of the tank must be ensured for its sturdiness, and there must be no gap between lid and its rim. At the end of the overflow pipe, the insect screen must be well secured at all times.

  vii. *Equipment and accessories*: All of the equipment and accessories installed in the rainwater-harvesting system, e.g., pump and valves must be checked at least once a year.

  viii. *Water quality*: The presence of mosquito larvae in stored rainwater must be checked for every 3 months. Accumulation of the sludge level must be monitored with the same frequency if rainwater is to be used for drinking purposes. An *E. coli* test should be performed at the initial development stage of the harvesting system to identify the risk of bacteriological contamination of the stored rainwater. In addition, this test should be performed when the system or a part of the system is altered.

The most commonly applied management practices in the sub-Saharan Africa include ridging, mulching, various types of furrowing and hoeing, and conservation tillage. Ridging—also known as furrow dikes, furrow damming, basin listing, basin tillage and micro-basins in different areas, can be designed as open or closed (tied) for holding water and facilitating infiltration in areas of low, erratic rainfall. In tied ridging, sometimes also called 'tied-furrows', ridge furrows are blocked with earth ties spaced at fixed distances to form a series of micro-catchment basins in the field.

17.9 Prospects of Rainwater Harvesting

It is believed that rainwater is harvested for mitigating the crisis of water prevailing in particular areas. Virtually this is the major and direct implication of rainwater harvesting, but there are various other indirect implications of rainwater-harvesting systems being installed in buildings. The prospective implications of rainwater harvesting in buildings are as follows:

- Decreases the pressure of using surface water and groundwater.
- Decreases the cost of water consumed.
- Increases groundwater availability.
- Decreases urban flooding and water logging.
- Harvested rainwater saves energy. A one-metre increase in groundwater level saves 0.40 kWh of electricity in extracting groundwater.
- Aids the development of energy-saving 'green' buildings.
- Helps to provide water during any crisis.
- Intervention in adaptation and decrease in vulnerability due to climate change.
- Renewable resource that does not pose any negative impacts on the environment.
- Salt-free source of water and its use in groundwater recharging decreases salt accumulation in soil.

17.10 Problems of a Rainwater-Harvesting System

Rainwater harvesting also has some negative implications in built environments. The following are those negative implications of rainwater harvesting in building development on surrounding environment.

- The availability of rainwater may be limited due to occurrence of long dry spells.
- Rainwater quality may not be consistent due to variability in air pollution and other sources of contamination.
- Regular consumption of mineral-free rainwater may cause nutritional deficiencies.
- The initial investment cost of the harvesting components is relatively high.
- Lack of proper operation and regular maintenance will restrict the desired output from the system.

17.11 Reasons to Upgrade Rain-Fed Agriculture

17.11.1 Large Scope for Poverty Alleviation

Agriculture plays a key role in economic development and poverty reduction. For instance, it has been shown that every 1% increase in agricultural yields translates into a 0.6%–1.2% decrease in the absolute poor (Wani et al., 2009). In sub-Saharan Africa (SSA), the majority of the poor make their living from agriculture. Here, majority of the poor (70% of the population accounting for 35% GDP) solely depend on agriculture as their source of livelihood. An investment in rain-fed agriculture in SSA would have a significant impact on poverty reduction; thus, agriculture is the engine of overall economic growth and consequently of broad-based poverty reduction and an investment in low-yielding rain-fed agriculture could have large impacts on poverty reduction.

17.11.2 Low Costs of Investment

The role of upgrading rain-fed agriculture is gaining increased attention with the high cost of expanding large-scale irrigation and environmental impacts of large dams; for instance, irrigated cereal production in sub-Saharan Africa, characterized by high marketing and transportation costs and limited marketing opportunities, might not be able to compete with subsidized food imports from the USA and Europe. In addition, the institutional infrastructure and experience required for irrigation operation, maintenance and management are lacking. Micro-credit schemes for water management investments in rain-fed agriculture have been suggested as a core strategy for enabling small-scale farmers to invest in water management in rain-fed agriculture (Wani et al., 2009).

17.11.3 Environmental Concerns Related to Large-Scale Irrigation

Diversion of water from rivers and lakes for agricultural purposes often adversely affects aquatic ecosystems, e.g. channel erosion, declines in biodiversity, introduction of invasive alien species, reduction of water quality, habitat fragmentation and reduced protection of flood plains and other inland and coastal fisheries. On the field scale, there are two major undesirable impacts of irrigation: salinization and waterlogging. Large-scale irrigation carries environmental risks and associated costs, which works in favour of investments in rain-fed agriculture.

17.11.4 Large Yield Gaps—High Potential

In developing countries, rain-fed grain yields are on average 1.5 t/ha, and increases in production have originated mainly from land expansion; In temperate regions rain-fed agriculture has some of the world's highest yields and even in tropical regions, agricultural yields in commercial rain-fed agriculture exceed 5–6 t/ha (Wani et al., 2009). In semi-arid regions in Africa and Asia, farmers' yields are two to four times lower than achievable yields for major rain-fed crops (Wani et al., 2009). The large yield gaps indicate a high potential for investments in rain-fed agriculture.

17.12 Conclusions

Water-harvesting techniques are useful in bridging short dry spells, but longer dry spells can lead to crop failure (Wang et al., 2009). This has made many farmers reluctant to use fertilizers, pesticides and labour in rain-fed settings. The cost of failure is higher, for the individual farmer who loses his/her income and for society. Risks to individual farmers can be mitigated by appropriate measures in rainwater harvesting, increasing the amount of rainwater that can be beneficially used by crops (i.e. effective precipitation). With the right incentives and measures to mitigate risks to individual farmers, water management in rain-fed agriculture holds a large potential to increase food production and reduce poverty while maintaining ecosystem services.

References

Adewoyin, A. A., Eludoyin, A. O., & Adetoro, O. O. I. (2022). *Extreme climatic conditions and effects on cropping activities at a farm settlement in southwestern Nigeria*. Preprint, https://doi.org/10.21203/rs.3.rs-1236423/v1.

Biazin, B., Sterk, G., Temesgen, M., Abdulkedir, A., & Stroosnijder, L. (2012). Rainwater harvesting and management in rain-fed agricultural systems in sub-Saharan Africa – A review. *Physics and Chemistry of the Earth, 47–48*, 139–151.

Biswas, S., Sahoo, S., Debsarkar, A., & Pal, M. (2021). Assessment of adoption potential of rooftop rainwater harvesting to combat water scarcity: A case study of North 24 Parganas district of West Bengal, India. *Arabian Journal of Geosciences, 14*(16), 1–16.

Brunner, U., & Haefner, H. (1986). The successful floodwater farming system of the Sabeans, Yemen Arab Republic. *Applied Geography, 6*(1), 77–86.

Bryan, K. (1929). Flood-water farming. *Geographical Review, 19*(3), 444–456.

Chen, Z., Wu, G., Wu, Y., Wu, Q., Shi, Q., Ngo, H. H., Saucedo, O. A. V., & Hu, H. Y. (2020). Water Eco-Nexus Cycle System (WaterEcoNet) as a key solution for water shortage and water environment problems in urban areas. *Water Cycle, 1*, 71–77.

Di Baldassarre, G., Wanders, N., AghaKouchak, A., Kuil, L., Rangecroft, S., Veldkamp, T. I., Garcia, M., van Oel, P. R., Breinl, K., & Van Loon, A. F. (2018). Water shortages worsened by reservoir effects. *Nature Sustainability, 1*(11), 617–622.

Droogers, P., Seckler, D., & Makin, I. (2001). *Estimating the Potential of Rain-Fed Agriculture* (Vol. 20). IWMI.

Eludoyin, A. O. (2016). Sustainability and water reclamation. In: Eslamian, S. (Ed.) *Urban Water Reuse Handbook* (Chapter 79, pp. 1077–1084). CRC Press, Taylor and Francis, New York, 1115p, ISBN 9781482229158.

Eludoyin, A. O. (2020). Accessibility to safe drinking water in selected urban communities in southwest Nigeria. *Water Productivity Journal, 1*(2), 1–10.

Eludoyin, A. O., & Olanrewaju, O. E. (2020). Water supply and quality in the Sub-Saharan Africa. In: Filho, W.L. et al (eds), *Encyclopedia of the UN Sustainable Development Goals, Clean Water and Sanitation* (pp. 1–17), Springer Nature, Switzerland AG. https://doi.org/10.1007/978-3-319-70061-8_166-1.

Eludoyin, A. O., Nevo, A. O., Abuloye, P. A., Eludoyin, O. M., & Awotoye, O. O. (2017). Climate events and impact on cropping activities of small-scale farmers in a part of southwest Nigeria. *Weather, Climate, and Society, 9*(2), 235–253.Eludoyin, A., Eludoyin, O., Martins, T., Oyinloye, M., & Eslamian, S. (2021). Water security using rainwater harvesting. In: Eslamian, S & Eslamian, F (eds), *Handbook of Water Harvesting and Conservation: Basic Concepts and Fundamentals* (Chapter 4, pp. 51–70), Wiley-Blackwell, New York. ISBN 9781119478942.

FAOSTAT. 2005. *Database*. Food and Agriculture Organization, Rome. Accessed November 2005. http://faostat.fao.org/.

Gleick, P. H. (1996). Basic water requirements for human activities: Meeting basic needs. *Water International, 21*(2), 83–92.

Güneralp, B., Güneralp, İ., & Liu, Y. (2015). Changing global patterns of urban exposure to flood and drought hazards. *Global Environmental Change, 31*, 217–225.

Haq, S. A. (2017). *Rainwater-Harvesting Technology. In: Harvesting Rainwater from Buildings*. Springer, Cham. https://doi.org/10.1007/978-3-319-46362-9_3.

Hoekstra, A. Y., Buurman, J., & Van Ginkel, K. C. (2018). Urban water security: A review. *Environmental Research Letters, 13*(5), 053002.

Idowu, M. A., Matthew, O. J., & Eludoyin, A. O. (2022). Evaluation of extreme climate over selected eco-climatic regions in Nigeria from observed and simulated (RegCM3) data. *International Journal of Global Warming, 26*(2), 222–232.

Kolarkar, A. S., Murthy, K. N. K., & Singh, N. (1983). 'Khadin—A method of harvesting water for agriculture in the Thar Desert. *Journal of Arid Environments, 6*(1), 59–66.

Lye, D. J. (2002). Health risks associated with consumption of untreated water from household roof catchment systems 1. *JAWRA Journal of the American Water Resources Association, 38*(5), 1301–1306.

Mackill, D. J., Nguyen, H. T., & Zhang, J. (1999). Use of molecular markers in plant improvement programs for rainfed lowland rice. *Field Crops Research, 64*(1–2), 177–185.

Martínez-Acosta, L., López-Lambraño, A. A., & López-Ramos, A. (2019). Design criteria for planning the agricultural rainwater harvesting systems: A review. *Applied Sciences, 9*(24), 5298.

Molden, D., Vithanage, M., De Fraiture, C., Faures, J. M., Gordon, L., Molle, F., & Peden, D. (2011). Water availability and its use in agriculture. In *Treatise on Water Science* (pp. 707–732). Elsevier, Oxford.

Omer, A., Elagib, N. A., Zhuguo, M., Saleem, F., & Mohammed, A. (2020). Water scarcity in the Yellow River Basin under future climate change and human activities. *Science of the Total Environment, 749*, 141446.

Onanuga, M. Y., Eludoyin, A. O., & Ofoezie, I. E. (2022). Urbanization and its effects on land and water resources in Ijebuland, southwestern Nigeria. *Environment, Development and Sustainability, 24*(1), 592–616.

Pérez Sánchez, J.M. (2019). Agricultural Terraces in Mexico. In: Varotto, M., Bonardi, L., Tarolli, P. (eds) World Terraced Landscapes: History, Environment, Quality of Life. Environmental History, vol 9. Springer, Cham. https://doi.org/10.1007/978-3-319-96815-5_10.

Reddy, P. P. (2016). Micro-catchment rainwater harvesting. In: *Sustainable Intensification of Crop Production*. Springer, Singapore. https://doi.org/10.1007/978-981-10-2702-4_14.

Rohitashw, K., Vijay, S., & Mahesh, K. (2011). Evaluation of evapotranspiration models for pea (Pisum sativum) in mid hill zone-India. *Universal Journal of Environmental Research and Technology, 1*(3), 329–337.

Sharma, S., Jackson, D. A., Minns, C. K., & Shuter, B. J. (2007). Will northern fish populations be in hot water because of climate change? *Global Change Biology, 13*(10), 2052–2064.

Słyś, D., & Stec, A. (2020). Centralized or decentralized rainwater harvesting systems: A case study. *Resources, 9*(1), 5. https://doi.org/10.3390/resources9010005.

Stroosnijder, L. (2003). Technologies for improving green water use efficiency in semi-arid Africa. In *Proceedings Water Conservation Technologies for Sustainable Dryland Agriculture in Sub-Saharan Africa* (pp. 92–102). Bloemfontein, South Africa, 8–11 April.

Velasco-Muñoz, J. F., Aznar-Sánchez, J. A., Batlles-delaFuente, A., & Fidelibus, M. D. (2019). Rainwater harvesting for agricultural irrigation: An analysis of global research. *Water, 11*(7), 1320.

Wani, S. P., Sreedevi, T. K., Rockström, J., & Ramakrishna, Y. S. (2009). Rainfed agriculture–past trends and future prospects. *Rainfed Agriculture: Unlocking the Potential, 7*, 1–33.

Yosef, B. A., & Asmamaw, D. K. (2015). Rainwater harvesting: An option for dry land agriculture in arid and semi-arid Ethiopia. *International Journal of Water Resources and Environmental Engineering, 7*(2), 17–28.

18

Optimization of Reservoir Operation for Irrigation

Laheab Abbas
Al-Maliki and
Sohaib Kareem
Al-Mamoori
University of Kufa

Khaled El-Tawil
Lebanese University

Nadhir Al-Ansari
*Lulea University
of Technology*

Fadi G. Comair
UNESCO IHP Council

Saeid Eslamian
*Isfahan University
of Technology*

18.1 Introduction

Irrigation water management is considered as an essential point to sustainable water use. Water used for irrigation is 70%–80% of the total consumptions of water. There is an urgent need for the methods and practices that reduce the excessive amount of water applied in irrigation without any decline in productivity.

It is necessary to manage the water resources efficiently to establish the maximum benefits of the crop productivity in light of the growing water crisis. To fulfil this requirement, a reservoir release should be integrated with the crop water distribution. Many factors and hydrological variables should be considered when modelling the reservoir operation for irrigation such as reservoir inflow, rainfall, evapotranspiration, reservoir storage and soil moisture (Eslamian et al., 2011; Valipour & Eslamian, 2014). The great challenge in optimizing reservoir operation is how to choose the variables and the constant in the model and this is depending on desired outputs itself.

Many optimization algorithms have been used and compared to optimize the reservoir operation. This area of research despite being mature, yet the development of modelling and programming techniques have made it very active till nowadays (Rani & Moreira, 2010). Many reviews have been conducted in the past from the early sixties to the present irrigation planning and management under uncertainty usually utilize stochastic dynamic programming (SDP) for its ability to deal with many optimization problems such as optimum interseasonal and intraseasonal water allocation, optimal seasonal acreage and optimum irrigation area. Also, it has the ability to integrate the optimization problem taking into account the uncertainty of demands, rainfall and runoff (Al-Mamoori & Al-Maliki, 2016; Dudley & Burt, 1973; Haddad et al., 2016). All of these models focused on optimizing the reservoir water release for irrigating the certain crop, that would result in optimal crop production.

Below is a briefed overview of some of the literature:

Dudley and Burt (1973) had developed an integrated SDP model for the optimal reservoir operation for intraseasonal and interseasonal irrigation demands considering three decisions classes: intertemporal water application rates, limiting the cropped areas for the season remained, and the optimal cropping area at the beginning of the season. Application of the model showed that it is more productive and beneficial (Dudley & Burt, 1973).

Vedula and Mujumdar (1992) developed a model for the optimal operating policy of a reservoir for irrigation under a multiple crops scenario using SDP. The model has been applied to the black soil's area of the Malaprabha Reservoir in Karnataka state, India.

Vedula and Kumar (1996) developed an integrated model to overcome these limitations. They proposed seasonal inputs for the model regarding the reservoir inflow and rainfall in the irrigated area, to decide the optimal reservoir operation policies and the multiple crops irrigation allocations. The model consists of two modules: the intraseasonal allocation model (module 1) and the seasonal allocation model (module 2). The model was applied to an existing reservoir in Karnataka state, India.

Mujumdar and Ramesh (1997) also developed a short-term real-time reservoir operation model for irrigation of multiple crops to overcome the limitations of earlier study he had conducted.

Georgiou et al. (2006) developed a non-linear discrete-time dynamic model to simulate a single-purpose reservoir operation during the irrigation season.

Nagesh Kumar et al. (2006) developed a genetic algorithm (GA) model that can be applied for optimal reservoir system operation to obtain the maximum benefits. The obtained model results indicate a success in utilizing the available water resources for optimal operating policy and optimal crop water allocations from an irrigation reservoir.

Consoli et al. (2008) developed a reservoir operation rule for irrigation in Pozzillo, Italy. He applied a combination of two methods, the multi-objective non-linear programming (LP) constraint method and step method. This combination helped the planner and the decision-makers to fix the used criteria and enhance the model performance.

Suiadee and Tingsanchali (2007) developed a combined simulation–GA optimization model to optimize the operation of Nam Oon Reservoir and Irrigation Project. Combining simulation with GA shows an excellent optimization efficient computation results.

Sudha et al. (2008) investigates whether significant improvements would be outcome from optimization of operation of the reservoir system. In their study, mixed integer linear programming (MILP) model is developed and five different management strategies are tested. The result indicates that a management strategy with deficit irrigation by supplying less water in non-critical growth period and maximum water during stress sensitive periods is a best viable solution for better performance. A MILP model, rather than a LP model, is used to ensure that the reservoir does not spill before reaching its capacity.

Georgiou and Papamichail (2008) developed a non-LP optimization model to determine the optimal reservoir release policies, the irrigation allocation to multiple crops and the optimal cropping pattern in irrigated agriculture. Giving importance to soil water balance. The model was applied to Havrias river in Northern Greece, and the results indicated that the used approach is general and can be applied to any reservoir to optimize water allocation and crop planning regardless of the constraints and the meteorological status.

Alaya et al. (2003) studied the optimal rules for Nebhana reservoir operation for irrigation in arid conditions using SDP technique. The most important finding for this study was that the SDP technique has a problem dealing with multiple-state variable or better known by the research community as the "curse of dimensionality".

Vedula (2002) presented a comprehensive and detailed overview of modelling reservoir operation for irrigation.

Azamathulla et al. (2008) developed and compared the GA and LP for the better performance in operating Chiller reservoir system in India. The most effective performance resulted from the GA model regarding both irrigation scheduling problems and the crops yield.

Adeyemo and Otieno (2010) applied the strategies of multi-objective differential evolution algorithm (MDEA) to maximize the irrigation benefit in the lower orange catchment of South Africa. MDEA with binomial crossover method is better in terms of the quantity and quality of non-dominated solutions generated.

Hosseinpourtehrani and Ghahraman (2011) proposed a Fuzzy-based model using non-LP to obtain optimal reservoir operation for irrigation of multiple crops.

Alizadeh and Mousavi (2013) explained the importance of using stochastic optimization methods for designing Hajiarab irrigation district in Iran. The results indicated that the achieved gross benefits were about two times greater than the net benefits.

Kumari and Mujumdar (2015) used the fuzzy state variables to develop a SDP model and applied it to irrigate multiple crops. The Fuzzy-State Stochastic Dynamic Programming (FSDP) model results were more acceptable than those of the classical stochastic dynamic models for the considered case study.

Bashiri-Atrabi et al. (2015) developed a metaheuristic technique called harmony search algorithm to optimize the reservoir operation with respect to flood control. This algorithm has proven to be effective in operation reservoir for flood management.

Gurav (2016) developed a multi-objective fuzzy linear programming model for the optimal cropping pattern planning utilizing the optimization software LINGO 13. The models were applied to a case study of the Jayakwadi Project in Maharashtra state, India and they show the ability to achieve an integrated irrigation planning with prime consideration for economic, social and environmental issues.

Garousi-Nejad et al. (2016) optimized the reservoir operation using firefly algorithm (FA) and compared the results with a commonly used optimization algorithm, the GA. The results showed that the FA won the lead in performance in terms of the convergence rate to the global optima and of the variance of the results about global optima when compared with the results of the GA.

Ren et al. (2017) developed a multi-objective fuzzy programming method for simultaneously optimizing the use of water and land resources for irrigation under uncertainty. The proposed model was then applied to a real-world case study in Wuwei City, Gansu Province, China and proved to be effective in saving the water resources (Adeyemo & Otieno, 2010).

Turgut et al. (2019) compared five different optimization algorithms (namely particle swarm, differential evolution, whale, crow search and master-slave algorithms) to optimize the Aswan High Dam (AHD) reservoir in Egypt. The master-slave algorithm surpasses the other ones and generates an operation policy for water balance to avoid the differences in water supply and water demand.

Sheibani et al. (2019) presented a reliability-based method for stochastic optimum design of the capacity and operation policy for both a reservoir and the downstream irrigation network. The results find out that combining the optimum values of the network area, reservoir capacity and the reliability index will result in the optimal operation of the reservoir system.

Dobson et al. (2019) performed a literature review and introduced a novel classification system of reservoir optimization approaches.

Alizadeh et al. (2018) proposed the different scenarios to obtain the optimal solutions for reservoir-irrigation district systems design using explicit stochastic hydro-economic optimization model. The reservoir system is affected by multiple interdependent sources of uncertainties and the considered variables such as reservoir size and capacity, the acreage area, reservoir operation obligations and irrigation management plans. The model was tested in the case study of Chamshir hydro-system in Iran and proved to be effective in preserving the irrigation water.

Paliwal et al. (2020) produced an operation optimization model for Mula reservoir in India based on the Jaya Algorithm. The developed model has proved to be more successful in solving the addressed problems than the other algorithms.

Valikhan-Anaraki et al. (2019) proposed an optimization model to reduce the irrigation deficiencies downstream Aydoghmoush Dam Reservoir based on a hybrid of the bat algorithm and particle swarm optimization (PSO) algorithm. The new model had an increasing convergence rate to avoid trapping in local optima. The results indicated that the hybrid optimization model was successful in reducing the computational time and increasing the convergence rate.

Banadkooki et al. (2020) compared five algorithms for reservoir optimization in Iran. The results showed that the superior performance goes to a new evolutionary algorithm, namely the crow algorithm (CA), to optimize the reservoir operation for irrigation management.

Ehteram et al. (2018) introduced an improved weed algorithm for optimizing the AHD reservoir operation to decrease the irrigation deficits in Egypt. The results suggested using the improved weed algorithm for solving the complex problems in water resources management.

Boudjerda et al. (2020) applied Dynamic Programming-Neural Network (DPNN) method to investigate the reservoir optimization. The application of DPNN method allowed increasing the satisfaction rate; in addition, the operation rule generated showed more reliable and resilience operation for the examined case study.

Karnatapu et al. (2020) introduced a hybrid model based on GA and non-LP to derive the optimal reservoir operating policies for a multi-purpose reservoir. The results prove the effectiveness of the model to indicate the optimal allocation of limited available water resources to any reservoir.

Raju et al. (2020) used LP to derive the optimal monthly releases for a single-purpose reservoir. The final model satisfied the downstream irrigation demands in addition to conserving a considerable amount of water from reduced spills.

Ashofteh et al. (2021) developed a bi-objective genetic programming (BO-GP) algorithm to optimize the operating rules of the Aidoghmoush reservoir (East Azerbaijan in northeastern Iran). The application results show a successful performance of the BO-GP algorithm in solving the bi-objective water supply problem with reservoir operation. This paper's results establish that the system vulnerability and its reliability range between 16%–41% and 46%–78%, respectively.

Kasiviswanathan et al. (2021) developed an artificial neural network (ANN) model for maximizing the total production with the optimal water releases from the reservoir irrespective of inflow to the reservoir. The proposed modelling framework has been demonstrated through a case example, Chittar river basin, India. The upper, lower and mean of forecasted inflow from the ANN model were used to arrive at the prediction interval of the depth of irrigation, total crop yield and area of irrigation. The results from this study suggested that the information on the uncertainty quantification helps in better understanding the reliability of the systems and for effective decision making.

This chapter highlights the models that have been developed specifically for optimization reservoir operation for purposes of irrigation, and will help decision-makers to compare and adopt the best approach to improving crop productivity and conserving water resources.

18.2 Reservoir Operation Optimization

The determination of reservoir releases on time is an essential aspect of operating the reservoir for irrigation. In other words, it is determining the appropriate operation policy that must be followed in each period, and that leads to the best objectives of the system in the long term. These objectives may be maximizing the annual yield of irrigated crops and thus maximizing the net benefits, or maximizing water use. The optimization model is a series of the decisions regarding reservoir releases during different periods of the year, and these decisions depend on the expected inputs for that period (e.g. flow, precipitation, evaporation, soil condition and the crop) (Vedula, 2002) (Figure 18.1).

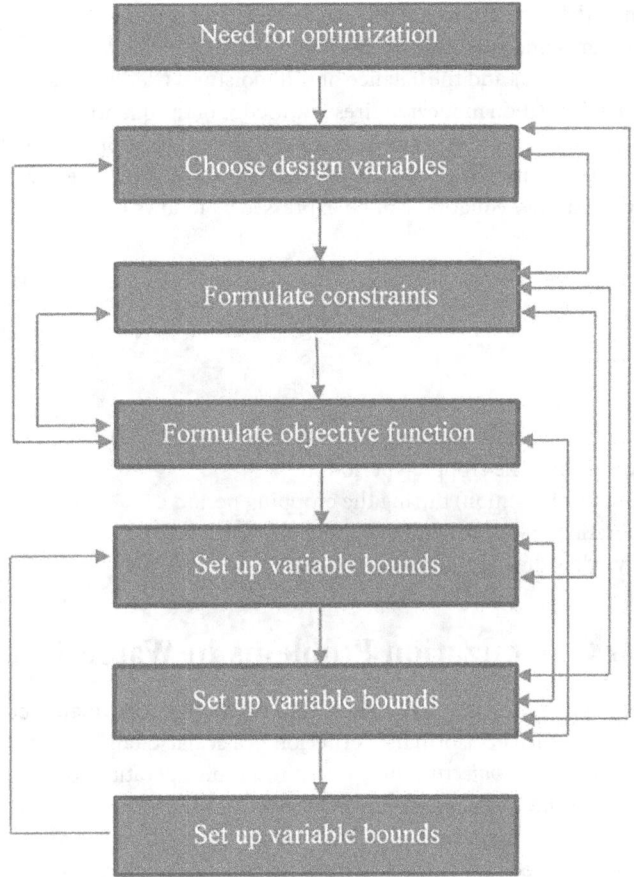

FIGURE 18.1　Optimal design procedure flowchart (Deb, 2012).

Since irrigation consumes the largest amount of released water, the optimum reservoir release is related to the allocation of irrigation in the area commanded by the reservoir during that period. Irrigation needs depend on the period of crop growth and soil moisture to achieve the maximum yield in the harvest period. Thus, the degree of detail that must be accounted for in the formulation of the optimization model depends on the need and physiology of the reservoir being modelled (Valipour et al., 2015).

Consequently, optimizing reservoir operation is of great importance in preserving water resources sustainability, and its importance is evident in two aspects: as a complement to the practical design of new reservoirs, and also as an improvement in the operation of old ones. In this chapter, the operation will be restricted to the quantities of water that must be extracted from the water sources (i.e. rivers), transported between the different systems of the reservoir and then released to the consumption points (i.e. irrigation). Literature definitely proved the ability to enhance reservoirs performance by utilizing the optimization models as well as guarantee efficacious design for new ones (Reddy & Kumar, 2007).

Generally, the most efficient optimization method for reservoir operation for irrigation is the non-linear optimization models. This type is formulated by combining reservoir-specific variables related to water availability with field variables related to the crop, in order to maximize the relative yield of the crops, taking into account various restrictions (Suiadee & Tingsanchali, 2007).

In the same manner, the restrictions are of two levels, the first being the restrictions related to the reservoir (i.e. water releases and reservoir storage), and the second the restrictions related to the farm (i.e. the water allocation to crops and the balance of soil moisture) (Reddy & Kumar, 2007).

The success of the maximization model requires knowledge of the quantities of water available during the irrigation season, the amounts of expected rainfall and the type of crop grown. Reservoirs operation is based on the continuity equation that determines the balance between the released and stored water for the reservoir. The continuity equation can be expressed as follows (Nagesh Kumar & Janga Reddy, 2007):

$$S_t + Q_t - R_t - L_t - \mathrm{OVF}_t = S_{t+1}, \tag{18.1}$$

where:

t: Cropping period
S_t: Initial active storage (i.e. at the beginning of the cropping period t),
Q_t: Reservoir inflow during the cropping period t,
R_t: Reservoir release (for irrigation) during the cropping period t,
L_t: Reservoir evaporation loss during the cropping period t, and
OVF_t: Reservoir overflow during the cropping period t.

18.3 Reservoir Optimization Problems in Water Resources

The first step in performing optimization problems is to derive the performance criterion F in terms of n variables (i.e. $x_1, x_2, \ldots, x_n$). This performance criterion is a scalar quantity that can take many forms and is commonly denoted as the *objective function*, in reservoir operation for irrigation purposes. The *objective function* refers to the water release in the system and the parameters that influence *objective function refer to as $x_1, x_2, \ldots, x_n$*; those parameters are either independent variables (e.g. time) or control parameters that can be modified. The most requisite optimization problem is to customize the parameters $x_1, x_2, \ldots, x_n$ in such a way as to ensure that the function F (the water release in this case) to be at its minimum quantity. This can be expressed mathematically as:

$$\text{minimize } F = f(x_1, x_2, \ldots, x_n). \tag{18.2}$$

A large number of parameters may influence the objective function, so, to simplify the notation, matrix notation is usually employed (Antoniou & Lu, 2007). Any optimization algorithm necessitates a sequence of steps to be performed. Many minimization algorithms for a certain objective function are available. Nevertheless, this chapter particularly concerned about the algorithms that result in the minimum amount of water release. The term *robust* is often used to indicate a reliable algorithm in mathematical programming terminology (Gill et al., 2019).

18.4 Optimization Algorithms Used for Optimizing Reservoir Operation for Irrigation

All algorithms in the optimization field are classified into exact or heuristic (Deb, 2012). The exact algorithms are formulated so they can find an optimal solution within a limited time. The problems optimizations in this class are simple and easy, so they can be carried out in unlimited (polynomial) time. The heuristic algorithms, on the other hand, are able to find the reasonable results in a short time because they don't inspect all the search space. Many approaches to mathematical programming were used in the last 50 years. Each branch consists of a set of optimization techniques that fit the category of a particular problem. The difference between these branches is directly related to the problem structure, the mathematical nature of the objective function, and the constraints. Ahmad et al. (2014) classified the reservoir optimization algorithms into four classes: LP, NLP, dynamic programming and computational intelligence (CI). Regardless of the multiplicity of algorithms used to operate the reservoir for irrigation, each method has the advantages and disadvantages that depend on the variables involved in the problem. Only as the rapid development of computer science helped facilitate all aspects of science, it also contributed to the emergence of improvement techniques that depend on CI, which is evolutionary computation. This technique provided the optimal solution with less computation time as well as the ability to link their results to the simulation model (Ahmad et al., 2014). The most used algorithms are mentioned below.

18.4.1 Particle Swarm Algorithm

The PSO algorithm is a metaheuristic algorithm based on the social behaviour of bird flocking and fish schooling. The concept of swarm intelligence is capable of solving complex mathematical problems existing in engineering (de Moura Meneses et al., 2009; Kennedy & Eberhart, 1995). Particles in this algorithm move in two movements: random and deterministic. During its random movement, the algorithm updates the position of the particle when it finds a better position than its previous one and continues to move until it finds the best global solution or after a certain number of attempts (Yang, 2009).

What characterizes this algorithm is that it has fewer variables to adjust, and these variables are discussed a lot in the literature; therefore, when comparing PSO with other optimization algorithms, it will have a preference (Sarkar et al. 2013). The basic form of a particle swarm algorithm is presented in Figure 18.2.

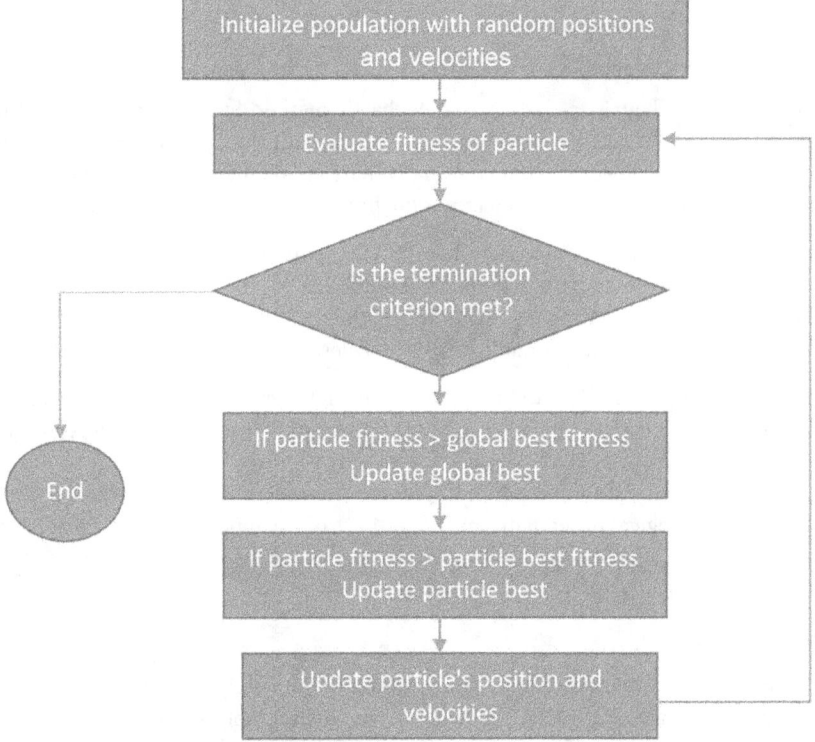

FIGURE 18.2 Basic flowchart of particle swarm algorithm (Jahandideh-Tehrani et al., 2020).

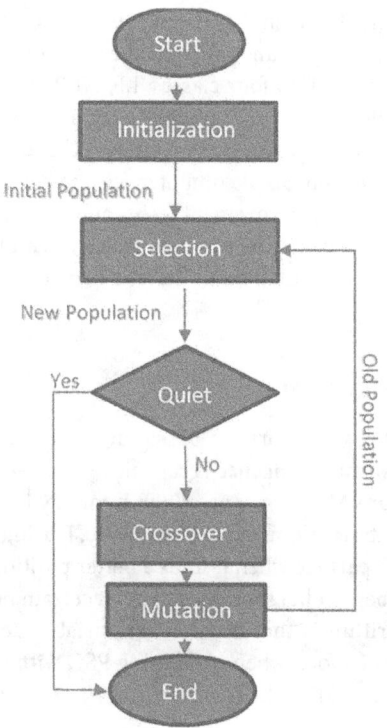

FIGURE 18.3   Basic genetic algorithms (GA) flowchart (Scrucca, 2013).

18.4.2 Genetic Algorithm

GAs are belonging to the heuristic class. It is a stochastic method based on principles of the natural competition between individuals for appropriating limited natural sources (natural selection) and has been used widely in engineering application especially in irrigation management (Deb, 2012; Festa, 2014; Gholizadeh, 2013). It has four main characteristics (Srinivasa Raju & Nagesh Kumar, 2004):

1. It works with a variable set coding, not the variable themselves.
2. It starts the search from a group of points, not one point.
3. It uses the decisive information, not derivatives.
4. It uses probabilistic, not deterministic transition rules.

Although GAs are random in nature, their results are much better than random local searches (trying different solutions to track the best one), because they also exploit historical information (Goldberg & Holland, 1988; Mathur & Nikam, 2009). Basic GA flowchart is shown in Figure 18.3.

Jahandideh-Tehrani et al. (2019) introduced the GA varieties since their emergence. Table 18.1 lists those varieties in chronological order, with the oldest (i.e. earliest appearance) ones placed at the top of the list.

18.4.3 Fuzzy Logic

Fuzzy logic algorithms have been used widely to optimize many water-resources-related problems such as in downscaling of climate variables, water quality modelling and water treatment, surface water hydrology applications, and reservoir operation (RO) (Kambalimath & Deka, 2020). These algorithms were complex for reservoir operation modelling whether it is a single or multi-purpose reservoir, yet

TABLE 18.1 Varieties of the GA (after Jahandideh-Tehrani et al., 2019)

Algorithm	Abbreviation	Year of Appearance	Reference
Binary-coded genetic algorithm	Binary-code GA	1985	Goldberg
Noisy genetic algorithm	NGA	1988	Fitzpatrick and Grefenstette
Constrained genetic algorithm	CGA	1989	Richardson et al.
Real-coded genetic algorithm	RCGA	1991	Wright
Self-adaptive genetic algorithm	Self-adaptive GA	1992	Bäck
Multi-objective genetic algorithm	MOGA	1995	Cieniawski et al.
Non-dominated sorting genetic algorithm	NSGA-II	1995	Srinivas and Deb
Hyper-cubic distributed genetic algorithm	HDGA	1997	Herrera and Lozano
Self-learning genetic algorithm	SLGA	1997	Han and May
Genetic-neuro fuzzy	GNF	1999	Seng and Khalid
Hybrid genetic algorithm and linear programming	GA-LP	2001	Cai et al.
Chaos genetic algorithm	Chaos GA	2002	Yuan et al.
Alternating fitness genetic algorithm	AFGA	2004	Zou and Lung
Self-organizing map-based multi-objective genetic algorithm	SBMOGA	2004	Kubota et al.
The non-dominated sorting genetic algorithm with support vector regression	SVR-NSGA-II	2004	Xu et al.
Non-dominated sorting genetic algorithm with artificial neural network	NSGA-II-ANN	2005	Nain and Deb
A combination of genetic algorithm and discrete differential dynamic programming	GA-DDDP	2005	Tospornsampan et al.
Genetic algorithm-based support vector machine	GA-SVM	2005	Lessmann et al.
Epsilon dominance non-dominated sorting genetic algorithm	Ɛ-NSGA-II	2005	Kollat and Reed
Macro-evolutionary multi-objective genetic algorithm	MMGA	2007	Chen et al.
GA-based fuzzy proportional derivative	GA-fuzzy PD	2007	Bagis and Karaboga
Adaptive neural network embedded genetic algorithm	Adaptive NN-GA	2009	Zou et al.
Grammatical evolution incorporated with parallel genetic algorithm	GEGA	2008	Chen et al.
Multi-tier interactive genetic algorithm	MIGA	2011	Wang et al.
Hybrid incremental dynamic programming and genetic algorithm	Hybrid IDP-GA	2012	Li et al.
Aggregation hybrid genetic algorithm	AHGA	2014	Huang

they became simpler by reducing the fuzzy rules (Sivapragasam et al., 2008). Consequently, a significant sum of investigations has been conducted and proved the effectiveness of this approach to solve the problems in water resources management. Figure 18.4 clarifies the basic steps in the fuzzy logic approach (Bogardi et al., 2004).

18.4.4 Firefly Algorithm

The FA belongs to the class of metaheuristic algorithms (Deb, 2012). It was inspired by the fireflies movement. The key influence parameters to the FA's efficiency are the light intensity and the adjacent firefly's attractiveness differences. There is also some effect for the scaling parameter in the distance between fireflies (Figure 18.5).

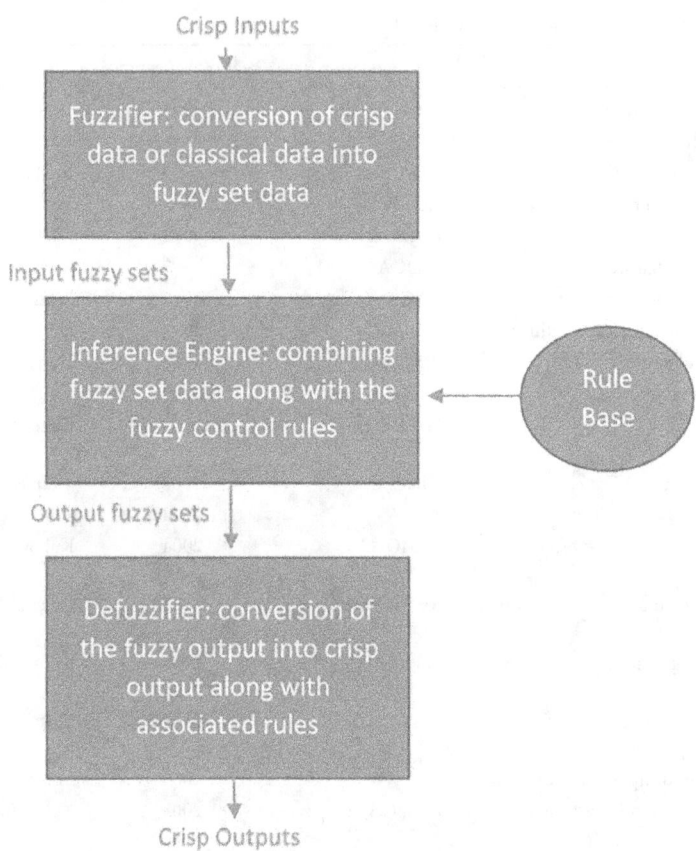

FIGURE 18.4 Fuzzy logic basic flowchart (Kambalimath & Deka, 2020).

18.4.5 Crow Algorithm

CAs are population-based metaheuristic algorithms that have been developed based on the intelligent behaviours of crow birds. In this algorithm, the populations simulate the behaviour of crows in searching for food, roving the environment that simulates the search area to find the best source of food that simulates the objective function. Thus, based on the following equation, the new position for the population would be defined, taking into account the constraints (Askarzadeh, 2016; Banadkooki et al., 2020).

18.5 Conclusions

The use of optimization algorithms for reservoir operation, in general, has proven to be a great success. Though there are a large and continually developing number of algorithms, the most crucial factor is how to formulate the problem in a way that guarantees to reach the appropriate decision regarding the water releases in the irrigation season. Some models in the literature dealt with the optimal irrigation allocation, while others dealt with the reservoir operation that achieves long-term objectives. However, the models that gave long-term solutions are the ones that combine the optimization of reservoir operation with the allocation of irrigation for each season. Based on the above, choosing the appropriate algorithm depends mainly on the objective function to be reached and the number of variables to be considered taking into account the constraints as the field conditions differ from one reservoir to another.

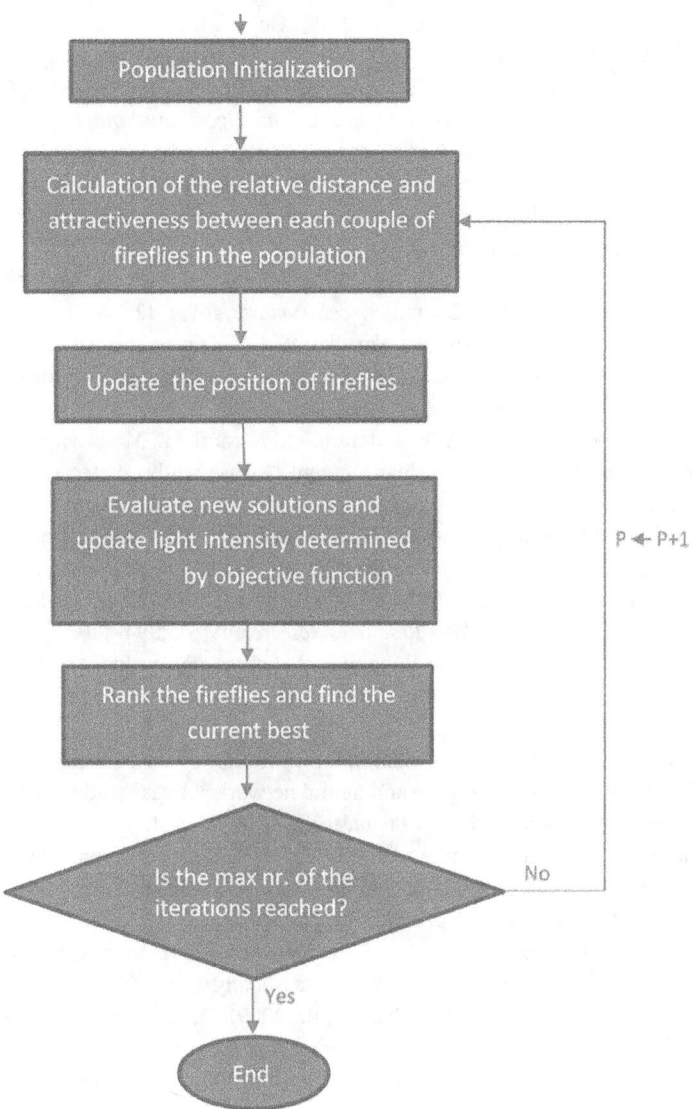

FIGURE 18.5   Firefly algorithm basic flowchart (Casciati & Elia, 2015).

References

Adeyemo, J. A., & Otieno, F. O. (2010). Maximum irrigation benefit using multiobjective differential evolution algorithm (MDEA). *OIDA International Journal of Sustainable Development, 1*(2), 39–44.

Ahmad, A., El-Shafie, A., Razali, S. F. M., & Mohamad, Z. S. (2014). Reservoir optimization in water resources: A review. *Water Resources Management, 28*(11), 3391–3405.

Al-Mamoori, S. K., & Al-Maliki, L. A. (2016). Evaluation of suitability of drainage water of al-Hussainia sector (Kut Iraq) to irrigate cotton crop. *Kufa Journal of Engineering, 7*(1), 67–78.

Alaya, A. B., Souissi, A., Tarhouni, J., & Ncib, K. (2003). Optimization of Nebhana reservoir water allocation by stochastic dynamic programming. *Water Resources Management, 17*(4), 259–272.

Alizadeh, H., & Mousavi, S. J. (2013). Stochastic order-based optimal design of a surface reservoir–irrigation district system. *Journal of Hydroinformatics, 15*(2), 591–606.

Alizadeh, H., Mousavi, S. J., & Ponnambalam, K. (2018). Copula-based chance-constrained hydro-economic optimization model for optimal design of reservoir-irrigation district systems under multiple interdependent sources of uncertainty. *Water Resources Research, 54*(8), 5763–5784.

Antoniou, A., & Lu, W.-S. (2007). *Practical Optimization: Algorithms and Engineering Applications*: Springer Science & Business Media, Germany.

Ashofteh, P.-S., Bozorg-Haddad, O., & Loáiciga, H. A. (2021). Application of bi-objective genetic programming for optimizing irrigation rules using two reservoir performance criteria. *International Journal of River Basin Management, 19*(1), 55–65.

Askarzadeh, A. (2016). A novel metaheuristic method for solving constrained engineering optimization problems: Crow search algorithm. *Computers & Structures, 169*, 1–12.

Azamathulla, H. M., Wu, F.-C., Ab Ghani, A., Narulkar, S. M., Zakaria, N. A., & Chang, C. K. (2008). Comparison between genetic algorithm and linear programming approach for real time operation. *Journal of Hydro-environment Research, 2*(3), 172–181.

Banadkooki, F. B., Adamowski, J., Singh, V. P., Ehteram, M., Karami, H., Mousavi, S. F., . . . Ahmed, E.-S. (2020). Crow algorithm for irrigation management: A case study. *Water Resources Management, 34*(3), 1021–1045.

Bashiri-Atrabi, H., Qaderi, K., Rheinheimer, D. E., & Sharifi, E. (2015). Application of harmony search algorithm to reservoir operation optimization. *Water Resources Management, 29*(15), 5729–5748.

Bogardi, I., Bardossy, A., & Duckstein, L. (1983). Regional management of an aquifer for mining under fuzzy environmental objectives. *Water Resources Research, 19*(6), 1394–1402.

Bogardi, I., Bardossy, A., Duckstein, L., & Pongracz, R. (2004). Fuzzy logic in hydrology and water resources. In R. V. Demicco & G. J. Klir (Eds.), *Fuzzy Logic in Geology* (Chapter 6, pp. 153-VIII): Burlington: Academic Press.

Boudjerda, M., Touaibia, B., & Mihoubi, M. (2020). Optimization of agricultural water demand using a hybrid model of dynamic programming and neural networks: A case study of Algeria. *International Journal of Environmental and Ecological Engineering, 14*(5), 121–124.

Casciati, S., & Elia, L. (2015). The potential of the firefly algorithm for damage localization and stiffness identification. In X.-S. Yang (Ed.), *Recent Advances in Swarm Intelligence and Evolutionary Computation* (pp. 163–178). Cham: Springer International Publishing. Switzerland.

Consoli, S., Matarazzo, B., & Pappalardo, N. (2008). Operating rules of an irrigation purposes reservoir using multi-objective optimization. *Water Resources Management, 22*(5), 551–564.

de Moura Meneses, A. A., Machado, M. D., & Schirru, R. (2009). Particle swarm optimization applied to the nuclear reload problem of a pressurized water reactor. *Progress in Nuclear Energy, 51*(2), 319–326.

Deb, K. (2012). *Optimization for Engineering Design: Algorithms and Examples*: PHI Learning Pvt. Ltd., New Delhi.

Dobson, B., Wagener, T., & Pianosi, F. (2019). An argument-driven classification and comparison of reservoir operation optimization methods. *Advances in Water Resources, 128*, 74–86.

Dudley, N. J., & Burt, O. R. (1973). Stochastic reservoir management and system design for irrigation. *Water Resources Research, 9*(3), 507–522.

Ehteram, M., P Singh, V., Karami, H., Hosseini, K., Dianatikhah, M., Hossain, M., . . . El-Shafie, A. (2018). Irrigation management based on reservoir operation with an improved weed algorithm. *Water, 10*(9), 1267.

Eslamian, S., Khordadi, M. J., & Abedi-Koupai, J. (2011). Effects of variations in climatic parameters on evapotranspiration in the arid and semi-arid regions. *Global and Planetary Change, 78*(3), 188–194.

Festa, P. (2014). A brief introduction to exact, approximation, and heuristic algorithms for solving hard combinatorial optimization problems. *Paper Presented at the 2014 16th International Conference on Transparent Optical Networks (ICTON)*, Graz, Austria.

Garousi-Nejad, I., Bozorg-Haddad, O., Loáiciga, H. A., & Mariño, M. A. (2016). Application of the firefly algorithm to optimal operation of reservoirs with the purpose of irrigation supply and hydropower production. *Journal of Irrigation and Drainage Engineering, 142*(10), 04016041.

Georgiou, P., & Papamichail, D. (2008). Optimization model of an irrigation reservoir for water allocation and crop planning under various weather conditions. *Irrigation Science, 26*(6), 487–504.

Georgiou, P., Papamichail, D., & Vougioukas, S. (2006). Optimal irrigation reservoir operation and simultaneous multi-crop cultivation area selection using simulated annealing. *Irrigation and Drainage: The Journal of the International Commission on Irrigation and Drainage, 55*(2), 129–144.

Gholizadeh, S. (2013). Structural optimization for frequency constraints. *Metaheuristic Applications in Structures and Infrastructures, 389*, 59 0f .

Gill, P. E., Murray, W., & Wright, M. H. (2019). *Practical Optimization.* The Woodlands, TX: SIAM.

Goldberg, D. E., & Holland, J. H. (1988). Genetic algorithms and machine learning. *Machine Learning, 3*(2), 95–99. https://doi.org/10.1023/A:1022602019183. Gurav, J. B. (2016). Optimal irrigation planning and operation of multi objective reservoir using fuzzy logic. *Journal of Water Resource and Protection, 8*(02), 226.

Haddad, O. B., Hosseini-Moghari, S.-M., & Loáiciga, H. A. (2016). Biogeography-based optimization algorithm for optimal operation of reservoir systems. *Journal of Water Resources Planning and Management, 142*(1), 04015034.

Hosseinpourtehrani, M., & Ghahraman, B. (2011). Optimal reservoir operation for irrigation of multiple crops using fuzzy logic. *Asian Journal of Applied Sciences, 4*(5), 493–513.

Jahandideh-Tehrani, M., Bozorg-Haddad, O., & Loáiciga, H. A. (2019). Application of non-animal-inspired evolutionary algorithms to reservoir operation: An overview. *Environmental Monitoring and Assessment, 191*(7), 439.

Jahandideh-Tehrani, M., Bozorg-Haddad, O., & Loáiciga, H. A. (2020). Application of particle swarm optimization to water management: An introduction and overview. *Environmental Monitoring and Assessment, 192*(5), 281

Kambalimath, S., & Deka, P. C. (2020). A basic review of fuzzy logic applications in hydrology and water resources. *Applied Water Science, 10*(8), 1–14.

Karnatapu, L. K., Annavarapu, S. P., & Nanduri, U. V. (2020). Multi-objective reservoir operating strategies by genetic algorithm and nonlinear programming (GA–NLP) hybrid approach. *Journal of the Institution of Engineers (India): Series A, 101*(1), 105–115.

Kasiviswanathan, K. S., Sudheer, K. P., Soundharajan, B.-S., & Adeloye, A. J. (2021). Implications of uncertainty in inflow forecasting on reservoir operation for irrigation. *Paddy and Water Environment, 19*(1), 99–111

Kennedy, J., & Eberhart, R. (1995). Particle Swarm optimization Conference. Proceedings of ICNN'95 - International Conference on Neural Networks, Perth, WA, Australia, 1995, pp. 1942–1948, vol. 4, doi: 10.1109/ICNN.1995.488968.

Kumari, S., & Mujumdar, P. (2015). Reservoir operation with fuzzy state variables for irrigation of multiple crops. *Journal of irrigation and drainage engineering, 141*(11), 04015015.

Mathur, Y., & Nikam, S. (2009). Optimal reservoir operation policies using genetic algorithm. *International Journal of Engineering and Technology, 1*(2), 184.

Mujumdar, P., & Ramesh, T. (1997). Real-time reservoir operation for irrigation. *Water Resources Research, 33*(5), 1157–1164.

Nagesh Kumar, D., & Janga Reddy, M. (2007). Multipurpose reservoir operation using particle swarm optimization. *Journal of Water Resources Planning and Management, 133*(3), 192–201.

Nagesh Kumar, D., Raju, K. S., & Ashok, B. (2006). Optimal reservoir operation for irrigation of multiple crops using genetic algorithms. *Journal of Irrigation and Drainage Engineering, 132*(2), 123–129.

Paliwal, V., Ghare, A. D., Mirajkar, A. B., Bokde, N. D., & Feijoo Lorenzo, A. E. (2020). Computer modeling for the operation optimization of Mula Reservoir, Upper Godavari Basin, India, using the Jaya Algorithm. *Sustainability, 12*(1), 84.

Raju, B. K., Chandre Gowda, C., & Karthika, B. (2020). Optimization of reservoir operation using linear programming. *International Journal of Recent Technology and Engineering, 8*(5), 1028–1032

Rani, D., & Moreira, M. M. (2010). Simulation–optimization modeling: A survey and potential application in reservoir systems operation. *Water Resources Management, 24*(6), 1107–1138.

Reddy, M. J., & Kumar, D. N. (2007). Optimal reservoir operation for irrigation of multiple crops using elitist-mutated particle swarm optimization. *Hydrological Sciences Journal, 52*(4), 686–701.

Ren, C., Guo, P., Tan, Q., & Zhang, L. (2017). A multi-objective fuzzy programming model for optimal use of irrigation water and land resources under uncertainty in Gansu Province, China. *Journal of Cleaner Production, 164*, 85–94. https://doi.org/10.1016/j.jclepro.2017.06.185

Sarkar, S., Roy, A., & Purkayastha, B. S. (2013). Application of particle swarm optimization in data clustering: A survey. *International Journal of Computer Applications, 65*(25), 38–46,

Scrucca, L. (2013). GA: A package for genetic algorithms in R. *Journal of Statistical Software, 53*(1), 1–37.

Sheibani, H., Alizadeh, H., & Shourian, M. (2019). Optimum design and operation of a reservoir and irrigation network considering uncertainty of hydrologic, agronomic and economic factors. *Water Resources Management, 33*(2), 863–879.

Sivapragasam, C., Sugendran, P., Marimuthu, M., Seenivasakan, S., & Vasudevan, G. (2008). Fuzzy logic for reservoir operation with reduced rules. *Environmental progress, 27*(1), 98–103.

Srinivasa Raju, K., & Nagesh Kumar, D. (2004). Irrigation planning using genetic algorithms. *Water Resources Management, 18*(2), 163–176.

Sudha, V., Venugopal, K., & Ambujam, N. K. (2008). Reservoir operation management through optimization and deficit irrigation. *Irrigation and Drainage Systems, 22*(1), 93–102.

Suiadee, W., & Tingsanchali, T. (2007). A combined simulation–genetic algorithm optimization model for optimal rule curves of a reservoir: A case study of the Nam Oon Irrigation Project, Thailand. *Hydrological Processes: An International Journal, 21*(23), 3211–3225.

Turgut, M. S., Turgut, O. E., Afan, H. A., & El-Shafie, A. (2019). A novel Master–Slave optimization algorithm for generating an optimal release policy in case of reservoir operation. *Journal of Hydrology, 577*, 123959.

Valikhan-Anaraki, M., Mousavi, S.-F., Farzin, S., Karami, H., Ehteram, M., Kisi, O., . . . Ahmed, A. N. (2019). Development of a novel hybrid optimization algorithm for minimizing irrigation deficiencies. *Sustainability, 11*(8), 2337.

Valipour, M., & Eslamian, S. (2014). Analysis of potential evapotranspiration using 11 modified temperature-based models. *International Journal of Hydrology Science and Technology, 4*(3), 192–207.

Valipour, M., Sefidkouhi, M. A. G., & Eslamian, S. (2015). Surface irrigation simulation models: A review. *International Journal of Hydrology Science and Technology, 5*(1), 51–70.

Vedula, S. (2002). Modelling reservoir operation for irrigation. In *Research Perspectives in Hydraulics and Water Resources Engineering* (pp. 317–334): World Scientific.

Vedula, S., & Kumar, D. N. (1996). An integrated model for optimal reservoir operation for irrigation of multiple crops. *Water Resources Research, 32*(4), 1101–1108.

Vedula, S., & Mujumdar, P. (1992). Optimal reservoir operation for irrigation of multiple crops. *Water Resources Research, 28*(1), 1–9. https://doi.org/10.1029/91WR02360

Yang, X.-S. (2009). Firefly algorithms for multimodal optimization. In O. Watanabe, & T. Zeugmann (Eds) *Stochastic Algorithms: Foundations and Applications*, pp. 169–178. Springer, Berlin, Heidelberg. SAGA 2009. Lecture Notes in Computer Science, vol 5792. https://doi.org/10.1007/978-3-642-04944-6_14.

19

Reducing Nitrate Leaching and Increasing Nitrogen Use Efficiency by Applying Nano-Fertilizers

Mehdi Jovzi
Agricultural Research Education and Extension Organization

Hamid Zare Abyaneh
Bu-Ali Sina University

Maryam Bayat Varkeshi
Malayer University

Saeid Eslamian
Isfahan University of Technology

19.1 Introduction

N fertilizer is considered as a limiting factor in obtaining high yield and quality (Zare Abyaneh et al., 2020). An adequate supply of nitrogen is essential for the optimum yield (Kiymaz and Ertek, 2015) and nitrogen is the major essential element for plant growth (Bhattacharyya et al., 2012; Lv et al., 2011). Quality crop production depends more on nitrogen as the most important macronutrient than on any other nutrient since its lack/shortage in soil has more adverse and observable consequences. Moreover, few soils are ever found that do not need nitrogen replenishment (Rahimizadeh et al., 2010; Albaji et al., 2015; Hooshmand et al., 2019; Neissi et al., 2020). Nitrogen is needed in plants for forming chlorophyll as well as increasing the crop protein and plant yield.

Application of too many nitrogen-containing fertilizers, however, is not economical and overdose of its application has become one of the significant sources for water contamination (Grignani et al., 2007; Silva et al., 2013). Nitrogen leaching into groundwater poses serious environmental hazards that result from the lack of oxygen available to organisms (Kim et al., 2000). Potato production has high potential for reactive N losses, which can occur in soluble form as NO_3^- with potential impacts on local and regional water quality (Padilla et al., 2018). Since agriculture aims to achieve sustainable crop production with higher yields and to maintain public health, improving the management practices that maximize crop yield and minimizing environmental impact while maintaining soil productivity are urgently needed (Du et al., 2016; Hou et al., 2007; Woli et al., 2016). Application of chemical fertilizers has long

DOI: 10.1201/9780429290114-25

been condemned because of their harmful effects on the environment and on the quality of agricultural products; hence, the researchers' quest for better alternatives. Fazlolahi and Eslamian (2013) performed nitrogen and phosphorus removal from municipal wastewater by three wetland plant species.

Proper management of water and fertilizers (Abbasi et al., 2012; Woli et al., 2016) or changes in the structure of fertilizers and utilization of new technologies can have positive results in reducing nitrogen leaching (Cui et al., 2006; Manjunatha et al., 2016; Souza et al., 2019). For example, controlled release urea (CRU), which provides a gradual N supply as crop requirements for a long period, has been extensively used (Ye et al., 2013) and CRU has been widely adopted as an effective mitigation alternative to improve the crops' productivity and reduce the labor/time inputs (Yang et al., 2011, 2012).

Zeolite as a slow-release fertilizer was used in some studies. A greenhouse test was conducted to evaluate spinach growth and spinach quality after application of zeolite pre-load with ammonium and potassium (Li et al., 2014). In this study, zeolite increased spinach yield compared to the conventional fertilizer with the same fertilizer amounts used. The application of natural zeolites has been reported to diminish the nutrient leaching, to increase crop water use efficiency (Coltorti et al., 2012; Gholamhoseini et al., 2013), and to decrease ammonia volatilization (Latifah Omar et al., 2010). The study of Colombani et al. (2015) suggested that natural zeolites charged with swine manure can be a viable option to retard the excess leaching of nutrients in agricultural lands and both the batch and column experiments should be performed together to cross-check and validate the obtained results.

Nano-fertilizers and slow-release fertilizers are appropriate alternatives to conventional fertilizers for the gradual and controlled supply of nutrients in soil. Alternative nano-fertilizers such as nano-chelate with chemical fertilizers not only reduce pollution but are also economical (Monreal, 2010). Studies with high levels of nitrogen input aimed at comparing fertilizer regimes have shown differences in nitrogen use efficiency (NUE). These effects were mainly associated with the differences in fertilizer types (Van Kempen et al., 1996). According to a study by Cui et al. (2006), nano-technology can be exploited to reduce the rate of fertilizer nutrient losses through leaching and to increase their availability to plants, which ultimately leads to reduced water and soil pollution (Johnson, 2006). The potato crop is known to have a low N uptake efficiency ranging between 40% and 60% (Rens et al., 2016, 2015). Thus, using coated and nano-urea fertilizers can increase the N uptake by reducing leaching and runoff, especially in the years with heavy rainfall on leaching-prone soils (Tian et al., 2018; Zareabyaneh and Bayatvarkeshi, 2015). According to Naderi and Danesh-Shahraki (2010), nano-fertilizers give rise to increased nutrient use efficiency, reduced soil pollution, reduced fertilizer application times, and in general, minimize the negative impacts of fertilizers. Suitable fertilizers can be produced by using nano-particles and nano-capsules. DeRosa et al. (2010) and Barmaki et al. (2010) reported that nano-fertilizer application may substantially increase the nutrient use efficiency and crop yield. Peyvandi et al. (2011b) compared the effects of nano-iron chelate and iron chelate on the Basilicum growth parameters and showed that the use efficiency of nano-chelate was higher than that of iron chelate. In a different study, the same authors reported that nano-iron chelate had significantly positive effects on the growth and activity of antioxidant enzymes than plain iron chelate did (Peyvandi et al., 2011a). Akhlaghi (2008) reported that sulfur-coated urea (SCU) as a slow-release fertilizer is more efficient and more beneficial to crops. Several researchers showed that SCU application significantly increases NUE in winter wheat (Fan et al., 2004; Lotfollahi et al., 2004; Malakouti et al., 2008). Lotfollahi et al. (2004) reported that SCU, compared to U, increased wheat yield and NUE when applied before planting. In a study aimed at enhancing NUE, Malakouti et al. (2008) carried out experiments in 22 wheat fields in 14 provinces in Iran during the period 2005–2004 and found that replacing U fertilizer with SCU not only led to increased NUE by 39% but also increased crop yield by 12%. According to Ryan and Hariq (1986), nitrogen fertilizers, especially SCU, increase NUE as a result of the reduced leaching and sublimation.

Ziaeyan and Keshavarz (2010) found that slow-release nitrogen fertilizers in potato cultivation are more economical than the other nitrogen fertilizers. Zvomuya et al. (2003) compared the U and SCU fertilizers and found that SCU increased both NUE and tuber yield. El-Gindy et al. (2000) studied the reaction of the potato crop to slow-release fertilizers in the different irrigation systems and showed that the effect of residual nitrogen in the slow-release fertilizers was higher than that of U. Zare Abyaneh and Bayat Varkeshi (2014) reported that soil nitrate increased while using nano-fertilizers. Their study showed that soil nitrate in nano-nitrogen chelate (NNC), sulfur-coated nano-nitrogen chelate (SNNC), and SCU fertilizers were 10.36%, 29.92%, and 23.95% more than the U fertilizer, respectively. It may, therefore, be concluded that it is essential to consider the use of nitrogen fertilizers with the high NUE not only for economic reasons but also for its multiple advantages of environmental control, reduced nitrate pollution in the groundwater and agricultural products, and enhanced public health.

The present study was, therefore, designed under greenhouse conditions to investigate the different levels of nitrogen in soil profile, the leaching of nitrogen into groundwater resources, and its effects on potato yield as a result of applying the four fertilizers, namely, NNC, SNNC, SCU, and U.

The aim of this study is the assessment of nitrogen-nano-fertilizer effect on nitrate leaching in soil under potato cultivation. Also, in this study, the yield of potato under different conditions of application of NNC, SNNC, SCU, and U was investigated.

19.2 Materials and Methods

19.2.1 Potato Crop

Potato with the scientific name of *Solanum tuberosum* L. is a major global food product, with several countries in Asia, North and South America, Europe, and Africa producing greater than 1 million metric tons (Mt) per year. In 2016, the top five leading nations for potato production were China (100 Mt), India (44 Mt), Russian Federation (31 Mt), Ukraine (22 Mt), and USA (20 Mt) (Souza et al., 2019).

The potato cropping season in the Hamedan province of Iran is around 95–110 days. According to the statistics published by the Ministry of Jihad Agriculture in 2014, Hamedan province accounts for 23.5% of the country's potato production. The province ranks the second in terms of area under potato cultivation with 25,500 ha and first in terms of the production with one million tons.

Due to its inherent physiological requirements, and because potatoes are frequently grown in coarse-textured soils, relatively large rates of N fertilizer are often applied. For example, recommended N rates (RNR) for the potatoes in the northern USA range from 180 to 390 kg N ha^{-1} y^{-1} for yield goals above 50 Mg tuber ha^{-1} (Lang et al., 1999; Rosen, 2018).

Thus, potato production has a high potential for the reactive N losses, which can occur in soluble form as NO_3^- with the potential impacts on local and regional water quality (Padilla et al., 2018).

19.2.2 Site Characteristics of Experimental

For the purpose of this study, a factorial experiment in CRD was performed in the years 2012 and 2013 in the greenhouse of the faculty of agriculture, Bu-Ali Sina university, Hamedan, Iran. The greenhouse was characterized by a temperature range of 13.8°C–51.9°C, and a relative humidity of 5%–57%. Potatoes were grown in 36 drainage lysimeters 55 cm in diameter and 90 cm in height (Figure 19.1). In order to evaluate the soil nitrate changes during the growing season, holes 15 cm across were created on the body of the lysimeters. Soil particles were prevented from entering the lysimeters by installing a polyethylene drainage tube at a distance of 3 cm from the bottom of the lysimeters, a sand filter layer 5 cm thick around the tube, and a geotextile filter layer 2 mm thick. In order to prepare the primary physical conditions, four heavy irrigations (10 L for every lysimeter) were carried out so that the lysimeter soil sank and its height reduced to 80 cm. Before planting, three soil samples were taken from each lysimeter, air dried, crushed, and analyzed to determine the soil physical and chemical properties. Similarly, samples of the irrigation water were taken for lab analysis and their properties were determined. Water and soil samples were collected in cleaned acid washed plastic bottles and sterilized plastic bags and stored in PE bags and maintained refrigerated during transportation to the laboratory. In the laboratory, the soil samples were homogenized at room temperature and a physical characterization was determined. The

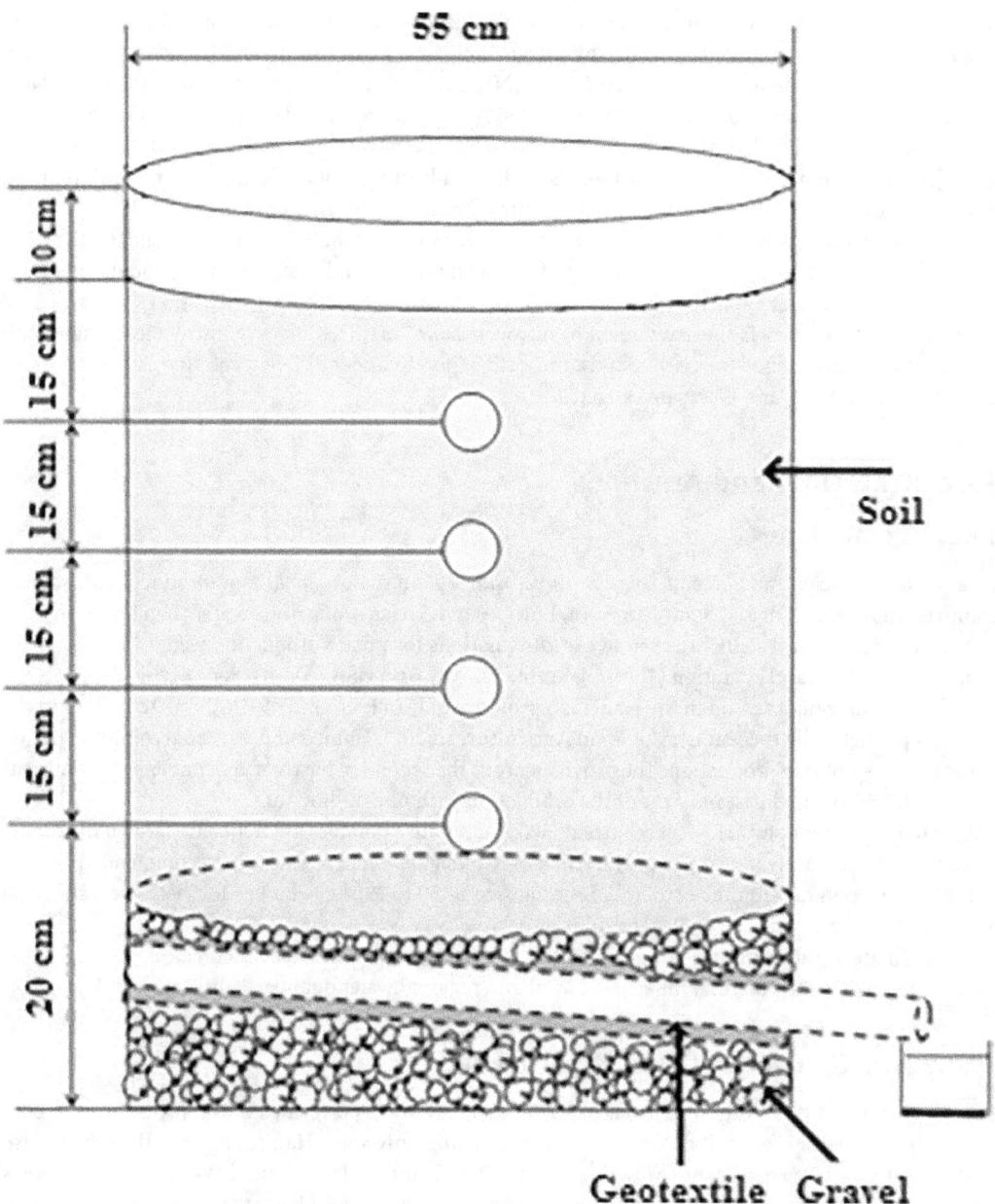

FIGURE 19.1 The lysimeter diagram and dimensions.

soil texture at the experimental site was sandy loam with 19% clay, 21% silt, and 60% sand. The collected samples have been analyzed to determine their chemical characteristics. Chemical parameters of water and soil were done by the standard methods.

Some of the chemical characteristics, including pH and electrical conductivity (EC) and sodium adsorption ratio (SAR), were determined using digital pH meter and digital EC meter, respectively, and anion and cation ions (positively and negatively charged ions) were analyzed using the titration method (Tables 19.1 and 19.2). The physical and chemical characteristics of the soil before planting and

TABLE 19.1 Initial Soil Chemical Characteristics of the Experimental Lysimeters.

Parameter	Amount
Na^+ (meq/L)	37
Mg^{2+} (meq/L)	28
Ca^{2+} (meq/L)	140
CO_2^{2+} (meq/L)	42
Cl^- (meq/L)	132
SO_4^{2-} (meq/L)	31
K^+ (mg/L)	689
P (mg/L)	50.8
N (%)	0.16
pH (--)	7.4
EC_e (dS/m)	1.7
SAR (--)	4.04
Nitrate (mg/L)	4.52

TABLE 19.2 Chemical Properties of Irrigation Water in Study Area.

Parameter	Amount
Na^+ (meq/L)	0.93
Mg^{2+} (meq/L)	1.9
Ca^{2+} (meq/L)	4.5
CO_2^{2+} (meq/L)	4.3
Cl^- (meq/L)	1.8
SO_4^{2-} (meq/L)	1.23
SAR (--)	0.52
EC (dS/m)	0.73
pH (--)	8.3

chemical properties of the irrigation water are reported in Tables 19.1 and 19.2, respectively. The EC values of the irrigation water are 0.73 and 1.7 dS/m, which are suitable for agricultural crops. Also, examination of the soil nutrient condition indicated that the soil used is nitrogen deficient but rich in phosphorus and potassium (Malakouti and Gheybi, 1997). Therefore, only nitrogen fertilizers were used in the treatments.

19.2.3 Specifications of Fertilizers Used

Nitrate was supplied by applying four fertilizers including NNC (27% nitrogen), SNNC (27% nitrogen), SCU (36% nitrogen), and U (46% nitrogen). Nano-chelate fertilizers (recently replacing the chemical fertilizers that are made through nano-technological techniques) that were produced by the self-assembling method were obtained from Khazra Co. This combination is registered under Patent No. US20120100372. Scanning Electron Microscopy (SEM) images containing information on topography, shape, size, and arrangement of particles in the surface soil were obtained using a Cam Scan MV2300 electron microscope at an operating voltage of 200 kV. Based on these images, the NNC size was 20–22 nm and that of SNNC was 44.07–83.89 nm (Figure 19.2). Also, transmission electron microscope (TEM) images from EM208 electron microscope for NNC and SNNC are given in Figure 19.3. These images show the nano-dimensions of the structure of NNC and SNNC fertilizers.

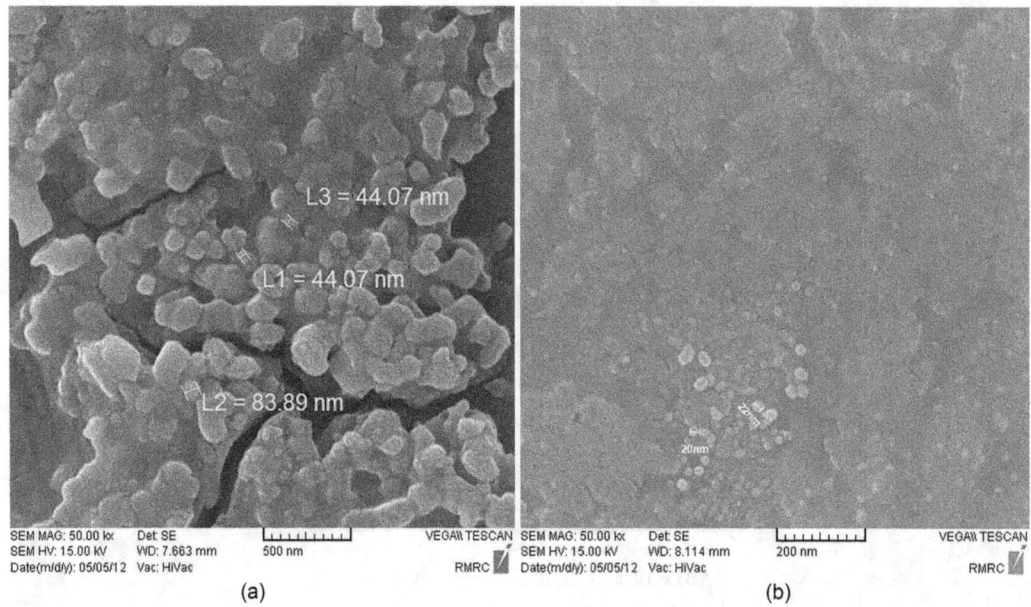

FIGURE 19.2   SEM images of NNC (a) and SNNC (b).

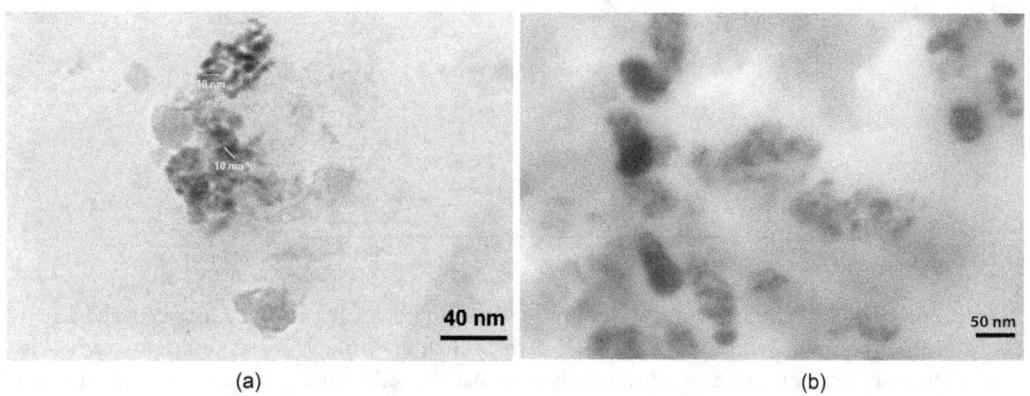

FIGURE 19.3   TEM images of NNC (a) and SNNC (b).

19.2.4 Potato Cultivation, Fertilizer Treatments, and Measurement of Parameters

Inside each lysimeter, six potatoes were cultivated. Considering the amount of nitrogen required for each tuber and the nitrogen content of the same input for all of the four treatments, the following levels of the fertilizers were applied: three levels of U: 100, 200, and 300 kg/ha; three levels of SCU: 113, 227, and 341 kg/ha; three levels of NNC: 127, 255, and 383 L/ha; and three levels of SNNC: 127, 255, and 383 L/ha. The three levels of nitrogen in each treatment were 46, 92, and 138 kg-N/ha.

Fertilizers were applied on the lysimeters' soil surface with irrigation water in two stages: at planting, and flowering stages. Irrigation period lasted 7 days and was performed in a total number of 16 stages, with 10 L of water gradually added in each stage to the soil surface. Since most of the nitrogen absorbed was in the form of nitrate (Mousavi Fazl and Faeznia, 2008), soil nitrate was measured at depths of 15, 30, 45, and 60 cm from planting time to the end of the growing season on a monthly basis. Soil samples were immediately analyzed in the laboratory. Drainage water samples were collected after irrigation and used to determine nitrate leaching using a spectrophotometer with U.V Jesco 7800 model at a wavelength of 400 μm (Mulvaney, 1996).

Yield was measured on the basis of potato weight using a balance with a precision of ±0.01 g and expressed per area of the lysimeters. To evaluate the experimental design, the treatments including U, SCU, NNC, and SNNC with three levels of nitrogen were performed based on a factorial experiment design in CRD with three replications.

The N-efficiency parameters of NUE, nitrogen uptake efficiency (UPE), and nitrogen utilization efficiency (UTE) were calculated for each treatment as follows (Aghaalikhani et al., 2012):

$$NUE = \frac{Y_E}{F_n} \tag{19.1}$$

$$UPE = \frac{L_n}{F_n} \tag{19.2}$$

$$UTE = \frac{Y_E}{L_n} \tag{19.3}$$

where Y_E is the tuber yield, F_n is the N supply, and L_n is the N uptake by the plant.

19.2.5 Nitrogen Balance

In the final confirmation of the research process, total nitrogen was calculated in each of the levels of fertilizer treatments in terms of kilograms of pure nitrogen per hectare, based on the measured values and compared with total injected nitrogen in the form of nitrogen balance equation (Brevé et al., 1997).

$$\Gamma_f = \Gamma_t + \Gamma_l + \Gamma_s + \Gamma_w + \Gamma_d \tag{19.4}$$

where Γ_f is the amount of nitrogen from the fertilization, Γ_t is the amount of nitrogen absorbed into the plant's underground organs, Γ_l is the amount of nitrogen absorbed into the plant's shoots, Γ_s is the amount of nitrogen remaining in the soil (minus primary soil nitrogen), Γ_w is the amount of leaching nitrogen (minus nitrogen in irrigation water), and Γ_d is the amount of nitrogen losses. Since the amount of mineral nitrogen is low, this component was removed (Hosseini et al., 2013). It should be noted that the study of sources showed that the potato tuber is a place of nitrogen accumulation in the underground organ, so the nitrogen absorbed in the tuber was considered as an underground organ (Kleinkopf et al., 1981; Malakouti, 1996).

19.2.6 Data Analysis

Data analysis was performed using the SAS 9.1 software and Duncan's multiple range tests at 5% level was used to compare the means.

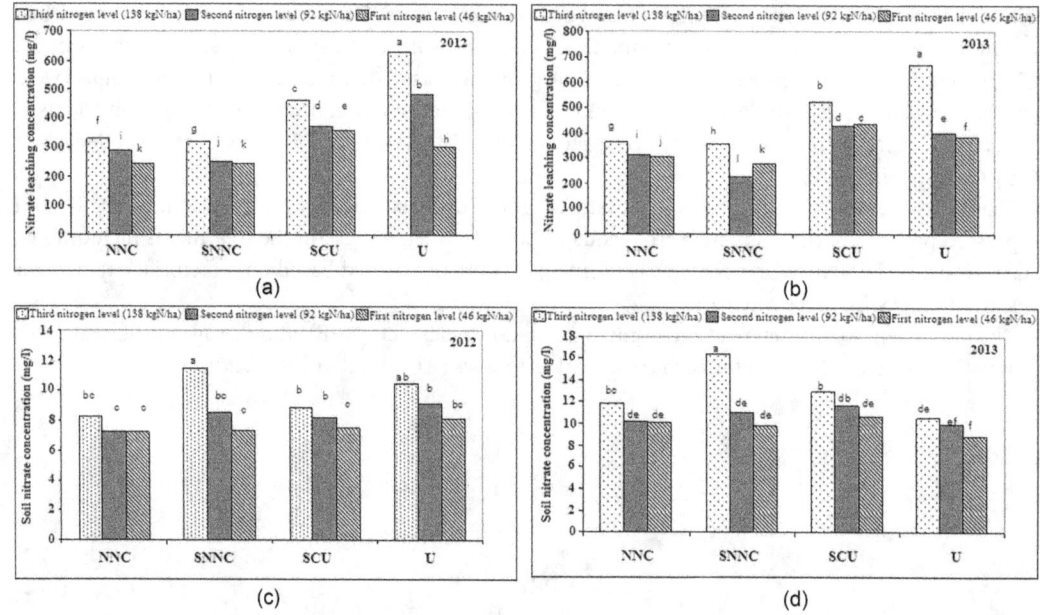

FIGURE 19.4 (a–d) Mean nitrogen concentration in leaching and soil at years of 2012 and 2013.

19.3 Results and Discussion

19.3.1 Soil Nitrate Concentration and Leaching

The mean values of the effect of different levels of fertilizer treatments on soil nitrate concentration and leaching during the growing season are shown in Figure 19.4. The vertical axis represents the nitrate concentration (mg/L) and the horizontal axis shows the average of three replicates of each nitrogen level in the four fertilizer treatments of NNC, SNNC, SCU, and U. In each case, different letters in each column indicate the significant differences between concentrations at 5% probability level. While the average values of nitrate leaching concentrations in most of the treatments are significantly different (designated by different letters above each column), the average soil nitrate concentrations are not significantly different. More specifically, the average soil nitrate concentrations for the first and second nitrogen levels of the NNC treatment are not significantly different from the first and second nitrogen levels of the SNNC and SCU treatments (note "de" above the columns).

As seen in Figure 19.4, soil nitrate concentration and nitrate leaching increase with increasing nitrogen levels in all of the four fertilizer treatments. Variations in the soil nitrate during the potato growing season showed that the highest level of soil nitrate occurred for the third level of each of the four fertilizer treatments so that by reducing the nitrogen input, soil nitrate available to plants also declined. The presence of nitrate in the soil during the crop growth period led to better plant nutrition perhaps due to its gradual delivery to the plant. The highest soil nitrate concentrations in the years 2012 and 2013 were observed in the third nitrogen level of the SNNC treatment with 11.50 and 16.41 mg/L, respectively. The lowest soil nitrate concentrations in the years 2012 and 2013 were observed in the first nitrogen level of the NNC (7.22 mg/L) and the U fertilizer treatments (8.65 mg/L), respectively.

It is also seen in Figure 19.4 that the average values of soil nitrate concentration with the NNC, SNNC, SCU, and U fertilizer treatments during the potato growing season in the years 2012 and 2013 were 7.57, 9.1, 8.2, 9.25 mg/L and 10.55, 12.42, 11.85, 9.56 mg/L, respectively. These results indicate that in the first year, soil nitrate concentrations in the SNNC and SCU fertilizers were 14.35% and 6.59%, respectively, and they were higher than the U fertilizer. But, soil nitrate concentration in the NNC fertilizer was 3.6% less than the U. In the second year, soil nitrate concentrations in NNC, SNNC, and SCU fertilizers were 10.36%, 29.92%, and 23.95%, respectively, and they were higher than that of the U fertilizer, which is reasonable given the low leaching rate with slow-release fertilizers compared to the others.

The value for nitrogen leaching with slow-release fertilizers in each of the three nitrogen levels was less than that for the U fertilizer. This result is consistent with the finding of El-Gindy et al. (2000) who reported that the slow-release fertilizers outperformed the U fertilizers due to the high residual effect of nitrogen in the potato cultivation. Also, Tian et al. (2018) reported that the CRU mitigated nitrate leaching under the cotton–garlic intercropping system in a 4-year field trial. In their study, the CRU treatments exhibited the higher NO_3^--N contents in 0–40 cm soil relative to the U treatments, but the NO_3^--N content in the U treatment was higher in the 60–100 cm soil layer, especially after 4 years of fertilization. The reason is that the NO_3^--N with a negative charge in the U treatment was less absorbed into the soil and irrigation caused more leaching of NO_3^--N from the upper soil layer (Zheng et al., 2016). Also, the similar results were reported by Shang et al. (2014). Their results showed that the total N contents of CRU treatments in the soil increased significantly in the 0–100 cm soil layers in the paddy fields.

Another point of importance in Figure 19.4 is the differences in the values of nitrate leaching among all of the three nitrogen levels of the treatments, which are significant at the 5% probability level. The highest nitrate leaching values in both years belong to the third level of U (629.23 and 670.78 mg/L) and the lowest nitrate leaching values in the years 2012 and 2013 belong to the second and first levels of SNNC (246.7 and 228.8 mg/L, respectively). With increasing nitrogen levels, the nitrate leaching value increased. Similar results have been reported by Bahmani et al. (2009). The mean values of nitrate leaching for the three nitrogen levels in the NNC, SNNC, SCU, and the U treatments are 308.21, 280.20, 431.85, and 479.49 mg/L, respectively. The use of slow-release fertilizers reduces nitrate leaching as is the case with the nano-chelate fertilizers. Zheng et al. (2016) reported that long-term application of CRU had a great potential to increase the wheat–corn yields and N use efficiency, to reduce application frequency, to improve soil fertility, to decrease the leaching of soil NO_3^--N and NH_4^+-N, and also to relieve soil pH decrease. Ryan and Hariq (1986) reported low nitrate leaching values due to the slow-release fertilizers such as SCU, which is consistent with the results of this research. Nitrate leaching reportedly reduces with increasing the surface to the volume ratio of nano-particles (Cui et al., 2006) or due to the chelating properties of the grafted nano-particles (Sikoa and Szmiat, 2001). The lower leaching of nutrients from the soil along with the reduced soil and water pollution makes the use of slow-release fertilizers more economical. Monreal (2010) claims that the use of nano-fertilizers can save $2,000 million, which would be otherwise lost to the low efficiency of other fertilizers.

19.3.2 Variations of Nitrate Concentration in Soil Depth

To investigate the effects of treatment on the nitrate levels, variations of nitrate at four different depths of 15, 30, 45, and 60 cm were measured as reported in Figure 19.5, which shows the nitrate distribution in the soil profile as a result of the fertilizer application. The differences observed may be attributed to differences in the structure of fertilizers. In the both years, the lowest amount of nitrate was observed in the all treatments at 15 cm soil depth, except for the NNC fertilizer treatment in 2013, which was obtained at 60 cm soil depth. These results are consistent with those of Bahmani et al. (2009) and Nabipoor et al. (2012). They stated that, in the case of U fertilizer, soil nitrate increased in the soil profile with increasing depth. Soil nitrate distribution with both the U and NNC fertilizers exhibited a reverse trend similar to that observed with the two sulfur fertilizers (SNNC and SCU). This could be due to the combination of two elements (nitrogen and sulfur) in SNNC and SCU, which is the same in both treatments despite the fact that their structures are quite different evidenced by the amount of nitrate measured at different soil depths. In the SNNC and SCU treatments, the highest nitrate concentration was observed at a depth of 30 cm beyond which it decreased up to a soil depth of 60 cm. According to Mohamadi and Faeznia (2001), the major portion of the potato root system is accumulated in a 30-cm layer and, therefore,

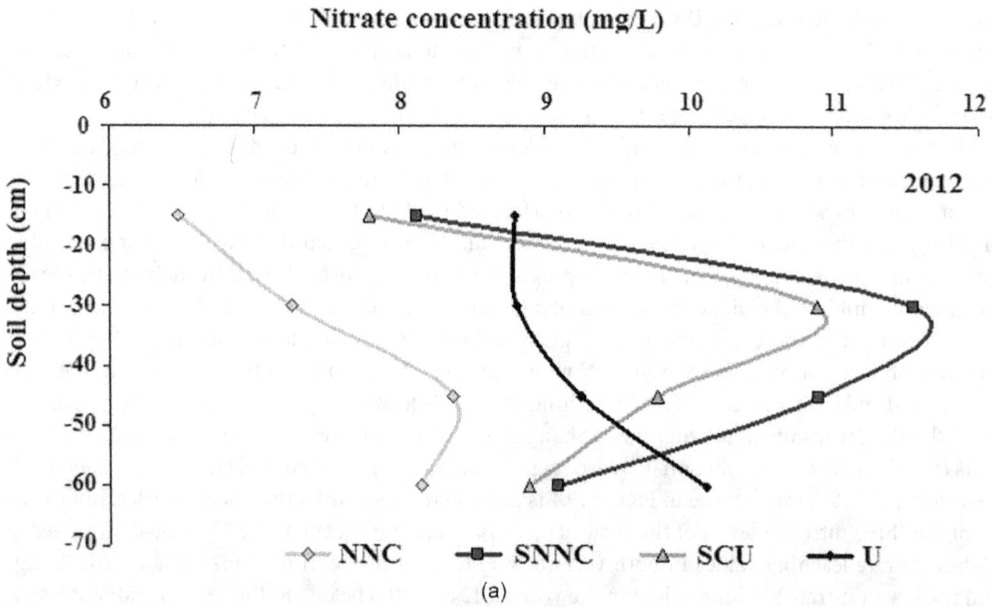

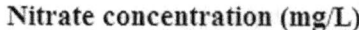

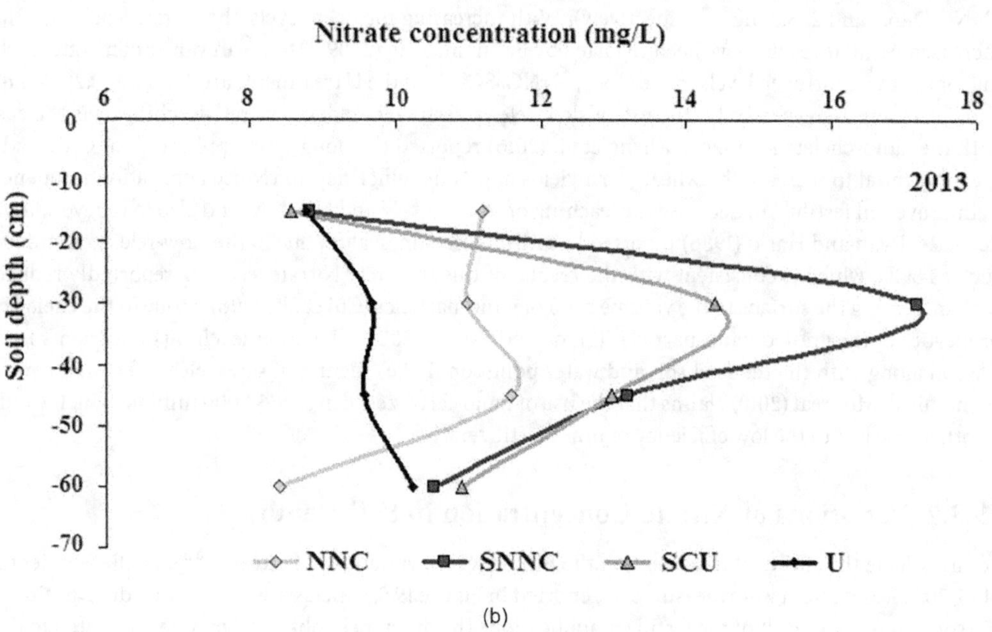

Nitrate concentration (mg/L)

(a)

Nitrate concentration (mg/L)

(b)

FIGURE 19.5 (a and b) Variations of nitrate concentration in soil depth under NNC, SNNC, SCU, and U application.

maximum nitrogen uptake by plants is expected to occur at this depth, which leads to an increased yield. At lower depths, nitrate usually leaches away and will not be available to the root.

Maximum nitrate concentration in the NNC treatment was observed at a depth of 45 cm, but its minimum was at a depth of 60 cm. The low nitrate level at this depth of the soil profile indicates the decreased nitrate leaching and losses. This understanding is confirmed by the results shown in Figure 19.4. Nitrate concentrations in the U treatment increased up to a depth of 60 cm. This means a lack of nitrogen for deep root development and high nitrate leaching at the lower layers. This observation is in agreement with the findings of Nabipoor et al. (2012) who reported a high rate of the U nitrate leaching in the soil profile.

19.3.3 Potato Yield in Different Fertilizer Treatments

Potato yield was measured to study the effect of nitrate movement in the soil profile on yield. The mean values of potato yield in the different treatments with the different levels of nitrogen are shown in Figure 19.6. Clearly, the highest potato yield is observed for the third nitrogen level in the SNNC

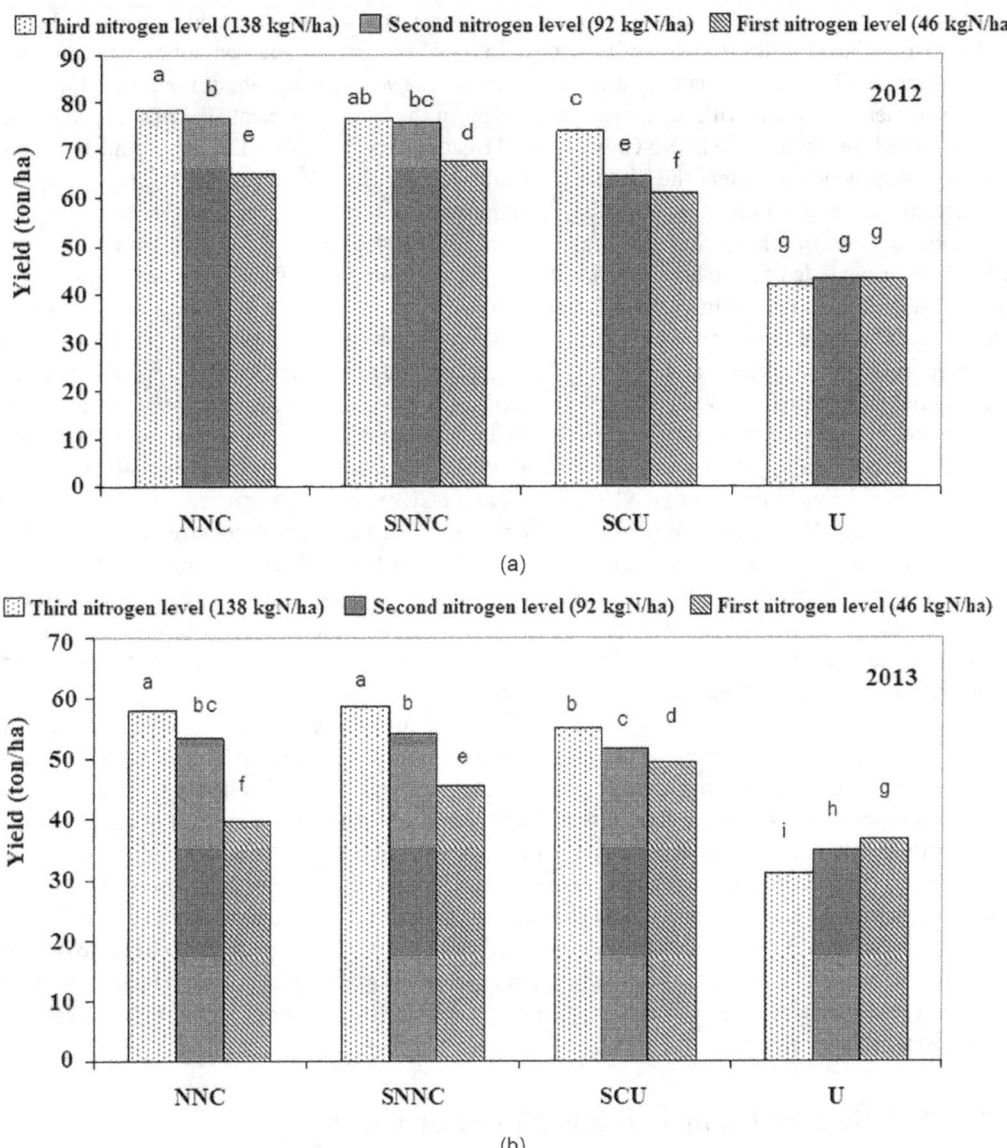

FIGURE 19.6 (a and b) Mean potato yield in fertilizer treatments with different nitrogen levels in the years 2012 and 2013.

treatment (in 2012 and 2013, 78.8 and 58.61 ton/ha, respectively) that is not significantly different from that observed for the third nitrogen level in the NNC treatment. The lowest yield is observed for the third nitrogen level in the U treatment which in the years 2012 and 2013 are equal to 42.2 and 31.05 ton/ha, respectively. In the U treatment, yield decreases linearly with the increasing levels of nitrogen input. This could be due to the increased nitrate leaching by the increasing levels of nitrogen, which is in agreement with the results shown in Figure 19.4. Guarda et al. (2004) reported the cases of reduced yield in wheat due to the high nitrogen levels, and Jiang et al. (2005) presented those of rice yield due to the lower nitrogen physiological efficiency and environmental pollution. However, the yield in the NNC treatment in this study increased significantly with increasing the nitrogen fertilizer. This indicates that in the NNC treatment, compared to the U one, nitrogen was efficiently converted to protein and the other materials required for increasing yield. In other words, by slowly releasing nitrogen, NNC provided food available in the growing season so that the final yield increased (Naderi and Danesh-Shahraki, 2010). This is while nitrogen in the U treatment increased not only in yield but also in nitrate leaching (as a negative trait). Thus, there are fewer concerns about the product quality and environmental hazards with the increased fertilizer in the NNC treatment. The mean yield of the three nitrogen levels in the NNC, SNNC, SCU, and U treatments were 61.83, 63.22, 59.32, and 39.61 ton/ha, respectively, which represent the highest yields of potato in the sulfur treatments. The results are presented in Figure 19.6, indicating the effect of high nitrate availability to the root system of plants on the potato yield. In other words, the high levels of nitrate at a depth of 30 cm in both the SNNC and SCU treatments led to the higher yields due to the accumulation of potato roots at this depth. Consumption of the fertilizer in the SNNC treatment led to an increase in yield equal to 2.25% compared to the NNC, and that in the SCU treatment led to a potato yield increase of 49.76% compared to the U treatment. On the other hand, the sulfur element as a nutrient in fertilizers improved the plant nutrition and might have contributed to the increased yield. However, the major increase in the potato yield observed in this study is due to the SNNC, SCU, NNC, and U treatments. Madani et al. (2009) stated that potato has the potential for yields in excess of 100 ton/ha, while yields of over 40 ton/ha are desirable. Since the yield in the three NNC, SNNC, and SCU treatments exceeded 40 ton/ha, in this study the NNC, SNNC, and SCU fertilizers can be recommended as the preferred treatments over the U treatment. This recommendation is further confirmed by the lower nitrate leaching associated with the three fertilizers. This is also in agreement with the findings of DeRosa et al. (2010) and Barmaki et al. (2010) who reported the increased yield by nano-fertilizer application, and those of El-Gindy et al. (2000); Fan et al. (2004); Lotfollahi et al. (2004); and Malakouti et al. (2008) who reported the increased yields with the SCU application. Tian et al. (2018) reported that CRU improved the crop yields under cotton–garlic intercropping system in a 4-year field trial. All of the results of their study indicated that CRU had a great potential to increase the crops yield, to improve the soil fertility, and to reduce the risk of groundwater pollution through nitrate leaching. However, the yields declined with the SCU application in some studies such as Babaakbari (2005) who investigated wheat cultivation and reported significantly lower yields with SCU than that with the U application. Gascho and Snyder (1976) investigated the application of SCU during the primary growth stages of sugarcane and observed increased growth rates, but the yield was fewer than that due to the application of ammonium sulfate used as the fertilizer. Nourgholi Poor et al. (2009) investigated the effect of the different sources of nitrogen on wheat yield and quality and found that SCU was not capable of supplying the nitrogen requirements of winter wheat as a substitute for U or ammonium nitrate. These studies indicate that further research is required with other crops.

19.3.4 N-Efficiency Parameters and Nitrate of Tubers

Table 19.3 presents a comparison of the average values of NUE, UPE, UTE, and tuber nitrate in each of the treatments during the years 2012 and 2013. Clearly, the tuber nitrate concentrations in these treatments are lower than the standard value of 250 mg/kg (World Health Organization, 1978). The

TABLE 19.3 Effects of Different Levels of Nitrogen in Fertilizer Treatments on N-Efficiency Parameters and Tuber Nitrate.

Year	Fertilizer	Nitrogen (kg/ha)	NUE (kg/ha)	UPE (kg/ha)	UTE (kg/ha)	Tuber Nitrate (mg/kg)
2012	NNC	138	509	0.367	1,387	170
		92	709	0.521	1,361	113
		46	1,046	0.678	1,543	80
	SNNC	138	5,111	0.291	1,758	198
		92	7,102	0.411	17,243	167
		46	1,056	0.519	2,033	127
	SCU	138	481	0.286	1,682	203
		92	596	0.316	1,888	187
		46	985	0.542	1,816	170
	U	138	273	0.177	1,547	216
		92	398	0.237	1,680	143
		46	696	0.409	1,702	104
2013	NNC	138	376	0.260	1,445	155
		92	392	0.309	1,595	111
		46	365	0.324	1,955	89
	SNNC	138	380	0.262	1,452	189
		92	500	0.337	1,481	151
		46	731	0.428	1,707	102
	SCU	138	356	0.258	1,382	217
		92	478	0.275	1,735	214
		46	793	0.476	1,668	213
	U	138	201	0.112	1,793	240
		92	322	0.164	1,967	144
		46	700	0.287	2,439	129

maximum tuber nitrate concentration of 240 mg/kg was obtained with 138 kg-N/ha supplied from U. The lowest tuber nitrate concentrations were due to the nano-fertilizer and SCU treatments. Also, Table 19.3 shows that the highest NUE belonged to the first nitrogen level of SCU while its lowest belonged to the third nitrogen level of U. It may, therefore, be concluded that nano-fertilizers and SCU are more efficient in cultivations with the nitrogen requirements. The changes in the UPE index are similar to those in the NUE index. In other words, the values of UPE and NUE decrease with increasing the fertilizer application while that of UTE increases with increasing nitrogen. Darwish et al. (2006) and Halitligil et al. (2002) experimented with potato and showed that NUE and UPE reduced with increasing the nitrogen level of U treatment, which is in line with the results reported in Table 19.3. It seems that a nonlinear relationship holds between the nitrogen fertilizer and tuber potato yield.

Based on the results obtained in this study, the order of the preferred types of fertilizers is SNNC>SCU>NNC>U for making nitrogen available to achieve the highest yield. However, when nitrate leaching and its effects on human health and the environment are in view, the priority changes to SNNC>NNC>SCU>U. Since no significant differences were observed in potato yields between the SCU and NNC treatments, the use of nano-chelate fertilizers is recommended. Thus, nano-technology can be successfully exploited to increase the food production, minimize costs, and protect the environment. The findings by Chinnamuthu and Boopathi (2009) also confirm this research's outcomes.

19.3.5 Nitrogen Balance

Nitrogen balance components in the different treatments and fertilizer levels at the end of the growing season by water, soil, and plant environment are shown in Figure 19.7. As stated in Section 19.2, components of the amount of nitrogen absorbed to the shoot, the amount of nitrogen remaining in the soil, the amount of leaching nitrogen, and the amount of nitrogen absorbed to the underground organ of the plant were used to draw the nitrogen balance. Nitrogen absorbed in the tuber was also considered as an underground organ (Kleinkopf et al., 1981; Malakouti, 1996). By subtracting these components from the nitrogen input to each treatment, the amount of nitrogen losses was obtained.

The results showed that the highest amount of leaching nitrogen in both years in the third level of U fertilizer is 50.5 kg/ha in the first year and 46.45 kg/ha in the second year, which is a negative factor indicating environmental concerns. It was mentioned in the interpretation of the results of Figure 19.4. The highest amount of leaching nitrogen in NNC, SNNC, and SCU treatments was similar to U in the third level of nitrogen (138 kg of pure nitrogen). So the highest amount of leaching in the first and second years of NNC fertilizer is 26.6 and 29.12, SNNC fertilizer is 25.6 and 28.77, SCU fertilizer

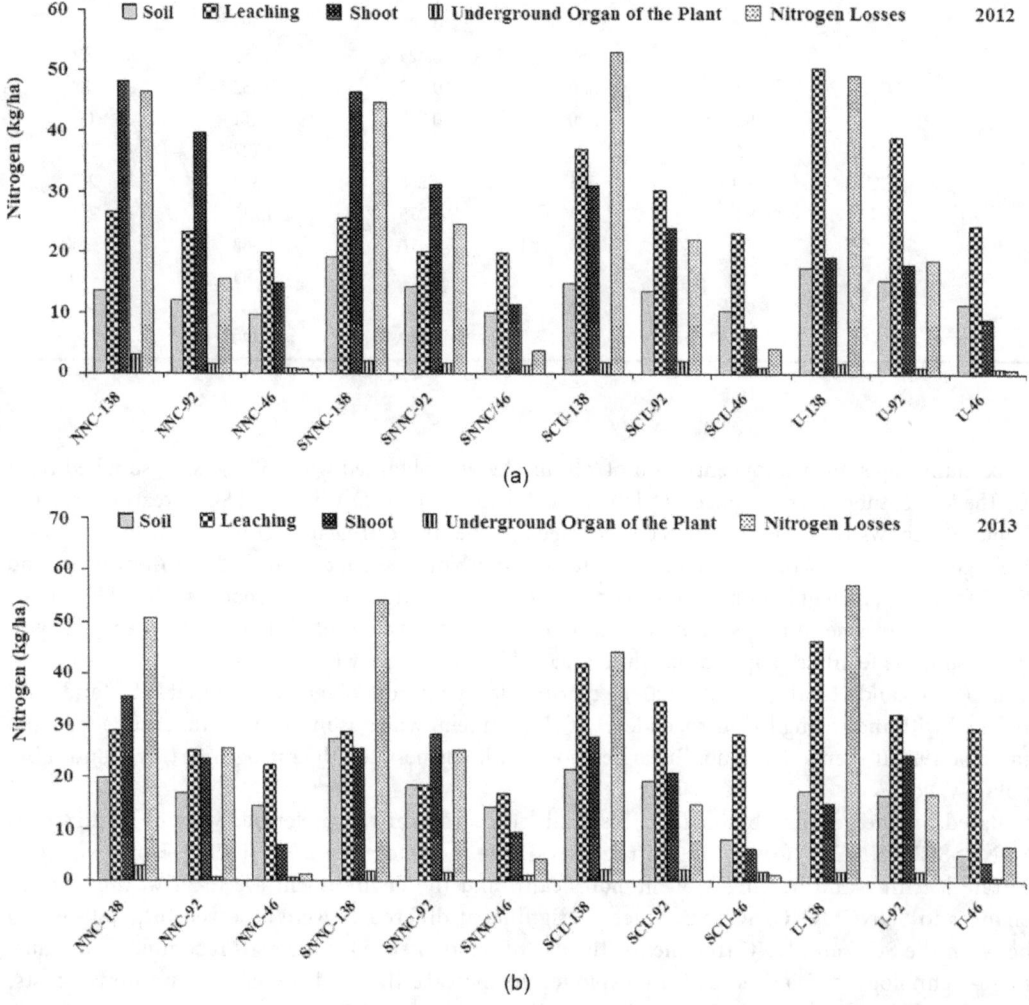

FIGURE 19.7 (a and b) Components of the nitrogen balance equation during 2 years of cultivation.

is 37.12 and 42.02 kg/ha, respectively. This indicates a significant difference in the severity of leaching nitrogen losses with increasing the level of fertilizer used in the U fertilizer treatment compared to the other treatments.

Shoot nitrogen levels in all of the three levels of nano-chelate and coated treatments were significantly higher than shoot nitrogen levels at the levels corresponding to U fertilizer treatment. Therefore, it can be expected that more nitrogen in the shoot will increase the amount of photosynthesis in the plant and consequently increase the yield; the tuber yield values shown in Figure 19.6 are in line with the results of Figure 19.7. In addition, the higher yield in NNC, SNNC, and SCU treatments is in line with the reports of Sharma (1992) and Varvel et al. (1997) that the increase in yield is due to the increase in photosynthesis due to the increase in leaf nitrogen. A noteworthy point in Figure 19.7 is the high nitrogen losses at high nitrogen consumption levels in all of the treatments. In other words, in all fertilizer treatments, the amount of nitrogen losses increases with increasing the level of nitrogen input. In the report of Bahmani et al. (2010) with increasing the level of nitrogen input, the amount of nitrogen losses increased, which is in line with this study.

19.4 Conclusions

The use of nano-technology in agriculture in recent years has attracted much attention globally. In this study, the reduction of nitrate leaching and increasing potato yield with emphasis on less soil and water pollution were investigated in the experiments with treatments that consisted of three nitrogen levels of NNC, SNNC, SCU, and U. The results showed that the slow-release fertilizers reduce the nitrate leaching and increase the plant available nitrogen in the soil during the growing season. Shoot nitrogen levels in all of three levels of nano-chelate and coated treatments were significantly higher than the shoot nitrogen levels at the levels corresponding to the U fertilizer treatment. Therefore, it can be expected that more nitrogen in the shoot will increase the rate of photosynthesis in the plant and lead to increased yield in the slow release fertilizer treatments compared to the U fertilizer treatment. This is evidenced by the reduced nitrate leaching by 35.72% as a result of applying NNC compared to the U fertilizer. The corresponding values for the SNNC and SCU fertilizers were 41.56% and 9.94%, respectively. In contrast, potato yields were 56.10%, 59.61%, and 49.76%, respectively, higher when the NNC, SNNC, and SCU fertilizers were applied than in the U treatment. Also, the maximum tuber nitrate concentration was observed when 138 kg-N/ha U was used. Comparison of N-efficiency parameters showed that the highest NUE value belonged to the first nitrogen level of SCU and its lowest belonged to the third nitrogen level of U. The comparison of different levels of nitrogen also indicated that low nitrogen levels in slow-release fertilizers were better than the high levels of U fertilizer, which is economical. Since, the results of this study were obtained in the greenhouse for potato, repeating this research method in the field conditions for other products is also recommended.

References

Abbasi, Y., Liaghat, A., Abbasi, F., 2012. Evaluation of nitrate deep leaching under maize furrow fertigation. *J. Water Soil (Agricultural Sci. Technol.)* 26, 842–853.

Aghaalikhani, M., Gholamhoseini, M., Dolatabadian, A., Khodaei-Joghan, A., Sadat Asilan, K., 2012. Zeolite influences on nitrate leaching, nitrogen-use efficiency, yield and yield components of canola in sandy soil. *Arch. Agron. Soil Sci.* 58, 1146–1169.

Akhlaghi, K., 2008. Formulation and manufacture of sealants for use in the production process sulfur-coated urea, in: *Iran Petrochemical Conference.* NPC, Tehran, Iran (in Persian).

Albaji, M., Nasab, S.B., Golabi, M., Nezhad, M.S., Ahmadee, M. 2015. Application possibilities of different irrigation methods in Hofel Plain. *Yüzüncü Yıl Üniversitesi Tarım Bilimleri Dergisi* 25 (1), 13–23.

Babaakbari, S.M., 2005. Nitrogen efficiency in two calcareous soils with different textures in Karaj wheat lands. M. Sc. Thesis, Faculty of Agriculture, Tarbiat Modarres University, Tehran, Iran (in Persian).

Bahmani, O., Boroumandnasab, S., Behzad, M., Naseri, A.A., 2009. Evaluation of potential nitrate and ammonium leaching under deficit irrigation in soil profile. *Iran. J. Irrig. Drain.* 3, 37–44.

Bahmani, O., Broumandnasab, S., Behzad, M., Naseri, A.A., 2010. Evaluation of potential nitrate and ammonium accumulation in the soil profile under irrigation and manure treatments with the LEACHM model. *Environ. Sci.* 7, 95–108.

Barmaki, S., Modares, M., Mehdizadeh, V., 2010. The application of nanotechnology in order to optimize the use of chemical fertilizers with emphasis on nano fertilizers, in: *The First Congress of the Fertilizer Challenge*, Tehran, Iran.

Bhattacharyya, P., Roy, K.S., Neogi, S., Adhya, T.K., Rao, K.S., Manna, M.C., 2012. Effects of rice straw and nitrogen fertilization on greenhouse gas emissions and carbon storage in tropical flooded soil planted with rice. *Soil Tillage Res.* 124, 119–130. https://doi.org/10.1016/j.still.2012.05.015.

Brevé, M.A., Skaggs, R.W., Parsons, J.E., Gilliam, J.W., 1997. DRAINMOD-N, a nitrogen model for artificially drained soil. *Trans. ASAE* 40, 1067–1075.

Chinnamuthu, C.R., Boopathi, P., 2009. Nanotechnology and agroecosystem. *Madras Agric. J.* 96, 17–31.

Colombani, N., Mastrocicco, M., Di Giuseppe, D., Faccini, B., Coltorti, M., 2015. Batch and column experiments on nutrient leaching in soils amended with Italian natural zeolitites. *Catena* 127, 64–71. https://doi.org/10.1016/j.catena.2014.12.022.

Coltorti, M., Giuseppe, D., Faccini, B., Passaglia, E., Malferrari, D., Mastrocicco, M., Colombani, N., 2012. ZeoLIFE, a project for water pollution reduction and water saving using a natural zeolitite cycle, in: 86° Congresso Società Geologica Italiana 2012. Società Geologica Italiana, pp. 853–853.

Cui, H., Sun, C., Liu, C., Jiang, Q., Gu, W., 2006. Applications of nanotechnology in agrochemical formulation, perspectives, challenges and strategies, institute of environment and sustainable development in agriculture. Chinese Acad. Agric. Sci. Beijing, China, 1–6.

Darwish, T.M., Atallah, T.W., Hajhasan, S., Haidar, A., 2006. Nitrogen and water use efficiency of fertigated processing potato. *Agric. Water Manag.* 85, 95–104. https://doi.org/10.1016/j.agwat.2006.03.012.

DeRosa, M.R., Monrea, C., Schnitzer, M., Walsh, R., Sultan, Y., 2010. Nanotechnology in fertilizers. *Nat. Nanotechnol.* 5, 91–91. https://doi.org/10.1038/nnano.2010.2.

Du, X., Chen, B., Meng, Y., Zhao, W., Zhang, Y., Shen, T., Wang, Y., Zhou, Z., 2016. Effect of cropping system on cotton biomass accumulation and yield formation in double-cropped wheat-cotton. *Int. J. Plant Prod.* 10, 29–44. https://doi.org/10.22069/ijpp.2016.2551.

El-Gindy, A.M., Mahmoud, M.H., El-Guibali, A.H., 2000. Response of plant to slow release fertilizers under drip irrigation systems. *Ann. Agric. Sci.* 1, 197–212.

Fan, X., Li, F., Liu, F., Kumar, D., 2004. Fertilization with a new type of coated urea: Evaluation for nitrogen efficiency and yield in winter wheat. *J. Plant Nutr.* 27, 853–865. https://doi.org/10.1081/PLN-120030675.

Fazlolahi, H., Eslamian, S.S., 2013. Nitrogen and Phosphorus removal from municipal wastewater by three wetland plant species. *J. River Eng.* 1(2), 14–20.

Gascho, G.J., Snyder, G.H., 1976. Sulfur-coated fertilizers for sugarcane: I. Plant response to sulfur-coated urea. *Soil Sci. Soc. Am. J.* 40, 119–122. https://doi.org/10.2136/sssaj1976.03615995004000010032x.

Gholamhoseini, M., Ghalavand, A., Khodaei-Joghan, A., Dolatabadian, A., Zakikhani, H., Farmanbar, E., 2013. Zeolite-amended cattle manure effects on sunflower yield, seed quality, water use efficiency and nutrient leaching. *Soil Tillage Res.* 126, 193–202. https://doi.org/10.1016/j.still.2012.08.002.

Grignani, C., Zavattaro, L., Sacco, D., Monaco, S., 2007. Production, nitrogen and carbon balance of maize-based forage systems. *Eur. J. Agron.* 26, 442–453. https://doi.org/10.1016/j.eja.2007.01.005.

Guarda, G., Padovan, S., Delogu, G., 2004. Grain yield, nitrogen-use efficiency and baking quality of old and modern Italian bread-wheat cultivars grown at different nitrogen levels. *Eur. J. Agron.* 21, 181–192. https://doi.org/10.1016/j.eja.2003.08.001.

Halitligil, M., Akin, A., Ylbeyi, A., 2002. Nitrogen balance of nitrogen-15 applied as ammonium sulphate to irrigated potatoes in sandy textured soils. *Biol. Fertil. Soils* 35, 369–378. https://doi.org/10.1007/s00374-002-0482-4.

Hooshmand, M., Albaji, M., Zadeh Ansari, N.A., 2019. The effect of deficit irrigation on yield and yield components of greenhouse tomato (Solanum lycopersicum) in hydroponic culture in Ahvaz region, Iran. *Scientia Horticulturae* 254, 84–90.

Hosseini, R.., Galeshi, S., Soltani, A., Kalateh, M., Zahed, M., 2013. The effect of nitrogen rate on nitrogen use efficiency index in wheat (Triticum Aestivum L.) cultivars. *Iran. J. F. Crop. Res.* 11, 300–306.

Hou, Z., Li, P., Li, B., Gong, J., Wang, Y., 2007. Effects of fertigation scheme on N uptake and N use efficiency in cotton. *Plant Soil* 290, 115–126. https://doi.org/10.1007/s11104-006-9140-1.

Jiang, L., Dong, D., Gan, X., Wei, S., 2005. Photosynthetic efficiency and nitrogen distribution under different nitrogen management and relationship with physiological N-use efficiency in three rice genotypes. *Plant Soil* 271, 321–328. https://doi.org/10.1007/s11104-004-3116-9.

Johnson, A., 2006. *Agriculture and Nanotechnology.* University of Wisconsin-Madison, Ward and Dutta, USA.

Kim, Y.S., Reid, F., Hansen, A., Zhang, Q., 2000. On-field crop stress detection system using multi-spectral imaging sensor. *Agric. Biosyst. Eng.* 1, 88–94.

Kiymaz, S., Ertek, A., 2015. Yield and quality of sugar beet (Beta vulgaris L.) at different water and nitrogen levels under the climatic conditions of Kırsehir, Turkey. *Agric. Water Manag.* 158, 156–165. https://doi.org/10.1016/j.agwat.2015.05.004.

Kleinkopf, G.E., Westermann, D.T., Dwelle, R.B., 1981. Dry matter production and nitrogen utilization by six potato cultivars 1. *Agron. J.* 73, 799–802.

Lang, N.S., Stevens, R.G., Thornton, R.E., Pan, W.L., Victory, S., 1999. Nutrient management guide: Central Washington irrigated potatoes. https://research.libraries.wsu.edu:8443/xmlui/handle/2376/6902.

Latifah Omar, O., Ahmed, O.H., Muhamad, A.M.N., 2010. Minimizing ammonia volatilization in water-logged soils through mixing of urea with zeolite and sago waste water. *Int. J. Phys. Sci.* 5, 2193–2197.

Li, Z., Zhang, Y., Li, Y., 2014. Zeolite as slow release fertilizer on spinach yields and quality in a greenhouse test. *J. Plant Nutr.* 36. https://doi.org/10.1080/01904167.2013.790429.

Lotfollahi, M., Malakouti, M., Safari, H., 2004. Nitrogen efficiency using SCU in light textured soil in Karaj. New methods of feeding wheat (Collection of articles of Malekooti et al.). Publ. Sana, Tehran, Iran, 751–759 (in Persian).

Lv, M., Li, Z., Che, Y., Han, F.X., Liu, M., 2011. Soil organic C, nutrients, microbial biomass, and grain yield of rice (Oryza sativa L.) after 18 years of fertilizer application to an infertile paddy soil. *Biol. Fertil. Soils* 47, 777–783. https://doi.org/10.1007/s00374-011-0584-y.

Madani, H., Farhadi, A., Pazoki, A., Changizi, M., 2009. Effect of different levels of nitrogen and zeolite on traits quantitative and qualitative of potato in Arak region. *New Find. Agric.* 4, 379–391.

Malakouti, M.J., 1996. Sustainable agriculture and increasing yield by optimizing fertilizer usage in Iran. Agriculture Education, Ministry of Jahad e Keshavarzi, Tehran, Iran (in Persian).

Malakouti, M.J., Bybordi, A., Lotfollahi, M., Shahabi, A.A., Siavoshi, K., Vakil, R., Ghaderi, J., Shahabifar, J., Majidi, A., Jafarnajadi, A.R., Dehghani, F., Keshavarz, M.H., Ghasemzadeh, M., Ghanbarpouri, R., Dashadi, M., Babaakbari, M., Zaynalifard, N., 2008. Comparison of complete and sulfur coated urea fertilizers with pre-plant urea in increasing grain yield and nitrogen use efficiency in wheat. *J. Agric. Sci. Technol.* 10, 173–183.

Malakouti, M.J., Gheybi, M.N., 1997. Determination of nutrients critical level of strategic products and correct fertilizer recommendation. Agricultural Education, Tehran, Iran (in Persian).

Manjunatha, S.B., Biradar, D.P., Aladakatti, Y.R., 2016. Nanotechnology and its applications in agriculture: A review. *J. Farm Sci.* 29, 1–13.

Mohamadi, A., Faeznia, F., 2001. Effects of water stress on growth and yield of two potato cultivars. Res. Report, Agric. Res. Center, Semnan, Shahrood, Iran (in Persian).

Monreal, C.M., 2010. Nano fertilizers for increased N and P use efficiencies by crops. *Summ. Inf. Curr. Provid. to MRI Concern. Appl. Round* 5, 12–13.

Mousavi Fazl, S.H., Faeznia, F., 2008. Effect of different moisture regimes and nitrogen on the yield and nitrate concentrations in potato tubers. *Iran. J. Soil Res.* 22, 243–250.

Mulvaney, R.L., 1996. Nitrogen—inorganic forms. *Methods Soil Anal. Part 3 Chem. Methods* 5, 1123–1184.

Nabipoor, M., Emami, H., Astaraee, A.R., 2012. Effect of nitrate fertilizer source, and irrigation on nitrate leaching, and its distribution in the soil profile, in: *The First National Conference of Water Management in Agriculture. Institute of Soil and Water Research*, Tehran, Iran (in Persian).

Naderi, M., Danesh-Shahraki, A., 2010. Nanofertilizers and Their Role in Sustainability of Agriculture, in: The 1st Iranian Fertilizer Challenges Congress. Half a Century of the Fertilizer Consumption. Tehran, Iran (in Persian).

Neissi, L., Albaji, M., Boroomand Nasab, S., 2020. Combination of GIS and AHP for site selection of pressurized irrigation systems in the Izeh plain, Iran. *Agric. Water Manage.* 231, 106004.

Nourgholi Poor, F., Bagheri, Y.R., Lotfollahi, M., 2009. Soil genesis in the lands of Islamic Azad University of Khoraskan (Isfahan). *J. Res. Agric. Sci.* 4, 120–129.

Padilla, F.M., Gallardo, M., Manzano-Agugliaro, F., 2018. Global trends in nitrate leaching research in the 1960–2017 period. *Sci. Total Environ.* 643, 400–413. https://doi.org/10.1016/j.scitotenv.2018.06.215.

Peyvandi, M., Mirza, M., Kamali Jamakani, Z., 2011a. The effect of Nano Fe Chelate and Fe Chelate on the growth and activity of some Antioxidant. *New Cell. Mol. Biotechnol. J.* 2, 25–32.

Peyvandi, M., Parande, H., Mirza, M., 2011b. Comparison of Nano Fe Chelate with Fe Chelate effect on growth parameters and Antioxidant enzymes activity of Ocimum Basilicum. *New Cell. Mol. Biotechnol. J.* 1, 89–98.

Rahimizadeh, M., Kashani, A., Zare Faizabadi, A., Koocheki, A., Nassiri Mahallati, M., 2010. Investigation of nitrogen use efficiency in wheat-based double cropping systems under different rate of nitrogen and return of crop residue. *Electron. J. Crop Prod.* 3, 125–142.

Rens, L., Zotarelli, L., Alva, A., Rowland, D., Liu, G., Morgan, K., 2016. Fertilizer nitrogen uptake efficiencies for potato as influenced by application timing. *Nutr. Cycl. Agroecosyst.* 104, 175–185. https://doi.org/10.1007/s10705-016-9765-2.

Rens, L.R., Zotarelli, L., Cantliffe, D.J., Stoffella, P.J., Gergela, D., Fourman, D., 2015. Biomass accumulation, marketable yield, and quality of Atlantic potato in response to nitrogen. *Agron. J.* 107, 931–942. https://doi.org/10.2134/agronj14.0408.

Rosen, C.J., 2018. Potato Fertilization on Irrigated Soils. https://extension.umn.edu/crop-specific-needs/potato-fertilization-irrigated-soils#nitrogen–1075460.

Ryan, J., Hariq, S.N., 1986. Crop and laboratory evaluation of nitrogen release from sulfur coated urea osmocote. *Leban Sci. Collect.* 2, 5–15.

Shang, Q., Gao, C., Yang, X., Wu, P., Ling, N., Shen, Q., Guo, S., 2014. Ammonia volatilization in Chinese double rice-cropping systems: A 3-year field measurement in long-term fertilizer experiments. *Biol. Fertil. Soils* 50, 715–725. https://doi.org/10.1007/s00374-013-0891-6.

Sharma, R.P., 1992. Effect of planting material, nitrogen and potash on bulb yield of rainy season onion (Allium cepa L.). *Indian J. Agron.* 37, 868–869.

Sikoa, L., Szmiat, P., 2001. Nitrogen sources, mineralization rates and plant nutrient benefits from compost. Compost Util. Hortic. Crop. Syst. CRC Press. USA.

Silva, J.G., França, M.G.C., Gomide, F.T.F., Magalhaes, J.R., 2013. Different nitrogen sources affect biomass partitioning and quality of potato production in a hydroponic system. *Am. J. Potato Res.* 90, 179–185. https://doi.org/10.1007/s12230-012-9297-5.

Souza, E.F.C., Rosen, C.J., Venterea, R.T., 2019. Contrasting effects of inhibitors and biostimulants on agronomic performance and reactive nitrogen losses during irrigated potato production. *F. Crop. Res.* 240, 143–153. https://doi.org/10.1016/j.fcr.2019.05.001.

Tian, X., Li, C., Zhang, M., Li, T., Lu, Y., Liu, L., 2018. Controlled release urea improved crop yields and mitigated nitrate leaching under cotton-garlic intercropping system in a 4-year field trial. *Soil Tillage Res.* 175, 158–167. https://doi.org/10.1016/j.still.2017.08.015.

Van Kempen, P., Le Corre, P., Bedin, P., 1996. Phytotechnie. La pomme terre, INRA, Paris, France, 363–414.

Varvel, G.E., Schepers, J.S., Francis, D.D., 1997. Ability for in-season correction of nitrogen deficiency in corn using chlorophyll meters. *Soil Sci. Soc. Am. J.* 61, 1233–1239.

Woli, P., Hoogenboom, G., Alva, A., 2016. Simulation of potato yield, nitrate leaching, and profit margins as influenced by irrigation and nitrogen management in different soils and production regions. *Agric. Water Manag.* 171, 120–130. https://doi.org/10.1016/j.agwat.2016.04.003.

World Health Organization, 1978. Nitrates, Nitrites, and N-Nitroso Compounds - Environmental Health Criteria 5. United Nations Environment Programme and World Health Organization, Geneva, Switzerland.

Yang, Y.-C., Zhang, M., Zheng, L., Cheng, D.-D., Liu, M., Geng, Y.-Q., 2011. Controlled release urea improved nitrogen use efficiency, yield, and quality of wheat. *Agron. J.* 103, 479–485. https://doi.org/10.2134/agronj2010.0343.

Yang, Y., Zhang, M., Li, Y.., Fan, X., Geng, Y., 2012. Controlled release urea improved nitrogen use efficiency, activities of leaf enzymes, and rice yield. *Soil Sci. Soc. Am. J.* 76, 2307–2317. https://doi.org/10.2136/sssaj2012.0173.

Ye, Y., Liang, X., Chen, Y., Liu, J., Gu, J., Guo, R., Li, L., 2013. Alternate wetting and drying irrigation and controlled-release nitrogen fertilizer in late-season rice. Effects on dry matter accumulation, yield, water and nitrogen use. *F. Crop. Res.* 144, 212–224. https://doi.org/10.1016/j.fcr.2012.12.003.

Zare Abyaneh, H., Bayat Varkeshi, M., 2014. The effect of nanofertilizers on nitrate leaching and its distribution in soil profile with an emphasis on potato yield. *Nano Sci. Nano Technol. An Indian J.* 8, 198–207.

Zare Abyaneh, H., Jovzi, M., Pak, N.A.E., Albaji, M., 2020. Crop yield response to partial root drying compared with regulated deficit irrigation, in: *Handbook of Irrigation System Selection for Semi-Arid Regions*, First edition. CRC Press/Taylor & Francis Group, Boca Raton, FL, pp. 29–54. https://doi.org/10.1201/9781003050261-5.

Zareabyaneh, H., Bayatvarkeshi, M., 2015. Effects of slow-release fertilizers on nitrate leaching, its distribution in soil profile, N-use efficiency, and yield in potato crop. *Environ. Earth Sci.* 74, 3385–3393. https://doi.org/10.1007/s12665-015-4374-y.

Zheng, W., Sui, C., Liu, Z., Geng, J., Tian, X., Yang, X., Li, C., Zhang, M., 2016. Long-term effects of controlled-release urea on crop yields and soil fertility under wheat-corn double cropping systems. *Agron. J.* 108, 1703–1716. https://doi.org/10.2134/agronj2015.0581.

Ziaeyan, A.H., Keshavarz, P., 2010. Increasing nitrogen use efficiency in potato by application of slow release N-fertilizers. *Iran. J. Soil Res. (Formerly Soil Water Sci.)* 24, 107–116.

Zvomuya, F., Rosen, C.J., Russelle, M.P., Gupta, S.C., 2003. Nitrate leaching and nitrogen recovery following application of polyolefin-coated urea to potato. *J. Environ. Qual.* 32, 480–489. https://doi.org/10.2134/jeq2003.4800.

Index

For Product Safety Concerns and Information please contact our EU
representative GPSR@taylorandfrancis.com
Taylor & Francis Verlag GmbH, Kaufingerstraße 24, 80331 München, Germany

www.ingramcontent.com/pod-product-compliance
Lightning Source LLC
Chambersburg PA
CBHW080902170526
45158CB00008B/1963